Cavitation-Enhanced Drug Delivery and Immunotherapy

Cavitation-Enhanced Drug Delivery and Immunotherapy

Editors

Brandon Helfield
Shashank Sirsi
James Kwan
Michael Gray

MDPI

Basel • Beijing • Wuhan • Barcelona • Belgrade • Novi Sad • Cluj • Manchester

Editors

Brandon Helfield
Department of Physics,
Department of Biology
Concordia University
Montreal, QC, Canada

Shashank Sirsi
Erik Jonsson School of
Engineering and
Computer Science
The University of Texas
at Dallas
Richardson, TX, USA

James Kwan
Department of
Engineering Science
University of Oxford
Oxford, UK

Michael Gray
Institute of Biomedical
Engineering
University of Oxford
Oxford, UK

Editorial Office

MDPI
St. Alban-Anlage 66
4052 Basel, Switzerland

This is a reprint of articles from the Special Issue published online in the open access journal *Pharmaceutics* (ISSN 1999-4923) (available at: https://www.mdpi.com/journal/pharmaceutics/special_issues/Microbubbles_Delivery).

For citation purposes, cite each article independently as indicated on the article page online and as indicated below:

Lastname, A.A.; Lastname, B.B. Article Title. *Journal Name* **Year**, *Volume Number*, Page Range.

ISBN 978-3-0365-9434-7 (Hbk)
ISBN 978-3-0365-9435-4 (PDF)
doi.org/10.3390/books978-3-0365-9435-4

Contents

pharmaceutics

MDPI

Editorial

Cavitation-Enhanced Drug Delivery and Immunotherapy

Brandon Helfield [1,2], Shashank Sirsi [3], James Kwan [4] and Michael Gray [5,*]

1 Department of Physics, Concordia University, Montreal, QC H3G 1M8, Canada; brandon.helfield@concordia.ca
2 Department of Biology, Concordia University, Montreal, QC H3G 1M8, Canada
3 Erik Jonsson School of Engineering and Computer Science, The University of Texas at Dallas, Richardson, TX 75080, USA; shashank.sirsi@utdallas.edu
4 Department of Engineering Science, University of Oxford, Oxford OX1 3PJ, UK; james.kwan@eng.ox.ac.uk
5 Biomedical Ultrasonics, Biotherapies and Biopharmaceuticals Laboratory, University of Oxford, Oxford OX3 7LD, UK
* Correspondence: michael.gray@eng.ox.ac.uk

Citation: Helfield, B.; Sirsi, S.; Kwan, J.; Gray, M. Cavitation-Enhanced Drug Delivery and Immunotherapy. *Pharmaceutics* **2023**, *15*, 2207. https://doi.org/10.3390/pharmaceutics15092207

Received: 18 August 2023
Accepted: 23 August 2023
Published: 26 August 2023

Welcome to this special issue on Cavitation-Enhanced Drug Delivery and Immunotherapy—a rapidly evolving area that has been buoyed in recent years by the development of methods harnessing the activity of ultrasound-stimulated bubbles known as cavitation. When properly controlled, cavitation can help overcome physical barriers to drug delivery whilst providing readily measurable information for timely quantitative feedback and treatment guidance. Microbubble assisted therapies have demonstrated impressive advancements in clinical trials and pre-clinical areas, including applications in neurology, oncology, cardiology and beyond.

In this special issue we have eight original research contributions highlighting the breadth of targets for cavitation-enhanced therapies, including Imran et al. [1]. (pancreas), Lacerda et al. [2] (head and neck cancer), and Ahmed et al. [3] (brain), who all discuss progress in fighting long-challenging pathologies. Imran et al. demonstrated substantially enhanced drug uptake in a porcine pancreatic tumor model using the SonoTran system, which is currently in a clinical trial for targeted chemotherapy to liver tumors. Lacerda et al. explored the treatment parameter space for a head and neck cancer combination therapy, showing the relative importance of O_2 microbubbles (MBs), mitochondrial respiration inhibitors, and radiation dose rate. In a murine glioma model study, Ahmed et al. demonstrated that an ultrasound mediated blood-brain barrier opening treatment following administration of anti-PD-L1 markedly improved overall survival in comparison to giving anti-PD-L1 alone.

Some newer targets and applications are also featured in the special issue. LuTheryn [4] et al. describe the first use of cationic MBs for targeted treatment of biofilms, which are notoriously drug-resistant and typically are negatively charged. Cationic MBs were also used by He et al. [5] to safely deliver angiogenic microRNA with an eye toward treatment of cardiovascular disease. Kerneis et al. [6] demonstrated the safety of sonoporation to facilitate inner ear drug delivery. Benton et al. [7] showed the importance of perfluoropentane droplet concentration in the therapeutic context of cardiovascular reperfusion injury. Finally, Martinez et al. [8] present cavitation dose relationships to microbubble size distributions and exposure conditions, and they suggest that gas volume fraction may be used as a unifying factor describing bubble behavior across a range of sizes and concentrations.

This collection also features four reviews, beginning with Navarro-Becerra and Borden [9] writing on the design and application of targeted microbubbles, challenges in their clinical implementation, and opportunities for protocol optimization and reporting standardization. Chapla et al. [10] reviewed design strategies for microbubble-nanoparticle complexes, highlighting the wide variety of available constructs, while noting the need for near term large animal safety and efficacy studies on the path to clinical evaluation.

Honari and Sirsi [11] also discuss ultrasound-sensitive particles, with notably clear descriptions of mechanisms by which mechanical and biological barriers to drug delivery may be overcome. The collection concludes with a review by Maardalen et al. [12] highlighting the potential role of cavitation in modulating anti-tumor immunity and identifying critical areas for further study such as the immunological effects resulting from cavitation activity ranging from gentle oscillation to violent collapse.

Taken together, this collection highlights the exciting progress and prospects for cavitation-mediated therapies, and we hope the information herein both informs and stimulates the readers to make the next important steps to understand and clinically translate techniques for cavitation-enhanced drug delivery and immunotherapy.

Conflicts of Interest: The authors declare no conflict of interest.

References

1. Imran, K.M.; Tintera, B.; Morrison, H.A.; Tupik, J.D.; Nagai-Singer, M.A.; Ivester, H.; Council-Troche, M.; Edwards, M.; Coutermarsh-Ott, S.; Byron, C.; et al. Improved Therapeutic Delivery Targeting Clinically Relevant Orthotopic Human Pancreatic Tumors Engrafted in Immunocompromised Pigs Using Ultrasound-Induced Cavitation: A Pilot Study. *Pharmaceutics* **2023**, *15*, 1585. [CrossRef] [PubMed]
2. Lacerda, Q.; Falatah, H.; Liu, J.-B.; Wessner, C.E.; Oeffinger, B.; Rochani, A.; Leeper, D.B.; Forsberg, F.; Curry, J.M.; Kaushal, G.; et al. Improved Tumor Control Following Radiosensitization with Ultrasound-Sensitive Oxygen Microbubbles and Tumor Mitochondrial Respiration Inhibitors in a Preclinical Model of Head and Neck Cancer. *Pharmaceutics* **2023**, *15*, 1302. [CrossRef] [PubMed]
3. Ahmed, M.H.; Hernández-Verdin, I.; Quissac, E.; Lemaire, N.; Guerin, C.; Guyonnet, L.; Zahr, N.; Mouton, L.; Santin, M.; Petiet, A.; et al. Low-Intensity Pulsed Ultrasound-Mediated Blood-Brain Barrier Opening Increases Anti-Programmed Death-Ligand 1 Delivery and Efficacy in Gl261 Mouse Model. *Pharmaceutics* **2023**, *15*, 455. [CrossRef] [PubMed]
4. LuTheryn, G.; Ho, E.M.L.; Choi, V.; Carugo, D. Cationic Microbubbles for Non-Selective Binding of Cavitation Nuclei to Bacterial Biofilms. *Pharmaceutics* **2023**, *15*, 1495. [CrossRef] [PubMed]
5. He, S.; Singh, D.; Yusefi, H.; Helfield, B. Stable Cavitation-Mediated Delivery of miR-126 to Endothelial Cells. *Pharmaceutics* **2022**, *14*, 2656. [CrossRef]
6. Kerneis, S.; Escoffre, J.-M.; Galvin, J.J., III; Bouakaz, A.; Presset, A.; Alix, C.; Oujagir, E.; Lefèvre, A.; Emond, P.; Blasco, H.; et al. Sonoporation of the Round Window Membrane on a Sheep Model: A Safety Study. *Pharmaceutics* **2023**, *15*, 442. [CrossRef]
7. Benton, R.P.; Al Rifai, N.; Stone, K.; Clark, A.; Zhang, B.; Haworth, K.J. Impact of Perfluoropentane Microdroplets Diameter and Concentration on Acoustic Droplet Vaporization Transition Efficiency and Oxygen Scavenging. *Pharmaceutics* **2022**, *14*, 2392. [CrossRef]
8. Martinez, P.; Bottenus, N.; Borden, M. Cavitation Characterization of Size-Isolated Microbubbles in a Vessel Phantom Using Focused Ultrasound. *Pharmaceutics* **2022**, *14*, 1925. [CrossRef] [PubMed]
9. Navarro-Becerra, J.A.; Borden, M.A. Targeted Microbubbles for Drug, Gene, and Cell Delivery in Therapy and Immunotherapy. *Pharmaceutics* **2023**, *15*, 1625. [CrossRef] [PubMed]
10. Chapla, R.; Huynh, K.T.; Schutt, C.E. Microbubble–Nanoparticle Complexes for Ultrasound-Enhanced Cargo Delivery. *Pharmaceutics* **2022**, *14*, 2396. [CrossRef] [PubMed]
11. Honari, A.; Sirsi, S.R. The Evolution and Recent Trends in Acoustic Targeting of Encapsulated Drugs to Solid Tumors: Strategies beyond Sonoporation. *Pharmaceutics* **2023**, *15*, 1705. [CrossRef] [PubMed]
12. Maardalen, M.; Carlisle, R.; Coussios, C. Cavitation-Mediated Immunomodulation and Its Use with Checkpoint Inhibitors. *Pharmaceutics* **2023**, *15*, 2110. [CrossRef]

pharmaceutics

MDPI

Article

Improved Therapeutic Delivery Targeting Clinically Relevant Orthotopic Human Pancreatic Tumors Engrafted in Immunocompromised Pigs Using Ultrasound-Induced Cavitation: A Pilot Study

Khan Mohammad Imran [1,2], Benjamin Tintera [2], Holly A. Morrison [2], Juselyn D. Tupik [2], Margaret A. Nagai-Singer [2], Hannah Ivester [1], McAlister Council-Troche [2], Michael Edwards [3], Sheryl Coutermarsh-Ott [2], Christopher Byron [4], Sherrie Clark-Deener [4], Kyungjun Uh [5], Kiho Lee [5], Paul Boulos [6], Cliff Rowe [6], Christian Coviello [6] and Irving C. Allen [1,2,*]

1 Graduate Program in Translational Biology, Medicine and Health, Virginia Polytechnic Institute and State University, Roanoke, VA 24061, USA
2 Department of Biomedical Sciences and Pathobiology, Virginia-Maryland College of Veterinary Medicine, Blacksburg, VA 24061, USA
3 Department of Small Animal Clinical Sciences, Virginia-Maryland College of Veterinary Medicine, Blacksburg, VA 24061, USA
4 Department of Large Animal Clinical Sciences, Virginia-Maryland College of Veterinary Medicine, Blacksburg, VA 24061, USA
5 Division of Animal Science, College of Agriculture Food and Natural Resources, University of Missouri, Columbia, MO 65211, USA
6 OxSonics Therapeutics, Oxford Science Park, Oxford OX4 4GA, UK
* Correspondence: icallen@vt.edu

Citation: Imran, K.M.; Tintera, B.; Morrison, H.A.; Tupik, J.D.; Nagai-Singer, M.A.; Ivester, H.; Council-Troche, M.; Edwards, M.; Coutermarsh-Ott, S.; Byron, C.; et al. Improved Therapeutic Delivery Targeting Clinically Relevant Orthotopic Human Pancreatic Tumors Engrafted in Immunocompromised Pigs Using Ultrasound-Induced Cavitation: A Pilot Study. *Pharmaceutics* **2023**, *15*, 1585. https://doi.org/10.3390/pharmaceutics15061585

Academic Editor: Maria Carafa

Received: 10 March 2023
Revised: 3 May 2023
Accepted: 22 May 2023
Published: 24 May 2023

Abstract: Pancreatic tumors can be resistant to drug penetration due to high interstitial fluid pressure, dense stroma, and disarrayed vasculature. Ultrasound-induced cavitation is an emerging technology that may overcome many of these limitations. Low-intensity ultrasound, coupled with co-administered cavitation nuclei consisting of gas-stabilizing sub-micron scale SonoTran Particles, is effective at increasing therapeutic antibody delivery to xenograft flank tumors in mouse models. Here, we sought to evaluate the effectiveness of this approach in situ using a large animal model that mimics human pancreatic cancer patients. Immunocompromised pigs were surgically engrafted with human Panc-1 pancreatic ductal adenocarcinoma (PDAC) tumors in targeted regions of the pancreas. These tumors were found to recapitulate many features of human PDAC tumors. Animals were intravenously injected with the common cancer therapeutics Cetuximab, gemcitabine, and paclitaxel, followed by infusion with SonoTran Particles. Select tumors in each animal were targeted with focused ultrasound to induce cavitation. Cavitation increased the intra-tumor concentrations of Cetuximab, gemcitabine, and paclitaxel by 477%, 148%, and 193%, respectively, compared to tumors that were not targeted with ultrasound in the same animals. Together, these data show that ultrasound-mediated cavitation, when delivered in combination with gas-entrapping particles, improves therapeutic delivery in pancreatic tumors under clinically relevant conditions.

Keywords: pancreatic cancer; paclitaxel; gemcitabine; cetuximab; drug delivery; SonoTran Particles; sonoporation; passive acoustic mapping; large animal cancer model; focused ultrasound

1. Introduction

Diagnosis of pancreatic cancer is associated with a poor prognosis, with a five-year survival rate as low as 6%. This poor prognosis can be attributed to the delayed presentation of patients with pancreatic cancer due to typically benign, or even a complete lack of symptoms during early stages. Location of the primary tumor can dictate the timing, signs and symptoms of pancreatic cancer, ascribing a more favorable prognosis to tumors causing symptoms in their earlier stages [1]. At this time, complete surgical resection is the only potentially curative treatment for pancreatic cancer; however, many patients do not qualify for tumor resection. Pancreatic cancers are assessed and categorized on a spectrum of high resectability to low resectability/unresectable. Characteristics of the primary tumor such as no arterial or venous involvement and no distant metastases could indicate a primary tumor with high resectability, while metastasis to lymph nodes excluding those that drain the peripancreatic tissues or involvement of the inferior vena cava are typically unresectable [2]. At the time of diagnosis, only 15–20% of patients with pancreatic cancer have tumors that are resectable due to the majority of patients having locally advanced or metastatic disease [2,3]. Chemotherapy and/or radiotherapy is currently considered to be adjuvant therapy to surgery in patients with borderline resectable tumors or as the first line treatment for patients with metastatic disease [4,5]. Although tumors characteristically have leaky blood vessels, pancreatic tumors are typically considered to be poorly vascularized and have a large amount of extracellular matrix, higher interstitial fluid pressure inside the tumor, and a lack of convection, which can all impede drug delivery to the tumor [6–8]. In addition to more traditional chemo- and small molecule therapeutics, monoclonal antibodies have shown promise in treating a variety of solid tumors [9,10]. However, the large molecular size of these antibodies can hinder their ability to penetrate pancreatic tumors and the use of these antibody therapies has yet to be optimized [7]. In pancreatic cancer, chemo-resistance, radio resistance and immunosuppression are all additional concerns that hinder otherwise promising therapeutic approaches [11–14]. In summary, strategies that improve the delivery and distribution of therapeutics to the tumor may significantly impact clinical outcomes [15].

Focused Ultrasound (FUS) has emerged as a noninvasive and drug-agnostic approach with the potential to improve therapeutic delivery in a variety of applications. Indeed, ultrasound-induced cavitation is an emerging approach that has shown significant promise in pre-clinical cancer studies. Of specific relevance to pancreatic cancer and the stromal niche, the use of low-intensity ultrasound combined with the co-administration of cavitation nuclei has been shown to increase drug delivery to neoplastic masses that are traditionally characterized by limited drug diffusion from the bloodstream [16–19]. For example, a mouse model using contralateral, flank-engrafted HT-29 colorectal carcinoma cells revealed significantly increased tumor concentrations of the therapeutic antibody Cetuximab [20]. Cetuximab was co-administered with cavitation nuclei consisting of either ultrasound contrast agent (SonoVue) or gas-stabilizing particles (SonoTran Particles) [20]. Ultrasound resulted in cavitation that increased tumoral Cetuximab concentrations either 2.1-fold (SonoVue) or 3.6-fold (SonoTran Particles) in US-treated tumors compared to the untreated contralateral tumor [20]. Complementing these studies, pre-clinical large animal studies have found that the SonoTran Particles combined with a clinically ready therapeutic ultrasound system (SonoTran System) are safe and effective [21] and that FUS-induced cavitation caused minimal levels of tissue damage in tissues such as the liver [22–24]. Thus, the objective of the current work is to expand upon these prior studies and evaluate the effectiveness of the SonoTran Particles combined with SonoTran System in a clinically relevant large animal model of pancreatic cancer that better recapitulates the human patient and tumor.

The overwhelming majority of preclinical animal models of cancer are rodent based. While mice are a valuable resource for understanding basic cancer biology and gaining mechanistic insight, due to their small size and differences in anatomy and physiology compared to human patients, direct therapeutic translation is often limited. This is especially true in the development and testing of biomedical devices, which often require multiple, miniaturized systems and optimization protocols that lack human relevance. To overcome these limitations, pigs are often used as human surrogates due to their similarity in size, genome, metabolism, anatomy, and physiology [25]. Indeed, for studies targeting pancreatic cancer, the pig pancreas is highly similar to the human organ, and the anatomic location presents many of the same ultrasound imaging and targeting complications that are typical challenges for human applications. However, to date, FUS applications in the pig have been generally limited to studies in the healthy pancreas and morbidity and mortality studies to demonstrate safety. While critical for proof of concept, the healthy pancreas does not effectively model or reproduce the tumor microenvironment, especially the stromal niche, which has significantly altered tissue mechanics compared to the healthy tissue [26–28]. In order to effectively take advantage of the strengths of the pig model and minimize the limitations, we recently developed a novel immunocompromised pig model that is receptive to xenografts with human tumors and tissues [22,23]. Using these *RAG2/IL2RG* knockout pigs, we have successfully engrafted multiple human cell lines, generating subcutaneous brain, breast, and liver tumors [22]. Of specific relevance to the present study, we have also effectively generated subcutaneous tumors using the human Panc-1 cell line, which is a common model of human PDAC [23]. These cells demonstrated 100% subcutaneous engraftment in our *RAG2/IL2RG* knockout pigs [23]. The tumors demonstrated characteristics that were indistinguishable from Panc-1 tumors propagated in immunocompromised mice and demonstrated hallmark features that were highly similar to characteristics typically observed in human patients [27].

In the current study, we expand the use of our novel *RAG2/IL2RG* knockout pigs and surgically engraft Panc-1 human PDAC cells in specific locations of the pancreas to generate orthotopic tumors. These clinically and physiologically relevant tumor models allow us to robustly evaluate the effectiveness of ultrasound-induced cavitation nuclei consisting of gas-stabilized particles (SonoTran Particles) in drug delivery to pancreatic tumors. Here, we characterize the orthotopic pancreatic tumors generated in the pigs and detail the effectiveness of the SonoTran System targeting. Using this model, we further show that cavitation increases the pancreatic tumor concentration of systemically delivered Cetuximab, gemcitabine, and paclitaxel. Together, these data show that noninvasive ultrasound-mediated cavitation, when delivered in combination with gas-entrapping particles, significantly improves therapeutic delivery in pancreatic tumors under clinically relevant conditions.

2. Materials and Methods

2.1. Materials

Panc-1 cells (CRL-1469) were obtained from ATCC and cultured following supplier protocols. Abraxane (paclitaxel) was obtained from Bristol Myers Squibb, New York, NY, USA. Cetuximab is a monoclonal antibody against epidermal growth factor receptor (EGFR) and was provided by Eli Lilly and Company, Indianapolis, IN, USA. Gemcitabine was obtained from Adooq biosciences LLC, Irvine, CA, USA. SonoTran Particles were manufactured by OxSonics, Oxford, UK.

The drug doses calculated for this work were extrapolated from human clinical doses. These therapeutics are administered at a dose calculated using patient surface area. To calculate a dose (mg/kg) to use in this study, for each drug, the theoretical dose for a human, minipig or micropig was calculated from literature surface area reference values. The calculated mg/kg values for a human, minipig and micro pig are shown in Table 1. Due to the increasing trend of mg/kg dose values as the recipient becomes smaller, the calculated values for micropigs were rounded upwards to provide the dosing for the immune-compromised piglet model used in this study.

Table 1. Extrapolating from clinical dose to experimental (piglet) dose.

Cetuximab Human Dose (250 mg/m^2) [29]				
	Reference [30] Surface Area (m^2)	**Calculated Dose (mg)**	**Reference [30] Weight (kg)**	**Dose mg/kg**
Human	1.62	405	60	6.8
Minipig	1.14	285	40	7.1

Table 1. *Cont.*

Cetuximab Human Dose (250 mg/m^2) [29]				
	Reference [30] Surface Area (m^2)	**Calculated Dose (mg)**	**Reference [30] Weight (kg)**	**Dose mg/kg**
Micropig	0.74	185	20	9.3
Immune-compromised piglet Cetuximab dose used in this study				10.0
Gemcitabine human dose (1000 mg/m^2) [29]				
Human	1.62	1620	60	27.0
Minipig	1.14	1140	40	28.5
Micropig	0.74	740	20	37.0
Immune-compromised piglet Gemcitabine dose used in this study				40.0
Abraxane human dose (125 mg/m^2) [31]				
Human	1.62	202.5	60	3.4
Minipig	1.14	142.5	40	3.6
Micropig	0.74	92.5	20	4.6
Immune-compromised piglet Abraxane dose used in this study				5.0

2.2. *Immunocompromised Porcine Model Generation*

To generate immunocompromised pigs, we utilized the CRISPR/Cas9 system to knock-out *RAG2* and *IL2RG* genes. Single-guide RNAs were designed to disrupt *RAG2* and *IL2RG* and BLASTed against the whole pig genome for their specificity. In vitro fertilization and microinjection of the CRISPR/Cas9 system were performed as previously described [22,32,33]. The immunocompromised status of each pig was validated through genotyping to confirm *RAG2/IL2RG* double knockout prior to engraftment surgery (Supplemental Figure S1). The piglets were harvested from the sow via a sterile hysterectomy, as described previously [33]. Ear notches were collected for genotyping shortly after birth. Immediately at birth, the immunocompromised piglets were aseptically transferred to their respective germ-free isolators. The pigs remained in the germ-free isolators during the course of the study to reduce the risk of infection. Each study method complied with the NIH Guide for the Care and Use of Laboratory Animals and the Virginia Tech Institutional Animal Care and Use Committee (IACUC).

2.3. *Generation of Orthotopic Human Panc-1 Pancreatic Tumor*

Human pancreatic ductal epithelial carcinoma, Panc-1 cells (ATCC, crl-1469) were cultured in RPMI media containing 10% fetal bovine serum and 1% Normocin (Invivogen) and lifted from culture plates with Trypsin (0.25%) in EDTA. Panc-1 cells were resuspended in Matrigel (Corning) on ice after trypsinization at a concentration of 6×10^6 cells per 100 μL. Under general anesthesia, pigs underwent ventral midline laparotomy surgery to visualize the pancreas and engraft 6×10^6 cells at two or three sites, depending on organ accessibility. The pigs resumed their regular feeding schedule at the germ-free housing facility after the surgery and post-recovery. The pigs were monitored for clinical signs of illness and tumor progression for the remaining duration of the study. Pigs were treated 26–30 days after the injection of tumor cells and the weight range of the pigs at the time of treatment was 6–7 kg. Pigs were treated as shown in Table 2, with cavitation induced in randomly selected tumors in each pig.

2.4. *SonoTran System*

The SonoTran System (OxSonics Ltd., Oxford, UK) includes an innovative hand-held curvilinear dual-frequency ultrasound probe. The probe is uniquely capable of B-mode ultrasound imaging (imaging array of 128 elements at 4 MHz central frequency), transmitting a second-frequency therapeutic ultrasound pulse (therapy array of 64 elements at 500 KHz central frequency, 8000 cycles, and 0.5 Hz Pulse Repetition Frequency (PRF)) to initiate inertial cavitation from SonoTran Particles, and, in real time, concurrently receiving acoustic emissions back from them for Passive Acoustic Mapping (PAM). These parameters have been optimized for quantifiably improved drug delivered across the vast majority of tumor models used in earlier studies [20,34,35]. PAM images correspond to the spatial images of detected acoustic energy (inertial cavitation) and is displayed as a colormap overlaid on the contemporaneous grayscale B-mode ultrasound image to localize, monitor, and achieve real-time quantification of inertial cavitation. PAM images were produced using a modified, frequency-domain Robust Capon Beamformer algorithm [36]. In order to provide real-time feedback, the calculation was implemented on a Graphics Processing Unit (GPU) using Compute Unified Device Architecture (CUDA).

Table 2. General characteristics of the animal, tumor, and treatment plan.

Pig ID	Number of Tumors	Tumor Volume (cm^3)	Treated with SonoTran (Yes/No)	Drugs Injected
1	3	1.25, 0.7, and 2.5	No	None
2	0	NA	No	None
3	3	1.05, 0.7 and 0.54	No	None
4	2	1.0 and 2.1	No	SonoTran Particles
5	3	1.6, 1.09 and 1.25	No	SonoTran Particles
6	2	1.1 and 0.53	Yes	Gemcitabine, Paclitaxel, Cetuximab and SonoTran Particles
7	2	0.3 and 0.35	Yes	Gemcitabine, Paclitaxel, Cetuximab and SonoTran Particles
8	2	0.45 and 0.39	Yes	Gemcitabine, Paclitaxel, Cetuximab and SonoTran Particles

2.5. In Vivo SonoTran Treatment

All animals were sedated using 2–4 mg/kg of Telazol-Ketamine-Xylazine (TKX) and remained under anesthesia during SonoTran treatment. Pigs were scrubbed using betadine with a special focus on the abdomen, cleaned with water, then dried, and hair covering the area of interest was trimmed using an electric trimmer. Depilatory cream was then applied to completely remove any additional hair that could hinder ultrasound imaging.

SonoTran System imaging mode (B-mode) was utilized to visualize each tumor and identify appropriate acoustic windows to treat randomly selected tumors while the pigs were under anesthesia and in a supine position (Figure 1). The pigs were then injected intravenously with a bolus dose of Cetuximab (10 mg/kg), paclitaxel (5 mg/kg), and gemcitabine (40 mg/kg). The drug concentration was determined based on the clinical dose (mg/kg) for Cetuximab and paclitaxel. A lower clinical dose (40 mg/kg) of gemcitabine was used instead of a high clinical dose of 250 mg/kg. SonoTran System therapy mode was first activated to establish that the ultrasound parameters used were insufficient to induce cavitation in the absence of SonoTran Particles. The pigs were then infused with the SonoTran Particle suspension (1.48 mg/mL in 5% Glucose) at a flow rate of 3 mL per minute and mechanical pressure was applied with the probe to move bowel gas overlying the pancreas to improve visualization and cavitation monitoring. During treatment, the SonoTran System B-mode images were combined to create PAM images were used to visualize the tumor and ensure correct ultrasound-mediated cavitation targeting. The total treatment time from start of infusion of the Sonotran particles to the completion of the study was approximately 20 min. US treatment was performed immediately after (<5 min) drug injections were finished. The SonoTran Particle suspension infusion was ongoing. Each drug was administered approximately 5 min apart. Immediately post-treatment, the pigs were euthanized using euthasol intracardiac injection following our IACUC guidelines, and a necropsy was performed, gross morphology changes of organs by the treatment were recorded and organs were collected for further analysis.

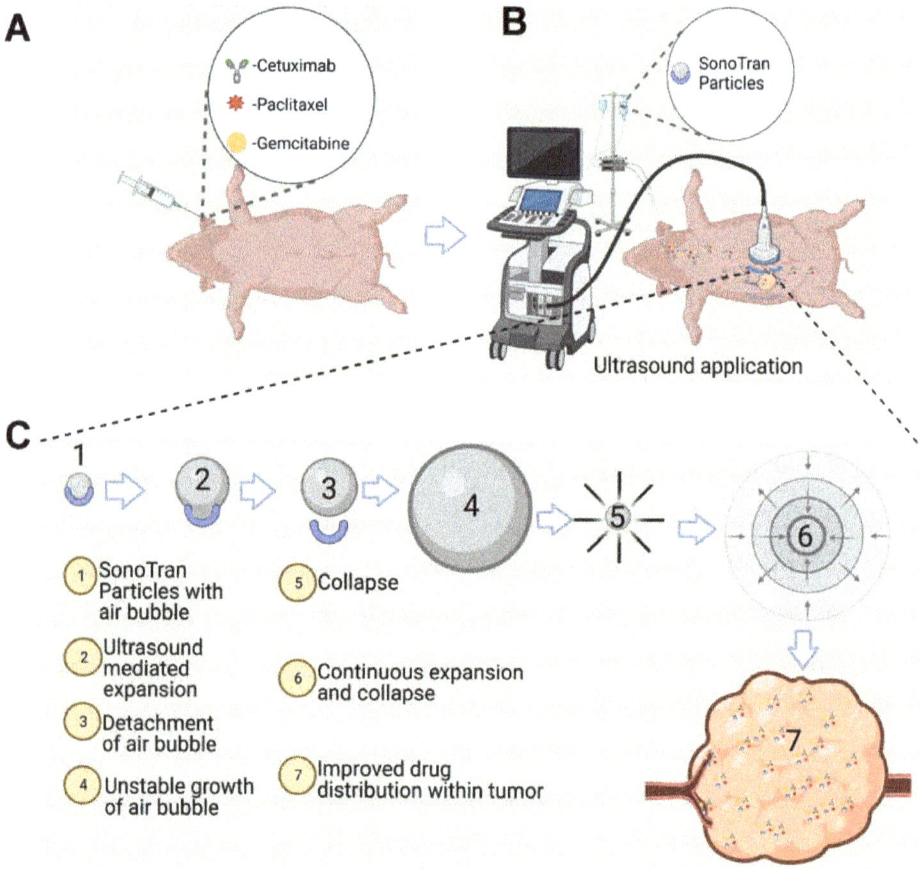

Figure 1. Schematic illustration of ultrasound-mediated cavitation using SonoTran Particles. (**A**) Cetuximab, paclitaxel, and gemcitabine are injected individually. (**B**) Infusion of gas-stabilized SonoTran Particles with concurrent ultrasound application targeting pancreatic tumor(s). (**C**) Schematics of air bubble expansion and collapse carried by the SonoTran Particles and FUS creating cavitation and improving drug distribution within the tumor.

2.6. Histopathology

Following necropsy, healthy pig pancreas, untargeted organs in the SonoTran treatment zone (i.e., gastrointestinal tract), untargeted tumors, and SonoTran targeted tumors were fixed in 10% formalin for at least 24 h prior to sectioning and staining. Tissue sections of the pig pancreas were subjected to hematoxylin and eosin (H&E) and trichrome stain to evaluate ablation, variation in collagen density and composition, damage to tumor surrounding structures (blood vessels and pancreatic ducts) in the ablation zone, and off-target damage [37]. All histopathology was evaluated by a board-certified veterinary pathologist to assess for regions of cellular damage and verify the expected degree of tissue ablation within and outside of the ablation zone.

2.7. Trichrome Image Scoring for Collagen Using FIJI

Trichrome-stained images were captured at 10X magnification from normal healthy pig pancreas, as well as SonoTran-treated and -untreated tumors. Images were taken from a total of 15 different representative regions of the pancreas. Images were subjected to color deconvolution into trichrome using FIJI and only the blue-stained regions (representing collagen in trichrome) were quantified. Images were converted into 8-bit images, duplicated, and thresholds were set to establish measurement parameters based on the mean and percent area, as previously described [38]. The percent area of analyzed particles in FIJI was calculated and used to generate the final data, as previously described [38].

2.8. Ultra-High-Performance Liquid Chromatography with Tandem Mass Spectrometry (UPLC-MS/MS)

Serum and tumor tissue was analyzed with mass spectrometry to quantify the amount of paclitaxel and gemcitabine. For mass spectrometry standards, 100 µL of serum from a healthy control pig was combined with 300 µL of the internal standard addition solution (ISAS, 1 µg/mL Gemcitabine + Paclitaxel-d5 in methanol + 1% acetic acid) in 2 mL polypropylene (PP) microcentrifuge tubes. Multiple reaction monitoring (MRM) transitions and specific mass spectrometry tuning parameters for the quantification of gemcitabine and paclitaxel are shown in Supplemental Figure S2. The protein precipitated samples were briefly shaken and then placed on a vortex table to extract for 5 min before being centrifuged (Eppendorf Microcentrifuge Model 5415R) at 16,100× *g* for 5 min. For analysis, 50 µL of the resulting supernatant solutions were then combined with 100 µL of deionized water in 2 mL amber autosampler vials with PP low-volume inserts. These were then briefly vortexed to homogenize before being placed in the refrigerated autosampler of the UPLC-MS/MS for analysis. For tissue analysis, 250 mg of tumor tissue was placed in 5 mL of acetonitrile and methanol before being mechanically homogenized. The sample was centrifuged, and the extract was removed before a secondary extraction of the homogenized tissue was performed to increase extraction efficiency. Methanol, acetic acid, and ISAS were added to the centrifuged sample post-primary extraction before being vortexed to re-homogenize prior to ultrasonic extraction for 20 min. The sample was centrifuged again at 16,100× *g* for 5 min and the supernatant was collected before being diluted, vortexed, and placed in the refrigerated autosampler of the UPLC-MS/MS for analysis. Sample extracts were subjected to chromatographic separation performed on a Waters I-Class UPLC system with an HSS T3 reverse phase column (Waters Acquity UPLC HSS T3, 100 mm length × 2.1 mm ID × 1.8 µm) and matching guard column (Waters Acquity UPLC HSS T3 VanGuard Pre-Column, 5 mm length × 2.1 mm ID × 1.8 µm) maintained at 40 °C. Four microliters of sample were injected onto the column using a refrigerated autosampler maintained at 5 °C. Mobile phase A consisted of 5 mM ammonium acetate ($NH_4CH_3CO_2$) + 0.05% acetic acid in H_2O, mobile phase B consisted of 5 mM ammonium formate + 0.1% formic acid in 95/5 Acetonitrile/H_2O, and mobile phase C was 100% MeOH. The mobile phase was delivered to the UPLC column at a flow rate of 0.4 mL per min.

2.9. Cetuximab ELISA

ELISA was used to quantify the amount of Cetuximab, which is an antibody-based therapeutic, in each specimen. The ELISA was not specific to Cetuximab but was suitable for any human antibody in pig blood and tissue samples. An untreated tumor from a pig that did not receive Cetuximab was used as a negative control. Tumor samples were weighed and homogenized in 1× cell culture lysis reagent (Promega, Hampshire, UK) and protease inhibitor cocktail (Abcam, Cambridge, UK) at a concentration of 1.25 mg/mL. Known quantities of Cetuximab were added to this tumor homogenate and serially diluted to generate a standard curve to quantify the amount of Cetuximab in the experimental specimens. For each ELISA, 50 µL of the primary antibody (Goat anti-human IgG-Fc specific, Sigma-Aldrich, St. Louis, MO, USA) was used to coat the wells of a 96-well high binding ELISA plate at a concentration of 1 µg/mL before being incubated overnight at 4 °C. Wells were washed at room temperature with wash buffer (PBST, 0.1% Tween 20 in 1× PBS), ensuring bubbles and liquid were removed from the wells at each wash. The wells were then blocked with 200 µL of 5% BSA in 1× PBST for at least one hour. The wells were washed again with wash buffer, and 50 μL of the standards or samples was added to the plate in either triplicate or duplicate, before being left to incubate for 1.5 h. The wells were washed again with wash buffer before being incubated with goat anti-human horseradish peroxidase-conjugated secondary antibody (Promega, Hampshire, UK) diluted to 1:2500 in 5% BSA in 1× PBS for 1.5 h at room temperature. After the wells were washed three times with wash buffer, TMB substrate (ThermoFisher Scientific, Oxford, UK) was added to the wells and incubated in the dark until a deep blue color had developed in the highest concentration standards. To stop the reaction, 50 μL of 1M H_2SO_4 was added per well and absorbance at 450 nm was determined using a plate reader (SpectraMax M5, Molecular Devices, San Jose, CA, USA). Absorbance values from control samples were used to generate a Four Parameter Logistic (4PL) Curve in AAT Bioquest website and sample absorbance values were extrapolated from this curve.

2.10. Statistical Analyses

Unless otherwise noted, data were analyzed using GraphPad Prism, version 9.0. Statistical significance was indicated as $p < 0.05$. All data are denoted as the mean ± SEM (ELISA and MS data show individual animal data points). A student's two-tailed paired t-test was performed when evaluating two experimental groups or an ANOVA was utilized with Tukey's post-test for complex datasets.

3. Results

3.1. Orthotopic Pancreatic Tumors Were Successfully Generated in the RAG2/IL2RG Deficient Immunocompromised Pigs

Generation of the *RAG2/IL2RG* knockout pigs was conducted as previously described [23,24,33]. Briefly, gRNAs targeting pig *RAG2* and *IL2RG* genes and Cas9 mRNA were injected into presumable zygotes following in vitro fertilization and cultured for five days. The injected embryos were transferred to surrogate gilts. Piglets were delivered through hysterectomy into a germ-free housing system, 114 days after in vitro fertilization. A total of eight piglets were utilized for the studies described. Genomic DNA from piglet ear notches revealed that both the *RAG2* and *IL2RG* genes were successfully disrupted in all eight piglets (Supplemental Figure S1). Abdominal surgery was performed under anesthesia to visualize and inject the Panc-1 human pancreatic cancer cells into the pancreas for xenograft implant. A total of 100 μL containing 6×10^6 Panc-1 cells was injected into specific regions of each pancreas, with each pig engrafted with either two or three tumors. Pigs were relocated to their germ-free isolators once they had fully recovered from anesthesia and they were clinically monitored throughout the remaining duration of the study. Every pig developed at least one tumor, with an orthotopic xenograft success rate of 94.44% (Table 2). Gross in situ tumors and excised formalin-fixed tumors are shown in Figure 2A,B respectively. Tumors were well tolerated without any illness and other abnormalities for the duration of this study and pigs were maintained for up to 45 days post-tumor engraftment, which corresponded to tumors reaching the average targeted treatment size of approximately 1 cm in diameter. The tumors were nodular and highly similar to tumors generated in immunocompromised mice, demonstrating multiple features that were consistent with tumors evaluated from human PDAC patients. The tumors were significantly stiffer than the surrounding pancreatic tissue and did not show regions of central necrosis, which is often found in mouse pancreatic tumor models.

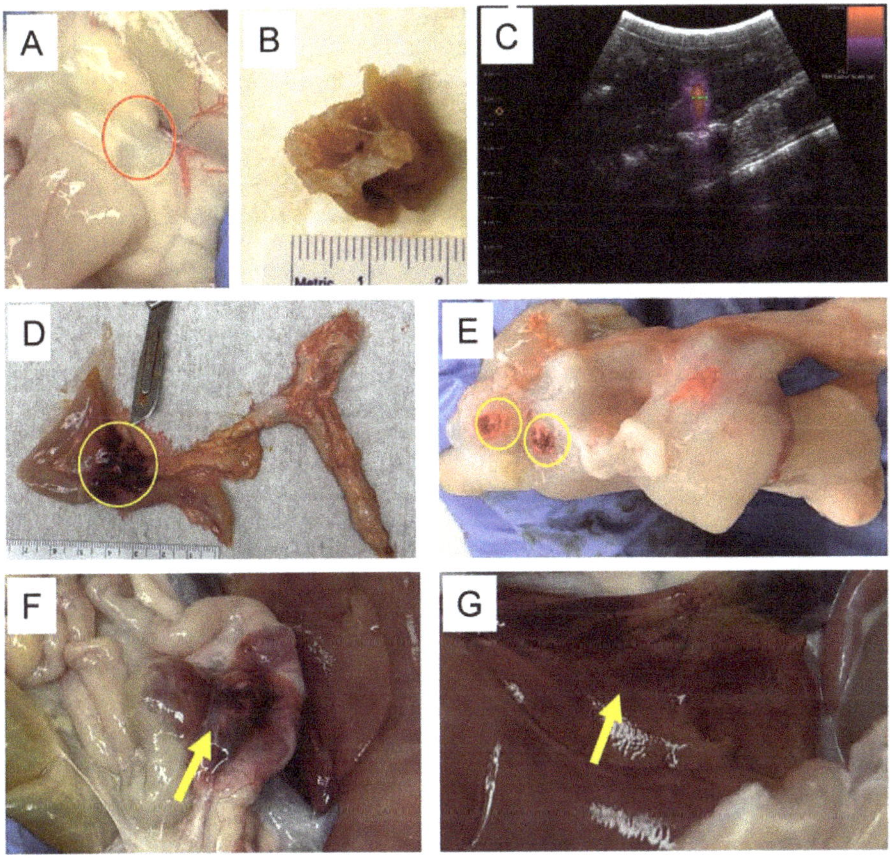

Figure 2. Panc 1 human pancreatic cancer cells were successfully engrafted in the pig pancreas and were effectively targeted using focused ultrasound. (**A**) Representative gross image of a human Panc-1 tumor in situ. (**B**) Excised human Panc-1 tumor. (**C**) US image of ongoing SonoTran therapy. Red color indicates active therapy targeting pancreatic tumor. (**D**) Gross image of human Panc-1 tumor grown in immunocompromised pig pancreas and hemorrhage from ultrasound-induced cavitation (indicated by yellow circle). (**E**) Gross image of Panc-1 tumor and hemorrhage following ultrasound-induced cavitation (indicated by yellow circles). (**F**) Off-target hemorrhage in the intestine (Yellow arrow). (**G**) Off-target hemorrhage in the liver (Yellow arrow).

3.2. FUS-Mediated Cavitation Using SonoTran Particles Effectively Targeted Orthotopic Panc-1 Tumors in the Pancreas, with Predominantly Superficial Off-Target Effects

For the current study, US was applied noninvasively through an abdominal window determined by using SonoTran System combined to SonoTran Particles. The pancreas was imaged, and tumors were detected and targeted within the ultrasound acoustic window in each pig (Figures 1 and 2). Prior to SonoTran treatment, we intravenously infused the pigs with Cetuximab (10 mg/kg), paclitaxel (5 mg/kg), and gemcitabine (40 mg/kg) (Figure 1A). Drug injections were then followed by infusion with SonoTran Particles (3 mL per minute) using an infusion pump to maintain a stable infusion rate during the US application (Figure 1B,C). SonoTran treatment was performed for around 20 min, and the FUS-mediated cavitation treatment phase was visualized using PAM images (the red area corresponding to the prediction of induced cavitation; Figure 2C). For the three pigs treated, inertial cavitation was efficiently induced in the pancreas. Cavitation energy measured by PAM displayed as the maximum energy value in Joules per FUS-pulse was used as a reference to cavitation activity. These "PAM max" values are plotted against time and peak rarefactional focal pressure (PRFP) in MPa (Supplemental Figure S3). Immediately upon the conclusion of SonoTran Treatment, pigs were euthanized, and a necropsy was performed. During the necropsy, both targeted and nontargeted tissues and organs were collected for subsequent analysis. Targeted tumors were readily identified by gross pathological lesions and regions of hemorrhage, which were grossly identified on all targeted and FUS treated tumors in the pancreas (Figure 2D,E). Grossly, off-target effects were also observed, almost exclusively in regions of the gastrointestinal tract that overlaid the pancreas within the acoustic window or injury to the liver in regions that were adjacent to the targeted region of the pancreas (Figure 2F,G). While these lesions were clearly identified grossly, histopathological assessments revealed that most of the off-target damage was superficial bruising or minor hemorrhaging (Figure 3). Hemorrhage was observed histopathologically in regions of the healthy pancreas near or adjacent to the tumors (Figure 3A). We also observed histopathologic evidence of hemorrhage in regions of the small intestine, especially in areas that overlay the pancreas and were within the acoustic window, and areas of the liver that were adjacent to the pancreas near the treatment zone (Figure 3B,C). Hemorrhaging in the nontargeted regions of the pancreas, intestine, and liver generally presented as small, punctate spherical patches spread over a wide area of the tissue (Figure 3A–C). No hemorrhaging or tissue damage was observed in the pancreas or tumors from animals that received the SonoTran Particles and therapeutics, but were not targeted with the US (control tumors, Figure 4A,B).

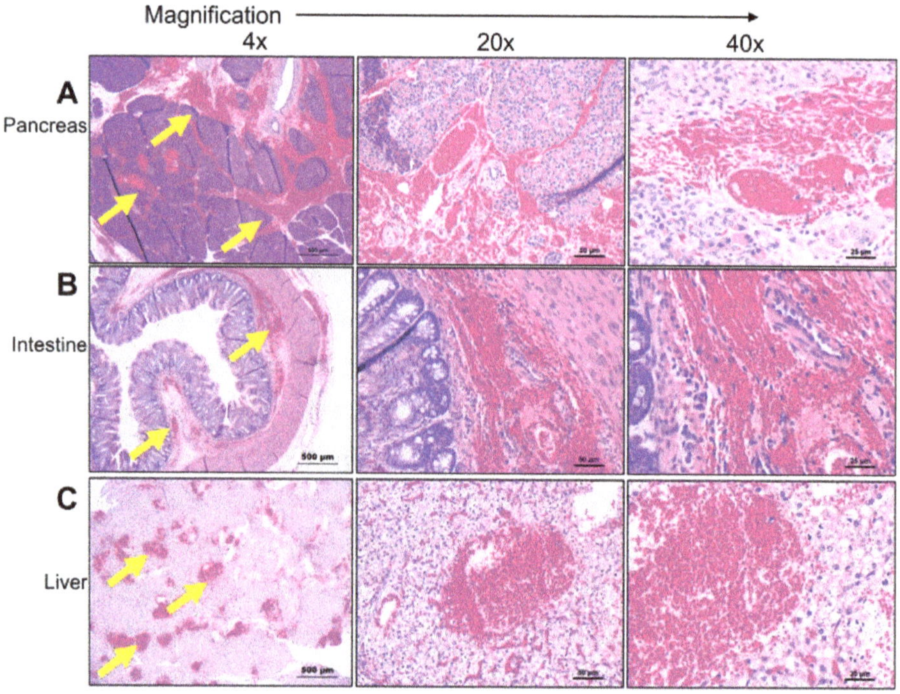

Figure 3. Histopathology assessments reveal ultrasound-mediated cavitation hemorrhaging in healthy organs within the treatment zone. Representative images of H&E-stained tissue sections were generated during histopathological evaluation of treatment effectiveness. (**A**) Treated immunocompromised pig pancreas showing minor to moderate hemorrhaging (yellow arrow) within the US treatment window of the healthy pancreas. (**B**,**C**) Additional off-target hemorrhaging was observed in the (**B**) intestine and (**C**) liver of treated animals. (**A**–**C**) Left to right- 4×, 20× and 40×, respectively. Scale bars are 500 µm for 4×, 50 µm for 20×, and 25 µm for 40× magnification.

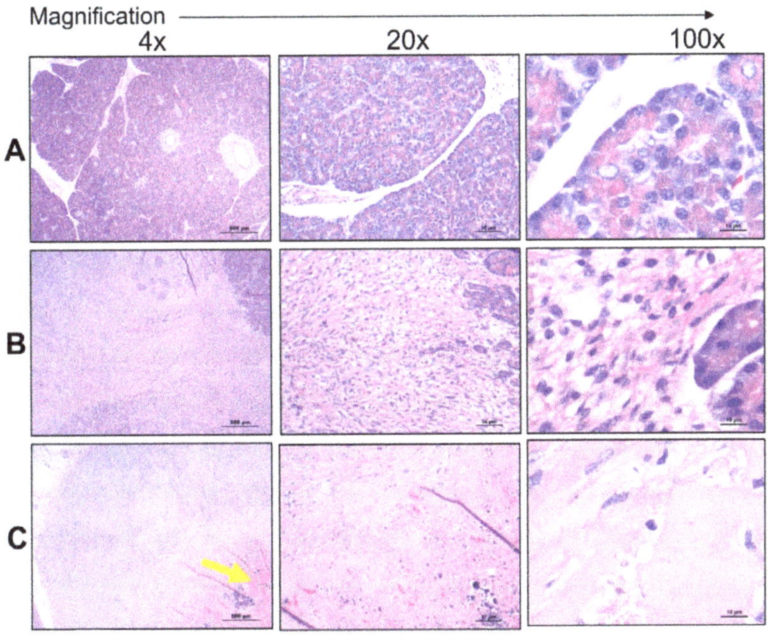

Figure 4. Effective targeting of human Panc-1 orthotopic porcine tumors in situ using ultrasound-mediated cavitation. Representative images of H&E-stained tissue sections were generated during the histopathological evaluation of treatment effectiveness in pigs with engrafted tumors. (**A**) Immunocompromised pig pancreas without ultrasound treatment. (**B**) Human Panc-1 tumors were effectively generated in the *RAG2/IL2RG*-deficient pigs and demonstrate histopathological features consistent with orthotopic cell line-based tumor models and human patient tumors. (**C**) Ultrasound-mediated cavitation resulted in clear margins defining the treatment zone with altered tumor structural features and areas of increased hemorrhage (yellow arrow). (**A–C**) Left to right- 4×, 20× and 100×, respectively. Scale bars are 500 μm for 4×, 50 μm for 20×, and 10 μm for 100× magnification.

3.3. Ultrasound-Induced Cavitation with SonoTran Particles Increases Drug Concentrations in Targeted Pancreatic Tumors

Collagenous stroma is a significant factor in chemoresistance, in part, by inhibiting drug penetration within the tumor [14,39,40]. Thus, we predict that the US and SonoTran therapy would significantly improve drug delivery. To evaluate this hypothesis, we evaluated the tumor deposition of three common cancer therapeutics: Cetuximab, gemcitabine, and paclitaxel (Figure 5). Each therapeutic was individually infused, intravenously, into pigs. Upon completion of the therapeutics injection, animals were infused with the SonoTran Particles, and a simultaneously selected tumor in each animal was targeted with FUS. Upon completion of SonoTran infusion and FUS treatment, the animals were euthanized, and drug concentrations were determined in the FUS-treated and untreated tumors in each animal using either ELISA for the antibody Cetuximab (Figure 5A) or mass spectrometric analysis (UPLC-MS/MS) to quantify gemcitabine and paclitaxel (Figure 5B,C). We observed a 4.77-fold trending increase in the concentration of Cetuximab in the FUS-targeted tumors compared with the tumors that were not targeted with FUS (Figure 5A, $p = 0.14$). Likewise, we observed a significant increase in gemcitabine in the FUS-targeted tumors compared with the tumors that were not targeted with FUS, with levels increasing from 1.58 to 2.34 μg/g of wet tissue (Figure 5B, $p < 0.05$). Similar to the increased concentrations of Cetuximab and gemcitabine, we also observed a consistent increase in paclitaxel in the FUS-targeted tumors, with levels increasing from 5.57 to 10.78 μg/g of wet tissue (Figure 5C, $p = 0.18$). However, there was more variability in the paclitaxel tumors treated with FUS (Figure 5C). Thus, these data were trending but did not achieve statistical significance. Taken together, our data suggest that ultrasound-induced cavitation with SonoTran Particles increases drug concentrations in targeted pancreatic tumors.

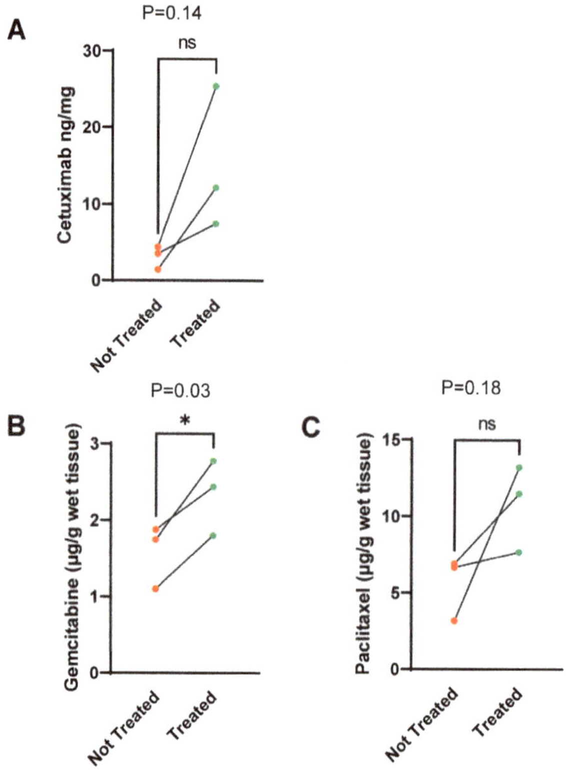

Figure 5. Ultrasound-mediated cavitation using cavitation nuclei increases the tumor delivery of cetuximab, gemcitabine, and paclitaxel. ELISA and mass spectrometry (UPLC-MS/MS) was used to quantify Cetuximab, gemcitabine, and paclitaxel, respectively. (**A**) ELISA quantification of Cetuximab comparing ultrasound targeted (treated) and untargeted (untreated) tumor tissue shows significant accumulation in the treated tumor. (**B**,**C**) MS analysis of gemcitabine and paclitaxel comparing treated and untreated tumor tissue reveals increased delivery of both drugs in the ultrasound-treated tumor compared to the untreated tumor. ns denotes not significant; * denotes $p < 0.05$.

3.4. FUS-Mediated Cavitation Using SonoTran Particles Increased Tissue Disruption and Reduced Stromal Collagen in the Tumor Microenvironment

The dense tumor stroma is a significant inhibitor of drug delivery in many different tumor types, including pancreatic tumors. As shown in Figure 4A, the healthy pig pancreas is composed of lobules separated and defined by connective tissue septae. Lobules of the pancreas contain clusters of exocrine acini cells. In the endocrine section of the pancreas, fully vascularized areas of several cell types called Islets of Langerhans are embedded within the pancreatic exocrine tissue. Cell cytoplasm and nuclei are clearly defined histopathology assessments of H&E-stained tissue (Figure 4A). Nodular Panc-1 tumors, with clear margins, were successfully engrafted in specific regions in the pig pancreas near anatomic landmarks to facilitate ultrasound targeting (Figure 4B). Histopathological evaluation of the tissue revealed that the tumors have no distinct lobules and cells were disarrayed or had a stream-like feature (Figure 4B 100×, right panel). Hemorrhaging around and within SonoTran targeted tumors was readily observed (Figure Figures 3A and 4C). Due to the mechanical force created by the continuous expansion and collapse of the bubbles stabilized in the SonoTran Particles (Figure 1C), and the application of higher cavitation energy to support increased drug delivery, we anticipated that the cavitation may reduce stromal density within treated tumors. To evaluate the effect of the therapy on the stroma density, we evaluated collagen content using trichrome staining and semi-quantification of density using FIJI image analysis (Figure 6). Consistent with pancreatic tumors in human patients, the Panc-1 tumors demonstrated a significant, 11-fold increase in collagen content compared to the normal, healthy pig pancreas (Figure 6A vs. Figure 6B, and quantification in Figure 6D). However, following FUS with SonoTran Particles, we observed a significant, 3.5-fold decrease in collagen compared to the untreated tumors (Figure 6B vs. Figure 6C and quantification in Figure 6D). Earlier studies have determined the micro-heterogeneity of collagen in human pancreatic tumors and showed that the amount of collagen is significantly associated with the biomechanical stiffness of tumors and inversely associated with vascular perfusion [37].

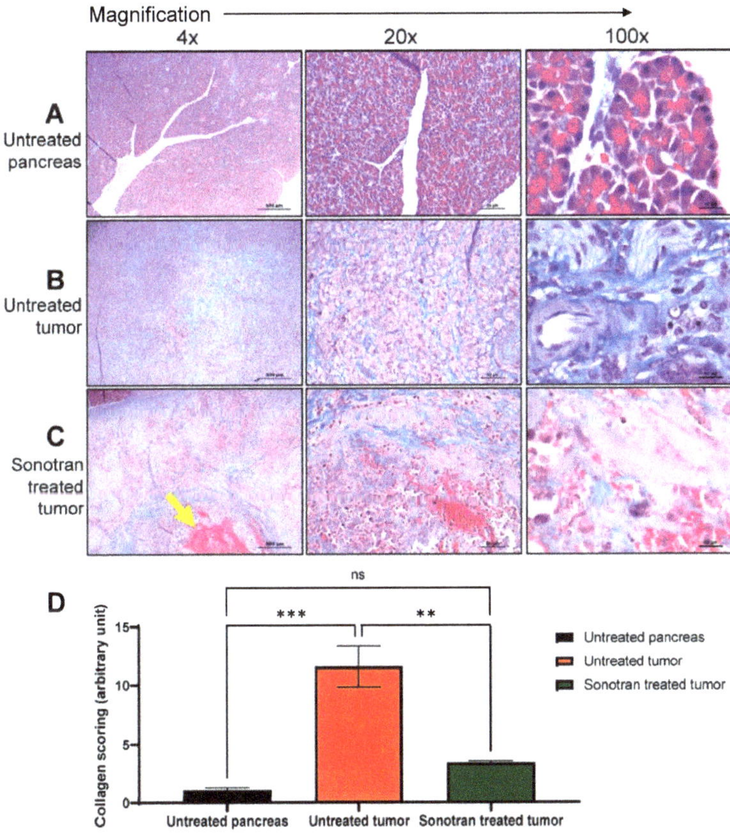

Figure 6. Trichrome stain reveals differences in collagen deposition and stroma among control, tumor and SonoTran-ultrasound-treated tumors. Representative tumor sections were stained with Trichrome for analysis of collagen deposition and tumor stroma assessments. (**A**) Immunocompromised pig pancreas without ultrasound treatment. (**B**) Orthotopic Panc-1 tumor in pig pancreas, without ultrasound treatment. (**C**) Panc-1 tumor treated with ultrasound-mediated cavitation using SonoTran Particles. (**D**) Quantification of trichrome images using FIJI image analysis reveals a reduction in collagen (blue staining in the images) after therapy. (**A–C**) Left to right- 4×, 20× and 100×, respectively. Scale bars are 500 µm for 4×, 50 µm for 20×, and 10 µm for 100× magnification. Areas of increased hemorrhage are noted with the yellow arrow. ns denotes not significant, ** denotes $p < 0.01$, *** denotes $p < 0.001$.

4. Discussion

Despite therapeutic progress over the last decade, pancreatic cancer has consistently ranks as one of the most lethal malignancies. A major driver of the dismal survival rate among patients is associated with the lack of effective chemotherapy treatment strategies and low response rates for highly promising immunotherapeutics. The lack of drug penetration into the solid tumor and poor drug distribution within the tumor microenvironment are typically cited as reasons for the failure of these therapeutics [7,41]. In pancreatic cancer, poor drug distribution is commonly driven by dense stroma formation and its physical/mechanical barriers. For example, the tumor interstitium can include a dense collagenous matrix that can constitute a total volume of up to 40% of the tumor [7]. These tumors can also have higher interstitial pressure inside the tumor compared to the periphery, with a lack of convection, and uneven or disarrayed distribution of vasculature-mediated blood supply can contribute to poor, nonuniform and uneven drug distribution [7,41–43]. Complementing these physical and environmental barriers, the development of chemoresistance is often relatively high for pancreatic cancer [26,42–46]. Beyond the physical complications associated with the tumor, many of the therapeutics used to treat pancreatic cancer also have significant limitations. For example, most drugs have high levels of detrimental side effects, off-target adverse events, and lower quality of life, due in part to the necessarily toxic nature of the drugs. This is partially driven by the systemic application of the therapeutics necessary to achieve the locally high doses necessary to impact tumor progression. Many of the current drugs used for pancreatic cancer also have relatively poor in vivo stability, which also necessitates high dosage ranges to compensate for drug breakdown. These issues are compounded by reduced local delivery to the tumor-associated with attenuated extravasation from the circulation and minimal penetration into the tumor, again driven in part by dense stroma formation [42,43,47]. Together, these represent significant challenges for pancreatic cancer drug delivery and refinements are critical to improving patient survival.

To address these issues, our results demonstrate that combining noninvasive focused ultrasound, coupled with co-administered cavitation nuclei consisting of gas-stabilizing SonoTran Particles, is effective at increasing local drug delivery for all three of the therapeutics tested. The greatest increase was observed for Cetuximab, for which our approach effectively doubled the local tumor concentration. Similarly, the combined use of focused ultrasound coupled with cavitation nuclei significantly increased gemcitabine, and we consistently observed a trend towards increased paclitaxel. As this was a pilot study with a limited number of animals, trending was observed and further studies are required to increase statistical power. The Cetuximab data are highly promising because it is a monoclonal antibody-based therapeutic that targets the extracellular domain of the epidermal growth factor receptor (EGFR). Prior Cetuximab studies have shown convincing results in preclinical mouse studies of pancreatic cancer [48,49]. However, the findings from human studies have been more mixed. For example, a Phase II study combining Cetuximab with gemcitabine in locally advanced or metastatic pancreatic cancer patients reported promising results [50]; however, a larger prospective phase II/III trial did not reveal any benefit or minimal benefit of Cetuximab [51,52]. Gemcitabine is one of the primary chemotherapy drugs used to treat pancreatic cancer and, as mentioned above in the Cetuximab trial, is commonly combined with other therapeutics. Of particular relevance to the current study, it is commonly administered with paclitaxel formulations, such as the Nab-paclitaxel (Abraxane®) combination used here. In early clinical trials, gemcitabine was proven effective as a single-agent therapeutic, improving symptoms and showing clinical benefits in approximately 20–30% of patients, with survival rates increasing by as much as 18% [53]. However, while these results are promising, the rise of chemotherapy-resistant tumors is still a significant challenge with gemcitabine, indicating the need for combination approaches [39]. Recently, clinical trials using combined gemcitabine and Nab-paclitaxel have produced impressive clinical effectiveness and safety profiles in both locally advanced and metastatic pancreatic cancer in patient populations [54]. Despite this progress, the overall survival rate for pancreatic cancer patients, especially those presenting with advanced stage, remains dismal, with poor five-year survival rates as low as 2–9% [55]. For each of these therapeutics, the suboptimal clinical effects and overall attenuated efficacy have been attributed to a combination of factors that includes high toxicity due, in part, to the elevated dosage necessary to effectively deliver the drugs to the pancreatic tumor and the molecular mechanisms of the drugs that can limit cellular uptake and tumor distribution. Based on our findings, combining ultrasound with co-administered cavitation nuclei has the potential to overcome these limitations by increasing the local concentration (i.e., drug infiltration into the tumor) of these therapeutics, thus enabling improved dosing and limiting the toxicity profiles of the therapeutics. Future studies can explore whether the cavitation-enhanced delivery of Cetuximab, gemcitabine, and paclitaxel results in improved therapeutic efficacy.

Our findings associated with improved drug delivery using ultrasound-induced cavitation are consistent with prior large-animal studies and rodent findings. Specifically related to the use of the SonoTran platform, prior studies have shown effective cavitation induction, detection, and display of cavitation events in real time using a passive acoustic mapping approach in various targeted tissues (e.g., liver) for up to one hour in large animal models [21]. These findings demonstrated the feasibility and safety of controlled cavitation in a large animal using a clinic-ready platform [21]. Expanding upon these data, a second study in rodents has also been reported, which utilized ultrasound-mediated cavitation nucleated by gas-entrapping nanoparticles, whereby the delivery of Cetuximab was found to be significantly improved in xenograft tumors [20]. Human HT-29 colorectal adenocarcinoma cells, which are strongly positive for EGFR, were engrafted bilaterally in the mouse flank and used as models to evaluate drug delivery [20]. Similar to the studies described here for the pig pancreatic cancer studies, Cetuximab was co-administered intravenously with cavitation nuclei (either SonoVue ultrasound contrast agent or SonoTran Particles) and one of the two tumors was exposed to focused ultrasound [20]. In the mouse study, cavitation increased Cetuximab concentrations either 2.1-fold with SonoVue or 3.6-fold with SonoTran Particles [20]. Together, these prior studies are highly similar to the findings we report in the current work. However, the prior studies were somewhat limited by the models utilized. For example, the prior large animal studies focused on monitoring cavitation in healthy tissues, which have vastly different physiology and mechanical properties compared to tumor tissue [22–24]. Likewise, the mouse studies are highly informative, albeit not highly relevant to in situ/orthotopic tumors or for direct clinical translation. Thus, the data generated here using our novel pig models with in situ relevant pancreatic tumors, surgically localized in areas of the pancreas that are difficult to both target and treat using traditional approaches, provide an ideal model for the evaluation of this drug delivery approach. In this model of human pancreatic cancer, targeted cavitation was readily observed and monitored in real time. Likewise, we observed significant increases in drug delivery in ultrasound-targeted tumors. It should be mentioned that off-target effects were clearly noted in the studies described here, especially hemorrhaging in the gastrointestinal tract that overlaid the pancreas. While this was pathologically determined to be superficial and predicted to have minimal clinical impact, the description of these events is critical for establishing the safety of the treatment strategy. It should also be noted that the tumors were treated with larger cavitation energies to maximize potential stromal penetration compared to prior treatments in large animals and mice in the described studies [20,21], which likely contributed to many of these off-target effects. Together, these studies are highly complementary to prior work in the field and further demonstrate the feasibility of using this approach to increase drug deposition in pancreatic tumors under highly relevant clinical conditions.

The effective deployment of our novel pig models described here is a significant innovation in drug delivery studies. Indeed, the lack of effective pre-clinical animal models of pancreatic cancer has contributed to the limited progress in drug development. In the biomedical device field, it is highly common to utilize pigs for the development and testing of human-relevant systems to evaluate safety and proof-of-concept studies in healthy tissues. Likewise, these studies are often complemented using mouse or rat models of either subcutaneous or orthotopic tumors and miniaturized devices with modified protocols to account for the small size of the animal models. This is particularly challenging in the mouse pancreas, which does not share many anatomical similarities with the human organ; additionally, its small size typically excludes practical orthotopic studies. Thus, the pig models deployed here minimize these limitations and provide us with a human-relevant model for therapeutic assessments. The ability to surgically engraft human tumors in the pancreas provides us with a highly robust, controllable, and reproducible model of pancreatic cancer. As noted in our results, the in situ pancreatic tumors that develop in these pigs are identical to tumors generated in orthotopic mice from Pan02 cell lines and demonstrate a variety of features that are consistent with tumors isolated from human patients [22–24]. While we have found this model to be ideal for biomedical device development and drug-targeting studies using human cell lines and patient-derived xenograft tumors, the lack of a functional immune system can limit some studies that require the immune niche. Beyond the pancreas, we have utilized these animals to engraft human liver, brain, breast, and sarcoma cell lines [22–24,56], which will be utilized in future studies to expand upon the findings related to ultrasound-induced cavitation and drug delivery.

The development of a dense collagenous stroma is one of the features of the porcine Panc-1 model that enables effective, realistic studies of drug delivery to pancreatic tumors. As described in the present work, the Panc-1 tumors demonstrate increased intra-tumoral collagen deposition, along with reduced and disarrayed vasculature (Figure 4). Following focused ultrasound treatment and cavitation, we observed significant disruption to this collagen matrix and increased hemorrhage in the targeted regions of each tumor. This is consistent with other prior studies that have reported damage to blood vessels through nanoparticle cavitation, which has been speculated to normalize the interstitial pressure inside of the tumor, leading to improved drug delivery and ultimately reducing the tumor burden [57]. Drugs can be distributed within a solid tumor through diffusion and convection. Small-molecule chemotherapeutics, such as 5-FU (130.078 Da), gemcitabine (263.198 Da), and paclitaxel (853.92 Da), enter the tumor predominantly through diffusion from the blood vessels, whereas, large molecules such as monoclonal antibodies, i.e., Cetuximab (134 kDa) predominantly enter the tumor through convection [7]. Intriguingly, mathematical models of drug infusion within solid tumors have shown that a molecule of 150 kDa would take up to a month to evenly distribute within a 1 cm^3 tumor [7,58,59], which significantly minimizes the clinical utility of such therapeutics. Thus, creating convection force surrounding the tumor can improve diffusion and convection-mediated small molecule drug delivery. Additionally, convection created by the expansion and collapse of the SonoTran Particles bubbles likely increased the distribution of the large molecule Cetuximab within the tumor. This process was likely augmented by the reduced collagen matrix observed in the tumors following treatment, which is expected to further improve drug deposition. The reduction in collagen demonstrates the reduction in the extracellular matrix. However, it should be noted that other physiological properties, such as poor vascularization and high interstitial fluid pressure, can also impact drug delivery and were not directly evaluated.

Significant effort has been focused on improving drug delivery and distribution to benefit pancreatic cancer patients [60–66]. We described successful use of focused ultrasound, coupled with co-administered cavitation nuclei consisting of gas stabilizing SonoTran Particles, to improve the in situ deposition of clinically relevant drugs for pancreatic cancer. We can conclude that the innovative use of human pancreatic cancer xenograft in large animals will provide a highly effective and clinically relevant model for device development studies. Future studies will explore potential mechanisms to mitigate the off-target impacts of the treatment by optimizing treatment parameters; optimize the magnitude of drug delivery to targeted tumors; and evaluate the ability of this approach to impact morbidity and mortality. In conclusion, this study is the first to show quantifiable increases in the levels of both small-molecule chemotherapeutics and large-molecule antibodies delivered to pancreatic tumors in a large animal model of clinical relevance to human patients.

Supplementary Materials: The following supporting information can be downloaded at: https://www.mdpi.com/article/10.3390/pharmaceutics15061585/s1, Supplemental Figure S1. Genotyping of piglets carrying modified *RAG2* and *IL2RG* genes; Supplemental Figure S2. MRM transitions and specific mass spectrometry tuning parameters for the quantification of gemcitabine and paclitaxel; Supplemental Figure S3. SonoTran-mediated cavitation treatment of pancreas tumors.

Author Contributions: Conceptualization, K.M.I., C.R., C.C. and I.C.A.; Data curation, K.M.I., M.C.-T., M.E., S.C.-O. and I.C.A.; Funding acquisition, I.C.A.; Investigation, K.M.I., B.T., H.A.M., J.D.T., M.A.N.-S., H.I., M.C.-T., M.E., S.C.-O., C.B., S.C.-D., K.U., K.L., P.B., C.R., C.C. and I.C.A.; Methodology, K.M.I., B.T., H.A.M., J.D.T., M.A.N.-S., H.I., M.C.-T., M.E., S.C.-O., C.B., S.C.-D., K.U., K.L., P.B., C.R., C.C. and I.C.A.; Writing—original draft, K.M.I., B.T. and I.C.A.; Writing—review & editing, K.M.I., B.T., H.A.M., J.D.T., M.A.N.-S., H.I., M.C.-T., M.E., S.C. O., C.B., S.C.-D., K.U., K.L., P.B., C.R., C.C. and I.C.A. All authors have read and agreed to the published version of the manuscript.

Funding: This research was supported by the Virginia Tech Foundation, The Focused Ultrasound Foundation, The Virginia Tech Institute for Critical Technology and Applied Sciences Center for Engineered Health, OxSonics Therapeutics, and the Virginia Maryland College of Veterinary Medicine (I.C.A.).

Institutional Review Board Statement: All animal experiments were approved and carried out in accordance with the Virginia Tech Institutional Animal Care and Use Committee (IACUC) under approved protocol 19-196 (approved 30 March 2020).

Informed Consent Statement: Not applicable.

Data Availability Statement: The data that support the findings of this study are reported in the manuscript and in supplemental data. Any other relevant replicate or data are available on request from the corresponding author.

Acknowledgments: We would like to thank Paige Eversole, Justin Markov Madanick, and Katie Orr for technical support on this project. We also appreciate the efforts of our animal support core for assistance with animal monitoring and care, especially Betsy Midkiff, Amy Rizzo, Rachel McNeill, Karen Hall, Kimberly Skroupa, Calvin Lau, and Kevin Pelzer. We would also like to acknowledge the Virginia Tech Animal Laboratory Services at the Virginia Maryland College of Veterinary Medicine for providing histopathology staining.

Conflicts of Interest: Dr. C. Rowe, Dr. P. Boulos, and Dr. C. Coviello are employees of OxSonics Therapeutics. Dr. C. Coviello is a founder and shareholder in OxSonics Therapeutics, which holds intellectual property relating to polymeric sonosensitive particles and ultrasound technology. The authors have no additional relevant affiliations or conflicting financial interests or any other personal relationships to disclose.

References

1. Kalser, M.H.; Barkin, J.; Macintyre, J.M. Pancreatic cancer. Assessment of prognosis by clinical presentation. *Cancer* **1985**, *56*, 397–402. [CrossRef] [PubMed]
2. Ryan, D.P.; Hong, T.S.; Bardeesy, N. Pancreatic adenocarcinoma. *N. Engl. J. Med.* **2014**, *371*, 1039–1049. [CrossRef] [PubMed]
3. Vareedayah, A.A.; Alkaade, S.; Taylor, J.R. Pancreatic adenocarcinoma. *Mo. Med.* **2018**, *115*, 230.
4. Khorana, A.A.; Mangu, P.B.; Berlin, J.; Engebretson, A.; Hong, T.S.; Maitra, A.; Mohile, S.G.; Mumber, M.; Schulick, R.; Shapiro, M. Potentially curable pancreatic cancer: American society of clinical oncology clinical practice guideline update. *J. Clin. Oncol.* **2017**, *35*, 2324–2328. [CrossRef] [PubMed]
5. Sohal, D.P.; Kennedy, E.B.; Cinar, P.; Conroy, T.; Copur, M.S.; Crane, C.H.; Garrido-Laguna, I.; Lau, M.W.; Johnson, T.; Krishnamurthi, S. Metastatic pancreatic cancer: ASCO guideline update. *J. Clin. Oncol.* **2020**, *38*, 3217–3230. [CrossRef] [PubMed]
6. Olive, K.P.; Jacobetz, M.A.; Davidson, C.J.; Gopinathan, A.; McIntyre, D.; Honess, D.; Madhu, B.; Goldgraben, M.A.; Caldwell, M.E.; Allard, D. Inhibition of Hedgehog signaling enhances delivery of chemotherapy in a mouse model of pancreatic cancer. *Science* **2009**, *324*, 1457–1461. [CrossRef]
7. Jain, R.K. Barriers to drug delivery in solid tumors. *Sci. Am.* **1994**, *271*, 58–65. [CrossRef]
8. Provenzano, P.P.; Hingorani, S. Hyaluronan, fluid pressure, and stromal resistance in pancreas cancer. *Br. J. Cancer* **2013**, *108*, 1–8. [CrossRef]

9. Huang, Z.-Q.; Buchsbaum, D.J. Monoclonal antibodies in the treatment of pancreatic cancer. *Immunotherapy* **2009**, *1*, 223–239. [CrossRef]
10. Tassev, D.V.; Cheung, N.-K.V. Monoclonal antibody therapies for solid tumors. *Expert Opin. Biol. Ther.* **2009**, *9*, 341–353. [CrossRef] [PubMed]
11. Andrén-Sandberg, Å. Pancreatic cancer: Chemotherapy and radiotherapy. *N. Am. J. Med. Sci.* **2011**, *3*, 1–12. [CrossRef] [PubMed]
12. Bear, A.S.; Vonderheide, R.H.; O'Hara, M.H. Challenges and opportunities for pancreatic cancer immunotherapy. *Cancer Cell* **2020**, *38*, 788–802. [CrossRef] [PubMed]
13. Looi, C.-K.; Chung, F.F.-L.; Leong, C.-O.; Wong, S.-F.; Rosli, R.; Mai, C.-W. Therapeutic challenges and current immunomodulatory strategies in targeting the immunosuppressive pancreatic tumor microenvironment. *J. Exp. Clin. Cancer Res.* **2019**, *38*, 162. [CrossRef] [PubMed]
14. Wang, Z.; Li, Y.; Ahmad, A.; Banerjee, S.; Azmi, A.S.; Kong, D.; Sarkar, F.H. Pancreatic cancer: Understanding and overcoming chemoresistance. *Nat. Rev. Gastroenterol. Hepatol.* **2011**, *8*, 27–33. [CrossRef]
15. Provenzano, P.P.; Cuevas, C.; Chang, A.E.; Goel, V.K.; Von Hoff, D.D.; Hingorani, S.R. Enzymatic targeting of the stroma ablates physical barriers to treatment of pancreatic ductal adenocarcinoma. *Cancer Cell* **2012**, *21*, 418–429. [CrossRef]
16. Ho, Y.-J.; Wang, T.-C.; Fan, C.-H.; Yeh, C.-K. Spatially uniform tumor treatment and drug penetration by regulating ultrasound with microbubbles. *ACS Appl. Mater. Interfaces* **2018**, *10*, 17784–17791. [CrossRef]
17. Kooiman, K.; Roovers, S.; Langeveld, S.A.; Kleven, R.T.; Dewitte, H.; O'Reilly, M.A.; Escoffre, J.-M.; Bouakaz, A.; Verweij, M.D.; Hynynen, K. Ultrasound-responsive cavitation nuclei for therapy and drug delivery. *Ultrasound Med. Biol.* **2020**, *46*, 1296–1325. [CrossRef]
18. Nesbitt, H.; Sheng, Y.; Kamila, S.; Logan, K.; Thomas, K.; Callan, B.; Taylor, M.A.; Love, M.; O'Rourke, D.; Kelly, P. Gemcitabine loaded microbubbles for targeted chemo-sonodynamic therapy of pancreatic cancer. *J. Control. Release* **2018**, *279*, 8–16. [CrossRef]
19. Xiao, N.; Liu, J.; Liao, L.; Sun, J.; Jin, W.; Shu, X. Ultrasound combined with microbubbles increase the delivery of doxorubicin by reducing the interstitial fluid pressure. *Ultrasound Q.* **2019**, *35*, 103–109. [CrossRef]
20. Grundy, M.; Bau, L.; Hill, C.; Paverd, C.; Mannaris, C.; Kwan, J.; Crake, C.; Coviello, C.; Coussios, C.; Carlisle, R. Improved therapeutic antibody delivery to xenograft tumors using cavitation nucleated by gas-entrapping nanoparticles. *Nanomedicine* **2020**, *16*, 37–50. [CrossRef]
21. Masiero, M.; Boulos, P.; Crake, C.; Rowe, C.; Coviello, C.M. Ultrasound-induced cavitation and passive acoustic mapping: SonoTran platform performance and short-term safety in a large-animal model. *Ultrasound Med. Biol.* **2022**, *48*, 1681–1690. [CrossRef] [PubMed]
22. Hendricks-Wenger, A.; Arnold, L.; Gannon, J.; Simon, A.; Singh, N.; Sheppard, H.; Nagai-Singer, M.A.; Imran, K.M.; Lee, K.; Clark-Deener, S. Histotripsy ablation in preclinical animal models of cancer and spontaneous tumors in veterinary patients: A review. *IEEE Trans. Ultrason. Ferroelectr. Freq. Control* **2021**, *69*, 5–26. [CrossRef] [PubMed]
23. Hendricks-Wenger, A.; Aycock, K.N.; Nagai-Singer, M.A.; Coutermarsh-Ott, S.; Lorenzo, M.F.; Gannon, J.; Uh, K.; Farrell, K.; Beitel-White, N.; Brock, R.M. Establishing an immunocompromised porcine model of human cancer for novel therapy development with pancreatic adenocarcinoma and irreversible electroporation. *Sci. Rep.* **2021**, *11*, 7584. [CrossRef]
24. Hendricks-Wenger, A.; Nagai-Singer, M.A.; Uh, K.; Vlaisavljevich, E.; Lee, K.; Allen, I.C. Employing Novel Porcine Models of Subcutaneous Pancreatic Cancer to Evaluate Oncological Therapies. In *Biomedical Engineering Technologies*; Springer: Berlin/Heidelberg, Germany, 2022; pp. 883–895.
25. Prather, R.S.; Lorson, M.; Ross, J.W.; Whyte, J.J.; Walters, E. Genetically engineered pig models for human diseases. *Annu. Rev. Anim. Biosci.* **2013**, *1*, 203–219. [CrossRef] [PubMed]
26. Hosein, A.N.; Brekken, R.A.; Maitra, A. Pancreatic cancer stroma: An update on therapeutic targeting strategies. *Nat. Rev. Gastroenterol. Hepatol.* **2020**, *17*, 487–505. [CrossRef]
27. Mahadevan, D.; Von Hoff, D.D. Tumor-stroma interactions in pancreatic ductal adenocarcinoma. *Mol. Cancer Ther.* **2007**, *6*, 1186–1197. [CrossRef]
28. Polani, F.; Grierson, P.M.; Lim, K.-H. Stroma-targeting strategies in pancreatic cancer: Past lessons, challenges and prospects. *World J. Gastroenterol.* **2021**, *27*, 2105–2121. [CrossRef]
29. Kullmann, F.; Hollerbach, S.; Dollinger, M.M.; Harder, J.; Fuchs, M.; Messmann, H.; Trojan, J.; Gäbele, E.; Hinke, A.; Endlicher, E.; et al. Cetuximab plus gemcitabine/oxaliplatin (GEMOXCET) in first-line metastatic pancreatic cancer: A multicentre phase II study. *Br. J. Cancer* **2009**, *100*, 1032–1036. [CrossRef]
30. Saadh, M.J.; Haddad, M.; Dababneh, M.F.; Bayan, M.F.; Al-Jaidi, B.A. A Guide for Estimating the Maximum Safe Starting Dose and Conversion it between Animals and Humans. *Syst. Rev. Pharm.* **2020**, *11*, 98–101.
31. Seufferlein, T.; Uhl, W.; Kornmann, M.; Algül, H.; Friess, H.; König, A.; Ghadimi, M.; Gallmeier, E.; Bartsch, D.K.; Lutz, M.P.; et al. Perioperative or only adjuvant gemcitabine plus nab-paclitaxel for resectable pancreatic cancer (NEONAX)—A randomized phase II trial of the AIO pancreatic cancer group. *Ann. Oncol.* **2023**, *34*, 91–100. [CrossRef]
32. Lei, S.; Ryu, J.; Wen, K.; Twitchell, E.; Bui, T.; Ramesh, A.; Weiss, M.; Li, G.; Samuel, H.; Clark-Deener, S. Increased and prolonged human norovirus infection in RAG2/IL2RG deficient gnotobiotic pigs with severe combined immunodeficiency. *Sci. Rep.* **2016**, *6*, 25222. [CrossRef] [PubMed]
33. Yuan, L.; Jobst, P.M.; Weiss, M. Gnotobiotic pigs: From establishing facility to modeling human infectious diseases. In *Gnotobiotics*; Elsevier: Amsterdam, The Netherlands, 2017; pp. 349–368.

34. Masiero, M.; Vojtasova, E.; Boulos, P.; Zammarchi, F.; van Berkel, P.; Crake, C.; Coviello, C.; Rowe, C. Ultrasound-induced cavitation enhances therapeutic efficacy of AXL-targeting ADC leading to improved survival in a human xenograft model of renal cancer. *Eur. J. Cancer* **2022**, *174*, S101–S102. [CrossRef]
35. Myers, R.; Coviello, C.; Erbs, P.; Foloppe, J.; Rowe, C.; Kwan, J.; Crake, C.; Finn, S.; Jackson, E.; Balloul, J.-M. Polymeric cups for cavitation-mediated delivery of oncolytic vaccinia virus. *Mol. Ther.* **2016**, *24*, 1627–1633. [CrossRef] [PubMed]
36. Coviello, C.; Kozick, R.; Choi, J.; Gyöngy, M.; Jensen, C.; Smith, P.P.; Coussios, C.-C. Passive acoustic mapping utilizing optimal beamforming in ultrasound therapy monitoring. *J. Acoust. Soc. Am.* **2015**, *137*, 2573–2585. [CrossRef] [PubMed]
37. Vincent, P.; Wang, H.; Nieskoski, M.; Gunn, J.R.; Marra, K.; Hoopes, P.J.; Samkoe, K.S.; Doyley, M.M.; Hasan, T.; Pogue, B.W. High-resolution ex vivo elastography to characterize tumor stromal heterogeneity in situ in pancreatic adenocarcinoma. *IEEE Trans. Biomed. Eng.* **2020**, *67*, 2490–2496. [CrossRef]
38. Manresa, M.C.; Miki, H.; Miller, J.; Okamoto, K.; Dobaczewska, K.; Herro, R.; Gupta, R.K.; Kurten, R.; Aceves, S.S.; Croft, M. A Deficiency in the Cytokine TNFSF14/LIGHT Limits Inflammation and Remodeling in Murine Eosinophilic Esophagitis. *J. Immunol.* **2022**, *209*, 2341–2351. [CrossRef]
39. Amrutkar, M.; Gladhaug, I.P. Pancreatic cancer chemoresistance to gemcitabine. *Cancers* **2017**, *9*, 157. [CrossRef]
40. Zeng, S.; Pöttler, M.; Lan, B.; Grützmann, R.; Pilarsky, C.; Yang, H. Chemoresistance in pancreatic cancer. *Int. J. Mol. Sci.* **2019**, *20*, 4504. [CrossRef]
41. Sriraman, S.K.; Aryasomayajula, B.; Torchilin, V.P. Barriers to drug delivery in solid tumors. *Tissue Barriers* **2014**, *2*, e29528. [CrossRef]
42. Patra, J.K.; Das, G.; Fraceto, L.F.; Campos, E.V.R.; Rodriguez-Torres, M.d.P.; Acosta-Torres, L.S.; Diaz-Torres, L.A.; Grillo, R.; Swamy, M.K.; Sharma, S. Nano based drug delivery systems: Recent developments and future prospects. *J. Nanobiotechnol.* **2018**, *16*, 71. [CrossRef]
43. Zhao, Z.; Ukidve, A.; Kim, J.; Mitragotri, S. Targeting strategies for tissue-specific drug delivery. *Cell* **2020**, *181*, 151–167. [CrossRef] [PubMed]
44. Mhaidly, R.; Mechta-Grigoriou, F. Fibroblast heterogeneity in tumor micro-environment: Role in immunosuppression and new therapies. *Semin. Immunol.* **2020**, *48*, 101417. [CrossRef] [PubMed]
45. Mouratidis, P.X.; Ter Haar, G. Latest Advances in the Use of Therapeutic Focused Ultrasound in the Treatment of Pancreatic Cancer. *Cancers* **2022**, *14*, 638. [CrossRef] [PubMed]
46. Wang, S.; Li, Y.; Xing, C.; Ding, C.; Zhang, H.; Chen, L.; You, L.; Dai, M.; Zhao, Y. Tumor microenvironment in chemoresistance, metastasis and immunotherapy of pancreatic cancer. *Am. J. Cancer Res.* **2020**, *10*, 1937–1953.
47. Li, C.; Wang, J.; Wang, Y.; Gao, H.; Wei, G.; Huang, Y.; Yu, H.; Gan, Y.; Wang, Y.; Mei, L. Recent progress in drug delivery. *Acta Pharm. Sin. B* **2019**, *9*, 1145–1162. [CrossRef]
48. Larbouret, C.; Robert, B.; Bascoul-Mollevi, C.; Penault-Llorca, F.; Ho-Pun-Cheung, A.; Morisseau, S.; Navarro-Teulon, I.; Mach, J.P.; Pèlegrin, A.; Azria, D. Combined cetuximab and trastuzumab are superior to gemcitabine in the treatment of human pancreatic carcinoma xenografts. *Ann. Oncol.* **2010**, *21*, 98–103. [CrossRef]
49. Tai, C.J.; Wang, H.; Wang, C.K., Tai, C.J.; Huang, M.T.; Wu, C.H.; Chen, R.J.; Kuo, L.J.; Wei, P.L.; Chang, Y.J.; et al. Bevacizumab and cetuximab with conventional chemotherapy reduced pancreatic tumor weight in mouse pancreatic cancer xenografts. *Clin. Exp. Med.* **2017**, *17*, 141–150. [CrossRef]
50. Xiong, H.Q.; Rosenberg, A.; LoBuglio, A.; Schmidt, W.; Wolff, R.A.; Deutsch, J.; Needle, M.; Abbruzzese, J.L. Cetuximab, a monoclonal antibody targeting the epidermal growth factor receptor, in combination with gemcitabine for advanced pancreatic cancer: A multicenter phase II Trial. *J. Clin. Oncol.* **2004**, *22*, 2610–2616. [CrossRef]
51. Cascinu, S.; Berardi, R.; Labianca, R.; Siena, S.; Falcone, A.; Aitini, E.; Barni, S.; Di Costanzo, F.; Dapretto, E.; Tonini, G.; et al. Cetuximab plus gemcitabine and cisplatin compared with gemcitabine and cisplatin alone in patients with advanced pancreatic cancer: A randomised, multicentre, phase II trial. *Lancet Oncol.* **2008**, *9*, 39–44. [CrossRef]
52. Philip, P.A.; Benedetti, J.; Corless, C.L.; Wong, R.; O'Reilly, E.M.; Flynn, P.J.; Rowland, K.M.; Atkins, J.N.; Mirtsching, B.C.; Rivkin, S.E.; et al. Phase III study comparing gemcitabine plus cetuximab versus gemcitabine in patients with advanced pancreatic adenocarcinoma: Southwest Oncology Group-directed intergroup trial S0205. *J. Clin. Oncol.* **2010**, *28*, 3605–3610. [CrossRef]
53. Heinemann, V. Gemcitabine: Progress in the treatment of pancreatic cancer. *Oncology* **2001**, *60*, 8–18. [CrossRef]
54. Philip, P.A.; Lacy, J.; Portales, F.; Sobrero, A.; Pazo-Cid, R.; Manzano Mozo, J.L.; Kim, E.J.; Dowden, S.; Zakari, A.; Borg, C.; et al. Nab-paclitaxel plus gemcitabine in patients with locally advanced pancreatic cancer (LAPACT): A multicentre, open-label phase 2 study. *Lancet Gastroenterol. Hepatol.* **2020**, *5*, 285–294. [CrossRef] [PubMed]
55. McGuigan, A.; Kelly, P.; Turkington, R.C.; Jones, C.; Coleman, H.G.; McCain, R.S. Pancreatic cancer: A review of clinical diagnosis, epidemiology, treatment and outcomes. *World J. Gastroenterol.* **2018**, *24*, 4846–4861. [CrossRef] [PubMed]
56. Brock, R.M.; Beitel-White, N.; Coutermarsh-Ott, S.; Grider, D.J.; Lorenzo, M.F.; Ringel-Scaia, V.M.; Manuchehrabadi, N.; Martin, R.C.G.; Davalos, R.V.; Allen, I.C. Patient Derived Xenografts Expand Human Primary Pancreatic Tumor Tissue Availability for ex vivo Irreversible Electroporation Testing. *Front. Oncol.* **2020**, *10*, 843. [CrossRef]
57. Larina, I.V.; Evers, B.M.; Ashitkov, T.V.; Bartels, C.; Larin, K.V.; Esenaliev, R.O. Enhancement of Drug Delivery in Tumors by Using Interaction of Nanoparticles with Ultrasound Radiation. *Technol. Cancer Res. Treat.* **2005**, *4*, 217–226. [CrossRef]
58. Chauhan, V.P.; Stylianopoulos, T.; Boucher, Y.; Jain, R.K. Delivery of molecular and nanoscale medicine to tumors: Transport barriers and strategies. *Annu. Rev. Chem. Biomol. Eng.* **2011**, *2*, 281–298. [CrossRef]

59. Jain, R.K. Delivery of molecular medicine to solid tumors. *Science* **1996**, *271*, 1079–1080. [CrossRef] [PubMed]
60. Carlisle, R.; Choi, J.; Bazan-Peregrino, M.; Laga, R.; Subr, V.; Kostka, L.; Ulbrich, K.; Coussios, C.-C.; Seymour, L.W. Enhanced tumor uptake and penetration of virotherapy using polymer stealthing and focused ultrasound. *J. Natl. Cancer Inst.* **2013**, *105*, 1701–1710. [CrossRef] [PubMed]
61. Kinoshita, M.; McDannold, N.; Jolesz, F.A.; Hynynen, K. Targeted delivery of antibodies through the blood–brain barrier by MRI-guided focused ultrasound. *Biochem. Biophys. Res. Commun.* **2006**, *340*, 1085–1090. [CrossRef]
62. Lyon, P.C.; Gray, M.D.; Mannaris, C.; Folkes, L.K.; Stratford, M.; Campo, L.; Chung, D.Y.; Scott, S.; Anderson, M.; Goldin, R. Safety and feasibility of ultrasound-triggered targeted drug delivery of doxorubicin from thermosensitive liposomes in liver tumours (TARDOX): A single-centre, open-label, phase 1 trial. *Lancet Oncol.* **2018**, *19*, 1027–1039. [CrossRef] [PubMed]
63. Lyons, B.; Hettinga, J.; Balkaran, J.; Collins, A.; Maardalen, M.; Katti, P.; Mannaris, C.; Bau, L.; Smith, C.; Gray, M. Gas-stabilizing solid cavitation nuclei for systemic or transdermal ultrasound-enhanced drug and vaccine delivery and immunomodulation. *J. Acoust. Soc. Am.* **2021**, *150*, A54. [CrossRef]
64. Staruch, R.M.; Hynynen, K.; Chopra, R. Hyperthermia-mediated doxorubicin release from thermosensitive liposomes using MR-HIFU: Therapeutic effect in rabbit Vx2 tumours. *Int. J. Hyperth.* **2015**, *31*, 118–133. [CrossRef] [PubMed]
65. Tak, W.Y.; Lin, S.-M.; Wang, Y.; Zheng, J.; Vecchione, A.; Park, S.Y.; Chen, M.H.; Wong, S.; Xu, R.; Peng, C.-Y. Phase III HEAT Study Adding Lyso-Thermosensitive Liposomal Doxorubicin to Radiofrequency Ablation in Patients with Unresectable Hepatocellular Carcinoma LesionsRFA and Lyso-Thermosensitive Liposomal Doxorubicin in HCC. *Clin. Cancer Res.* **2018**, *24*, 73–83. [CrossRef] [PubMed]
66. Mondal, J.; Lakkaraju, R.; Ghosh, P.; Ashokkumar, M. Acoustic cavitation-induced shear: A mini-review. *Biophys. Rev.* **2021**, *13*, 1229–1243. [CrossRef] [PubMed]

pharmaceutics

Article

Improved Tumor Control Following Radiosensitization with Ultrasound-Sensitive Oxygen Microbubbles and Tumor Mitochondrial Respiration Inhibitors in a Preclinical Model of Head and Neck Cancer

Quezia Lacerda [1,2], Hebah Falatah [1,2,3,4], Ji-Bin Liu [1], Corinne E. Wessner [1,2], Brian Oeffinger [2], Ankit Rochani [5,6], Dennis B. Leeper [7], Flemming Forsberg [1], Joseph M. Curry [8], Gagan Kaushal [5], Scott W. Keith [9], Patrick O'Kane [1], Margaret A. Wheatley [2] and John R. Eisenbrey [1,*]

1 Department of Radiology, Thomas Jefferson University, Philadelphia, PA 19107, USA
2 School of Biomedical Engineering, Science and Health Systems Drexel University, Philadelphia, PA 19104, USA
3 College of Applied Medical Sciences, King Saud Bin Abdulaziz University for Health Sciences, Jeddah 22384, Saudi Arabia
4 King Abdullah International Medical Research Center, Jeddah 22384, Saudi Arabia
5 Department of Pharmaceutical Sciences, Thomas Jefferson University, Philadelphia, PA 19107, USA
6 Department of Pharmaceutical Sciences, Wegmans School of Pharmacy, St. John Fisher University, Rochester, NY 14618, USA
7 Department of Radiation Oncology, Thomas Jefferson University, Philadelphia, PA 19107, USA
8 Department of Otolaryngology, Thomas Jefferson University, Philadelphia, PA 19107, USA
9 Division of Biostatistics, Department of Pharmacology, Physiology, and Cancer Biology, Thomas Jefferson University, Philadelphia, PA 19107, USA
* Correspondence: john.eisenbrey@jefferson.edu; Tel.: +1-215-503-5188

Citation: Lacerda, Q.; Falatah, H.; Liu, J.-B.; Wessner, C.E.; Oeffinger, B.; Rochani, A.; Leeper, D.B.; Forsberg, F.; Curry, J.M.; Kaushal, G.; et al. Improved Tumor Control Following Radiosensitization with Ultrasound-Sensitive Oxygen Microbubbles and Tumor Mitochondrial Respiration Inhibitors in a Preclinical Model of Head and Neck Cancer. *Pharmaceutics* **2023**, *15*, 1302. https://doi.org/10.3390/pharmaceutics15041302

Academic Editors: Xiangyang Shi, Shashank Sirsi, James Kwan, Michael Gray and Brandon Helfield

Received: 16 January 2023
Revised: 10 March 2023
Accepted: 17 April 2023
Published: 21 April 2023

Abstract: Tumor hypoxia (oxygen deficiency) is a major contributor to radiotherapy resistance. Ultrasound-sensitive microbubbles containing oxygen have been explored as a mechanism for overcoming tumor hypoxia locally prior to radiotherapy. Previously, our group demonstrated the ability to encapsulate and deliver a pharmacological inhibitor of tumor mitochondrial respiration (lonidamine (LND)), which resulted in ultrasound-sensitive microbubbles loaded with O_2 and LND providing prolonged oxygenation relative to oxygenated microbubbles alone. This follow-up study aimed to evaluate the therapeutic response to radiation following the administration of oxygen microbubbles combined with tumor mitochondrial respiration inhibitors in a head and neck squamous cell carcinoma (HNSCC) tumor model. The influences of different radiation dose rates and treatment combinations were also explored. The results demonstrated that the co-delivery of O_2 and LND successfully sensitized HNSCC tumors to radiation, and this was also enhanced with oral metformin, significantly slowing tumor growth relative to unsensitized controls ($p < 0.01$). Microbubble sensitization was also shown to improve overall animal survival. Importantly, effects were found to be radiation dose-rate-dependent, reflecting the transient nature of tumor oxygenation.

Keywords: radiotherapy; tumor hypoxia; contrast-enhanced ultrasound; oxygen-loaded microbubbles; drug delivery; head and neck cancer; lonidamine; metformin

1. Introduction

Radiation therapy remains a preferred treatment approach in patients with head and neck squamous cell carcinoma (HNSCC), with nearly 75% receiving radiation therapy [1]. HNSCC development has historically been caused by tobacco use and excessive alcohol consumption. However, in recent decades, human papillomavirus (HPV) infection has become an increasingly recognized causative factor for oropharyngeal tumors (80%), and HNSCC has become the most common HPV-related malignancy [2–4]. The radiation dose is often limited by the tolerance of the associated normal tissue located in the radiation field. The conventional clinical radiation dose for HNSCC consists of fractionated doses of 1.5–2 Gy of external-beam radiation five times a week for up to seven weeks [5–7]. Roughly two-thirds of the biological damage produced by X-rays is caused by free radicals from the reactive oxygen species (ROS) induced by ionizing radiation. Furthermore, the damage created by free radicals in DNA can be more readily repaired in a hypoxic environment [8,9].

Tumor hypoxia has been identified as the key mechanism of radiation therapy resistance in tumors due to the limited production of ROS. The irregular, angiogenic vasculature that forms in tumors cannot supply sufficient oxygen to the rapidly dividing cells, resulting in an oxygen consumption rate in tumor cells that is greater than the amount of oxygen supplied by the blood flow in these tumors [8,10,11]. Healthy tissue generally exhibits oxygen partial pressures (pO_2) ranging from 40 to 60 mmHg, while many tumors exhibit partial pressures between 2 and 18 mmHg [12]. Several studies have shown that tumor hypoxia is directly associated with a poor prognosis in patients with advanced HNSCC [13–15]. Molecular oxygen is a well-known radiosensitizer, and studies have shown that relatively low levels (<20 mmHg) of oxygen are required for radiosensitization [10,16].

To overcome hypoxia-associated radiotherapy resistance, systemic approaches have been investigated to improve tumor energetics and oxygenation prior to radiotherapy. Recent studies explored the ability to overcome hypoxia in solid tumors prior to photodynamic therapy using nanoparticles delivered along with mitochondria-associated oxidative phosphorylation disrupting drugs [17–19]. One such drug, metformin, is a biguanide that is generally used for the treatment of type 2 diabetes and has been shown to reduce cancer-related morbidity and mortality. Metformin is an attractive adjunct given its 50–60% absolute oral bioavailability while also inhibiting mitochondrial respiration [20,21]. However, this and other systemic approaches lack tumor specificity. Consequently, more recent work has focused on localized approaches for overcoming tumor hypoxia, including ultrasound-triggered delivery of bioactive drugs and gases prior to radiotherapy [22].

Our group has demonstrated the effectiveness of using surfactant-shelled microbubbles with oxygen cores ($SE61O_2$) to deliver oxygen to hypoxic tumors prior to radiation therapy [6,23–26]. The feasibility of this approach was first demonstrated in breast tumor xenografts, where oxygenation significantly improved both tumor control and animal survival ($p < 0.03$) [24]. However, that study's limitation was that the duration of oxygenation (less than 3 min) hindered clinical adoption. In an effort to prolong oxygenation, we recently showed the ability to deliver both O_2 and lonidamine (LND), a pharmacological mitochondrial respiration inhibitor, in an HNSCC tumor model, which prolonged the duration of tumor oxygenation for up to 5 min [6]. Furthermore, when investigating LND bioavailability, SE61 proved to be an optimal targeted delivery vehicle, improving LND biodistribution in tissue and plasma.

Building on this prior work, the study presented here aimed to determine whether the combination of O_2 microbubbles with mitochondrial respiration inhibitors (both co-encapsulated LND and oral metformin) would improve tumor treatment response and survival by improving radiosensitization. LND has been shown to not only inhibit aerobic glycolysis in tumor cells but also increase aerobic glycolysis in normal cells and display a selective effect on tumors by producing intracellular acidosis [27,28]. Additionally, given the limited duration of oxygenation with this platform, the influence of radiation dose rate was also explored in this model. These are essential steps in understanding the mechanisms of radiotherapy sensitization and demonstrating suitability for clinical translation.

2. Materials and Methods

2.1. Microbubble Fabrication

Surfactant-shelled oxygen microbubbles ($SE61O_2$) were fabricated in accordance with our previously reported methods [6]. Briefly, LND (Sigma-Aldrich; St. Louis, MO, USA) was incubated with water-soluble vitamin E (Tocopheryl polyethylene glycol succinate; TPGS) (Eastman Chemical Company; Kingsport, TN, USA) micelles while continuously stirring for 48 h at 37 °C. The micelle solution (with or without LND) was mixed with sorbitan monostearate (Montane 60 PHA Premium), a gift from Seppic (Paris, France), after being autoclaved for 35 min. The mixture of the two surfactants was then purged with perfluorocarbon (PFC) gas for one minute (octafluoropropane, Advanced Specialty Gasses; Reno, NV, USA), followed by sonication while continuously purging with PFC gas. The resultant microbubbles were separated using gravity. After washing (to separate the unincorporated surfactant), the microbubbles were collected and diluted 1:1 (*v/v*) with 10% (*w/v*) glucose to provide lyoprotection. The microbubbles were then transferred in 2 mL aliquots into 10 mL lyophilization vials. Samples were flash frozen in a −20 °C bath, lyophilized using a freeze-dryer under vacuum for at least 20 h, and capped under vacuum. Prior to use, microbubbles were charged with a sterile-filtered (0.2 μm filter) gas of choice, PFC, oxygen, or nitrogen (Airgas, Radnor, PA, USA) through the vial stoppers using a syringe needle and reconstituted immediately by injecting 2 mL of 0.5 X phosphate-buffered saline (PBS).

2.2. In Vitro Acoustic Characterization

Acoustic enhancement and stability were quantified in vitro in a closed-loop flow phantom setup (ATS Laboratories, CIRS, Norfolk, VA, USA). Microbubbles were injected and insonated using a 10L4 transducer and an Acuson Sequoia ultrasound scanner (Siemens Healthineers, Issaquah, WA, USA) at room temperature. Imaging of the microbubbles flowing through the embedded vessel was performed in cadence pulse sequencing mode every 30 s for 10 min (n = 3, for microbubbles charged with O_2 and N_2 with and without LND) at low mechanical index (MI) imaging to visualize enhancement (MI = 0.12). The mean enhancement returned to the transducer was determined by drawing regions of interest on the contrast imaging plane; all analyses were performed in ImageJ (NIH, Bethesda, MD, USA). Enhancement (in dB) was determined every 30 s and plotted against time.

2.3. Light Microscopy

To access visual changes to microbubble physical characteristics, an Olympus 1X71 microscope (Olympus Corporation, Tokyo, Japan) was used to image drug-loaded and -unloaded SE61. Images were processed using the Olympus cellSens Standard software (Olympus Corporation, Tokyo, Japan). Images were obtained using a magnification factor of 64×.

2.4. Microbubble Lonidamine Loading Quantification

The quantification of LND loading was performed as previously described [6]. In brief, all quantifications used a Dionex Ultimate 3000 HPLC system (ThermoFisher, Waltham, MA, USA) attached to a Thermo Orbitrap mass spectrometer. An XBridge C18 column (4.6 mm × 150 mm, 3.5 μm; Waters) was used for all separations. Freeze-dried sample vials were reconstituted in methanol, vortexed for 15–20 s, filtered with a syringe filter, and analyzed by HPLC analysis. The mobile phase was composed of 50% solvent A (0.1% formic acid in water) and 50% solvent B (0.1% formic acid in acetonitrile) by the isocratic method at a flow rate of 0.2 mL/min [29]. Each injection ran for 17 min at a volume of 5 μL with the compartment temperature at 4 °C and the column temperature at 30 °C.

2.5. Cell Line and Reagents

A well-characterized HNSCC cell line, CAL27 (ATCC, Manassas, VA, USA), was maintained in growth media consisting of Dulbecco's modified Eagle's medium (DMEM), High Glucose, Gluta MAX™ (Gibco™, Waltham, MA, USA) supplemented with 10% fetal bovine serum (FBS; Corning, NY, USA) and 1% penicillin–streptomycin (ATCC, Manassas, VA, USA) and kept at 37 °C in 5% CO_2.

2.6. Implantation and Tumor Growth

Animal experiments were carried out in accordance with Thomas Jefferson University's Institutional Animal Care and Use Committee (IACUC; approved 15 June 2020). HNSCC tumors were generated in athymic nude mice (The Jackson Lab, Bar Harbor, ME, USA) (split evenly by gender) subcutaneously on the right flank with 5×10^5 CAL27 cells (ATCC, Manassas, VA, USA) and 100 μL matrigel (Corning, NY, USA). Once tumors reached 100 mm^3, animals were randomly assigned to control and treatment groups with metformin pre-treatment (OM) (Table 1). Animals were monitored twice per week using digital caliper measurements until sacrifice was required (tumor mass > 10% body weight or animal showing signs of distress).

Table 1. Treatment groups.

GROUP	TREATMENT	ULTRASOUND	5 GY	OM	n:(F1)	n:(F2)
1	SE61O_2/LND	Yes	Yes	Yes	7	3
2	SE61N_2/LND	Yes	Yes	Yes	6	2
3	SE61O_2	Yes	Yes	Yes	5	3
4	SE61O_2/LND	Yes	Yes	No	8	3
5	SE61O_2	Yes	Yes	No	5	3
6	OM	Yes	Yes	Yes	7	1
7	None	No	No	No	6	2

2.7. In Vivo Acoustic Analysis

Each microbubble group listed in Table 1 received a 0.1 mL bolus injection of agent followed by a 0.05 mL saline flush with ultrasound triggering at the tumor. Continuous B-mode and contrast mode imaging was acquired using a 10L4 transducer and an Acuson Sequoia (Siemens Healthineers, Issaquah, WA, USA) ultrasound scanner. Following peak enhancement, a 4-second destructive pulse was employed to trigger and destroy microbubbles within the tumor environment (MI = 1.4). This was followed by 10 s of low MI imaging to visualize microbubble reperfusion (MI = 0.12) and repeated for 75 s.

2.8. Treatment Administration

The 61 tumor-bearing mice were randomly assigned to the groups listed in Table 1. Therapy response was compared with and without O_2 microbubble pre-sensitization. Animals in groups receiving metformin pre-treatment had it added to their drinking water at a concentration of 1 mg/mL [21,30,31]. The mice received 300 mg/kg/day on average. Following the same protocol previously reported for injectable anesthetics [6,24], animals from each group were anesthetized using a mixture of ketamine (75 mg/kg) and acepromazine (1 mg/kg). SE61 groups received microbubbles through a 24-gauge tail vein angiocatheter. Following microbubble administration, animals receiving radiation therapy were placed 40 cm from the radiation source and covered with 4 mm lead shielding, exposing the right flank and tail. Immediately following ultrasound triggering, animals received 5 Gy (for an estimated 25% tumor inhibition) using an X-RAD 320 biological irradiator (Precision X-Ray, Madison, CT, USA) with a beam quality of 320 kV with two different added filtrations. Overall, 44 animals received 5 Gy with a 2 mm aluminum filter (F1), and 17 animals received 5 Gy with a 1.5 mm aluminum, 0.25 mm copper, and 0.75 mm tin filter (F2). These filters corresponded to radiation dose rates of 3.59 and 1.36 Gy/min, respectively. Following treatment, animals were monitored until sacrifice was required per the criteria mentioned above.

2.9. Data Analysis

Significant differences between acoustic properties were determined using a one-way ANOVA. Tumor volumes for each animal were plotted against time after treatment. A simple survival analysis was conducted in Prism (GraphPad Software, San Diego, CA, USA) using Kaplan–Meier curves from the day of treatment until the day each animal was sacrificed. To analyze differences between groups and filters, log-rank tests were used. Differences in growth rate parameters were established using a Generalized Estimating Equations (GEE) model of exponential growth for robust standard error estimation accounting for the autoregressive correlations among the longitudinal measures within animals. A significance level of $\alpha = 0.05$ was used, and statistical analysis was performed in Prism and SAS v 9.4 (SAS Institute, Cary, NC, USA), with plots of the models created in R (Wirtschaftsuniversität, Vienna, AT, USA).

3. Results

3.1. Microbubble Lonidamine Loading and Physical Characterization

Following characterization with HPLC, LND encapsulation within the microbubble showed an average of 25.7 ± 1.5 µg/mL, with a total of 705.7 ± 80.8 µg LND encapsulated per batch. Though the encapsulation efficiency was low (1.71%), it was able to surpass the minimum required loading of 4.8 µg/mL based on the minimum effective LND dose in tissue [28,32]. The population and size distribution LND microbubbles have been previously reported, at an average of (2.7 ± 0.2) × 10^9 MB/mL [6]. Light microscopy images showing the visual representation of the microbubble sizes to display visible changes in the microbubble morphology after adding LND are shown in Figure 1.

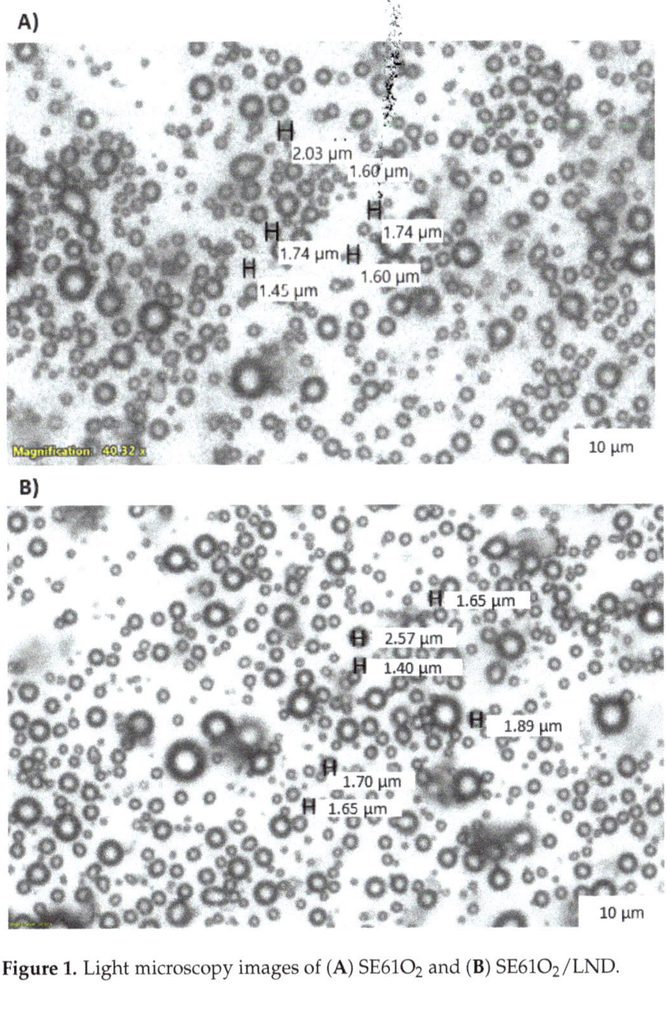

Figure 1. Light microscopy images of (**A**) SE61O_2 and (**B**) SE61O_2/LND.

3.2. Microbubble Imaging In Vitro and In Vivo

The ability to non-invasively destroy SE61 microbubbles with or without LND and with or without O_2 was confirmed in a pulsatile flow phantom setup, where agents were insonated in a single plane with higher intensity ultrasound pulses (MI = 1.4) (Figure 2A). All samples showed strong signals within the lumen and remained stable for around 10 min. Stability curves were constructed with the normalized enhancement of each region of interest in the contrast imaging plane every 30 s (shown in Figure 2B). All samples had 80% retained signal for over 7 min. No significant differences were observed in agent stability with or without LND or between O_2 and N_2 cores (p = 0.99).

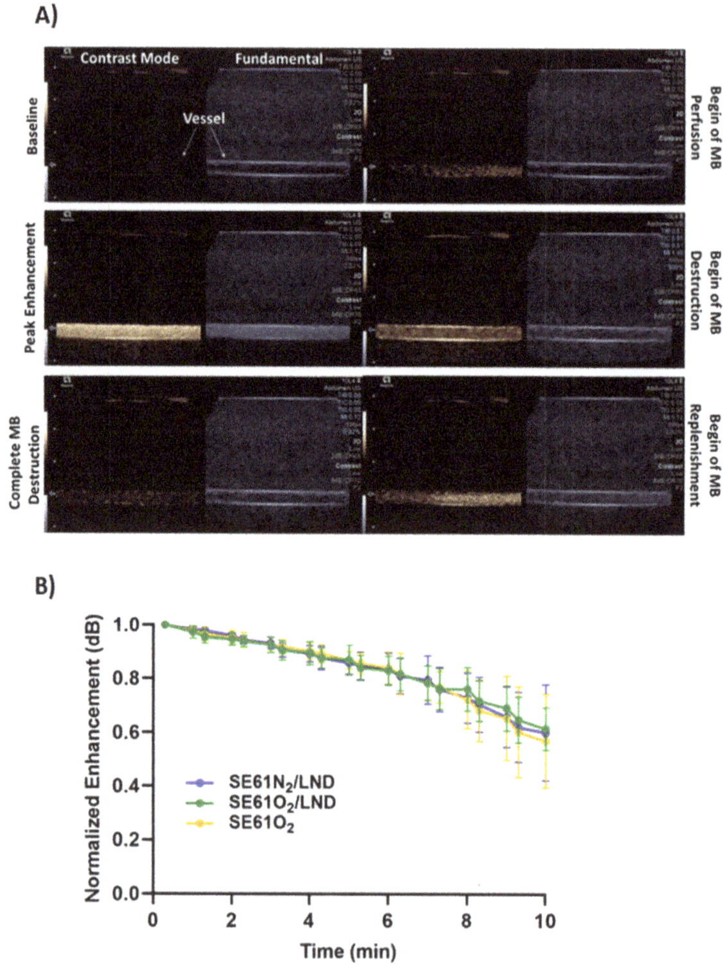

Figure 2. (**A**) Illustrative ultrasound images of SE61O_2/LND in a closed-loop flow phantom showing contrast mode (**left**) and fundamental (B-mode) (**right**). Image depth markers correspond to 0.3 cm increments with the flow direction from right to left. (**B**) Microbubble stability curves of SE61O_2/LND (green), SE61O_2 (yellow), and SE61N_2/LND (blue) at a non-destructive MI (MI = 0.12). No statistically significant differences among formulations were observed in vitro (p = 0.99).

Imaging and tolerability in the HNSCC tumor model in vivo showed that all microbubbles samples, whether oxygenated, charged with nitrogen, and/or drug-loaded (SE61O_2, SE61O_2/LND, and SE61N_2/LND), were well tolerated following intravenous injection. These agents demonstrated perfusion into the tumor environment with strong enhancement on contrast-specific imaging (Figure 3). Following administration, at peak enhancement, a destructive pulse sequence was employed to destroy microbubbles within the tumor (MI = 1.4) followed by 10 s of low MI imaging to visualize reperfusion (MI = 0.12). This flash-replenishment sequence was repeated for 75 s, confirming the ability to destroy these agents locally as well as their stability in the systemic circulation (Figure 3). Images show the time course of SE61O_2/LND from prior to injection (i.e., baseline) over complete perfusion to complete destruction and reperfusion after the destructive pulses.

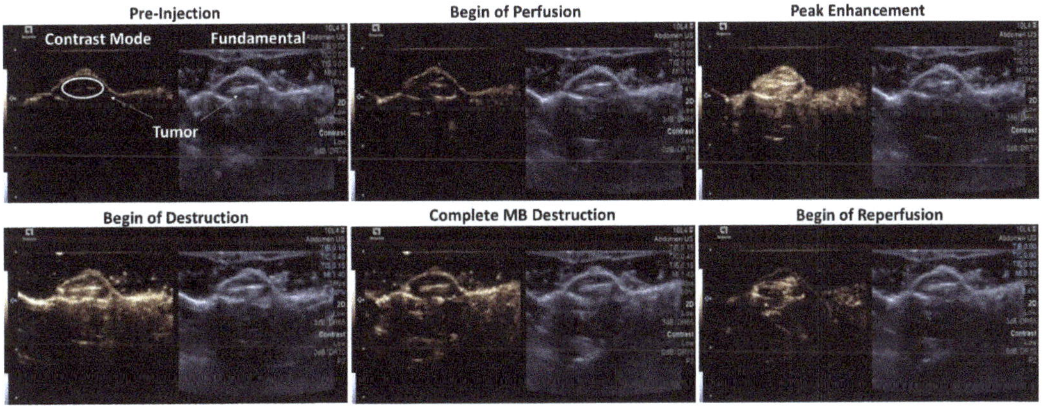

Figure 3. Illustrative ultrasound images of SE61O_2/LND in an HNSCC tumor model generated in immunocompromised mice in dual B-mode/cadence pulse sequencing mode. Images show pre-injection of microbubbles (MB), start of MB perfusion, peak enhancement post-injection, start of MB destruction, complete MB destruction, and start of reperfusion (image depth corresponds to 0.3 cm increments). The left side of the display is the nonlinear imaging mode (contrast), and the right is the conventional B-mode (grayscale imaging).

3.3. Therapy Experiments

Therapeutic enhancement evaluated in an HNSCC tumor model is shown in Figure 4 comparing tumor growth over time. The results demonstrate that despite some expected variability due to differences in tumor size and growth rate, the experimental group that received SE61O_2/LND with OM pre-treatment followed by 5 Gy delivered with F1 had improved tumor growth control compared to animals in the SE61O_2 + US + 5 Gy, OM + US + 5 Gy, and the no treatment groups. These results were further confirmed after fitting to a regression model curve, shown in Figure 5. Both O_2 microbubbles (shown in dark green and orange) receiving OM pre-treatment and 5 Gy showed a statistically significant improvement in tumor control relative to the other controls ($p < 0.01$) other than N_2 microbubbles (blue in the plot) and LND microbubbles without OM (purple in the plot). For LND-loaded bubbles treated with ultrasound and 5 Gy radiation, the presence of O_2 (versus N_2) showed approximately 20 days of improvement in tumor control, although this was not statistically significant ($p = 0.11$). However, no differences were observed in tumor growth volumes between animals receiving OM pre-treatment, 5 Gy radiation, and O_2 microbubbles with or without LND ($p = 0.83$) and SE61O_2/LND with or without OM ($p = 0.17$). These results indicate that while the addition of each mitochondrial tumor respiration inhibitor alone improves tumor control, no additional benefit is provided by combining multiple inhibitors.

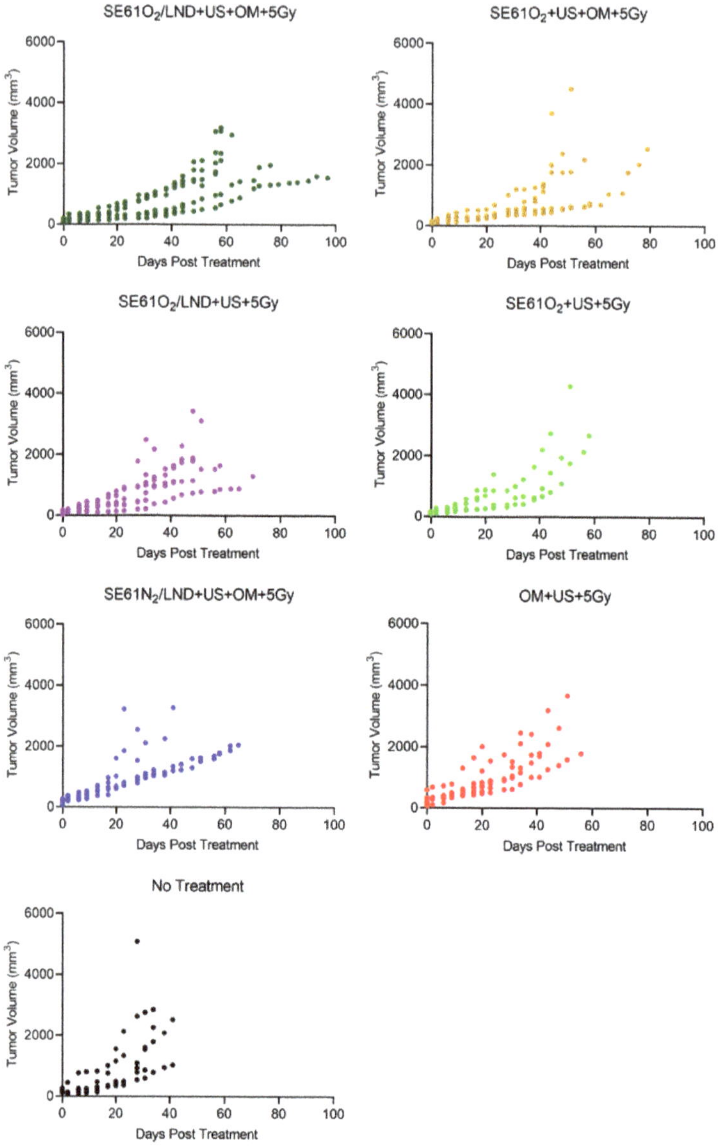

Figure 4. Tumor growth over time per animal for each group showing raw tumor volumes (mm^3) from the day of treatment until sacrifice was required. Results show the influence of ultrasound (US), SE61 microbubbles, radiation therapy (5 Gy), gas (oxygen or nitrogen), and tumor mitochondrial respiration inhibitors (metformin (OM) and lonidamine (LND)) on an HNSCC tumor model.

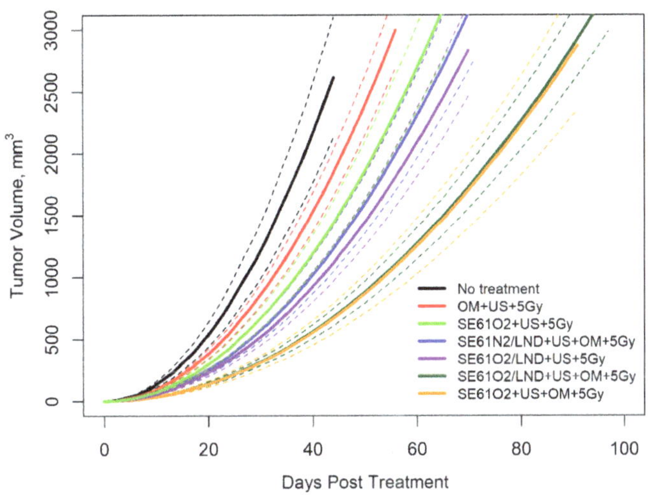

Figure 5. Tumor volumes plotted against time with fitted exponential growth curves showing tumoral response to therapy, with 95% confidence bands shown as dashed lines. Tumor volumes are plotted to the day of treatment to show the influence of ultrasound (US), SE61 microbubbles, radiation therapy (5 Gy), gas (oxygen or nitrogen), and tumor mitochondrial respiration inhibitors (metformin (OM) and lonidamine (LND)).

The influence of radiation dose rate on animal survival is shown in Figure 6A,B. Similarly, animals treated with O_2 LND-loaded microbubbles, OM pre-treatment, and 5 Gy radiation (dark green in the plot) showed an improvement in median animal survival (63 days) compared to all control groups ($p = 0.0064$), except for animals that received O_2 microbubbles with OM pre-treatment and 5 Gy of radiation (orange in the plot) ($p = 2873$) (Figure 6A). Additionally, the experimental group (dark green plot) receiving SE61O_2/LND with OM pre-treatment had a probability of survival of 86% at 58 days post-treatment, which was statistically significant compared to all other groups ($p = 0.003$). Interestingly, animals treated with SE61O_2/LND with OM pre-treatment showed an improvement in animal survival compared to animals treated with SE61O_2/LND without OM pre-treatment ($p = 0.0108$). However, there was no difference in median animal survival for animals that received O_2 microbubbles, metformin pre-treatment, and 5 Gy radiation (without LND), reaching 59 days ($p = 0.29$). Additionally, animals that received SE61O_2/LND and OM pre-treatment followed by 5 Gy delivered with F1 (3.59 cGy/min) showed an improved animal survival ($p = 0.0008$), with a median survival of 63 days, when compared to animals that received the same treatment with 5 Gy delivered with F2 (1.36 cGy/min), reaching 29 days (Figure 6B). This demonstrates that while the platform provides a significant improvement in tumor control with co-encapsulated LND and OM, a relatively high radiation dose rate is still required given the limited duration of oxygenation.

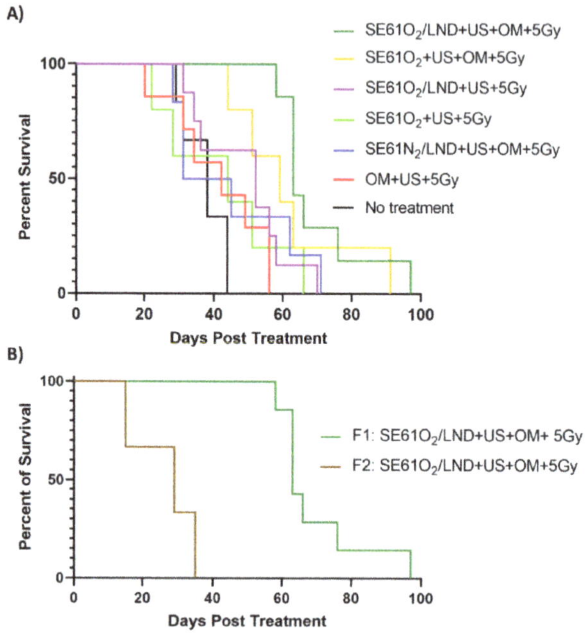

Figure 6. Survival proportion of animals following treatment until day of sacrifice. (**A**) Animal survival of all groups; treated groups received radiation with Filter 1 (F1; 3.59 cGy/min). (**B**) A comparison of the experimental group ($SE61O_2$/LND) with ultrasound triggering (+ US) and metformin pre-treatment (OM) that received radiation with Filter 1 (dark green plot) and animals treated with the same treatment but that received radiation with Filter 2 (F2; 1.36 cGy/min) (brown plot).

4. Discussion

Disrupting ultrasound-sensitive microbubbles containing oxygen locally may overcome tumor hypoxia prior to radiotherapy. Earlier in vivo studies by our group demonstrated the feasibility of disrupting oxygen microbubbles within a murine breast tumor model and a model of metastatic breast cancer in the brain, which raised the mean tumor pO_2 to as much as 20 mmHg. This increase in oxygenation significantly improved tumor control and animal survival [24,26]. The sensitization capability of oxygen-loaded microbubbles before radiation therapy has also been proven by others [33–36]. Other groups explored the ability to enhance the efficacy of treatment by decreasing tumors' oxygen consumption with mitochondrial oxidative phosphorylation inhibitors or agents to depress the expression of programmed death-ligand 1 (PD-L1) in tumors [37–41].

More recently, our group investigated the ability to deliver oxygen in combination with a pharmacological tumor mitochondrial respiration inhibitor, LND, to HNSCC tumors to prevent cells from upregulating oxidative phosphorylation, which in turn extended the duration of oxygenation in tumors [6]. That work demonstrated that O_2, when delivered along with LND, raised the pO_2 by nearly 30 mmHg in HNSCC solid tumors, with levels remaining elevated for up to 5 min [6]. This work showed the therapeutic efficacy of co-delivering O_2 microbubbles with mitochondrial respiration inhibitors, namely LND or metformin. The stability of this platform was confirmed in vitro in a pulsatile flow phantom. Stability curves showed that all samples, independent of gas and/or drug content, had 80% retained signal for over 7 min (cf., Figure 2).

Imaging and tolerability examined in vivo revealed that all microbubble samples, whether oxygenated, charged with nitrogen, and/or drug-loaded ($SE61O_2$, $SE61O_2$/LND, and $SE61N_2$/LND), were well tolerated following intravenous injection, with no adverse reaction in any animal. The ability to improve tumor control and survival by sensitizing

HNSCC tumors to radiotherapy was explored in this model. Regression modeling demonstrated that both O_2 microbubbles (with and without LND) receiving OM pre-treatment and 5 Gy showed a statistically significant improvement in tumor control relative to the other controls ($p < 0.01$), except for the N_2 microbubbles ($p = 0.11$) and LND microbubbles without OM pre-treatment ($p = 0.17$). This demonstrates that both metformin alone (light green versus orange growth curves in Figure 5) and LND alone (light green versus purple growth curves in Figure 5) had an added effect on radiosensitization in this model. We attribute the more pronounced effects of metformin to the higher intratumoral concentrations due to its superior bioavailability (>50% orally [21]). Notably, the addition of multiple mitochondrial respiration inhibitors (orange versus dark green growth curves in Figure 5) did not provide additional benefits, presumably due to both targeting similar mechanisms for inhibiting aerobic glycolysis.

Microbubble destruction (without O_2) has been shown to increase radiosensitivity by inducing cell apoptosis due to increased ceramide production induced by the shear stress generated from inertially cavitating microbubbles [24,42–44]. These findings are reflected by the improved radiosensitization of animals receiving N_2 microbubbles (although tumor control was further improved with the addition of O_2). Additionally, we hypothesize that the limited improvement in tumor control with the addition of LND is due to the relatively low drug concentration (~2.55 µg/injection, considerably less than prior preclinical experiments using intraperitoneal injection [27]) and due to the application of radiation over a short enough duration whereby both O_2 agents provided sufficient tumor oxygenation based on prior work [6]. An interesting finding in this work was the influence of radiation dose rate on therapeutic benefits, shown by delivering 5 Gy with two different filtrations (3.59 cGy/min vs. 1.36 cGy/min). Animals placed in our experimental group receiving SE61O_2/LND, OM pre-treatment, and 5 Gy delivered with F1 showed improved tumor growth control and survival with a median survival of 63 days when compared to animals that received the same treatment with 5 Gy with F2 (29 days; $p = 0.0008$) (Figure 6B). This demonstrates that while the platform provides an improvement in tumor control, a relatively high radiation dose rate is still required given the limited duration of oxygenation currently available with this microbubble platform.

Though this study produced encouraging results, several limitations exist. The therapeutic efficacy of this microbubble platform was only evaluated in immunocompromised mice, which prevents an analysis of the immune system's role in treatment response. The experimental goal of this study was to show tumor burden and animal survival; for this reason, we have not yet analyzed acute cellular response to radiation therapy with SE61 microbubbles. Future work will focus on immunocompetent animals to better assess the clinical relevance of this platform. We also anticipate expanding this work to other solid tumors easily accessible to ultrasound and currently treated with radiotherapy such as breast cancer, prostate cancer, and soft tissue sarcoma. Additionally, the radiation dose rate should be better refined to reflect changes in tumor oxygenation duration with the addition of LND. Finally, to overcome the relatively low drug concentration of LND, future work will focus on optimizing the LND dose delivered.

5. Conclusions

This work demonstrated that the co-delivery of O_2 via ultrasound-sensitive microbubbles combined with a tumor mitochondrial respiration inhibitor improved radiosensitivity and animal survival in an HNSCC model. Additionally, this work highlights the importance of dose rate in the context of the limited duration of oxygenation currently available with this microbubble platform.

Author Contributions: Conceptualization, Q.L., J.-B.L., M.A.W., J.R.E.; methodology, Q.L., J.-B.L., M.A.W., J.R.E., H.F., C.E.W., B.O., A.R., F.F., G.K., J.M.C., P.O., S.W.K.; software, Q.L., S.W.K., J.R.E.; validation, Q.L., J.R.E., M.A.W.; formal analysis, Q.L., J.R.E., M.A.W., S.W.K., A.R.; resources, J.R.E., M.A.W., J.M.C., D.B.L., F.F., G.K.; data curation, Q.L., H.F., C.E.W., B.O., J.R.E.; writing—original draft preparation, Q.L. and J.R.E.; writing—review and editing, H.F., J.-B.L., C.E.W., B.O., A.R., D.B.L., F.F., J.M.C., G.K., S.W.K., P.O., M.A.W.; visualization, Q.L., C.E.W., S.W.K.; supervision, M.A.W., J.M.C., G.K., J.R.E.; funding acquisition, J.R.E., M.A.W. All authors have read and agreed to the published version of the manuscript.

Funding: Funding for this work was provided in part by the United States National Institute of Health R01 EB026881.

Institutional Review Board Statement: The animal study protocol was approved by the Institutional Animal Care and Use Committee of Thomas Jefferson University (protocol 02138, approved 15 June 2020).

Informed Consent Statement: Not applicable.

Data Availability Statement: Data will be made available on request.

Acknowledgments: Montane 60 PHA Premium was received as a generous gift from Seppic (Paris, France). The Acuson Sequoia ultrasound scanner was provided by Siemens Healthineers (Issaquah, WA, USA).

Conflicts of Interest: C.E.W.: Clinical consultant, Bracco Diagnostics, and speaking bureau, Canon Medical Imaging USA. F.F.: Equipment support from Butterfly, Siemens, GE Healthcare, and Canon; drug support from Bracco, GE HealthCare, and Lantheus Medical Imaging; member of Lantheus Medical Imaging Scientific and Sonothera Advisory Boards; consultant for EXACT Therapeutics; lecturer for GE HealthCare; and research grants from Canon. J.R.E.: Equipment support from Siemens, GE HealthCare, and Canon; drug support from Bracco and Lantheus Medical Imaging; member of Lantheus Medical Imaging Scientific Advisory Board; and research grants from GE HealthCare. All other authors declare that they have no known competing financial interests or personal relationships that could have appeared to influence the work reported in this paper.

References

1. Reid, P.; Staudacher, A.H.; Marcu, L.G.; Olver, I.; Moghaddasi, L.; Brown, M.P.; Bezak, E. Characteristic differences in radiation-induced DNA damage response in human papillomavirus-negative and human papillomavirus-positive head and neck cancers with accumulation of fractional radiation dose. *Head Neck* **2021**, *43*, 3086–3096. [CrossRef]
2. Lechner, M.; Liu, J.; Masterson, L.; Fenton, T.R. HPV-associated oropharyngeal cancer: Epidemiology, molecular biology and clinical management. *Nat. Rev. Clin. Oncol.* **2022**, *19*, 306–327. [CrossRef] [PubMed]
3. Avril, D.; Foy, J.-P.; Bouaoud, J.; Grégoire, V.; Saintigny, P. Biomarkers of radioresistance in head and neck squamous cell carcinomas. *Int. J. Radiat. Biol.* **2022**, 1–11. [CrossRef] [PubMed]
4. Mirghani, H.; Amen, F.; Tao, Y.; Deutsch, E.; Levy, A. Increased radiosensitivity of HPV-positive head and neck cancers: Molecular basis and therapeutic perspectives. *Cancer Treat. Rev.* **2015**, *41*, 844–852. [CrossRef] [PubMed]
5. Tolentino, E.d.S.; Centurion, B.S.; Ferreira, L.H.C.; de Souza, A.P.; Damante, J.H.; Rubira-Bullen, I.R.F. Oral adverse effects of head and neck radiotherapy: Literature review and suggestion of a clinical oral care guideline for irradiated patients. *J. Appl. Oral Sci.* **2011**, *19*, 448–454. [CrossRef]
6. Lacerda, Q.; Rochani, A.; Oeffinger, B.; Liu, J.B.; Wessner, C.E.; Tahmasebi, A.; Falatah, H.; Lee, P.; Leeper, D.B.; Forsberg, F.; et al. Tumoral oxygenation and biodistribution of Lonidamine oxygen microbubbles following localized ultrasound-triggered delivery. *Int. J. Pharm.* **2022**, *625*, 122072. [CrossRef]
7. Langendijk, J.A.; Doornaert, P.; Leeuw, I.M.V.-D.; Leemans, C.R.; Aaronson, N.K.; Slotman, B.J. Impact of Late Treatment-Related Toxicity on Quality of Life Among Patients With Head and Neck Cancer Treated with Radiotherapy. *J. Clin. Oncol.* **2008**, *26*, 3770–3776. [CrossRef]
8. Hall, E.J.; Giaccia, A.J. *Radiobiology for the Radiologist*, 8th ed.; Wolters Kluwer: Philadelphia, PA, USA, 2019.
9. Lehnert, S. *Biomolecular Action of Ionizing Radiation*; CRC Press: Boca Raton, FL, USA, 2007. [CrossRef]
10. Rockwell, S.; Dobrucki, I.T.; Kim, E.Y.; Marrison, S.T.; Vu, V.T. Hypoxia and radiation therapy: Past history, ongoing research, and future promise. *Curr. Mol. Med.* **2009**, *9*, 442–458. [CrossRef]
11. Muz, B.; de la Puente, P.; Azab, F.; Azab, A.K. The role of hypoxia in cancer progression, angiogenesis, metastasis, and resistance to therapy. *Hypoxia* **2015**, *3*, 83–92. [CrossRef]
12. Brown, J.M.; Wilson, W.R. Exploiting tumour hypoxia in cancer treatment. *Nat. Rev. Cancer* **2004**, *4*, 437–447. [CrossRef]

13. Nordsmark, M.; Bentzen, S.M.; Rudat, V.; Brizel, D.; Lartigau, E.; Stadler, P.; Becker, A.; Adam, M.; Molls, M.; Dunst, J.; et al. Prognostic value of tumor oxygenation in 397 head and neck tumors after primary radiation therapy. An international multi-center study. *Radiother. Oncol.* **2005**, *77*, 18–24. [CrossRef] [PubMed]
14. Rudat, V.; Stadler, P.; Becker, A.; Vanselow, B.; Dietz, A.; Wannenmacher, M.; Molls, M.; Dunst, J.; Feldmann, H.J. Predictive value of the tumor oxygenation by means of pO_2 histography in patients with advanced head and neck cancer. *Strahlenther. Onkol.* **2001**, *177*, 462–468. [CrossRef] [PubMed]
15. Brizel, D.M.; Dodge, R.K.; Clough, R.W.; Dewhirst, M.W. Oxygenation of head and neck cancer: Changes during radiotherapy and impact on treatment outcome. *Radiother. Oncol. J. Eur. Soc. Ther. Radiol. Oncol.* **1999**, *53*, 113–117. [CrossRef] [PubMed]
16. Overgaard, J. Hypoxic Radiosensitization: Adored and Ignored. *J. Clin. Oncol.* **2007**, *25*, 4066–4074. [CrossRef] [PubMed]
17. Zhou, Z.; Chen, J.; Liu, Y.; Zheng, C.; Luo, W.; Chen, L.; Zhou, S.; Le, Z.; Shen, J. Cascade two-stage tumor re-oxygenation and immune re-sensitization mediated by self-assembled albumin-sorafenib nanoparticles for enhanced photodynamic immunotherapy. *Acta Pharm. Sin. B* **2022**, *12*, 4204–4223. [CrossRef] [PubMed]
18. Zhou, Z.; Liu, Y.; Song, W.; Jiang, X.; Deng, Z.; Xiong, W.; Shen, J. Metabolic reprogramming mediated PD-L1 depression and hypoxia reversion to reactivate tumor therapy. *J. Control. Release* **2022**, *352*, 793–812. [CrossRef]
19. Liu, Y.; Jiang, Y.; Zhang, M.; Tang, Z.; He, M.; Bu, W. Modulating Hypoxia via Nanomaterials Chemistry for Efficient Treatment of Solid Tumors. *Acc. Chem. Res.* **2018**, *51*, 2502–2511. [CrossRef]
20. Galluzzi, L.; Kepp, O.; Heiden, M.G.V.; Kroemer, G. Metabolic targets for cancer therapy. *Nat. Rev. Drug Discov.* **2013**, *12*, 829–846. [CrossRef]
21. Luo, Q.; Hu, D.; Hu, S.; Yan, M.; Sun, Z.; Chen, F. In vitro and in vivoanti-tumor effect of metformin as a novel therapeutic agent in human oral squamous cell carcinoma. *BMC Cancer* **2012**, *12*, 517. [CrossRef]
22. Chowdhury, S.M.; Lee, T.; Willmann, J.K. Ultrasound-guided drug delivery in cancer. *Ultrasonography* **2017**, *36*, 171–184. [CrossRef]
23. Eisenbrey, J.R.; Albala, L.; Kramer, M.R.; Daroshefski, N.; Brown, D.; Liu, J.B.; Stanczak, M.; O'Kane, P.; Forsberg, F.; Wheatley, A.M. Development of an ultrasound sensitive oxygen carrier for oxygen delivery to hypoxic tissue. *Int. J. Pharm.* **2015**, *478*, 361–367. [CrossRef] [PubMed]
24. Eisenbrey, J.R.; Shraim, R.; Liu, J.B.; Li, J.; Stanczak, M.; Oeffinger, B.; Leeper, D.B.; Keith, S.W.; Jablonowski, L.J.; Forsberg, F.; et al. Sensitization of Hypoxic Tumors to Radiation Therapy Using Ultrasound-Sensitive Oxygen Microbubbles. *Int. J. Radiat. Oncol. Biol. Phys.* **2018**, *101*, 88–96. [CrossRef] [PubMed]
25. Oeffinger, B.E.; Vaidya, P.; Ayaz, I.; Shraim, R.; Eisenbrey, J.R.; Wheatley, M.A. Preserving the Integrity of Surfactant-Stabilized Microbubble Membranes for Localized Oxygen Delivery. *Langmuir ACS J. Surf. Colloids* **2019**, *35*, 10068–10078. [CrossRef] [PubMed]
26. Delaney, L.J.; Ciraku, L.; Oeffinger, B.E.; Wessner, C.E.; Liu, J.B.; Li, J.; Nam, K.; Forsberg, F.; Leeper, D.B.; O'Kane, P.; et al. Breast Cancer Brain Metastasis Response to Radiation After Microbubble Oxygen Delivery in a Murine Model. *J. Ultrasound Med. Off. J. Am. Inst. Ultrasound Med.* **2019**, *38*, 3221–3228. [CrossRef]
27. Nath, K.; Guo, L.; Nancolas, B.; Nelson, D.S.; Shestov, A.A.; Lee, S.C.; Roman, J.; Zhou, R.; Leeper, D.B.; Halestrap, A.P.; et al. Mechanism of antineoplastic activity of lonidamine. *Biochim. Biophys. Acta* **2016**, *1866*, 151–162. [CrossRef]
28. Floridi, A.; Paggi, M.G.; Marcante, M.L.; Silvestrini, B.; Caputo, A.; De Martino, C. Lonidamine, a selective inhibitor of aerobic glycolysis of murine tumor cells. *J. Natl. Cancer Inst.* **1981**, *66*, 497–499.
29. Rochani, A.K.; Wheatley, M.; Oeffinger, B.E.; Eisenbrey, J.R.; Kaushal, G. LC-MS based stability-indicating method for studying the degradation of lonidamine under physical and chemical stress conditions. *Res. Pharm. Sci.* **2020**, *15*, 312–322. [CrossRef]
30. Ueno, S.; Kimura, T.; Yamaga, T.; Kawada, A.; Ochiai, T.; Endou, H.; Sakurai, H. Metformin enhances anti-tumor effect of L-type amino acid transporter 1 (LAT1) inhibitor. *J. Pharmacol. Sci.* **2016**, *131*, 110–117. [CrossRef]
31. Zhao, D.; Long, X.-D.; Lu, T.-F.; Wang, T.; Zhang, W.-W.; Liu, Y.-X.; Cui, X.-L.; Dai, H.-A.; Xue, F.; Xia, Q. Metformin decreases IL-22 secretion to suppress tumor growth in an orthotopic mouse model of hepatocellular carcinoma. *Int. J. Cancer* **2015**, *136*, 2556–2565. [CrossRef]
32. Patel, R.; Lacerda, Q.; Oeffinger, B.E.; Eisenbrey, J.R.; Rochani, A.K.; Kaushal, G.; Wessner, C.E.; Wheatley, M.A. Development of a Dual Drug-Loaded, Surfactant-Stabilized Contrast Agent Containing Oxygen. *Polymers* **2022**, *14*, 1568. [CrossRef] [PubMed]
33. McEwan, C.; Owen, J.; Stride, E.; Fowley, C.; Nesbitt, H.; Cochrane, D.; Coussios, C.C.; Borden, M.; Nomikou, N.; McHale, A.P.; et al. Oxygen carrying microbubbles for enhanced sonodynamic therapy of hypoxic tumours. *J. Control. Release* **2015**, *203*, 51–56. [CrossRef] [PubMed]
34. Reusser, T.D.; Song, K.-H.; Ramirez, D.; Benninger, R.K.; Papadopoulou, V.; Borden, M.A. Phospholipid Oxygen Microbubbles for Image-Guided Therapy. *Nanotheranostics* **2020**, *4*, 83–90. [CrossRef] [PubMed]
35. Fix, S.M.; Papadopoulou, V.; Velds, H.; Kasoji, S.K.; Rivera, J.N.; Borden, M.A.; Chang, S.; Dayton, P.A. Oxygen microbubbles improve radiotherapy tumor control in a rat fibrosarcoma model—A preliminary study. *PLoS ONE* **2018**, *13*, e0195667. [CrossRef] [PubMed]
36. Feshitan, J.A.; Legband, N.D.; Borden, M.A.; Terry, B.S. Systemic oxygen delivery by peritoneal perfusion of oxygen microbubbles. *Biomaterials* **2014**, *35*, 2600–2606. [CrossRef] [PubMed]

37. Liu, Y.; Zhou, Z.; Hou, J.; Xiong, W.; Kim, H.; Chen, J.; Zheng, C.; Jiang, X.; Yoon, J.; Shen, J. Tumor Selective Metabolic Reprogramming as a Prospective PD-L1 Depression Strategy to Reactivate Immunotherapy. *Adv. Mater.* **2022**, *34*, 2206121. [CrossRef]
38. Li, M.; Shao, Y.; Kim, J.H.; Pu, Z.; Zhao, X.; Huang, H.; Xiong, T.; Kang, Y.; Li, G.; Shao, K.; et al. Unimolecular Photodynamic O_2-Economizer To Overcome Hypoxia Resistance in Phototherapeutics. *J. Am. Chem. Soc.* **2020**, *142*, 5380–5388. [CrossRef]
39. Xiong, W.; Qi, L.; Jiang, N.; Zhao, Q.; Chen, L.; Jiang, X.; Li, Y.; Zhou, Z.; Shen, J. Metformin Liposome-Mediated PD-L1 Downregulation for Amplifying the Photodynamic Immunotherapy Efficacy. *ACS Appl. Mater. Interfaces* **2021**, *13*, 8026–8041. [CrossRef]
40. Zheng, C.; Luo, W.; Liu, Y.; Chen, J.; Deng, H.; Zhou, Z.; Shen, J. Killing three birds with one stone: Multi-stage metabolic regulation mediated by clinically usable berberine liposome to overcome photodynamic immunotherapy resistance. *Chem. Eng. J.* **2023**, *454*, 140164. [CrossRef]
41. Zhang, R.; Yang, Y.; Dong, W.; Lin, M.; He, J.; Zhang, X.; Tian, T.; Yang, Y.; Chen, K.; Lei, Q.-Y.; et al. D-mannose facilitates immunotherapy and radiotherapy of triple-negative breast cancer via degradation of PD-L1. *Proc. Natl. Acad. Sci. USA* **2022**, *119*, e2114851119. [CrossRef]
42. Czarnota, G.J.; Karshafian, R.; Burns, P.N.; Wong, S.; Al Mahrouki, A.; Lee, J.W.; Caissie, A.; Tran, W.; Kim, C.; Furukawa, M.; et al. Tumor radiation response enhancement by acoustical stimulation of the vasculature. *Proc. Natl. Acad. Sci. USA* **2012**, *109*, E2033–E2041. [CrossRef]
43. Daecher, A.; Stanczak, M.; Liu, J.B.; Zhang, J.; Du, S.; Forsberg, F.; Leeper, D.B.; Eisenbrey, J.R. Localized microbubble cavitation-based antivascular therapy for improving HCC treatment response to radiotherapy. *Cancer Lett.* **2017**, *411*, 100–105. [CrossRef] [PubMed]
44. Eisenbrey, J.R.; Forsberg, F.; Wessner, C.E.; Delaney, L.J.; Bradigan, K.; Gummadi, S.; Tantawi, M.; Lyshchik, A.; O'Kane, P.; Liu, J.-B.; et al. US-triggered Microbubble Destruction for Augmenting Hepatocellular Carcinoma Response to Transarterial Radioembolization: A Randomized Pilot Clinical Trial. *Radiology* **2021**, *298*, 450–457. [CrossRef] [PubMed]

pharmaceutics

Article

Low-Intensity Pulsed Ultrasound-Mediated Blood-Brain Barrier Opening Increases Anti-Programmed Death-Ligand 1 Delivery and Efficacy in Gl261 Mouse Model

Mohammed H. Ahmed [1,2,*], Isaias Hernández-Verdin [1], Emie Quissac [1], Nolwenn Lemaire [1], Coralie Guerin [3], Lea Guyonnet [3], Noël Zahr [4], Laura Mouton [1], Mathieu Santin [1], Alexandra Petiet [1], Charlotte Schmitt [5], Guillaume Bouchoux [5], Michael Canney [5], Marc Sanson [6], Maïté Verreault [1], Alexandre Carpentier [5,6] and Ahmed Idbaih [5,*]

1 Institut du Cerveau—Paris Brain Institute—ICM, Inserm, CNRS, AP-HP, Hôpital de la Pitié Salpêtrière, Sorbonne Université, F-75013 Paris, France
2 School of Cancer & Pharmaceutical Sciences, King's College London, London SE1 9NH, UK
3 Cytometry Department, Institute Curie, F-75006 Paris, France
4 Pharmacokinetics and Therapeutic Drug Monitoring Unit, Inserm, CIC-1901, UMR ICAN 1166, AP-HP, Hôpital de la Pitié Salpêtrière, Sorbonne Université, F-75013 Paris, France
5 CarThera, Institut du Cerveau et de la Moelle Épinière (ICM), F-75013 Paris, France
6 Institut du Cerveau—Paris Brain Institute—ICM, Inserm, CNRS, AP-HP, DMU Neurosciences, Service de Neurologie 2-Mazarin, Hôpital de la Pitié Salpêtrière, Sorbonne Université, F-75013 Paris, France
* Correspondence: mohammed.ahmed@kcl.ac.uk (M.H.A.); ahmed.idbaih@aphp.fr (A.I.); Tel.: +44-(0)-20-7836-5454 (M.H.A.); +33-01-42-16-03-85 (A.I.); Fax: +33-01-42-16-04-18 (A.I.)

Citation: Ahmed, M.H.; Hernández-Verdin, I.; Quissac, E.; Lemaire, N.; Guerin, C.; Guyonnet, L.; Zahr, N.; Mouton, L.; Santin, M.; Petiet, A.; et al. Low-Intensity Pulsed Ultrasound-Mediated Blood-Brain Barrier Opening Increases Anti-Programmed Death-Ligand 1 Delivery and Efficacy in Gl261 Mouse Model. *Pharmaceutics* **2023**, *15*, 455. https://doi.org/10.3390/pharmaceutics15020455

Academic Editors: Brandon Helfield, Shashank Sirsi, James Kwan and Michael Gray

Received: 8 December 2022
Revised: 20 January 2023
Accepted: 24 January 2023
Published: 30 January 2023

Abstract: Therapeutic antibodies targeting immune checkpoints have shown limited efficacy in clinical trials in glioblastoma (GBM) patients. Ultrasound-mediated blood–brain barrier opening (UMBO) using low-intensity pulsed ultrasound improved drug delivery to the brain. We explored the safety and the efficacy of UMBO plus immune checkpoint inhibitors in preclinical models of GBM. A blood–brain barrier (BBB) opening was performed using a 1 MHz preclinical ultrasound system in combination with 10 µL/g microbubbles. Brain penetration of immune checkpoint inhibitors was determined, and immune cell populations were evaluated using flow cytometry. The impact of repeated treatments on survival was determined. In syngeneic GL261-bearing immunocompetent mice, we showed that UMBO safely and repeatedly opened the BBB. BBB opening was confirmed visually and microscopically using Evans blue dye and magnetic resonance imaging. UMBO plus anti-PDL-1 was associated with a significant improvement of overall survival compared to anti-PD-L1 alone. Using mass spectroscopy, we showed that the penetration of therapeutic antibodies can be increased when delivered intravenously compared to non-sonicated brains. Furthermore, we observed an enhancement of activated microglia percentage when combined with anti-PD-L1. Here, we report that the combination of UMBO and anti-PD-L1 dramatically increases GL261-bearing mice's survival compared to their counterparts treated with anti-PD-L1 alone. Our study highlights the BBB as a limitation to overcome in order to increase the efficacy of anti-PD-L1 in GBM and supports clinical trials combining UMBO and in GBM patients.

Keywords: SonoCloud; GL261 mouse model; ultrasound-mediated drug delivery; glioblastoma; immune checkpoint inhibitors

1. Introduction

Glioblastoma (GBM) is the most malignant primary brain tumor in adults, with a median overall survival of less than 18 months after initial diagnosis [1]. Despite significant efforts in the neuro-oncology field to develop new therapeutic alternatives, temozolomide (approved in 2005) remains the gold standard chemotherapy in GBM treatment [2]. For over five decades, research has been focused on developing new anti-cancer therapies for GBM, including anti-neoplastic agents [3], molecular targeted drugs [4], immunotherapeutic approaches [5], and angiogenesis inhibiting compounds [6]; however, the prognosis of patients has hardly improved [7]. Blood-brain barrier (BBB), which is specific to the blood vessels in the central nervous system (CNS), prevents most systemic therapeutic compounds from reaching the brain parenchyma and GBM cells [8] although it is disrupted in some areas (i.e., blood–tumor barrier).

Several innovative strategies have been studied to enhance the delivery of chemotherapeutic agents and antibodies to the brain [8]. Ultrasound mediated BBB opening (UMBO) using low-intensity pulsed ultrasound (LIPU) has now been studied in preclinical [9] and clinical settings [10]. LIPU is delivered to the brain simultaneously with an I.V. of micron-sized bubbles for a few minutes, allowing the microbubbles to oscillate. Microbubble oscillation produces mechanical stretching on vessel walls that allows a transient BBB opening [11]. UMBO has shown a good safety profile for BBB opening in recurrent GBM patients [10,12] and is now being studied in dozens of clinical trials using a range of transcranial [13] or implantable ultrasound devices [10] for treating both primary and secondary brain tumors as well as neurodegenerative diseases [14].

The choice of therapeutic agents to deliver after UMBO is crucial and remains a point of discussion among researchers. Direct stimulation of the immune system with immune checkpoint inhibitors (ICI, e.g., PD-1/PD-L1) showed promising effects alone or with other chemotherapies in multiple cancers. PD-L1 proteins are expressed as surface molecules by cancerous cells, such as GBM cells [15], and provide a tumor escape mechanism when bound to PD-1 proteins at the surface of activated T-lymphocytes leading to their exhaustion [16]. Despite their promise in other cancers, nivolumab (anti-PD-1) has shown no additional efficacy over bevacizumab in phase III clinical trials in recurrent GBM patients [17]. Similarly, Avelumab (anti-PD-L1) in combination with molecular targeted drugs did not improve the outcome of GBM patients [18]. In the present study, we evaluated the effect of anti-PD-L1 and anti-CTLA-4 alone and in combination with UMBO in syngeneic Gl261 mouse models.

2. Materials and Methods

2.1. Low-Intensity Pulsed Ultrasound Preclinical Device

The pre-clinical ultrasound system (CarThera, Paris, France) was identical to that described in other studies (Supplementary Material, Figure S1A) [9].The system consisted of a 1 MHz, 10-mm diameter acoustic transducer that was coupled to the head of the mouse at 15-mm from the transducer. Sonication was performed for 120 s using a 25,000-cycle burst at a 1 Hz pulse repetition frequency and an acoustic pressure of 0.3 MPa as measured in water.

2.2. Calibration of Low-Intensity Pulsed Ultrasound Device

The ultrasound transducer (CarThera®) used in this study was calibrated on a regular basis (Figure 1A,B). The aim was to map the ultrasound field and to determine the electrical set point that the generator uses during the experiments to obtain the targeted acoustic pressure in situ. The calibration was performed using degassed water at room temperature with a 200 µm needle hydrophone (HNC0200, ONDA). A 2D acoustic field was scanned at 5 mm from the transducer surface with a 3-axis computer-controlled motorized positioning system (UMS, Precision Acoustics, UK). The 3D acoustic field was computed from this pattern with the Rayleigh integral. The hydrophone was positioned at the spatial peak pressure determined from the 3D acoustic field. The ratio between the active electrical power drawn by the transducer measured with an oscilloscope and the square of the spatial peak acoustic pressure was measured. This ratio is used as a calibration coefficient by the generator during subsequent experiments: the active electrical power needed to obtain the targeted pressure is calculated by the generator using this coefficient at the beginning of manipulation, and the generator adjusts its set point to obtain the specified active electrical power measured internally.

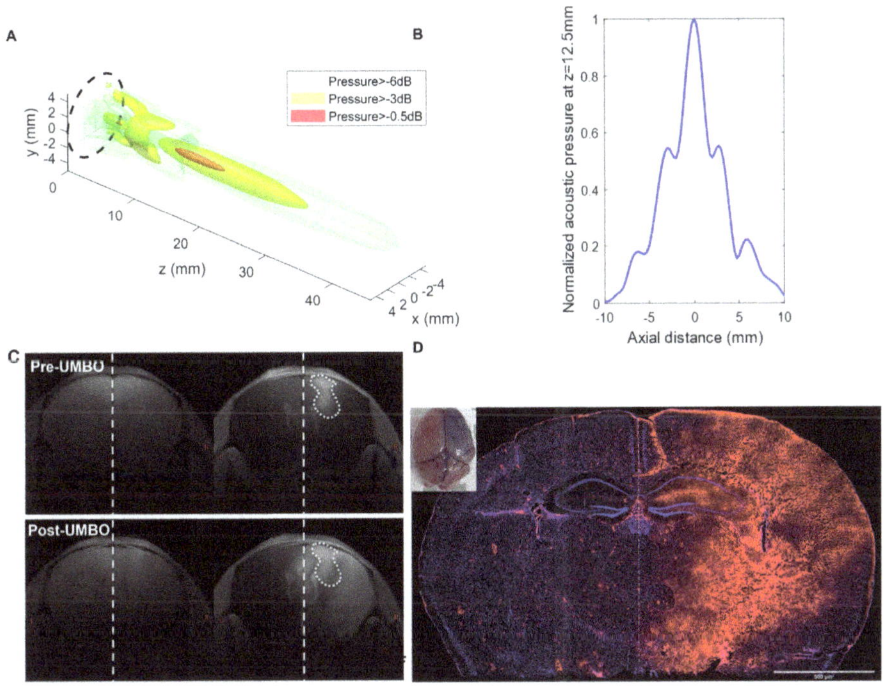

Figure 1. UMBO parameters are effective for BBB opening in healthy and Gl261 bearing mice. (Panel **A**,**B**): 3D structure of the ultrasound waves generated by the calibrated low-intensity ultrasound system. (Panel **C**): T1W MRI showed a marked Gd contrast enhancement within an hour following UMBO. An enhancement of Gd following UMBO in mice ($n = 2$, analysis of the Gd enhancement is shown in Supplementary Material, Figure S1C). The two top T1-MRI images were obtained before UMBO (pre-Gd left MRIs; and post-Gd). The two bottom T1-MRI images were obtained after UMBO (pre-Gd left MRIs; and post-Gd). UMBO is effective in C57BL/6 mice. right MRIs). (Panel **D**): Evans blue staining was enhanced in sonicated brain hemisphere compared to the control hemisphere visually and by fluorescence ((Panel **D**); Evans blue in red, DAPI in blue).

2.3. *Ultrasound-Mediated Blood–Brain Barrier Opening (UMBO)*

UMBO was delivered to both UMBO and UMBO plus anti-PD-L1 groups. Anti-PD-L1 (6E11 Genentech) was administered intraperitoneally in two independent experiments ($n = 8$) at a dose of 200 µg sixty minutes before UMBO application. Mice were maintained under anesthesia with isoflurane (2%, 2 L/min O_2). For each UMBO application, 10 mL/kg SonoVue® was injected through the intravenous route less than 10 s before the start of the ultrasound application. For each session, UMBO was validated using an additional control mouse. Each control mouse was injected intravenously with a solution of 2.7% Evans blue (Sigma, E2129) in phosphate buffer saline (PBS) at a dose of 4 mL/kg ten minutes post-sonication. All mice received 10 mL/kg warm saline injection in each treatment protocol before anesthesia to prevent any possible hypovolemia or hypothermia effect. I.P injection of anti-PD-L1 injection was given 60 min before sonication to ensure anti-PD-L1 absorption (Supplementary Material, Figure S1B). UMBO test mice were sacrificed 15 min following Evans' blue injection, and their brain was harvested. The passage of Evans blue was assessed both visually and by ZEISS Axio-Scan fluorescence imaging of cryo-sectioned brains.

2.4. Cell Culture and In Vivo Studies

GL261 cells were cultured in Dulbecco's modified essential medium (DMEM) supplemented with 10% fetal bovine serum and 1% penicillin/streptomycin. Cells were passaged twice weekly according to their confluence. The animal ethics committee at the Ministry of Higher Education and Research in Paris approved all protocols involving live mice (protocol #17503 and #26137). C57BL/6 mice were purchased from Charles River and were given a week of acclimation before starting any experiment.

GL261 was transduced with a *luciferase/mKate2* vector as described before [19]. GL261-luciferase (1.4×10^5 cells/2 µL) were inoculated into the right caudate nucleus-putamen (AP +1 mm, DV −0.25 mm, ML −0.15 mm) of 7–8 weeks old C57BL/6 females using a stereotactic injection frame (David Kopf Instruments, Angeles, CA, USA). Mice were imaged using the IVIS Spectrum (PerkinElmer, Waltham, MA, USA) 10 min following a 2 mg subcutaneous injection of luciferin (Sigma Aldrich (MO, USA), L9504). The growth of GL261-luciferase cells was confirmed by two IVIS imaging one week apart of intracranial cell injection. We observed that mice with bioluminescence values lower than 5×10^5 photon/second would not develop GBM tumors during the characterization of tumor growth in our mouse models. Therefore, we have included mice with bioluminescence values over 5×10^5 photon/second. Mice were randomly placed into treatment arms once they passed the bioluminescence cutoff value.

Animals were treated with 200 µg of anti-CTLA-4 (Bristol-Myers Squibb, New York, NY, USA, G1-XAS-Ab), anti-PD-L1 (Genentech, New York, NY, USA, 6E11), IgG1 (BXCELL, West Lebanon, NH, USA, BE0083), and InVivoPure pH 6.5 Dilution Buffer (BXCELL, IP0065) for four doses (Supplementary Material, Figure S1B). Unless stated otherwise, animals were sacrificed when they showed signs of tumor-associated illness (20% body weight loss or changes in behavior or posture).

2.5. Pharmacokinetic (PK) Analysis of Therapeutic Antibodies with and without UMBO

The PK analysis was performed using an identical molecular weight with similar conformational structure IgG1 isoform. Thirty-six mice were used in the pharmacokinetic experiment. Mice were separated into control and UMBO groups. Six time points were selected as follows: 0.15, 0.3, 3, 6, 24, 48, and 96 h. Each mouse received a 200 µg of nivolumab (Bristol-Meyers Squibb, New York, NY, USA) I.V. injection 10 min following the BBB opening. 100 µL of blood was collected through cardiac puncture using a pre-heparinized syringe. Serum was collected by centrifugation of the blood at 3500 rpm for 10 min. All samples (plasma and brain) were then analyzed using an ultra-performance liquid chromatography (UPLC) system coupled to mass spectrometry (LC-MS/MS; MS-8060, Shimadzu, Nakagyo-ku, Kyoto, Japan). Peak integration and quantification were performed using LabSolutions Insight LC-MS software. Nivolumab was quantified with signature peptide ASGGITFSNSGMHWVR by nano-surface and molecular orientation limited proteolysis (Shimadzu, Japan) [20].

2.6. MRI Data Acquisition

Two GL261-bearing mice were used in the experiment. Two sessions per mouse were completed in 2 consecutive days to decrease any distress effect of long isoflurane exposure. MRI acquisitions were performed using a preclinical 11.7 T MRI scanner (Biospec, Bruker BioSpin, Ettlingen, Germany) equipped with a CryoProbe dedicated to mouse brain imaging (Biospec, Bruker BioSpin, Germany). Gd enhancement volume was estimated on T_1w MRI pre and post BBB opening manually on FSLeyes. Animals were anesthetized with 1% isoflurane in O_2 (2 L/min). Respiratory rate and body temperature were monitored while mice were restrained. For each animal, the protocol consisted of: (i) acquiring pre-gadolinium enhancement anatomical T_1-weighted (T_1w) images using a multi-slice multi echo (MSME) sequence with the following parameters: repetition time (T.R.) = 400 ms, echo time (T.E.) = 5 ms (one single echo), four averages, 14 slices, and resolution = $60 \times 60 \times 500$ µm^3, (ii) following injection of a total volume of 100 µL of gadolinium (Gd; DOTAREM®,Guerbet, Aulnay-sous-Bois, France) at 0.5 mM and at physiological temperature in the tail vein of the mouse outside the MRI scanner, and (iii) acquiring post-gadolinium T_1w images using the same sequence. The MGE sequence was acquired with the following parameters: T.R. = 80 ms, ten echoes ranging from T.E. = 2.7 ms to 35.1 ms (echo spacing = 3.6 ms), and isotropic resolution of $60 \times 60 \times 60$ μm^3.

2.7. *mRNA Sequencing*

Six mice with a confirmed tumor of comparable size (as measured by bioluminescence imaging) were included in this experiment. Mice were divided into two groups (UMBO group and vehicle group). The vehicle group was treated with inVivoPure pH 6.5 Dilution Buffer (BXCELL, IP0065). Two treatment sessions (days 21 and 24) were applied in this experiment. Mice were sacrificed 24 h after the last treatment by cervical dislocation, and the right hemisphere was stored in 5 mL RNALater (Thermofisher AM7020). Lysing Matrix D (MBio, 6913050) was used to homogenize the collected brain tissues. mRNA was extracted using Maxwell RSC simply RNA automated RNA purification kit (Promega, AS1340). RNA quality was analyzed using high-sensitivity RNA chips. For RNA sequencing, NovaSeq 6000 sequencer (200 cycles, 800 million reads) and reagent kit. The reads (202 bp length, 100 million input reads) were mapped with the STAR v2.7.2a (default parameters) software to the reference genome (version GRCm38) on new junctions and known annotations. Mapping parameters were obtained from STAR outputs obtaining around 90% of unique mapped reads for all samples. Read counts from STAR were used as input for differential expression analysis using DESeq2. Furthermore, normalized counts were obtained using the variance-stabilizing transformation (VST) method from DESeq2 to be used as input for gene set variation analysis (GSVA) [21] to evaluate signature enrichment of microglia expression (Slc2a5, Siglech, P2ry12, Gpr34, P2ry13, Olfml3, Tmem119, and Fcrls) [22], microglia sensome (96 genes) [23] or antigen presentation related genes (Ciita, Psme2b, Erap1, Irf1, Tapbp, Psme2, Psme1, Pdia3, Psme3, Tap1, B2m, Calr, Tap2, Hspa1a, H2-Ab1k, H2-K1, and H2-D1) [24]. For heatmaps representation (ComplexHeatmap R package), VST gene expression values were first quantile normalized and $\log^2$ transformed, then converted to Z-scores by subtracting the average expression value of gene i (Gi) from the gene expression within sample x(Sx); the resulting value was divided by the SD of Gi; the formula is: Z-scoreGiSx − (Expression GiSx-μGi)/σGi. Raw data were uploaded to Gene Expression Omnibus with accession number GSE220909.

2.8. *Immunohistochemistry (IHC)*

A 150 KDa rat IgG2 antibody targeting PD-L1 was used in our IHC staining (BXCELL, #BE0101). A goat anti-rat secondary IgG (H+L) antibody (BA-9400) was used to detect the anti-PD-L1. Iba1 protein was detected using 1:1000 (Abcam, #ab178846). Mouse brains were fixed overnight in 4% paraformaldehyde (PFA), then immersed in 30% sucrose overnight for cryoprotection. Next, brains were stored in Tissue-Tek® O.C.T and stored at −80 °C10 μm cryosections were harvested using Leica CM1950 cryostat. Slides were stored at −80 °C until analysis.

2.9. *Quantitative Digital Droplet Polymerase Chain Reaction (ddPCR)*

GL261 tumor-bearing mice four weeks following cell inoculation were used in the ddPCR experiment. A single UMBO treatment was completed, and 30 min later, blood (100 μL) was collected in heparinized tubes through cardiac puncture. Whole blood DNA was extracted automatically using Maxwell® Blood DNA Purification Kit (AS1010). QX200 ddPCR EvaGeen® was utilized to detect *mKate2* and *Luciferase* genes in the extracted DNA. Primer3Plus web interface was used to design *mKate2*, and *Luciferase* primers and primers were purchased from Life Technologies. The following forward (FR) and reverse (RV) primers were used: luciferase-FR, TCCACGATGAAGAAGTGCTC; *luciferase*-RV, AGGCTACAAACGCTCTCATC; *mKate2*-FR, GGTGAGCGAGCTGATTAAGG; and mKate2-RV, GGGTGTGGTTGATGAAGGTT.

2.10. Flow Cytometry

Twenty mice with a confirmed tumor of comparable sizes were included in this experiment. Mice were separated into four groups: UMBO group, anti-PD-L1 (Genentech, 6E11) group, UMBO plus anti-PD-L1, and vehicle group (n = 5/group). One treatment session was delivered in this experiment. Mice were perfused using cold distilled phosphate buffer saline (DPBS) ~16 h after treatment. Tumor-bearing hemispheres were isolated immediately and stored in 2 mL ice-cold Hanks' balanced salt solution. According to the manufacturer's protocol, the right hemisphere was isolated and mixed in the enzyme mix solution from the adult brain dissociation kit (Miltenyi Biotec, Cologne, Germany, #130-107-677). Cells gentleMACS® Octo Dissociator with Heaters (#130-096-427) and gentleMACS C Tubes (#130-093-237) were used to perform mice brain dissociation. The number of dissociated cells was calculated using Scepter® 3.0 Handheld Cell Counter. Flow cytometry was applied as described in Supplementary Material, Figure S2.

2.11. Statistical Tests

Statistical analysis was performed using Prism software (GraphPad Software, San Diego, CA, USA). Data are shown as mean values plus and minus standard error of the mean (SEM). Statistical significance of differences between groups was verified using appropriate statistical tests. Significance levels were denoted with asterisks: * for $p \leq 0.05$; ** for $p \leq 0.01$; *** for $p \leq 0.001$, and **** for $p \leq 0.0001$.

3. Results

3.1. Repeated UMBO Is Safe and Effective in Immunocompetent Mice

UMBO parameters were previously optimized [25] in healthy mice and we evaluated UMBO parameters and treatment frequency in GL261 bearing mice. T_1w MRI (Figure 1C) showed a marked gadolinium contrast enhancement within an hour following the UMBO (Figure 1C). Furthermore, biweekly UMBO (four sonications in total without drugs) was evaluated in the GL261-bearing mice. Mouse weight was unaffected (Figure 2A) and no significant difference in the overall survival (OS) between UMBO and non-treated groups was observed (Figure 2B). Overall, the UMBO parameters used for repeated BBB opening were safe and well-tolerated in GL261-bearing mice.

Pilot study with no UMBO using a GL261-luciferase orthotopic GBM mouse model was performed. This experiment aimed to determine the effect of anti-PD-L1 and anti-CTLA-4 in our GBM mouse model and select the best candidates to combine with UMBO. Anti-PD-L1 antibody alone showed a small regression in tumor growth (Figure 2C) with a limited improvement of survival in GL261-bearing mice (Figure 2D), Anti-CTLA-4 treatment did not reduce tumor growth (Figure 2C) or animal survival (Figure 2D).

Furthermore, using immunofluorescence, we were able to detect PD-L1 expression on the surface of the GL261 cell line using anti-PD-L1 (Genentech, 6E11). RT-PCR was used to evaluate the quantitative expression of PD-L1 in the GL261 cell line and we have used the Nfpp10 cell line as a positive control. PD-L1 expression was significantly higher in the GL261 cell line compared to our control (Supplementary Material, Figure S3). Overall, this makes the anti-PD-L1 antibody the best candidate to combine UMBO in the GL261 GBM mouse model.

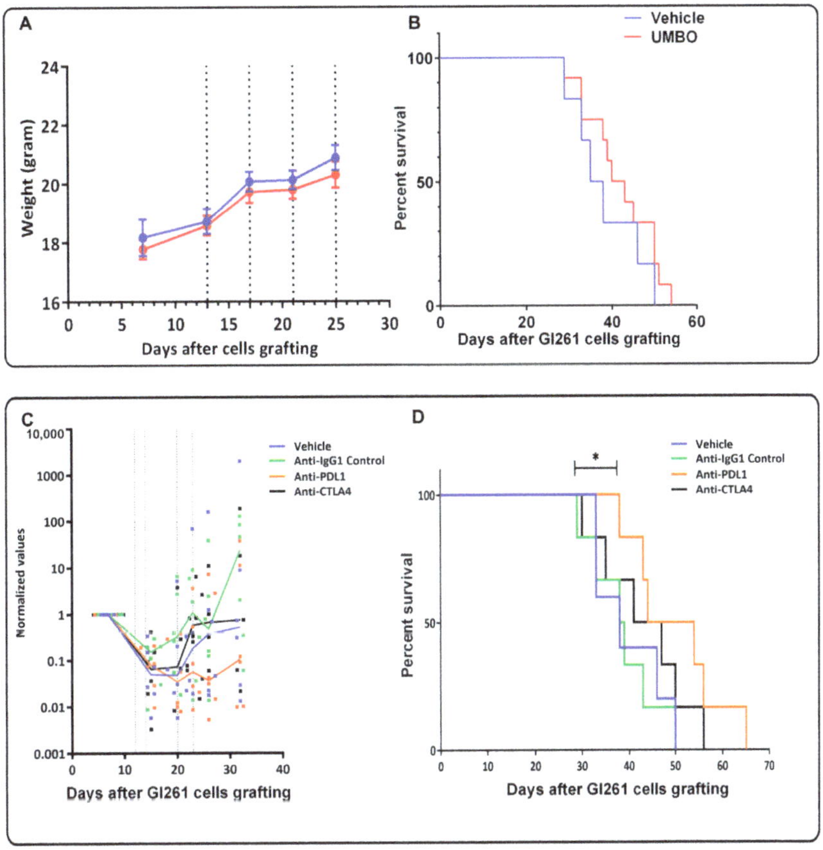

Figure 2. UMBO parameters are safe and anti-PD-L1 increased survival of GL261 bearing mice. Repeated UMBO alone does not affect OS (Panel **B**) or body weight in GL261-bearing mice compared to the non-treated (Panel **A**). (Panel **C**,**D**): Animals were treated with 200 µg of anti-PDL-1, Anti-IgG1 as (a control), anti-CTL-4, or vehicle for four doses. Bioluminescence measures normalized to the first measured value performed on day 7 after cell inoculation in GL261-bearing mice model. Each colored dot represents values for one animal and the line represents the median value for the group. Bioluminescence signal was measured weekly. Dotted lines represent the days of treatments. (Panel **D**). Kaplan Meier curves in GL261-bearing mice. Anti-PD-L1 alone improved OS in Gl261 bearing mice. Anti-CTLA-4 did not improve OS in GL261-bearing mice. * for $p \leq 0.05$.

3.2. *UMBO Dramatically Increased the Efficacy of anti-PD-L1 in GL261-Bearing Mice*

We then investigated the combined effect of UMBO with anti-PD-L1 in the GL261-bearing mice. Mice with comparable bioluminescence values were divided into five groups: (i) UMBO group, (ii) anti-PD-L1 group, (iii) UMBO plus anti-PD-L1 group, (iv) IgG1 group, and (v) IgG1 plus UMBO group. Mice that had received an anti-PD-L1 antibody plus UMBO showed 76% long-term survivals (13/17) compared to 26% in anti-PD-L1 alone (4/15) and 0% in control groups (0/16). The results from two independent experiments (Exp) were summarized in Table 1. UMBO plus anti-PD-L1 showed a significant regression in tumor growth following four sessions of treatments compared to controls ($p < 0.01$) (Figure 3B). Furthermore, Anti-PD-L1 alone showed a limited regression of tumor growth ($p < 0.05$) compared to control (Figure 3B). Kaplan–Meier estimate showed a significant difference in OS of UMBO's plus anti-PD-L1 treated mice versus anti-PD-L1 alone (Figure 3C). Furthermore, a higher significant difference (Figure 3C) was observed in UMBO plus anti-PD-L1 treated mice compared to anti-IgG1 plus UMBO treated mice.

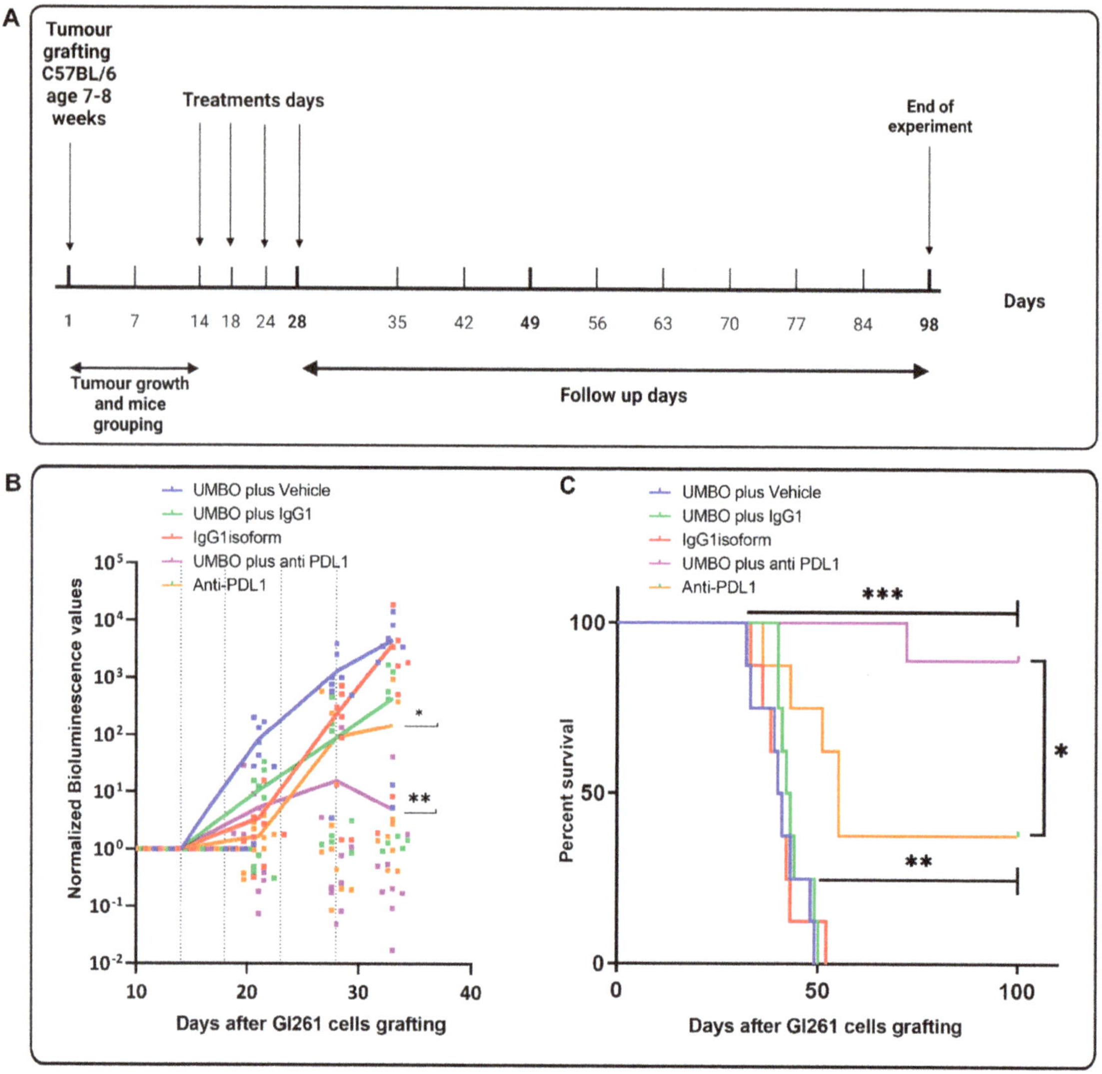

Figure 3. UMBO increased the efficacy of anti-PD-L1 in GL261-bearing mice. (Panel **A**): shows a timeline of the treatment protocol used in the survival experiments. Cell grafting was done on day 1 while treatments were applied on days 14, 18, 24, and 28. Mice were followed up for 100 days. (Panel **B**,**C**): (Panel **B**): bioluminescence normalized to the first measured value after cell inoculation in GL261-bearing mice. Each dot represents values for one animal and the line represents the median value for the group. Bioluminescence signal was measured weekly. Dotted lines represent the days of treatments. Anti PD-L1 plus UMBO suppressed tumor growth over time compared to controls. ** for $p \leq 0.01$. Repeated UMBO alone does not affect OS (Panel **C**) or body weight (Supplementary Material, Figure S4) in GL261-bearing mice compared to the non-treated (Panel **C**). (Panel **C**): UMBO plus an-ti-PD-L1 increased OS (* $p < 0.05$) compared to anti-PD-L1 alone and (*** $p < 0.0001$) compared to control groups.

Table 1. Summary of long-term survivals and statistical analyses.

		Exp. 1	Exp. 2
No of Long-term Survivals > 100 days	UMBO plus vehicle	0/8	0/8
	UMBO plus IgG1	0/8	0/8
	IgG1 isoform	0/8	0/8
	UMBO plus anti-PD-L1	7/9	6/8
	Anti-PD-L1	3/8	1/7
Percentage	UMBO plus anti-PD-L1	77%	75%
	Anti-PD-L1	37%	14%
Statistic	UMBO plus anti-PD-L1/UMBO alone	$p = 0.0009$	$p = 0.0012$
	anti-PD-L1/IgG1 isoform	$p = 0.0057$	$p = 0.0191$
	UMBO plus anti PD-L1/anti-PD-L1	$p = 0.0480$	$p = 0.0360$

3.3. UMBO Increased the Penetration of anti-PD-1 and anti-PD-L1 Antibodies into the Brain Parenchyma

IHC staining of anti-PD-L1 (BXCELL, BE0101) confirmed UMBO's trend to deliver anti-PD-L1 to the right hemisphere brain parenchyma (Figure 4A). Furthermore, an already clinically optimized method [21] was used to compare a size matched IgG1 antibody's pharmacokinetics with and without UMBO. Three C57BL/6 mice per time point (six-time points) per group were used in the analysis. We observed a comparable serum concentration of nivolumab in control and UMBO-treated mice.

Interestingly, higher concentrations of nivolumab were detected in mice brains treated with nivolumab plus UMBO. As expected, we detected a negligible concentration ($\leq$0.2 µg/200 mg brain) of nivolumab in control mice brains (Figure 4B). The maximum concentration (C_{max}) of nivolumab in normal brain tissue was detected at 24 h and started to decline and reach a negligible concentration at 96 h. Therefore, a regimen of biweekly antibody administration was performed. The brain to plasma concentration ratio shows that UMBO enhanced the ratio of nivolumab passage across the BBB at 3, 24, 48 h but not at 96 h (Figure 4C).

A

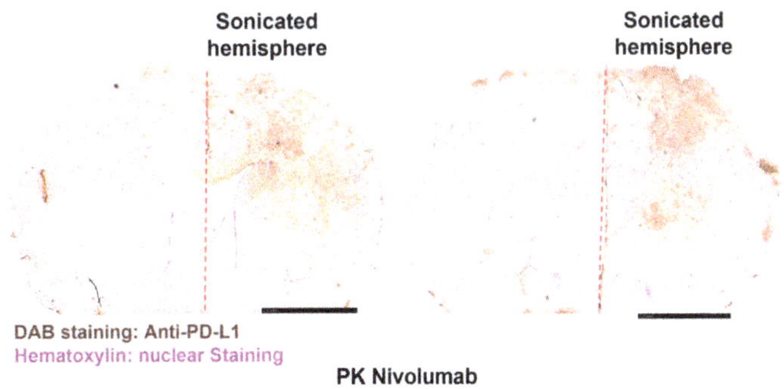

Figure 4. *Cont.*

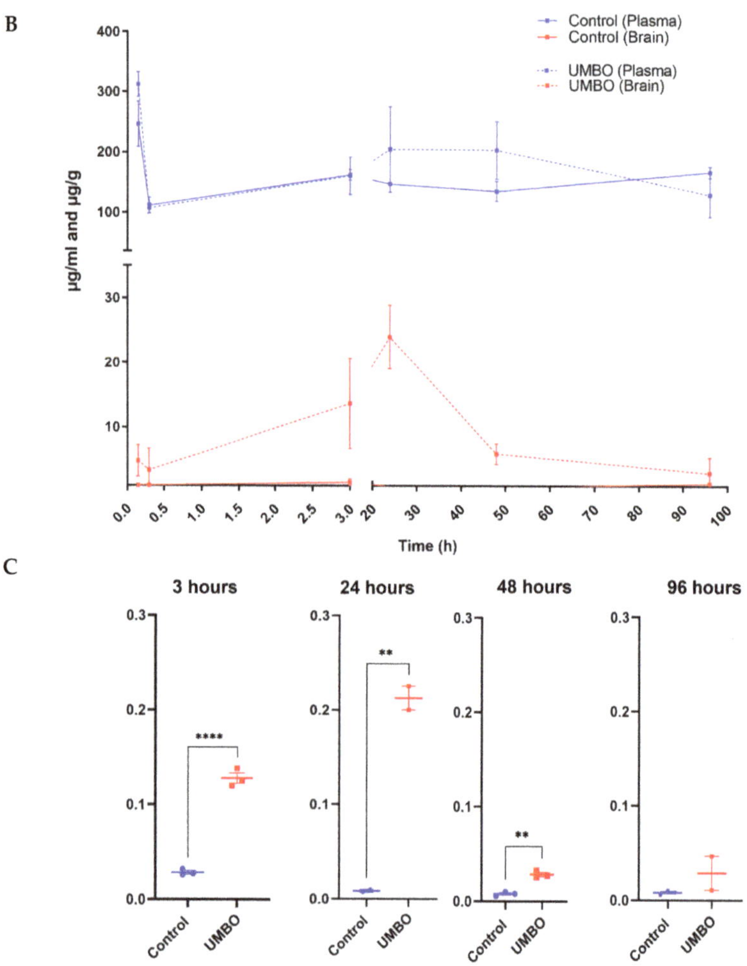

Figure 4. UMBO increased the delivery of ICIs to the brain parenchyma. (Panel **A**): IHC staining of anti-PD-L1 (BXCELL) (Panel **B**): PK of Nivolumab concentration in the C57BL/6 mice blood and brain. (Panel **C**): Brain/plasma ratio of nivolumab concentration over time. UMBO enhanced the brain/plasma ratio of nivolumab compared to control mice. ** for $p \leq 0.01$. **** for $p \leq 0.0001$.

3.4. UMBO plus Anti-PD-L1 Activates Microglia and Modulates Microglial Phenotype

We studied the immune cell populations in our treatment groups using flow cytometry. Interestingly, we observed that UMBO plus anti-PD-L1 significantly enhanced the percentage of activated microglia compared to anti-PD-L1 treatment alone (Figure 5B). UMBO alone was not associated with a significant enhancement of activated microglia percentage compared to the vehicle group; however, a trend was observed ($p = 0.150$). On the other hand, we did not observe any significant changes in the percentage of $CD8^+$ and $CD4^+$ T-lymphocytes or $CD206^+$ macrophages in all groups. Immunofluorescence staining of microglia in the anti-PD-L1 plus UMBO treated group confirmed this finding and showed a phenotype of activated microglia. IHC showed a double nucleus staining of Iba1 in the UMBO plus anti-PD-L1 treated GL261-bearing mice suggesting a possible induction of microglia cell division (Figure 5G).

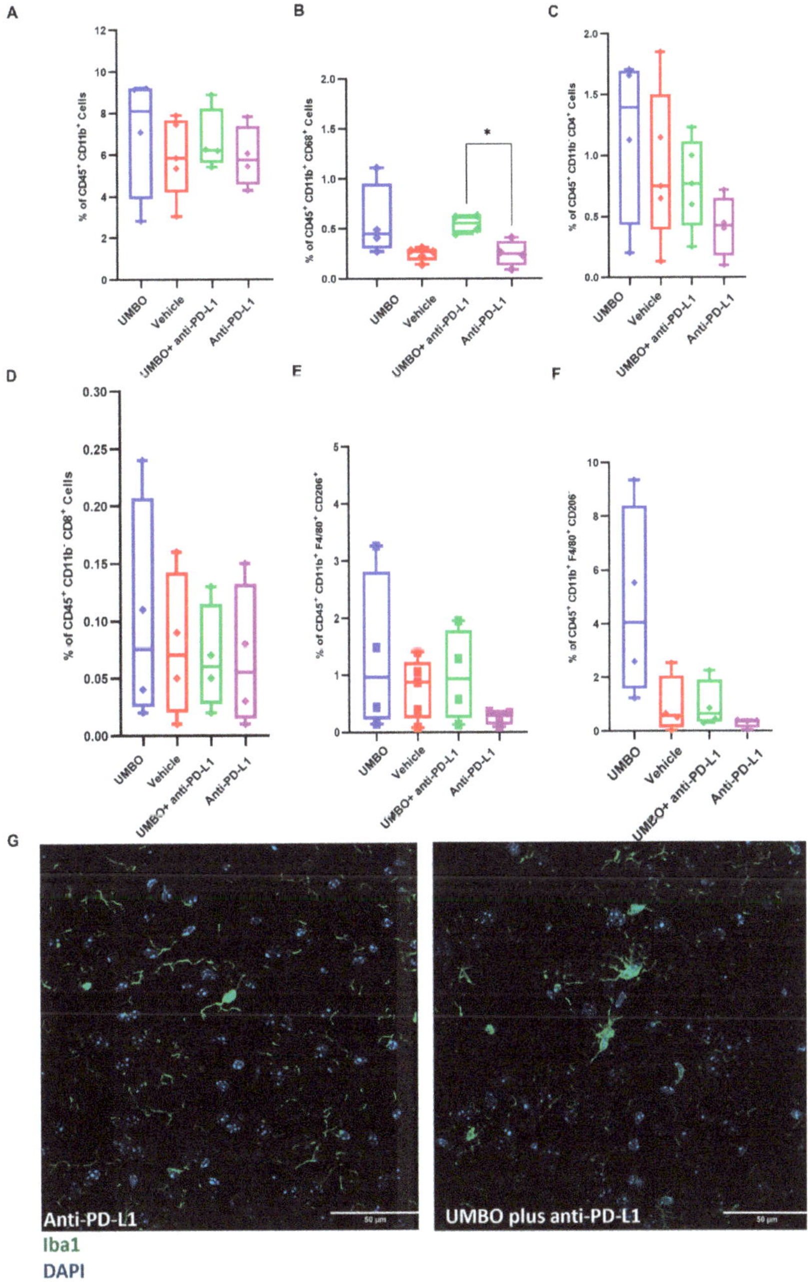

Figure 5. UMBO plus anti-PD-L1 activates microglia and modulates microglial phenotype. (Panel **A**): Flow cytometry analysis of the percentage of total microglia in all different groups (n = 4–5). (Panel **B**): UMBO plus anti-PD-L1 significantly enhanced (* $p < 0.05$) the percentage of CD68$^+$ cells than an-ti-PD-L1 alone. (Panel **C**,**D**): UMBO plus anti-PD-L1 did not influence CD4$^+$ and CD8$^+$ T-lymphocytes percentages compared to other groups. (Panel **E**,**F**): No significant difference in CD206$^+$ and CD206$^-$ macrophages in all groups. UMBO plus anti-PD-L1 did not modulate macrophages' expression. (Panel **G**): Green: Iba1 microglia/macrophages Blue: DAPI nuclear staining; microglia staining in anti-PD-L1 plus UMBO (right photos) treated group confirm a phenotype of activated microglia.

Using bulk RNA sequencing, we analyzed whether UMBO modulates antigen presentation related genes compared to the vehicle treated group. UMBO did not influence antigen presentation (Figure 6A) or affect microglia gene expression (Figure 6B). Microglial ability to sense changes in the cellular environment was recently termed as microglia sensome [23]. We used the same gene signature to evaluate microglial sensome with and without UMBO. Interestingly, UMBO significantly induced the expression of gene signatures for microglial sensome compared to control (Figure 6B). Additionally, we evaluated whether UMBO enhanced circulating tumor DNA release to the bloodstream. GL261-bearing mice with significant tumors were used in this experiment. We observed a significant elevation in the number of copies for both *luciferase* (Figure 6D) and *mKate2* (Figure 6E) in the UMBO treated group compared to the control.

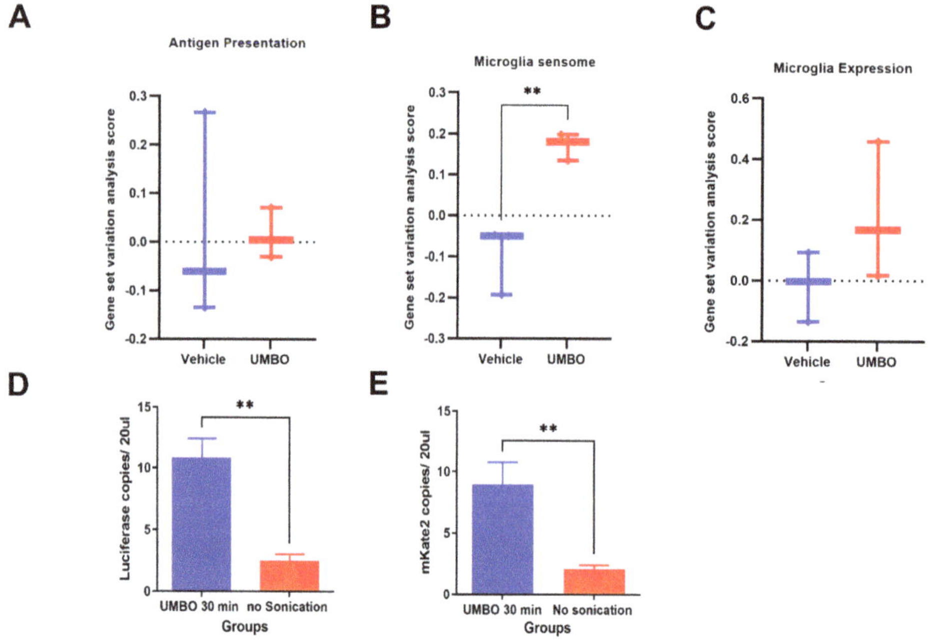

Figure 6. Repeated UMBO is associated with microglia's gene signature in the GL261 mouse model. (Panel **A**): UMBO did not influence antigen presentation gene signature compared to the non-treated. (Panel **C**): UMBO alone is associated with a significant (** $p < 0.01$) enrichment of microglia sensome gene signature (Panel **D**) compared to vehicle ($n = 3$). No significant difference was observed in microglial gene expression (Panel **B**). Heat maps of microglia gene signature and antigen presentation genes used in panels A and C are presented in Supplementary Material, Figure S5. (Panel **D**,**E**) represent the ddPCR analysis of *Luciferase* DNA (Panel **D**) and *mKate2* DNA (Panel **E**) in the blood 30 min following UMBO.

4. Discussion

UMBO and several innovative strategies continuously evolve to overcome the BBB by increasing drug delivery [8]. Immunotherapies, including ICIs and cell therapies, have revolutionized multiple solid tumors' treatments through activating the general antitumor immune response. The CheckMate-143 phase three clinical trial failed to demonstrate any higher efficacy of nivolumab over bevacizumab and showed that nivolumab did not improve OS in GBM patients. Several reasons might explain the low efficacy of ICIs in GBM: (i) low tumor mutation load, (ii) lack of predictor of response and lack of selection of patients, (iii) low penetration of ICIs within the brain parenchyma, (iv) low peripheral priming, (v) local immunosuppression, and (vi) low penetration of T-lymphocytes [26].

We explored the BBB as the limitation for antibody and lymphocytes penetration and priming and evaluated the potential of UMBO to deliver large therapeutics to the brain and the possible modulation of the immune microenvironment in GBM mouse models. We confirmed the limited efficacy of ICIs in the Gl261-bearing and Nfpp10-bearing mouse models. Consistent with our data, Reardon et al. reported limited efficacy of anti-PD-L1 and anti-CTLA-4 in GL261-bearing mice with a different treatment regimen [27].

To the best of our knowledge, this is the first research article that reports a dramatic increase in the OS of GL261-bearing mice when treated with UMBO plus anti-PD-L1. Indeed, 76% of GL261-bearing mice treated with anti-PD-L1 plus UMBO survived longer than 100 days compared to 26% of mice treated with anti-PD-L1 alone. We investigated the mechanisms involved in the anti-tumor effect. We hypothesized that the BBB was responsible for the limited efficacy by blocking anti-CTLA-4 and anti-PD-L1 from reaching the GBM tumor. A recent study reported enhanced efficacy of antibody conjugates following their delivery to GBM tumors [28]. Furthermore, focused ultrasound enhanced the delivery of intranasal anti-PD-L1 but not OS in GL261-bearning mice [29] and a very recent clinical study showed that UMBO enhanced the delivery of trastuzumab in brain metastases [30].

In our setting, we reported that UMBO enhanced antibody concentration up to 28-fold compared to control. UMBO was optimized to disturb one hemisphere; however, in our PK analysis, we used a whole-brain homogenization method; therefore, local concentrations of nivolumab could have been higher. Consistent with our data, a study has shown that UMBO enhanced the delivery of bevacizumab [31] ~149 KDa and anti-PD1 [32] compared to non-sonicated brains in a glioma mouse model. UMBO plus 200 μg of the anti-PD-L1 biweekly treatment regimen was used to maintain the higher concentration of anti-PD-L1 within the brain parenchyma. Immune checkpoint blockade with anti-PD-L1 was performed on day 14 post-inoculation to allow for T-lymphocytes exhaustion [33].

UMBO stimulates detectable peripheral circulation of GL261 DNA [34]. Zhu et al., investigated the possibility of using UMBO for liquid biopsies in GBM models. They detected that green fluorescent protein mRNA 20 minutes following UMBO in the GL261-GFP expressing mouse model [35] supports the passage of tumor material from the brain to blood flow stream. Consistent with our timepoint, another research group described that UMBO increases cell-free DNA in time-dependent matter [36]. The priming effect of circulating DNA could activate naïve T-lymphocytes through their exposure to new antigens. BBB protects the tumor from T-lymphocytes infiltration and immune activation. Therefore, detecting GL261 tumors in the peripheral circulation might activate the global antitumor effect. Further functional analysis of lymphocyte activation should be performed to evaluate any priming effect of UMBO.

Our results showing microglia activation in the UMBO plus anti-PD-L1 treated GL261-bearing mice suggest a possible mechanism for the observed enhanced therapeutic efficacy of anti-PD-L1. Our flow cytometry analysis confirms published data that showed a higher ratio of Iba-1 staining in sonicated regions compared to the non-sonicated [37]. PD-L1 is expressed on the cell surface of both GL261 and microglia [38]. A possible effect on microglia phenotype might be related to the combined effect of UMBO and anti-PD-L1 delivery to the brain parenchyma. Activated microglia might have a cytotoxic effect against GL261 tumor cells [39]; therefore, further investigation of the activated microglia role in GBM should be addressed.

To date, there is no clear evidence of the effect of UMBO on T-lymphocytes passage to the brain. We did not observe any significant elevation in the percentage of $CD8^+$ and $CD4^+$ T-lymphocytes at one timepoint (~16 h). This effect might be related to the timing of sample collection as we only evaluated our treatment regimen at one timepoint. We observed a delayed antitumor effect in UMBO and anti-PD-L1 group which can be related to a delayed effect on T-lymphocytes. Furthermore, we have not analyzed any subpopulations of $CD8^+$ T lymphocytes i.e., $PD\text{-}1^+$ $CD8^+$ T-lymphocytes.

Syngeneic mice models and especially the GL261 mouse model used in our experiments is a limitation of the current study. The GL261 mouse model: (i) has a high mutation load which is not consistent with GBM patients and (ii) a variability in terms of responses to ICIs in vivo [33]. Another limitation is the inability to demonstrate functional analysis of the role of UMBO in priming naïve T-lymphocytes through their exposure to new antigens. Additional functional analysis of the effect of UMBO plus anti-PD-L1 would explain the dramatic effect on OS that was observed in our study.

5. Conclusions

Our study showed statistically significant increased brain penetration and efficacy of anti-PD-L1 in GL261-bearing mice when delivered by UMBO. We have also provided clear evidence of the possible safe and effective delivery of large therapeutic agents using UMBO. Further investigations are needed to confirm the impact of UMBO on brain penetration and efficacy of chemotherapeutic agents and anti-PD-L1 to overcome the resistance of GBM to the current treatments.

Supplementary Materials: The following supporting information can be downloaded at: https://www.mdpi.com/article/10.3390/pharmaceutics15020455/s1, Figure S1: Schematic representation of LIPU procedure. (A) Graphical representation of LIPU generator set up. (B) Experimental timeline for anti-PD-1 treatment with BBB disruption. (C) A quantitative analysis of Gd enhancement pre and post UMBO in Gl261-bearing mice. (** $p < 0.001$); Figure S2: A representative gating strategy to identify immune cell subsets in Gl261 bearing brains following treatments. Mice were perfused using cold distilled phosphate buffer saline (DPBS) ~16 h after treatment. Tumor-bearing hemispheres were isolated, dissociated. Samples were acquired on a spectral flow cytometer (Aurora, Cytek) and analyzed by FlowJo software (FlowJo, LLC). Briefly, cells were selected based on their morphology, doublets, and dead cells were excluded using (Biolegend, #423107) while tumor cells were excluded based on their *mKate* expression. Monocytes (Ly6C$^+$ Ly6G$^-$) and neutrophils (Ly6C$^+$ Ly6G$^+$) were excluded from non-tumoral live cells using Ly-6C (Biolegend, #128036) and Ly-6G (Biolegend, #127617). Microglia were identified based on their expression of CD11b$^+$ and CD45^{low} using CD45 (Biolegend, #103131) and CD11b (Biolegend, #101255). Activated microglia were identified as CD68$^+$ using (Biolegend, #137003). F4/80 marker (Biolegend, #123117) was used to determine macrophages in the CD45^{high} CD11b$^+$ cell population. CD206 marker (Biolegend, #141729) was used to distinguish between subpopulations of macrophages. Lymphocytes CD4$^+$ (Biolegend, #100541) and CD8$^+$ (Biolegend, #100737) were identified on the CD45$^+$ CD11b$^-$ fraction of non-tumoral live cells. The percentage of each subpopulation was calculated and used for comparisons; Figure S3: Quantitative expression of PD-L1 in GL261 cell line and we have used Nfpp10 cell line as a positive control (left figure). Immunofluorescence of PD-L1 expression in GL261 cell line. (* $p < 0.05$); Figure S4: None of the treatments affected the body weight of GL261 bearing mice. Analysis of body weight shows no significant changes on body weight following treatments. Dotted lines represent the days of treatments; Figure S5: Heat maps of microglia gene signature (left figure) and antigen presentation genes (right figure) used in data analysis in Figure 6A,C.

Author Contributions: Conceptualization, M.H.A., A.I., M.V., I.H.-V., E.Q., N.L., C.G., L.G., N.Z., L.M., M.S. (Marc Sanson), A.P., C.S., G.B., M.C., M.S. (Mathieu Santin), M.V., A.C. and A.I.; methodology, M.H.A., A.I., M.V., I.H.-V., E.Q., N.L., C.G., L.G., N.Z., L.M., M.S. (Marc Sanson), A.P., C.S., G.B., M.C., M.S. (Mathieu Santin), M.V., A.C. and A.I.; software, I.H.-V., C.G. and L.G.; validation, formal analysis, M.H.A., A.I., M.V. and I.H.-V.; resources, L.M., M.S. (Mathieu Santin) and A.P.; data curation, M.C.; writing—original draft preparation, M.H.A. and A.I.; writing—review and editing, M.V., I.H.-V., E.Q., N.L., C.G., L.G., N.Z., L.M., M.S. (Marc Sanson), A.P., C.S., G.B., M.C., M.S. (Mathieu Santin), M.V., A.C. and A.I.; visualization, I.H.-V., E.Q., N.L., C.G., L.G., N.Z., L.M., M.S. (Mathieu Santin), A.P., C.S., G.B., M.C., M.S.(Marc Sanson), M.V., A.C. and A.I.; supervision, M.H.A., A.I. and A.C.; project administration, M.V.; funding acquisition, M.H.A. and A.I. All authors have read and agreed to the published version of the manuscript.

Funding: This work was supported by the European Union's Horizon 2020 research and innovation program under the Marie Skłodowska-Curie grant agreement #766069 (GLIO-TRAIN).

Institutional Review Board Statement: The animal ethics committee at the Ministry of Higher Education and Research in Paris approved all protocols involving live mice (protocol #17503 approved in 19 October 2020 and protocol #26137 approved in 6 April 2020).

Informed Consent Statement: Not applicable.

Data Availability Statement: Raw data will be available upon request from the authors.

Acknowledgments: We acknowledge Genentech for providing anti-PD-L1 (Clone 6E11) antibody and we appreciate our valuable discussions with Michel Mallat and his comments regarding microglia. We acknowledge the Salk Institute for Biological Studies for providing us with the Nfpp10 cells.

Conflicts of Interest: AI reports grants and travel funding from CarThera. A.C. is a paid consultant of CarThera and has an ownership interest in CarThera. C.S., G.B. and M.C. are employees of CarThera. All other authors have no conflict of interest.

References

1. Ostrom, Q.T.; Bauchet, L.; Davis, F.G.; Deltour, I.; Fisher, J.L.; Langer, C.E.; Pekmezci, M.; Schwartzbaum, J.A.; Turner, M.C.; Walsh, K.M.; et al. The epidemiology of glioma in adults: A "state of the science" review. *Neuro Oncol.* **2014**, *16*, 896–913. [CrossRef] [PubMed]
2. Pace, A.; Dirven, L.; Koekkoek, J.A.F.; Golla, H.; Fleming, J.; Ruda, R.; Marosi, C.; Rhun, E.L.; Grant, R.; Oliver, K.; et al. European Association for Neuro-Oncology (EANO) guidelines for palliative care in adults with glioma. *Lancet Oncol.* **2017**, *18*, e330–e340. [CrossRef] [PubMed]
3. Atiq, A.; Parhar, I. Anti-neoplastic Potential of Flavonoids and Polysaccharide Phytochemicals in Glioblastoma. *Molecules* **2020**, *25*, 4895. [CrossRef] [PubMed]
4. Touat, M.; Idbaih, A.; Sanson, M.; Ligon, K.L. Glioblastoma targeted therapy: Updated approaches from recent biological insights. *Ann. Oncol. Off. J. Eur. Soc. Med. Oncol.* **2017**, *28*, 1457–1472. [CrossRef] [PubMed]
5. Weenink, B.; French, P.J.; Sillevis Smitt, P.A.E.; Debets, R.; Geurts, M. Immunotherapy in Glioblastoma: Current Shortcomings and Future Perspectives. *Cancers* **2020**, *12*, 751. [CrossRef]
6. Wang, N.; Jain, R.K.; Batchelor, T.T. New Directions in Anti-Angiogenic Therapy for Glioblastoma. *Neurother. J. Am. Soc. Exp. NeuroTherapeutics* **2017**, *14*, 321–332. [CrossRef]
7. Lara-Velazquez, M.; Al-Kharboosh, R.; Jeanneret, S.; Vazquez-Ramos, C.; Mahato, D.; Tavanaiepour, D.; Rahmathulla, G.; Quinones-Hinojosa, A. Advances in Brain Tumor Surgery for Glioblastoma in Adults. *Brain Sci.* **2017**, *7*, 166. [CrossRef]
8. Drean, A.; Goldwirt, L.; Verreault, M.; Canney, M.; Schmitt, C.; Guehennec, J.; Delattre, J.Y.; Carpentier, A.; Idbaih, A. Blood-brain barrier, cytotoxic chemotherapies and glioblastoma. *Expert Rev. Neurother.* **2016**, *16*, 1285–1300. [CrossRef]
9. Zhang, D.Y.; Dmello, C.; Chen, L.; Arrieta, V.A.; Gonzalez-Buendia, E.; Kane, J.R.; Magnusson, L.P.; Baran, A.; James, C.D.; Horbinski, C.; et al. Ultrasound-mediated Delivery of Paclitaxel for Glioma: A Comparative Study of Distribution, Toxicity, and Efficacy of Albumin-bound Versus Cremophor Formulations. *Clin. Cancer Res. Off. J. Am. Assoc. Cancer Res.* **2020**, *26*, 477–486. [CrossRef]
10. Idbaih, A.; Canney, M.; Belin, L.; Desseaux, C.; Vignot, A.; Bouchoux, G.; Asquier, N.; Law-Ye, B.; Leclercq, D.; Bissery, A.; et al. Safety and Feasibility of Repeated and Transient Blood–Brain Barrier Disruption by Pulsed Ultrasound in Patients with Recurrent Glioblastoma. *Clin. Cancer Res.* **2019**, *25*, 3793–3801. [CrossRef]
11. Sheikov, N.; McDannold, N.; Vykhodtseva, N.; Jolesz, F.; Hynynen, K. Cellular mechanisms of the blood-brain barrier opening induced by ultrasound in presence of microbubbles. *Ultrasound Med. Biol.* **2004**, *30*, 979–989. [CrossRef]
12. Carpentier, A.; Canney, M.; Vignot, A.; Reina, V.; Beccaria, K.; Horodyckid, C.; Karachi, C.; Leclercq, D.; Lafon, C.; Chapelon, J.Y.; et al. Clinical trial of blood-brain barrier disruption by pulsed ultrasound. *Sci. Transl. Med.* **2016**, *8*, 343re342. [CrossRef]
13. Martínez-Fernández, R.; Máñez-Miró, J.U.; Rodríguez-Rojas, R.; Del Álamo, M.; Shah, B.B.; Hernández-Fernández, F.; Pineda-Pardo, J.A.; Monje, M.H.G.; Fernández-Rodríguez, B.; Sperling, S.A.; et al. Randomized Trial of Focused Ultrasound Subthalamotomy for Parkinson's Disease. *N. Engl. J. Med.* **2020**, *383*, 2501–2513. [CrossRef]
14. Rezai, A.R.; Ranjan, M.; D'Haese, P.-F.; Haut, M.W.; Carpenter, J.; Najib, U.; Mehta, R.I.; Chazen, J.L.; Zibly, Z.; Yates, J.R.; et al. Noninvasive hippocampal blood−brain barrier opening in Alzheimer's disease with focused ultrasound. *Proc. Natl. Acad. Sci. USA* **2020**, *117*, 9180–9182. [CrossRef]
15. Hao, C.; Chen, G.; Zhao, H.; Li, Y.; Chen, J.; Zhang, H.; Li, S.; Zhao, Y.; Chen, F.; Li, W.; et al. PD-L1 Expression in Glioblastoma, the Clinical and Prognostic Significance: A Systematic Literature Review and Meta-Analysis. *Front. Oncol.* **2020**, *10*, 1015. [CrossRef]
16. Azoury, S.C.; Straughan, D.M.; Shukla, V. Immune Checkpoint Inhibitors for Cancer Therapy: Clinical Efficacy and Safety. *Curr. Cancer Drug Targets* **2015**, *15*, 452–462. [CrossRef]
17. Reardon, D.A.; Brandes, A.A.; Omuro, A.; Mulholland, P.; Lim, M.; Wick, A.; Baehring, J.; Ahluwalia, M.S.; Roth, P.; Bähr, O.; et al. Effect of Nivolumab vs Bevacizumab in Patients With Recurrent Glioblastoma: The CheckMate 143 Phase 3 Randomized Clinical Trial. *JAMA Oncol.* **2020**, *6*, 1003–1010. [CrossRef]
18. Awada, G.; Ben Salama, L.; De Cremer, J.; Schwarze, J.K.; Fischbuch, L.; Seynaeve, L.; Du Four, S.; Vanbinst, A.M.; Michotte, A.; Everaert, H.; et al. Axitinib plus avelumab in the treatment of recurrent glioblastoma: A stratified, open-label, single-center phase 2 clinical trial (GliAvAx). *J. Immunother. Cancer* **2020**, *8*, e001146. [CrossRef]

19. Plessier, A.; Le Dret, L.; Varlet, P.; Beccaria, K.; Lacombe, J.; Mériaux, S.; Geffroy, F.; Fiette, L.; Flamant, P.; Chrétien, F.; et al. New in vivo avatars of diffuse intrinsic pontine gliomas (DIPG) from stereotactic biopsies performed at diagnosis. *Oncotarget* **2017**, *8*, 52543–52559. [CrossRef]
20. Iwamoto, N.; Yokoyama, K.; Takanashi, M.; Yonezawa, A.; Matsubara, K.; Shimada, T. Application of nSMOL coupled with LC-MS bioanalysis for monitoring the Fc-fusion biopharmaceuticals Etanercept and Abatacept in human serum. *Pharmacol. Res. Perspect.* **2018**, *6*, e00422. [CrossRef]
21. Hänzelmann, S.; Castelo, R.; Guinney, J. GSVA: Gene set variation analysis for microarray and RNA-Seq data. *BMC Bioinform.* **2013**, *14*, 7. [CrossRef] [PubMed]
22. Haage, V.; Semtner, M.; Vidal, R.O.; Hernandez, D.P.; Pong, W.W.; Chen, Z.; Hambardzumyan, D.; Magrini, V.; Ly, A.; Walker, J.; et al. Comprehensive gene expression meta-analysis identifies signature genes that distinguish microglia from peripheral monocytes/macrophages in health and glioma. *Acta Neuropathol. Commun.* **2019**, *7*, 20. [CrossRef] [PubMed]
23. Maas, S.L.N.; Abels, E.R.; Van De Haar, L.L.; Zhang, X.; Morsett, L.; Sil, S.; Guedes, J.; Sen, P.; Prabhakar, S.; Hickman, S.E.; et al. Glioblastoma hijacks microglial gene expression to support tumor growth. *J. Neuroinflamm.* **2020**, *17*, 120. [CrossRef] [PubMed]
24. Schmidt, J.; Smith, A.R.; Magnin, M.; Racle, J.; Devlin, J.R.; Bobisse, S.; Cesbron, J.; Bonnet, V.; Carmona, S.J.; Huber, F.; et al. Prediction of neo-epitope immunogenicity reveals TCR recognition determinants and provides insight into immunoediting. *Cell Rep. Med.* **2021**, *2*, 100194. [CrossRef] [PubMed]
25. Dréan, A.; Lemaire, N.; Bouchoux, G.; Goldwirt, L.; Canney, M.; Goli, L.; Bouzidi, A.; Schmitt, C.; Guehennec, J.; Verreault, M.; et al. Temporary blood-brain barrier disruption by low intensity pulsed ultrasound increases carboplatin delivery and efficacy in preclinical models of glioblastoma. *J. Neuro-Oncol.* **2019**, *144*, 33–41. [CrossRef]
26. Beccaria, K.; Canney, M.; Bouchoux, G.; Desseaux, C.; Grill, J.; Heimberger, A.B.; Carpentier, A. Ultrasound-induced blood-brain barrier disruption for the treatment of gliomas and other primary CNS tumors. *Cancer Lett.* **2020**, *479*, 13–22. [CrossRef]
27. Reardon, D.A.; Gokhale, P.C.; Klein, S.R.; Ligon, K.L.; Rodig, S.J.; Ramkissoon, S.H.; Jones, K.L.; Conway, A.S.; Liao, X.; Zhou, J.; et al. Glioblastoma Eradication Following Immune Checkpoint Blockade in an Orthotopic, Immunocompetent Model. *Cancer Immunol. Res.* **2016**, *4*, 124–135. [CrossRef]
28. Guo, H.; Wang, R.; Wang, D.; Wang, S.; Zhou, J.; Chai, Z.; Yao, S.; Li, J.; Lu, L.; Liu, Y.; et al. Deliver anti-PD-L1 into brain by p-hydroxybenzoic acid to enhance immunotherapeutic effect for glioblastoma. *J. Control. Release Off. J. Control. Release Soc.* **2020**, *320*, 63–72. [CrossRef]
29. Ye, D.; Yuan, J.; Yue, Y.; Rubin, J.B.; Chen, H. Focused Ultrasound-Enhanced Delivery of Intranasally Administered Anti-Programmed Cell Death-Ligand 1 Antibody to an Intracranial Murine Glioma Model. *Pharmaceutics* **2021**, *13*, 190. [CrossRef]
30. Meng, Y.; Reilly, R.M.; Pezo, R.C.; Trudeau, M.; Sahgal, A.; Singnurkar, A.; Perry, J.; Myrehaug, S.; Pople, C.B.; Davidson, B.; et al. MR-guided focused ultrasound enhances delivery of trastuzumab to Her2-positive brain metastases. *Sci. Transl. Med.* **2021**, *13*, eabj4011. [CrossRef]
31. Liu, H.L.; Hsu, P.H.; Lin, C.Y.; Huang, C.W.; Chai, W.Y.; Chu, P.C.; Huang, C.Y.; Chen, P.Y.; Yang, L.Y.; Kuo, J.S.; et al. Focused Ultrasound Enhances Central Nervous System Delivery of Bevacizumab for Malignant Glioma Treatment. *Radiology* **2016**, *281*, 99–108. [CrossRef]
32. Lee, H.; Guo, Y.; Ross, J.L.; Schoen, S., Jr.; Degertekin, F.L.; Arvanitis, C. Spatially targeted brain cancer immunotherapy with closed-loop controlled focused ultrasound and immune checkpoint blockade. *Sci. Adv.* **2022**, *8*, eadd2288. [CrossRef]
33. Aslan, K.; Turco, V.; Blobner, J.; Sonner, J.K.; Liuzzi, A.R.; Núñez, N.G.; De Feo, D.; Kickingereder, P.; Fischer, M.; Green, E.; et al. Heterogeneity of response to immune checkpoint blockade in hypermutated experimental gliomas. *Nat. Commun.* **2020**, *11*, 931. [CrossRef]
34. Lun, M.; Lok, E.; Gautam, S.; Wu, E.; Wong, E.T. The natural history of extracranial metastasis from glioblastoma multiforme. *J. Neuro-Oncol.* **2011**, *105*, 261–273. [CrossRef]
35. Zhu, L.; Cheng, G.; Ye, D.; Nazeri, A.; Yue, Y.; Liu, W.; Wang, X.; Dunn, G.P.; Petti, A.A.; Leuthardt, E.C.; et al. Focused Ultrasound-enabled Brain Tumor Liquid Biopsy. *Sci. Rep.* **2018**, *8*, 6553. [CrossRef]
36. Zhang, D.Y.; Gould, A.; Happ, H.C.; Youngblood, M.W.; Dmello, C.; Kang, S.J.; Canney, M.; Stupp, R.; Carvill, G.L.; Sonabend, A.M. Ultrasound-mediated blood–brain barrier opening increases cell-free DNA in a time-dependent manner. *Neuro-Oncol. Adv.* **2021**, *3*, vdab165. [CrossRef]
37. Sinharay, S.; Tu, T.-W.; Kovacs, Z.I.; Schreiber-Stainthorp, W.; Sundby, M.; Zhang, X.; Papadakis, G.Z.; Reid, W.C.; Frank, J.A.; Hammoud, D.A. In vivo imaging of sterile microglial activation in rat brain after disrupting the blood-brain barrier with pulsed focused ultrasound: [18F]DPA-714 PET study. *J. Neuroinflamm.* **2019**, *16*, 155. [CrossRef]
38. Chen, Q.; Xu, L.; Du, T.; Hou, Y.; Fan, W.; Wu, Q.; Yan, H. Enhanced Expression of PD-L1 on Microglia After Surgical Brain Injury Exerts Self-Protection from Inflammation and Promotes Neurological Repair. *Neurochem. Res.* **2019**, *44*, 2470–2481. [CrossRef]
39. Li, Y.; Zhang, R.; Hou, X.; Zhang, Y.; Ding, F.; Li, F.; Yao, Y.; Wang, Y. Microglia activation triggers oligodendrocyte precursor cells apoptosis via HSP60. *Mol. Med. Rep.* **2017**, *16*, 603–608. [CrossRef]

pharmaceutics

MDPI

Article

Cationic Microbubbles for Non-Selective Binding of Cavitation Nuclei to Bacterial Biofilms

Gareth LuTheryn [1,2,*], Elaine M. L. Ho [2,3], Victor Choi [4] and Dario Carugo [1,*]

1 Nuffield Department of Orthopaedics, Rheumatology and Musculoskeletal Sciences (NDORMS), The Botnar Research Centre, University of Oxford, Windmill Road, Oxford OX3 7HE, UK
2 Faculty of Engineering and Physical Sciences, University of Southampton, University Road, Southampton SO17 1BJ, UK; elaine.ho@rfi.ac.uk
3 Artificial Intelligence and Informatics, The Rosalind Franklin Institute, Harwell Campus, Didcot OX11 0QX, UK
4 Institute of Biomedical Engineering, Department of Engineering Science, University of Oxford, Parks Road, Oxford OX1 3PJ, UK; victor.choi@eng.ox.ac.uk
* Correspondence: gareth.lutheryn@ndorms.ox.ac.uk (G.L.); dario.carugo@ndorms.ox.ac.uk or udriplab@gmail.com (D.C.)

Abstract: The presence of multi-drug resistant biofilms in chronic, persistent infections is a major barrier to successful clinical outcomes of therapy. The production of an extracellular matrix is a characteristic of the biofilm phenotype, intrinsically linked to antimicrobial tolerance. The heterogeneity of the extracellular matrix makes it highly dynamic, with substantial differences in composition between biofilms, even in the same species. This variability poses a major challenge in targeting drug delivery systems to biofilms, as there are few elements both suitably conserved and widely expressed across multiple species. However, the presence of extracellular DNA within the extracellular matrix is ubiquitous across species, which alongside bacterial cell components, gives the biofilm its net negative charge. This research aims to develop a means of targeting biofilms to enhance drug delivery by developing a cationic gas-filled microbubble that non-selectively targets the negatively charged biofilm. Cationic and uncharged microbubbles loaded with different gases were formulated and tested to determine their stability, ability to bind to negatively charged artificial substrates, binding strength, and, subsequently, their ability to adhere to biofilms. It was shown that compared to their uncharged counterparts, cationic microbubbles facilitated a significant increase in the number of microbubbles that could both bind and sustain their interaction with biofilms. This work is the first to demonstrate the utility of charged microbubbles for the non-selective targeting of bacterial biofilms, which could be used to significantly enhance stimuli-mediated drug delivery to the bacterial biofilm.

Keywords: microbubble; cavitation nuclei; Biofilm; microbubble targeting; cationic microbubble; drug delivery; ultrasound drug delivery

Citation: LuTheryn, G.; Ho, E.M.L.; Choi, V.; Carugo, D. Cationic Microbubbles for Non-Selective Binding of Cavitation Nuclei to Bacterial Biofilms. *Pharmaceutics* **2023**, *15*, 1495. https://doi.org/10.3390/pharmaceutics15051495

Academic Editor: Clive Prestidge

Received: 31 March 2023
Revised: 6 May 2023
Accepted: 12 May 2023
Published: 13 May 2023

1. Introduction

Biofilms present a multifaceted challenge to achieving a positive therapeutic outcome in chronic infections, principally due to their competency as a physicochemical barrier to treatment [1,2]. Biofilms are typically implicated in severe chronic infection and disease such as in chronic wounds and cystic fibrosis, which represent an estimated global annual expenditure of $281 and $7.5 billion, respectively [3]. Moreover, biofilms represent the world's leading cause of antimicrobial resistance, which a UK report anticipated would lead to 10 million deaths worldwide by 2050 and a $100 trillion burden to the global economy [4]. For these reasons, research pertaining to the development of novel antibiofilm technologies and pharmaceutical agents has gained substantial momentum. Growing research suggests that targeting the delivery of antimicrobial agents to biofilms may increase their therapeutic potential [5–7] by increasing their local concentration to overcome the biofilm's innate tolerance and reducing the negative systemic side-effects of antimicrobial therapy [8,9]. To achieve greater control over drug delivery and enhance therapeutic efficacy, the use of ultrasound responsive agents has rapidly become an important area of research. Microbubbles (MBs) are small, gas-filled and shelled spherical particles, typically between 0.5 μm and 10 μm in diameter [10,11]. The composition, application, and versatility of MBs, as both ultrasound contrast agents in diagnostics and cavitation nuclei for drug delivery, have been comprehensively reviewed elsewhere [11–18]. Briefly, the composition of a microbubble shell is a key factor in determining the MB stability, biophysical effects in response to ultrasound exposure, and therapeutic outcomes. Enhancing the localisation of MBs to the target treatment site is acutely linked to their therapeutic efficacy. Some strategies employed for targeted drug delivery and gene therapy with functionalised MBs rely on the use of biochemical ligand-receptor targeting, acoustic radiation forces, electrostatic charge, and magnetic manipulation [19–22].

Due to the heterogeneity of the extracellular polymeric substances (EPS) that constitute the extracellular matrix of the biofilm, the identification of an appropriate, commonly expressed target receptor has proved challenging [7]. Recent work has demonstrated that microbubbles functionalised with an Affimer protein can be utilised to selectively target the *Staphylococcus aureus* biofilm virulence factor, clumping factor A [23]. One potential ligand target present in the extracellular matrix of clinically significant biofilms, such as *Pseudomonas aeruginosa*, are lectins [24]; they are carbohydrate-binding proteins that play an important role in mediating initial bacterial adhesion to surfaces for biofilm formation and subsequent polysaccharide cross-linking in the extracellular matrix of biofilms [25]. A significant limitation of this method of biofilm targeting, however, is that lectin receptors have a very weak natural affinity for their carbohydrate ligands, making them a difficult target in an already impervious structure [6,26]. The affinity that a ligand has for the intended receptor also suffers from high variability across biological systems; this is a widespread issue for ligand-receptor mediated specific targeting [27]. Importantly, authors have previously outlined that such means of targeting may not be directly translatable to other strains of the same bacterial species [6,23,25]. This naturally raises the issue of how successfully this means of targeting can be implemented clinically, where multi-species biofilms are abundant and specific strains often unidentified. Due to the challenges in current research on specific ligand-receptor mediated targeting of bacterial biofilms, it was the aim of this research to validate the use of a non-selective means of targeting the bacterial biofilm. Regardless of constituent species or specific strains, bacterial biofilms exhibit a net negative charge. This is owed to the ubiquitous presence of negatively charged extracellular DNA and polysaccharide constituents of the biofilm extracellular matrix, as well as highly negatively charged bacterial cell membrane components such as teichoic acids and lipopolysaccharides [28,29]. It has been demonstrated for example, that *P. aeruginosa* biofilms have a zeta potential (i.e., a measure of the electrostatic charge at the biofilm's surface) in the range of -20 mV [30].

The incorporation of cationic lipids allows MBs to electrostatically interact with negatively charged molecules; this method has already been used to bind nucleic acids to MBs for gene delivery [31,32]. An interesting point to note is that MBs augmented with a charge have not been used as an explicit means to facilitate MB interaction with biofilms. Recent work in this area has shown that cationic microbubbles can selectively localise in tumour vasculature [33], which demonstrates the utility of electrostatic interactions as a means of targeting regardless of the specific biomarkers present. In applications requiring intravascular administration of MBs, non-specific interactions of MBs with endothelial cells or blood constituents before reaching their target site would however be undesirable. Therefore, the use of cationic MBs for non-selectively targeting of specific sites or features is not well explored. For non-intravenous therapies however, such as the treatment of superficial wound infections, utilising positively charged MBs can provide the means to induce a prolonged MB-biofilm interaction unhindered. This application has been explored in conjunction with the delivery of therapeutic gases such as nitric oxide (NO), which has shown to induce dispersal in *P. aeruginosa* biofilms [34,35]. The use of a non-selectively targeted MB has demonstrated improved treatment efficacy by reducing the diffusion distance between this highly reactive gas and the biofilm. While utility and success of this cationic MB formulation and therapeutic approach have been demonstrated in other work [36], the present study aims to outline the formulation and testing underpinning its successful implementation. Herein, we hypothesised that cationic microbubbles could therefore be utilised as a non-selective means of targeting bacterial biofilms to achieve greater local concentration of MBs associated with and proximal to the biofilm and in turn increase the efficacy of ultrasound-mediated antibiofilm therapy.

2. Materials and Methods

2.1. Production of MBs

Room air microbubbles (RAMBs) and NO-filled MBs (NOMBs) were produced as previously described [36,37]. Briefly, 1,2-distearoylphosphatidylcholine (DSPC) (850365P, Avanti, Sigma-Aldrich, Alabaster, AL, USA) and polyoxyethylene (40) stearate (PEG40s) (P3440, Sigma-Aldrich, St. Loui, MO, USA) dissolved in chloroform were combined in a 20 mL capacity and 23 mm inner diameter glass vial (15394769, Fisherbrand™, Fisher Scientific, Dublin, Ireland) in a 9:1 molar ratio, using a 1 mL Luer lock glass syringe (1MR-GT, S.G.E Gas Tight Syringe, Supelco, UK) to achieve a final lipid concentration of 4 mg/mL. The chloroform was evaporated to obtain a dry lipid film; this was then rehydrated with 5 mL of degassed 0.01 M sterile phosphate-buffered saline (PBS) (P4417, Sigma-Aldrich, Gillingham, UK). Each vial was placed on a stirring hotplate (Fisherbrand™, Isotemp™) for 30 min at a temperature of 90 °C and at 700 rpm. Using a 120 W, 3.175 mm diameter tip sonicator (20 kHz, Fisher Scientific FB120, Pittsburgh, PA, USA), the lipid suspension was homogenously dispersed for 150 s at 40% power (48 W), with the sonicator tip fully immersed in the liquid. Microbubbles were subsequently formed by placing the sonicator tip at the liquid–air interface of the homogenised lipid suspension for 30 s at 70% power (84 W). Upon completion of the second sonication step, the vial was placed immediately into an ice bath to rapidly cool the MB suspension.

For the generation of nitric oxide microbubbles, the same sonication steps were performed; however, the dry lipid film was reconstituted with PBS that was sparged with pure nitrogen for 20 min [38]. Thereafter, a constant flow of nitrogen was used to purge air from the headspace of the vial. The eNO generator (NitricGen Inc., Madison, WI, USA) was used to flush 40 ppm NO at 1.5 L/min through the sparged PBS during the first sonication step [39]. During the second sonication step, the flow of NO was maintained with the needle placed at the interface between the fluid and the sonicator tip.

Cationic MBs (+) were produced by adding the cationic phospholipid 1,2-distearoyl-sn-glycero-3-ethylphosphocholine (DSEPC) (Avanti Polar Lipids, Alabaster, AL, USA) dissolved in chloroform to the DSPC and PEG40s mixture in quantities determined by the desired molar ratio, final lipid concentration, and sample volume as described in other research [36,37,40,41] (see Table 1).

Table 1. Stock concentrations, molar ratio compositions, and volumes used for neutral and cationic MBs.

MB Composition	DSPC	PEG40s	DSEPC
Stock concentrations (mg/mL)	25	10	10
	Molar ratio		
Uncharged	9	0.5	0
Cationic (+)	9	0.5	1
Preparation volumes (μL) to produce 2 mL of MB suspension with a final lipid concentration of 4 mg/mL			
Uncharged	210	75	0
Cationic (+)	190	68	57

2.2. Development of Microfluidic Flow Cells for Creating Surfaces with Augmented Electrostatic Charge

Flow cells were designed using Solidworks (Solidworks 2018, Dassault Systèmes, France) and converted to DXF files, which were compatible with the software of a laser cutting machine (Lasercut 5.3, HPC Laser Ltd., Yorkshire, UK). The flow cells were designed to fit a 75 mm × 25 mm glass slide. The inlet and outlet ports were laser cut (LS1690, HPC Laser Ltd., Yorkshire, UK) from a layer of 600 mm thick clear acrylic. Double-sided tape (0.14 mm thick, 3M™ Double Coated Polyester Tape 9731, Self-Adhesive Supplies, Reading, UK) was laid over the acrylic, and the channels were laser cut from the tape. A glass substrate was then fixed over the double-sided tape (Figure 1). Untreated borosilicate glass coverslips (0CON-161, 75 × 25 × 0.17 mm glass coverslips, Logitech, Glasgow, UK) were used as an uncharged substrate in the flow cell. Quartz coverslips (Alfa Aesar™ Quartz Microscope Slides, Fisher Scientific UK Ltd.) were instead used for creating flow cells with a substrate having a negative electrostatic charge, as quartz becomes relatively anionic when in contact with a solution (e.g., PBS) at pH 7.4 [42]. To ensure the flow cell was fluid-tight, water containing a red food dye was pumped through the flow cell at increasing flow rates until failure (i.e., corresponding to visible leakage) occurred. Two parallel channels were used to increase the throughput of the device and allow comparative studies to be performed with biofilms grown under the same conditions. A high aspect-ratio design was used for the channel cross-section to provide relatively uniform wall shear stress over the substrate surface.

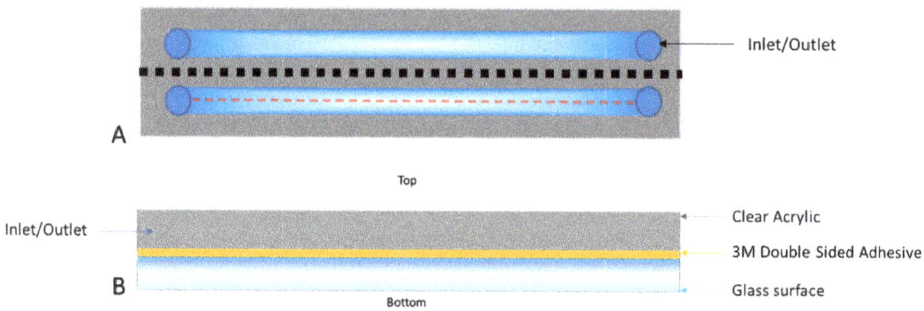

Figure 1. Schematic diagram of microfluidic flow cell with main features highlighted. The flow cell measures 75 × 25 mm (length × width), and the fluid channel is 0.14 mm high and 3.5 mm wide. (**A**) Top view of the flow cell with a dashed red line indicating the centreline of the channel, which is approximately 1.7 mm equidistant from the lateral walls of the channel. (**B**) Cross-sectional view of (**A**) taken at the dotted black line, with 'top' and 'bottom' orientation indications (figure not to scale).

2.3. Assessment of MB Charge

The charge associated with each MB shell composition was assessed using a dynamic light scattering apparatus (DLS; Zetasizer Nano ZS, Malvern Instruments, Malvern, UK); MBs were prepared as outlined in the above protocol except for the second sonication step. The lipid vesicles obtained from this process were diluted immediately prior to measurement (30 µL in 970 µL) in 10 mM HEPES buffer. Samples were centrifuged at 10,000 rpm for 30 s to remove any formed bubbles which would not be compatible with the measurement technique. Measurements were performed using the Smoluchowski algorithm for up to 100 runs, with three measurements conducted per sample.

2.4. Determination of MB Binding Characteristics under Static Conditions

To determine if the incorporation of a positive charge into the MB shell can promote contact with negatively charged surfaces, an artificially charged environment was created for controlled preliminary testing. Firstly, a flow cell with an uncharged glass coverslip was primed with PBS. Tubing (4 mm inner diameter, Masterflex Transfer Tubing, Platinum-

Cured Silicone, WZ-95802-03, Cole Parmer, Saint Neots, UK) was fitted to the inlet and outlet of the flow cell. Approximately 1 mL of each undiluted MB suspension was manually injected into the flow cell via the tubing until the suspension was visible in the outlet tubing. To prevent back flow of the MB suspension, the outlet tubing was then clamped to create a seal. A total of 10 'before interaction' images were acquired at 50 mm intervals from the inlet (left) to the outlet hole (right), along the centreline of the flow cell channel. Images were captured under bright field microscopy (Olympus, IX71), with a 50× (Olympus, LMPLFLN) objective lens using a CCD camera (Hamamatsu ORCA-ER, C4742-80) and analysed in accordance with Appendix A. The centreline of the channel was approximately 1.7 mm equidistant from the channel side walls, as indicated in Figure 1, and was selected for MB imaging purposes, as the wall shear stress acting over this region of the glass surface would be relatively uniform. Each MB suspension was allowed to interact with the charged or uncharged surface for 60 s; this interaction was facilitated by reversing the orientation of the flow cell. This meant the neutral or charged surface was positioned to be the 'top' of the flow cell, allowing MBs to passively float towards and interact with the surface. After 60 s, the device was reoriented so the neutral or charged surface was again positioned at the bottom of the flow cell; a further 10 'post interaction' images were then captured and analysed in the same way as previously described. A 60 s interaction period was selected as a representative clinically viable treatment timescale for a topical administration; this also aligns with the time taken for the majority of the MBs in suspension to passively float up towards the surface. This procedure was repeated for three independent vials of each MB composition with an uncharged and charged surface. The percentage change in MB interaction with each surface was calculated using Equation (1); where V_1 corresponds to the MB concentration (in MB/mL) present before contact with the charged or uncharged surface was induced (i.e., glass surface facing upwards), and V_2 corresponds to the concentration of MBs that remained in contact with the charged or uncharged surface after induced interaction was stopped (i.e., glass surface facing downwards).

Equation (1): Percentage change in MB concentration.

$$\begin{bmatrix} \textit{Change in} \\ \textit{MB concentration } (\%) \end{bmatrix} = \frac{V_2 - V_1}{V_1} \times 100 \tag{1}$$

2.5. Assessment of MB Non-Selective Binding Strength under Flow

The mean wall shear stress exerted by the fluid flow over the channel substrate was varied by adjusting the flow rate of fluid through the channel. The volumetric flow rate required to generate a given mean wall shear stress was calculated by first determining the pressure drop along the channel length (for a given wall shear stress value). The pressure drop was calculated from Equation (2) [43], where ΔP = pressure drop along the channel (Pa), τ = mean wall shear stress (Pa), P_W = wetted channel perimeter (m) = 2 × channel width + 2 × channel depth (assuming the channel was completely filled, P_W = 0.00728 m), L = channel length (m) = 0.0655 m, and A = cross-sectional area of the channel (m^2) = 4.9×10^{-7} m^2.

Equation (2): Pressure drop along the flow cell channel.

$$\Delta P = \frac{\tau \times P_w \times L}{A} \tag{2}$$

The required volumetric flow rate was subsequently calculated from the pressure drop along the channel using Hagen–Poiseuille's law, which was corrected for calculating volumetric flow rate within a rectangular channel [44], where Q = volumetric flow rate (mL/min), ΔP = pressure drop along the channel length (Pa), D_H = channel hydraulic diameter (m) = 4 × cross-sectional area / wetted perimeter = 2.7×10^{-4} m, μ = dynamic viscosity of water = 0.00089 kg/ms, L = channel length (m), and 0.2 is the correction factor for a rectangular channel with a width of 3.5 mm and depth of 0.14 mm. The calculated volumetric flow rates required to achieve mean wall shear stress values of 0.1, 0.2, 0.5, and 1 Pa are shown in Table 2.

Table 2. Volumetric flow rates of PBS corresponding to mean wall shear stress values to be exerted on MBs bound to the glass surface. The volumetric flow rates for each required mean wall shear stress were calculated from the pressure drop along the channel using Equations (2) and (3).

Mean Wall Shear Stress (Pa)	0.1	0.2	0.5	1
Pressure drop along channel (Pa)	97.31	194.63	486.57	973.14
Volumetric flow rate (mL/min)	0.06	0.13	0.32	0.65

Equation (3): Hagen–Poiseuille's law corrected for estimating flow rate in a rectangular channel.

$$Q = \frac{\Delta P \times \pi \times D_H^4}{128 \times \mu \times L \times 0.2} \times \frac{10^6}{60} \quad (3)$$

The MB concentration was assessed after increasing the mean wall shear stress acting on the glass substrate to determine the MB binding strength to uncharged and anionic substrates. The MB suspension was drawn into a 5 mL syringe (BD Luer Lock, 309649) and manipulated by inversion to ensure homogeneous distribution of MBs. A second 10 mL Luer Lock syringe (BD Luer Lock, SYR912) was filled with sterile 0.01 M PBS. Both syringes were connected to the inlets of a three-way valve (Masterflex Fitting, polycarbonate, Three-Way, Stopcock with Male Luer Lock, UY-30600-02, Cole Parmer, UK), with the outlet connected to the flow cell through a segment of tubing. The tubing and flow cell were primed with PBS before the MB suspension was pumped into the flow cell at a flow rate of 0.01 mL/min until the channel was filled. When the channel was full of MB suspension in a 1:1 volumetric ratio with PBS, the flow cell was placed with the glass surface facing up for 60 s to allow interaction of the MBs with the uncharged/charged surface. PBS was then pumped through the channel at 0.01 mL/min for 90 s to remove any unbound or weakly associated MBs.

After visual inspection confirmed MBs were stationary within the channel, ten images were acquired at 5 mm intervals along the centreline of the channel. This ensured the full length of the channel occupied by MBs was analysed, with a direct comparison of the same areas of the channel before and after application of flow. PBS was pumped into the flow cell for 15 s at a time, with flow rates of 0.06, 0.13, 0.32, and 0.65 mL/min respectively, to achieve wall shear stress values shown in Table 2. After the application of each flow rate, ten images at the same 5 mm positions along the centreline of the channel were acquired after visual confirmation that MBs had become stationary. The procedure was repeated in triplicate with both MB compositions in uncharged and charged flow cells. Segmentation of individual MBs was not possible for this experiment as there was significant clustering of the MBs. Therefore, in an adaptation of the method outlined in Appendix A, the total percentage area covered by the MBs in each image was determined by image thresholding and binarisation in ImageJ (NIH). The total area covered by MBs obtained through thresholding was calculated for each set of 10 images at each flow rate and for each MB composition. For each formulation, the mean area covered by MBs derived from 10 images at each flow rate was normalised by dividing the mean area by the initial area covered by MBs. This is represented by Equation (4); where $\bar{x}$ corresponds to the mean percentage area of MBs present in the channel after flow rates of 0.06, 0.13, 0.32, or 0.65 mL/min, respectively, were applied, and the mean percentage area of MBs present in the channel prior to the application of flow is denoted as μ. The results of Equation (4) equate the initial percentage coverage for each formulation to 100%, whilst providing the proportional change in MBs due to the application of various wall shear stress values.

Equation (4): Normalisation of percentage area covered by MBs.

$$[Normalised\ MB\ Area(\%)] = \frac{\bar{x}}{\mu} \times 100 \quad (4)$$

2.6. Assessment of Non-Selective Interaction of MBs with P. aeruginosa Biofilms

P. aeruginosa (PAO1) biofilms were grown in Ibidi® dishes (μ-Dish 35 mm, glass bottom, Thistle Scientific) for 24 h in wound constituent medium (WCM) at 37 °C as described previously [36]. Biofilms were washed three times with sterile PBS to remove any planktonic or weakly attached cells from the dish growth area. Biofilms were live/dead stained with 2.5 μM Syto9 (S34854, Invitrogen™, ThermoFisher Scientific) and 9 μM propidium iodide (P3566, Invitrogen™, ThermoFisher Scientific) for 5 min and stored under foil to prevent exposure to light. Excess stain was removed by washing with sterile PBS and visual confirmation of biofilm's presence was performed with fluorescence microscopy before the application of MBs (using an EVOS M5000 optical microscope).

Cationic DSPC RAMB+ and NOMB+ and uncharged DSPC RAMB and NOMB suspensions were diluted 1:5 (by volume) in sterile PBS and 1 mL of MB suspension was gently pipetted onto the previously live/dead stained biofilm. Each MB formulation was tested in triplicate using three independent biofilm samples (n = 18); the total MB-biofilm interaction time was 60 s for each biofilm. After 60 s, the Ibidi® dish was held at a 45° angle and washed from the top with 1 mL of sterile PBS three times; this ensured removal of MBs not interacting strongly with the biofilm. Each biofilm was imaged with the EVOSM5000 microscope (using a 20× Plan Fluor EVOS AMEP42924 objective); green fluorescent Syto9 was assessed under light emission at 510 nm, gain of 5, light source intensity of 0.5, brightness of 0.5, exposure time of 20 ms, and using a green fluorescent protein (GFP) LED cube (AMEP4653, Invitrogen™, ThermoFisher Scientific). Red fluorescent propidium iodide was assessed under light emission at 593 nm, gain of 5, light source intensity of 0.5, brightness of 0.5, exposure time of 20 ms, 20× magnification objective, and using a red fluorescent protein (RFP) LED cube. MBs were observed in brightfield to assess their localisation pattern, and biofilm association was calculated as percentage area using ImageJ in an adaptation of the steps outlined in Appendix A. Briefly, the brightfield images of MBs were isolated from the combined biofilm-MB images. An automatic threshold was applied to the image before a mask of each MB was created to eliminate the gas core. Based on the total area of the image, the percentage area occupied by MBs was then calculated by the software.

2.7. Statistical Analysis

All data were assessed for normal distribution. For normally distributed data, both paired and independent t-tests have been used to compare data depending on the relationship between variables. For multiple treatment conditions, a one-way ANOVA was used to identify any significant difference between groups. All data were analysed and plotted using Prism 8.4.3 (GraphPad), with a threshold value for significance of <0.05 where * = $p < 0.05$, ** = $p < 0.005$, *** = $p < 0.0005$, and **** = $p < 0.0001$.

3. Results

The mean size, concentration, and zeta potential of each MB formulation immediately after production are shown in Table 3; the results are averaged from three replicates with standard deviation of the mean shown where applicable. There was no statistically significant difference between the size or concentration of each formulation of MBs, consistent with previous work on neutral and cationic MBs [45]. Full data are shown in the Appendix A (Figure A2).

Table 3. The mean (n = 3) diameter with standard deviation, mean concentration, and mean zeta potential with standard deviation of each MB formulation immediately after production.

Microbubble Formulation	Mean MB Diameter (μm)	Mean MB Concentration (MB/mL)	Zeta Potential (mV)
Uncharged RAMB	3.48 ± 2.28	5.63×10^7	−2.44 ± 2.60
Uncharged NOMB	2.85 ± 2.72	1.71×10^8	
Cationic RAMB	5.25 ± 2.70	1.96×10^8	19.74 ± 0.95
Cationic NOMB	3.22 ± 1.74	2.69×10^8	

3.1. Microbubble Binding Characteristics in Static Conditions

In testing which involved no flow of fluid through the channel after MB administration, cationic room air MBs (RAMBs$^+$) showed a significant ($p < 0.05$) increase in surface interaction compared to their uncharged counterparts (Figure 2). There was approximately a 40% increase in surface-associated MBs. The results for room air MBs (RAMBs) demonstrate that when the charged lipid is removed, the increase in surface interaction is lost. With NO as the gas core of the MB, any effect of the charged lipid in promoting interaction is removed. For both NOMBs$^+$ (−25%) and NOMBs (−65%), there is an apparent net repulsion of the MBs from a negatively charged surface.

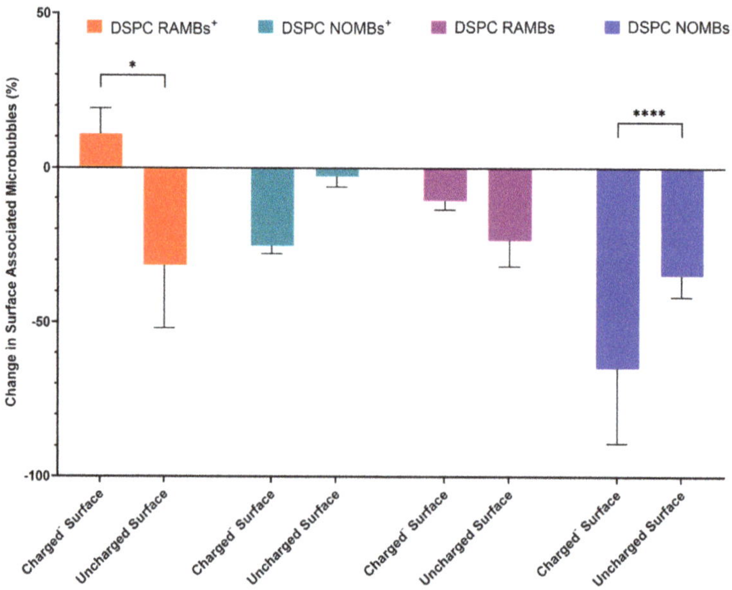

Figure 2. Assessment of the change in surface-associated cationic RAMBs$^+$ and NOMBs$^+$ and their respective uncharged equivalents after a 60 s interaction with either a negatively charged or uncharged surface. The data show that the inclusion of the positively charged DSEPC lipid in the shell of DSPC RAMB$^+$, significantly increased the quantity of MBs that can maintain contact with a negatively charged surface by 40%. For both NOMB and NOMB$^+$ suspensions, there was a consistent net loss of MBs that interacted with either surface, but this loss increased significantly in the presence of a negatively charged surface. Approximately 65% of NOMBs failed to remain in contact with the negatively charged surface, compared to a loss of 40% from the uncharged surface. This apparent repulsion of NOMBs from the negatively charged surface is mitigated somewhat by the presence of the cationic DSEPC lipid in NOMBs$^+$, of which only 25% were lost from the negatively charged surface and <10% were lost from an uncharged surface. Error bars represent the standard deviation of the mean. * = $p < 0.05$, and **** = $p < 0.0001$.

3.2. Assessment of MB Non-Selective Binding Strength in Dynamic Conditions

A large proportion of RAMBs+ (>75%) was able to sustain an effective interaction with a negatively charged surface up to wall shear stress values of 0.5 Pa; in contrast, when the charged surface was removed, there was a substantial loss of RAMBs+ (>40%) exposed to the same wall shear stress (Figure 3). This attests to the relative strength of electrostatic interaction of RAMBs+ with the negatively charged surface. In uncharged RAMBs and RAMB+ suspensions, there was no statistically significant difference in their ability to sustain an interaction with either a charged or uncharged surface. Contrary to the findings of static binding experiments, NOMBs+ appear better able to maintain an interaction with the artificially charged surface than RAMBs+; however, there was no significant difference in between the charged and uncharged NOMBs or RAMBs' abilities to sustain an interaction with the negatively charged surface under increasing wall shear stress levels (Figure 4). At 1 Pa, >90% of RAMBs irrespective of formulation were flushed out of the system, indicative that the non-selective interaction of room–air MBs with surfaces is non-permanent and easily reversible. It should be noted in contrast that at 1 Pa, 72.9% of NOMBs+ were still in contact with the negatively charged surface, compared to 41.4% with the uncharged surface.

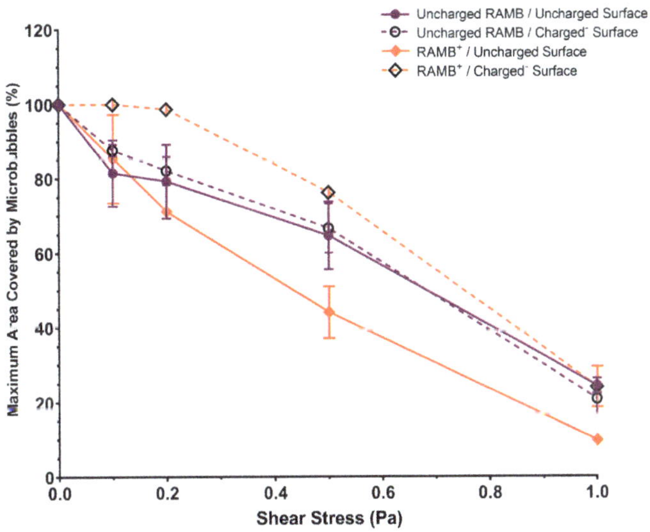

Figure 3. Percentage of maximum area covered by MBs for uncharged RAMB (9:0.5 DSPC:PEG40s) and cationic RAMB+ (9:0.5:1 DSPC:PEG40s:DSEPC). The total area covered by each MB composition was assessed over 10 images, captured after the application of incremental wall shear stress values from 0 to 1 Pa. Solid lines represent MB association with the uncharged surface; dashed lines represent MB association with the negatively charged surface. RAMBs+ demonstrated the highest affinity for the negatively charged surface, with no substantial loss of MBs in contact with the surface until wall shear stress values > 0.2 Pa. All values were normalised by taking the percentage value of the area covered by MBs at 0 Pa as the baseline maximum percentage area covered (100%). Error bars represent the standard deviation of the mean.

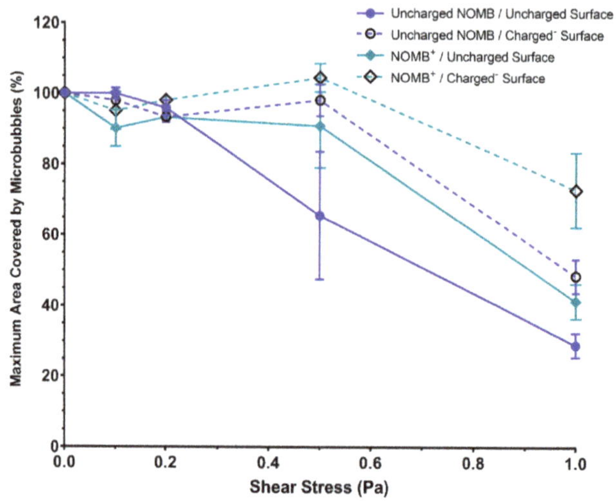

Figure 4. Percentage of maximum area covered by MBs for uncharged NOMBs (9:0.5 DSPC:PEG40s) and cationic NOMB$^+$ (9:0.5:1 DSPC:PEG40s:DSEPC). The total area covered by each MB composition was assessed over 10 images, captured after the application of incremental wall shear stress values from 0 to 1 Pa. Solid lines represent MB association with the uncharged surface; dashed lines represent MB association with the negatively charged surface. Uncharged NOMBs interacted with the uncharged surface as expected, with a rapid decline in MBs associated with the surface in response to wall shear stress > 0.2 Pa. For both NOMB and NOMB$^+$ suspensions, there was an accumulation of MBs on the charged surface up to 0.5 Pa, with only 50% and 30% loss respectively in maximum coverage at 1 Pa. All values were normalised by taking the percentage value of the area covered by MBs at 0 Pa as the baseline maximum percentage area covered (100%). Error bars represent the standard deviation of the mean.

3.3. Assessment of MB Interaction with *P. aeruginosa* Biofilms

Images of biofilms captured with fluorescence microscopy and MBs captured with brightfield and epifluorescence microscopy have been overlaid in ImageJ to provide a comprehensive visual representation of the interaction between MBs and the *P. aeruginosa* biofilm (Figure 5). Visual inspection of the interaction between both cationic MBs$^+$ and uncharged MB suspensions with *P. aeruginosa* biofilms and the glass growth substrate indicated that there was a low level of innate MB association to both strata irrespective of net charge. As observed in Figure 5a, clusters of MBs were clearly associated with the glass surface of the growth area, but there was no specific interaction with the distinct areas of biofilm growth. In contrast, DSPC$^+$ RAMBs exhibited a significantly ($p < 0.001$) increased quantity of MBs (38%) (Figure 6), which appeared to be more highly localised and non-selectively bound to areas of biofilm (Figure 5b). From visual inspection, there also appeared to be a positive correlation between the quantity of bound MBs and depth of biofilm growth; specifically, in areas where the biofilm is denser, there was a visibly discernible increase in the aggregation of MBs (Figure 5b).

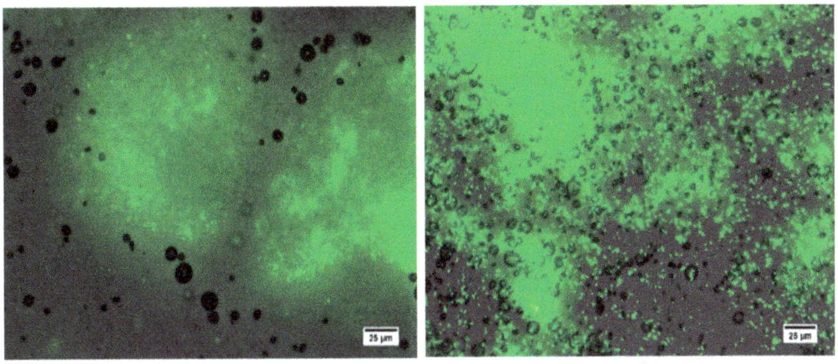

(a) (b)

Figure 5. Visual assessment of the interaction between uncharged DSPC RAMBs (**a**) and cationic DSPC RAMBs$^+$ (**b**) with a *P. aeruginosa* biofilm (stained with Syto9, green). Visual inspection confirms, there is an increased level of non-selective binding of cationic DSPC RAMBs$^+$ to defined areas of biofilm growth, compared to a low level of residual uncharged MB interaction with the glass growth surface and little to no association of uncharged MBs to defined areas of biofilm growth. Scale bar = 25 µm.

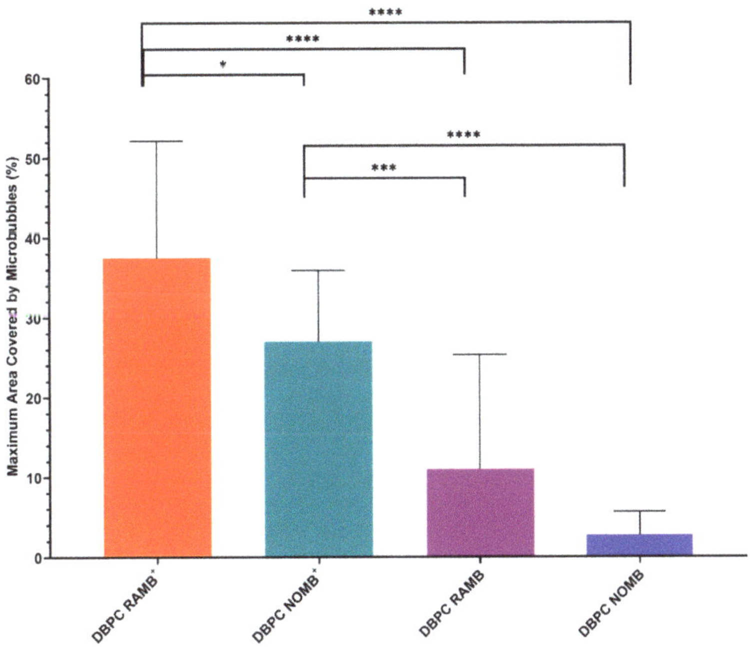

Figure 6. The maximum average percentage area covered by each MB formulation, assessed by analysing three independent biofilm samples for each of the four MB formulations tested. After administration of MBs to biofilms and subsequent washing to remove all MBs with a weak or no association to the biofilms, the average percentage coverage of the remaining MBs was 37% for RAMBs$^+$, 27% for NOMBs$^+$, 11% for uncharged RAMBs, and 3% for uncharged NOMB suspensions; error bars report the standard deviation of the mean. Quantification of MB coverage was assessed by determining the percentage area that MB formulations covered in each image, which incorporated both MB association to the *P. aeruginosa* biofilms and glass substrate of the Ibidi® dish growth area. * = $p < 0.05$, *** = $p < 0.0005$, and **** = $p < 0.0001$.

In this experiment, the innate level of uncharged MB interaction with biofilms and growth surface was on average 10% for uncharged RAMBs and 2% for uncharged NOMBs (Figure 6). Sharing consistency with binding experiments conducted in flow cells with an artificially charged surface (Figure 2), RAMBs$^+$ exhibited a significantly ($p < 0.001$) increased level of association to the biofilm (38%) compared to all other formulations; NOMBs$^+$ were the second most effective (28%). The incorporation of cationic charge into the NOMBs$^+$ shell was shown to significantly increase ($p < 0.005$) their association with the *P. aeruginosa* biofilms and glass substrate, when compared to both uncharged RAMBs and NOMBs. There was an average decrease in non-selective NOMBs$^+$ binding of 7% compared to RAMBs$^+$, but an average increase in NOMBs$^+$ interaction of 16% and 24%, respectively, over uncharged RAMBs and NOMBs (Figure 6).

4. Discussion

Due to the challenges in current research on specific ligand-receptor mediated targeting of bacterial biofilms, it was the aim of this research to validate the use of a non-selective means of targeting the bacterial biofilm. We hypothesised that cationic microbubbles could therefore be utilised as a non-selective means of targeting bacterial biofilms to achieve greater local concentration of MBs associated with and proximal to the biofilm and in turn increase the efficacy of ultrasound-mediated antibiofilm therapy. In order to assess this research aim, a microfluidic flow cell was designed where a glass or quartz surface could be bonded to an acrylic manifold (Figure 1). To provide an effective level of mechanical support to the 170 μm thick glass surfaces used in these experiments (required to allow high magnification microscopy), the manifold selected was a 6 mm thick clear acrylic. This allowed each flow cell to be optically transparent, produced rapidly and precisely by laser cutting, and be cost effective at a material cost of £0.83 per flow cell. Aqueous solutions in contact with a solid surface can create charge on the surface depending on the composition of the solid material and chemical properties of the solution; zeta potential can be used as a measure of the relative charge created at this surface [46]. Borosilicate glass coverslips bonded to the acrylic manifold generate a surface that does not create or sustain a charge, making them a suitable representation of an uncharged surface. Conversely, a quartz surface in contact with PBS at pH 7.4 has been shown to have a zeta potential of approximately −40 mV [47], which provides a valuable artificial control for a surface with a negative electrostatic charge. The artificially anionic and uncharged environments that were created in the flow cell formed an essential proof of concept to determine the ability of cationic MBs to interact with a negatively charged surface. It was important to first characterise this interaction in an environment that is static and can be controlled to best establish if there could be an effect before moving into more complex biological testing on biofilms. This characterisation of interaction focused on inducing contact between RAMBs$^+$, RAMBs, NOMBs$^+$, and NOMBs and either a negatively charged or uncharged surface. Then, it was studied by inverting the device, creating ideal conditions for weakly associated MBs to passively float away from the surface. As this method is carried out using smooth surfaces with augmented charge, it strictly accounts for MB association induced by electrostatic forces only. This is an important assessment as it provides a baseline for the efficacy of incorporating charge as a means of non-selective targeting without overestimation in the data caused by MB trapping at the surface due to variations in topography intrinsic to biological samples such as biofilms. The zeta potential measurements in this work were carried out on uncharged (DSPC:PEG40s) and cationic (DSPC:PEG40s:DSEPC) MB formulations. As outlined in previous research [40], the measurement was performed on samples that underwent only the first sonication step of the MB production process. Microbubbles are inherently buoyant and will float during measurements in a DLS apparatus, subsequently reducing the accuracy of the measurement. After the first sonication step the sample contains a dispersion of nanoscale lipid vesicles (i.e., microbubble precursors) that are neutrally buoyant, thereby making the measurement possible. Whilst this approach makes the measurement easier to obtain and increases its accuracy, it has been shown that the electrostatic charge of microbubbles and of their precursor lipid vesicles of the same formulation remain consistent [40]. Similarly, the purpose of this measurement was to determine the relative charge created by the MB shell formulation only; therefore, the sonication step was only performed in air. The DSPC:PEG40s:DSEPC cationic MB formulation was shown to have a zeta potential of 19.74 ± 0.95 mV (Table 3), which is congruent with other research that has produced cationic MBs with a zeta potential of approximately 25 mV using similar sonication method, lipid composition, and molar ratios employed in this work [40]. An indicative assessment of cationic MB charge has been carried out elsewhere [41]; this work showed that the most frequently reported range of MB charge was 15 mV to 40 mV; however, a higher reported charge of 60 mV was also previously determined for a DSTAP:DSPC:PEG40s lipid-shelled MB used in gene delivery to skeletal muscle.

The data for static interaction of RAMBs$^+$ provided an insight into the efficacy of charge as a means of non-selective binding; there was a significant increase of >40% in RAMBs$^+$ associated with the negatively charged surface compared to the uncharged surface, which affirms that the observed effect is due to interaction of the cationic MB and the anionic surface (Figure 2). Uncharged RAMBs displayed no significant percentage increase in MBs associated with the charged surface compared to the uncharged surface. This provides evidence that the non-selective binding effect is explicitly caused by the incorporation of DSEPC in the microbubble shell. Given this evidence that cationic MBs containing DSEPC promote electrostatic interactions with anionic surfaces, it appears anomalous that there was a greater loss of NOMBs$^+$ from the negatively charged surface compared to the uncharged surface. It was hypothesised that a NO-specific mediated interaction was causing interference with the electrostatic interactions, consequently reducing the binding efficacy of the charged MBs. Evidence for this effect is clear in the data for the interaction of uncharged NOMBs with negatively charged surfaces; there was a highly statistically significant difference between the number of NOMBs$^+$ lost from the negatively charged surface compared to the uncharged surface. It would therefore appear that the presence of DSEPC effectively mitigates the repulsive effect seen in MBs with an NO core from anionic surfaces, since 40% more DSPC NOMBs$^+$ remained in contact with the charged surface than DSPC NOMBs (Figure 2). In all experiments carried out in this research involving NOMBs and negatively charged surfaces, there has been a demonstrably weaker attraction between NOMB and negatively charged substrates compared to RAMBs, despite the incorporation of a cationic moiety into the NOMB shell. Although NO is a highly reactive molecule due to the presence of an unpaired electron, it is not an anionic molecule in either the gas or solution phase [48]. However, under physiological and environmental conditions, the bioactivity of NO is limited by its oxidation to the stable anionic products nitrite (NO^{2-}) and nitrate (NO^{3-}), with nitrate being the prevailing oxidation product formed [49]. As demonstrated in other research, it is highly likely that the repulsion of NOMBs from negatively charged surfaces seen in these experiments is caused by the accumulation of the anionic products NO^{2-} and NO^{3-} in solution [50–52]. The exception to this effect seen in experiments conducted under flow to assess NOMB binding strength (Figure 4) is likely due to the flow of PBS facilitating removal of oxidised products of NO, thereby reducing the concentration of negative ions in the fluid and mitigating their interference with the MB binding process. Building on this assumption, it would seem plausible that the loss of NO from NOMBs by oxidation would render their utility as a means of inducing dispersal in biofilms invalidated. However, in delivering a solution that invariably contains the oxidative products of NO, we provide bacterial cells within the biofilm with an important source of nitrate that can be recycled to form NO via bacterial nitrate reductases [53].

Research in this area has explored the antimicrobial effects of nitrite in solution [49]; this has been demonstrated in cystic fibrosis models of *P. aeruginosa* infection, where a 3-log reduction in bacteria was seen after a four-day prolonged exposure to 15 mM nitrite [54]. Further research in vitro has shown that the conversion of nitrite to NO is potentiated in the presence of ascorbic acid, which is produced by clinically important species of bacteria such as *P. aeruginosa* and *S. aureus* and results in enhance bactericidal activity [55,56]. Importantly, the impact of providing bacterial biofilms with sources NO^{3-} that can be reduced endogenously to NO by aerobically grown *P. aeruginosa* biofilms has been characterised by Rodgers et al. [57], who have shown that low concentrations (100 nM to 1 μM) of endogenous NO can induce biofilm dispersal. The presence of nitrite and nitrate in the effluent of biofilms usually only occurs in mature biofilms (i.e., >5 days old), at which point the endogenous production of NO induces dispersal of the biofilm architecture [57]. Therefore, by providing biofilms with an abundant source of nitrite and nitrate in the NOMB solution, there will be a corresponding upregulation in denitrification and endogenous production of NO to more rapidly induce a major biofilm dispersal event. The expedited endogenous production of NO in biofilms for dispersal may prove to be an interesting avenue of exploration for future work in this area. The relative strength of the electrostatic interaction between cationic MBs and an anionic surface was assessed by exposing MBs in the flow cell to increasing levels of wall shear stress from 0.1 to 1 Pa by altering the volumetric flow rate through the channel (Table 2). For physiological reference, mean wall shear stress in large veins is typically <0.1 Pa, whilst it is higher in arterioles at up to 8 Pa [58]. Previous research has demonstrated that wall shear stress values up to 0.06 Pa could promote microbubble binding; however, there is a decline in attachment after this point [25]. We hypothesised that in a similar manner, we would observe a decline in surface-associated MBs in response to increasing wall shear stress levels. It was confirmed that the dissociation of MBs from both charged and uncharged surfaces was less pronounced at wall shear stress values up to 0.5 Pa, but occurred more rapidly from 0.5 Pa to 1 Pa for all formulations (Figure 3). Phospholipid-coated αvβ3-targeted MBs used in endothelial cell targeting by Langeveld et al. remained attached to their target under flow up to shear stress values of approximately 0.2 Pa [59]. There was a significant reduction in the number of bound MBs from approximately 0.5 Pa, which is consistent with levels of attachment and loss of cationic MBs under flow in this work. Interestingly, Langeveld et al. also compared how the homogeneity of the ligand distribution affected MB binding. They showed that MBs with a more homogenous ligand distribution had a higher binding efficacy than those with a heterogenous distribution [59]; future work in this area could investigate how the arrangement of lipids in a cationic MB may also affect MB binding efficacy. In our investigation, both $RAMBs^+$ and $NOMBs^+$ were able to remain in contact or continue accumulating on the negatively charged surface up to wall shear stress values of 0.2 Pa and 0.5 Pa, respectively. This result would suggest that in comparison to studies that have assessed the binding strength of ligand-targeted MBs under increasing wall shear stress levels, the electrostatic interaction induced here can withstand a greater level of wall shear stress before initial detachment begins [27,60]. These findings are directly concurrent with work carried out by Edgeworth et al., who demonstrated that MBs electrostatically attached to a surface remained bound up to wall shear stress values of 0.66 Pa [61]. However, in the present study, $RAMBs^+$ and $NOMBs^+$ were reduced by approximately 90% and 30%, respectively, of their maximum coverage surface at a wall shear stress of 1 Pa (Figures 3 and 4); whereas, ligand-receptor targeted MBs have been shown to withstand wall shear stress levels of up to 50 Pa [61]. It is important to note, however, that this may not be an accurate representation of the MB binding and detachment profiles that would occur in vivo. As most studies can only estimate the local site density of the target receptor, there may be a significant over- or under-estimation in the data [27].

Though the microfluidic flow cells provided a valid and essential initial assessment of cationic MB interactions with charged surfaces to be able to understand the interaction between cationic MBs and biofilms more comprehensively, they were not fit for purpose. Due to the difficulty in analysing biofilm architecture and MBs simultaneously and a lack of translational applicability in terms of how MB suspensions would be applied clinically, biofilms were instead grown in Ibidi® dishes that feature a 170 μm thick glass coverslip growth area. This provided a means to grow biofilms on a surface where MB suspensions could be applied topically under clinically relevant conditions. The data present the percentage area occupied by MBs in microscopy images (Figure 6) which were retained after a three-step washing process to remove all weakly associated MBs. These data were averaged across three independent biofilms to demonstrate consistency in MB binding regardless of biofilm topography. The data shared a strong correlation with the effects observed in flow cell experiments; RAMBs$^+$ exhibited a significantly ($p < 0.001$) increased level of non-selective association with the biofilm (38%) compared to all other formulations, with NOMBs$^+$ being the second most effective (28%) (Figure 6). Caudwell et al. showed that the *S. aureus* specific binding Affimer AClfA1 conjugated with MBs, could achieve a nine-fold increase in MBs bound to the biofilm over non-targeted control MBs [23]. The authors note that utilising MBs in biofilm targeting has some limitations; principally, MBs are inherently unstable over time and can be destroyed during injection/administration processes [23]. Moreover, though there is great value in the specificity of the ligand for targeting biofilms, a potential limitation of this interaction is that this specificity may be highly variable between strains of the same bacterial species. However, this work sought to overcome such a limitation by utilising the non-selective nature of cationic charge to promote MB–biofilm interaction. In this work, the MB coverage as a proportion of the total area in the image was assessed, as opposed to limiting the analysis to only the percentage area of the biofilm covered by MBs. Though this includes MBs non-selectively interacting with the glass surface of the growth area, it provides a more accurate representation of what is likely to occur in vitro. By not eliminating MBs that were peripheral to the biofilm but not directly associated, we gain a better understanding of the innate level of MB retention due to varied surface topography. Moreover, even though they are not directly associated with the biofilm, proximal MBs may still have a significant impact on the overall efficacy of treatment due to the cavitation regimes induced upon exposure to ultrasound [13,62,63]. Specifically, fluid flows induced by microbubble cavitation (also known as microstreaming) may potentiate local drug delivery by imparting greater shear stress on the biofilm to increase permeability and enhance transport of chemical species [64,65]. As demonstrated by Pereno et al. [65], microstreaming velocities can peak in the order of 1000 μm/s and be maintained at 100 μm/s at distances > 1 mm from the source MBs cavitating in response to a 1 MHz ultrasound stimulus. It is important to note that the reduced number of surface-associated NOMBs in this experiment provides further support to the hypothesis developed from flow cell experiments; there is a net repulsion between MBs with a NO core and surfaces with a net negative charge, i.e., biofilms in this experiment. Evidence to support this can be seen in the significant increase ($p = <0.0001$) in NOMB$^+$ binding to biofilms compared to their uncharged counterparts; it can thus be hypothesised that the positive charge of the lipid in this instance is aiding the MB in overcoming the effect of NOMB repulsion from the biofilm. There is still a significant amount of NOMBs$^+$ bound to biofilms when compared to uncharged RAMBs ($p = <0.0005$) and NOMBs ($p = <0.0001$), which only cover a maximum surface area of 10% and 2%, respectively (Figure 6). These data, coupled with the significant difference ($p = <0.05$) between NOMBs$^+$ and RAMBs$^+$, add credence to the claim that the cause of this apparent repulsion is NO or more specifically its anionic oxidised products.

5. Conclusions

It was the aim of this research to determine if cationic MBs could be used as a means of non-selectively targeting the bacterial biofilm to achieve a greater local concentration of MBs associated with and proximal to the biofilm and in turn increase the efficacy of ultrasound-mediated antibiofilm therapy. The research aims were achieved by assessing the binding affinity and strength of cationic MBs to uncharged and charged surfaces under static and dynamic conditions within a flow cell capable of augmenting surface charge that was developed in this research to facilitate the study of these interactions in a controlled manner. Whereby, the addition of a cationic charge to the MB shell facilitated a consistent increase in the number of MBs that could bind to a negatively charged substrate. This interaction with a negatively charged substrate was shown to be sustained up to wall shear stress levels of 1 Pa, with cationic MBs remaining in consistently greater numbers compared to their uncharged counterparts. Subsequent testing demonstrated the efficacy of cationic MBs in significantly enhancing MB contact when applied directly to a *P. aeruginosa* biofilm. This corresponded to a significant 26% and 24% increase in the number of RAMBs^{+} and NOMBs^{+}, respectively, able to bind and sustain contact with the biofilm. This work demonstrated for the first time that cationic MBs have the ability to increase proximity, promote, and sustain contact with bacterial biofilms, validating their utility as a means of non-selectively targeting the biofilm.

Author Contributions: Conceptualization, D.C. and G.L.; methodology G.L., E.M.L.H. and D.C.; formal analysis, G.L. and E.M.L.H.; investigation, G.L., E.M.L.H. and V.C.; writing—original draft preparation, G.L., E.M.L.H. and D.C.; writing—review and editing, G.L., E.M.L.H. and D.C.; supervision, D.C. All authors have read and agreed to the published version of the manuscript.

Funding: We thank the Engineering and Physical Sciences Research Council (EPSRC) for funding this research through an EPSRC Doctoral Prize EP/N509747/1 (awarded to Gareth LuTheryn by the University of Southampton) and the EPSRC Programme Grant "Beyond Antibiotics" (EP/V026623/1).

Institutional Review Board Statement: Not applicable.

Informed Consent Statement: Not applicable.

Data Availability Statement: Data can be made available upon request.

Conflicts of Interest: The authors declare no conflict of interest.

Appendix A

Appendix A.1. Characterisation of MBs Size and Concentration

To assess the stability of uncharged and cationic MBs at environmental temperature, three vials each of air MBs (control) and NOMBs were prepared. A 10 μL sample was pipetted onto a Neubauer haemocytometer with a 0.2 mm thick cover slip placed on top; samples were taken from each vial at intervals of 10, 20, 40, and 60 min after the initial reading. Between the readings at each time point, the vials remained sealed with a polyethylene cap on a bench at room temperature (21 °C ± 2 °C). The MBs were observed using bright field microscopy (Olympus, IX71) with a 50× (Olympus, LMPLFLN) objective lens; ten images were acquired at each time point for all vials using a CCD camera (Hamamatsu ORCA-ER, C4742-80). The images were processed using ImageJ (Fiji) to determine the MB diameter, size distribution, and volume. The contrast of each image was normalised to differentiate between the white background and dark MB before the threshold was automatically calculated by ImageJ to differentiate the MBs from the surrounding fluid and identify individual MBs. The images taken had a width of 1344 pixels and a height of 1024 pixels; in this experiment, 700 pixels equated to 100 μm. The area in pixels of all particles was determined by using the set-scale function in ImageJ, using the dimensions of the picture with a known distance on each image. The circularity threshold was set between 0.7 and 1, which eliminated any debris from counting. MB concentration was determined using Equation (A1); where, the number of MBs was the total number of MBs in all the images (three vials, ten images per vial = 30 images per MB composition), b = length covered in field of view (μm) = 192 μm, h = height covered in field of view (μm) = 146 μm, t = thickness of MB suspension imaged on haemocytometer (μm) = 100 μm, number of images = 10 images taken per vial, and number of vials = 3 vials per composition. The post-production number of MBs per 5 mL vial was approximately 8.5×10^{8}.

Equation (A1): Calculation for determining concentration of MB/mL.

$$[\text{Mean MB}] = \frac{\text{No.of MBs}}{b \times h \times t \times \text{No.of images per vial} \times \text{No.of vials}} \tag{A1}$$

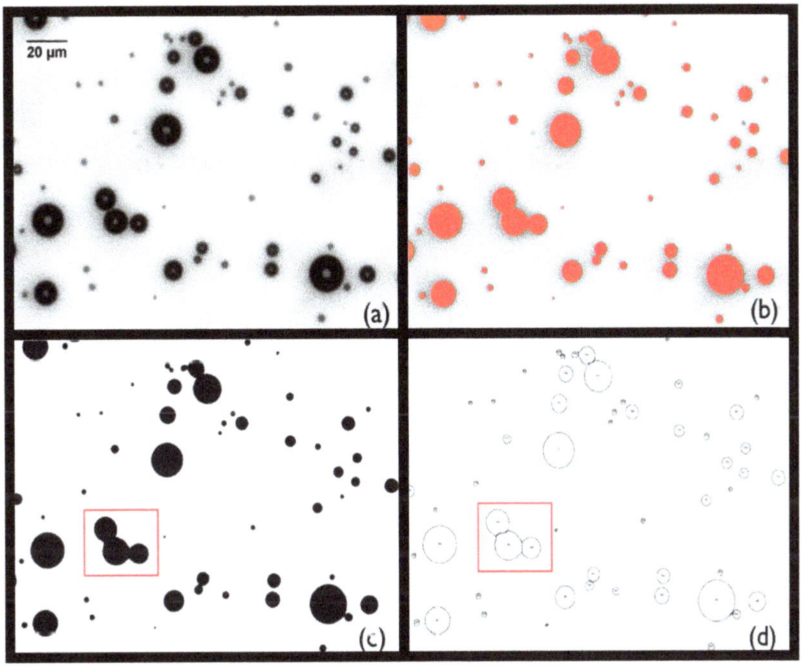

Figure A1. Image analysis protocol for determining size and concentration of MBs by optical microscopy. Images were taken with a 50× objective; the scale bar represents a 20 μm distance in each image. (**a**) Bright field image of MBs suspended in PBS. (**b**) Automatic threshold was applied to the image. (**c**) A mask was created based on the thresholded image. (**d**) Watershed algorithm was applied to count MBs which appeared to be touching as separate MBs (see red boxes in (**c**,**d**)). Only features with circularity > 0.7 were selected to remove any cropped MBs. The area and diameter of each MB were determined by calibrating a scale in which the number of pixels corresponds to a known distance in μm; specifically in these experiments, 700 pixels corresponded to 100 μm; therefore, the number of pixels per MB could be used to automatically calculate the dimensions of each MB in the image.

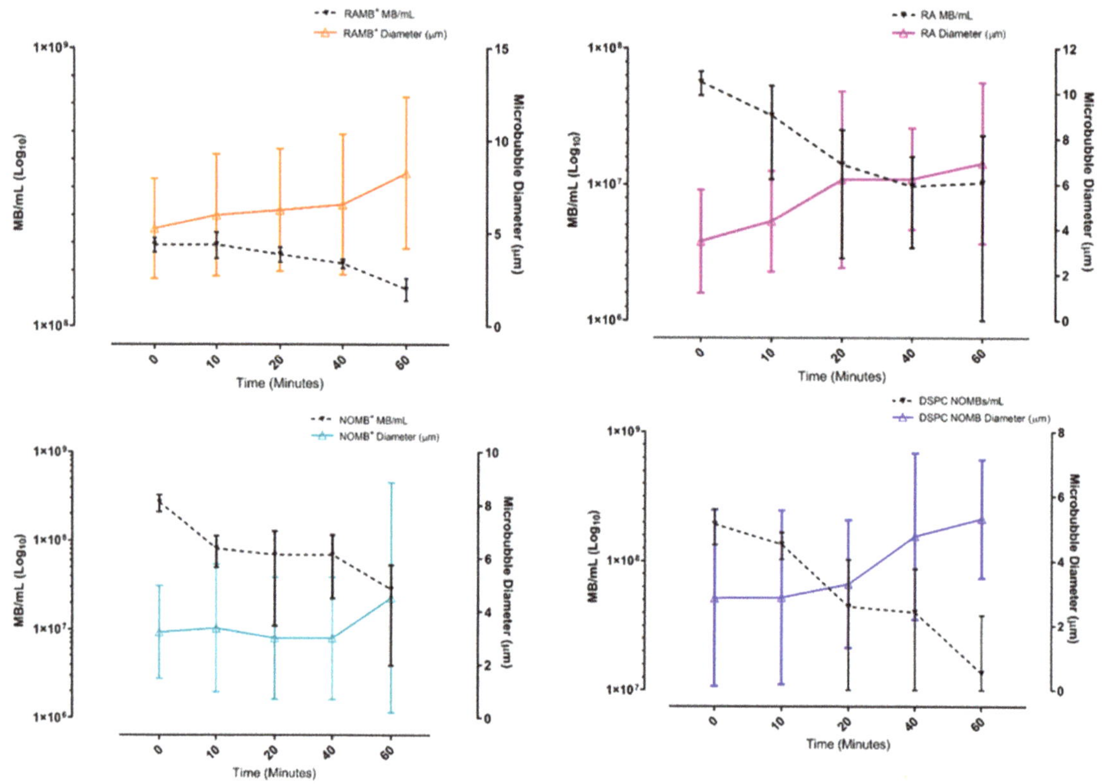

Figure A2. Size and concentration data for each MB formulation assessed at room temperature over a 60 min period post-production. (**Upper left**): RAMBs$^+$; (**upper right**): neutral RAMBs; (**lower left**): NOMBs$^+$; (**lower right**): neutral NOMBs.

Appendix A.2. Assessing Flow Cells for Leakage under Increasing Flow Rates

The flow cell was tested for any potential leakage, with both the quartz and borosilicate glass coverslip bonded to the acrylic manifold. Our assessment verified that the flow cells could maintain an effective sealing, i.e., did not leak, up to volumetric flow rates of 25 mL/min. The use of red food dye ensured any leakages from tubing or bleeding of fluid into the tape would be clearly visible (Figure A3). All flow rates for experimentation were kept below 1 mL/min, significantly minimising any risk of leakage during experimentation. This provided confidence that the flow cells could sustain continuous flow over prolonged periods of time, given that the typical flow rates used in these experiments were at a maximum of 1 mL/min. As the biocompatibility of the flow cells was assessed for the growth of *P. aeruginosa* biofilms, the leakage tests also verified that there was very little risk of biohazardous material escaping from the flow cell. Due to the limitations of the microscope used in the initial phase of this research, simultaneous visualisation of MBs and biofilms was not possible. Therefore, the flow cell was only utilised as a means of providing a smooth surface with an uncharged or anionic charge for the investigation of cationic MB interaction under (i) static conditions to assess the level of passive electrostatic interaction and (ii) flow to determine the relative binding strength of this interaction.

Figure A3. Example of a flow cell subjected to leakage testing, which failed upon exposure to a flow rate of 25 mL/min. There is indication of leakage of fluid and precipitation of the dye (red-orange areas) into the tape bonding the glass substrate to the acrylic manifold. Scale bar is 1 cm.

References

1. Ciofu, O.; Rojo-Molinero, E.; Macià, M.D.; Oliver, A. Antibiotic Treatment of Biofilm Infections. *APMIS* **2017**, *125*, 304–319. [CrossRef]
2. Carniello, V.; Peterson, B.W.; van der Mei, H.C.; Busscher, H.J. Physico-Chemistry from Initial Bacterial Adhesion to Surface-Programmed Biofilm Growth. *Adv. Colloid. Interface Sci.* **2018**, *261*, 1–14. [CrossRef]
3. Cámara, M.; Green, W.; MacPhee, C.E.; Rakowska, P.D.; Raval, R.; Richardson, M.C.; Slater-Jefferies, J.; Steventon, K.; Webb, J.S. Economic Significance of Biofilms: A Multidisciplinary and Cross-Sectoral Challenge. *NPJ Biofilms Microbiomes* **2022**, *8*, 1–8. [CrossRef] [PubMed]
4. O'Neil, J. Antimicrobial Resistance: Tackling a Crisis for the Health and Wealth of Nations. 2014. Available online: https://amr-review.org/sites/default/files/AMR%20Review%20Paper%20-%20Tackling%20a%20crisis%20for%20the%20health%20and%20wealth%20of%20nations_1.pdf (accessed on 11 May 2023).
5. Sharma, D.; Misba, L.; Khan, A.U. Antibiotics versus Biofilm: An Emerging Battleground in Microbial Communities. *Antimicrob. Resist. Infect. Control* **2019**, *8*, 76. [CrossRef] [PubMed]
6. Meiers, J.; Zahorska, E.; Röhrig, T.; Hauck, D.; Wagner, S.; Titz, A. Directing Drugs to Bugs: Antibiotic-Carbohydrate Conjugates Targeting Biofilm-Associated Lectins of Pseudomonas Aeruginosa. *J. Med. Chem.* **2020**, *63*, 11707–11724. [CrossRef] [PubMed]
7. Palmioli, A.; Sperandeo, P.; Polissi, A.; Airoldi, C. Targeting Bacterial Biofilm: A New LecA Multivalent Ligand with Inhibitory Activity. *ChemBioChem* **2019**, *20*, 2911–2915. [CrossRef]
8. Heta, S.; Robo, I. The Side Effects of the Most Commonly Used Group of Antibiotics in Periodontal Treatments. *Med. Sci.* **2018**, *6*, 6. [CrossRef]
9. De Gunzburg, J.; Ghozlane, A.; Ducher, A.; Le Chatelier, E.; Duval, X.; Ruppé, E.; Armand-Lefevre, L.; Sablier-Gallis, F.; Burdet, C.; Alavoine, L.; et al. Protection of the Human Gut Microbiome from Antibiotics. *J. Infect. Dis.* **2018**, *217*, 628–636. [CrossRef]
10. Sirsi, S.; Borden, M. Microbubble Compositions, Properties and Biomedical Applications. *Bubble Sci. Eng. Technol.* **2009**, *1*, 3–17. [CrossRef]
11. Lattwein, K.R.; Shekhar, H.; Kouijzer, J.J.P.; van Wamel, W.J.B.; Holland, C.K.; Kooiman, K. Sonobactericide: An Emerging Treatment Strategy for Bacterial Infections. *Ultrasound Med. Biol.* **2020**, *46*, 193–215. [CrossRef]
12. Unger, E.C.; Porter, T.; Culp, W.; Labell, R.; Matsunaga, T.; Zutshi, R. Therapeutic Applications of Lipid-Coated Microbubbles. *Adv. Drug Deliv. Rev.* **2004**, *56*, 1291–1314. [CrossRef] [PubMed]
13. Kooiman, K.; Vos, H.J.; Versluis, M.; de Jong, N. Acoustic Behavior of Microbubbles and Implications for Drug Delivery. *Adv. Drug Deliv. Rev.* **2014**, *72*, 28–48. [CrossRef]
14. Stride, E.; Saffari, N. Microbubble Ultrasound Contrast Agents: A Review. *Proc. Inst. Mech. Eng. H* **2003**, *217*, 429–447. [CrossRef]
15. Ibsen, S.; Schutt, C.E.; Esener, S. *Microbubble-Mediated Ultrasound Therapy: A Review of Its Potential in Cancer Treatment*; Dove Press: Macclesfield, UK, 2013; Volume 7, pp. 375–388.
16. Kooiman, K.; Roovers, S.; Langeveld, S.A.G.; Kleven, R.T.; Dewitte, H.; O'Reilly, M.A.; Escoffre, J.M.; Bouakaz, A.; Verweij, M.D.; Hynynen, K.; et al. Ultrasound-Responsive Cavitation Nuclei for Therapy and Drug Delivery. *Ultrasound Med. Biol.* **2020**, *46*, 1296–1325. [CrossRef]
17. LuTheryn, G.; Glynne-Jones, P.; Webb, J.S.; Carugo, D. Ultrasound-mediated Therapies for the Treatment of Biofilms in Chronic Wounds: A Review of Present Knowledge. *Microb. Biotechnol.* **2019**, *1751–7915*, 13471. [CrossRef] [PubMed]
18. Izadifar, Z.; Babyn, P.; Chapman, D. Ultrasound Cavitation/Microbubble Detection and Medical Applications. *J. Med. Biol. Eng.* **2018**, *39*, 259–276. [CrossRef]
19. Martin, M.J.; Chung, E.M.L.; Goodall, A.H.; Martina, A.D.; Ramnarine, K.V.; Fan, L.; Hainsworth, S.V.; Naylor, A.R.; Evans, D.H. Enhanced Detection of Thromboemboli with the Use of Targeted Microbubbles. *Stroke* **2007**, *38*, 2726–2732. [CrossRef]

20. Zhao, S.; Borden, M.; Bloch, S.H.; Kruse, D.E.; Ferrara, K.W.; Dayton, P.A. Increasing Binding Efficiency of Ultrasound Targeted Agents with Radiation Force. In Proceedings of the Proceedings—IEEE Ultrasonics Symposium, Montreal, QC, Canada, 23–27 August 2004; Volume 2, pp. 1114–1117.
21. Weller, G.E.R.; Villanueva, F.S.; Klibanov, A.L.; Wagner, W.R. Modulating Targeted Adhesion of an Ultrasound Contrast Agent to Dysfunctional Endothelium. *Ann. Biomed. Eng.* **2002**, *30*, 1012–1019. [CrossRef]
22. Owen, J.; Rademeyer, P.; Chung, D.; Cheng, Q.; Holroyd, D.; Coussios, C.; Friend, P.; Pankhurst, Q.A.; Stride, E. Magnetic Targeting of Microbubbles against Physiologically Relevant Flow Conditions. *Interface Focus* **2015**, *5*, 20150001. [CrossRef]
23. Caudwell, J.A.; Tinkler, J.M.; Johnson, B.R.G.; McDowall, K.J.; Alsulaimani, F.; Tiede, C.; Tomlinson, D.C.; Freear, S.; Turnbull, W.B.; Evans, S.D.; et al. Protein-Conjugated Microbubbles for the Selective Targeting of S. Aureus Biofilms. *Biofilm* **2022**, *4*, 100074. [CrossRef]
24. Sommer, R.; Wagner, S.; Rox, K.; Varrot, A.; Hauck, D.; Wamhoff, E.-C.; Schreiber, J.; Ryckmans, T.; Brunner, T.; Rademacher, C.; et al. Glycomimetic, Orally Bioavailable LecB Inhibitors Block Biofilm Formation of Pseudomonas Aeruginosa. *J. Am. Chem. Soc.* **2018**, *140*, 2537–2545. [CrossRef] [PubMed]
25. Johansson, E.M.V.; Crusz, S.A.; Kolomiets, E.; Buts, L.; Kadam, R.U.; Cacciarini, M.; Bartels, K.M.; Diggle, S.P.; Cámara, M.; Williams, P.; et al. Inhibition and Dispersion of Pseudomonas Aeruginosa Biofilms by Glycopeptide Dendrimers Targeting the Fucose-Specific Lectin LecB. *Chem. Biol.* **2008**, *15*, 1249–1257. [CrossRef] [PubMed]
26. Wagner, S.; Hauck, D.; Hoffmann, M.; Sommer, R.; Joachim, I.; Müller, R.; Imberty, A.; Varrot, A.; Titz, A. Covalent Lectin Inhibition and Application in Bacterial Biofilm Imaging. *Angew. Chem. Int. Ed.* **2017**, *56*, 16559–16564. [CrossRef] [PubMed]
27. Takalkar, A.M.; Klibanov, A.L.; Rychak, J.J.; Lindner, J.R.; Ley, K. Binding and Detachment Dynamics of Microbubbles Targeted to P-Selectin under Controlled Shear Flow. *J. Control Release* **2004**, *96*, 473–482. [CrossRef]
28. Renner, L.D.; Weibel, D.B. Physicochemical Regulation of Biofilm Formation. *MRS Bull.* **2011**, *36*, 347–355. [CrossRef]
29. Harper, R.A.; Carpenter, G.H.; Proctor, G.B.; Harvey, R.D.; Gambogi, R.J.; Geonnotti, A.R.; Hider, R.; Jones, S.A. Diminishing Biofilm Resistance to Antimicrobial Nanomaterials through Electrolyte Screening of Electrostatic Interactions. *Colloids. Surf. B Biointerfaces* **2019**, *173*, 392–399. [CrossRef] [PubMed]
30. Khelissa, S.O.; Abdallah, M.; Jama, C.; Barras, A.; Chihib, N.E. Comparative Study on the Impact of Growth Conditions on the Physiology and the Virulence of Pseudomonas Aeruginosa Biofilm and Planktonic Cells. *J. Food Prot.* **2019**, *82*, 1357–1363. [CrossRef] [PubMed]
31. Huang, C.; Zhang, H.; Bai, R. Advances in Ultrasound-Targeted Microbubble-Mediated Gene Therapy for Liver Fibrosis. *Acta Pharm. Sin. B* **2017**, *7*, 447–452. [CrossRef] [PubMed]
32. Carugo, D.; Owen, J.; Crake, C.; Lee, J.Y.; Stride, E. Biologicallyand Acoustically Compatible Chamber for Studying Ultrasound-Mediated Delivery of Therapeutic Compounds. *Ultrasound Med. Biol.* **2015**, *41*, 1927–1937. [CrossRef]
33. Diakova, G.B.; Du, Z.; Klibanov, A.L. Targeted Ultrasound Contrast Imaging of Tumor Vasculature with Positively Charged Microbubbles. *Invest. Radiol.* **2020**, *55*, 736–740. [CrossRef]
34. Barraud, N.; Storey, M.V.; Moore, Z.P.; Webb, J.S.; Rice, S.A.; Kjelleberg, S. Nitric Oxide-Mediated Dispersal in Single- and Multi-Species Biofilms of Clinically and Industrially Relevant Microorganisms. *Microb. Biotechnol.* **2009**, *2*, 370–378. [CrossRef]
35. Barraud, N.; Kelso, M.J.; Rice, S.A.; Kjelleberg, S. Nitric Oxide: A Key Mediator of Biofilm Dispersal with Applications in Infectious Diseases. *Curr. Pharm. Des.* **2015**, *21*, 31–42. [CrossRef] [PubMed]
36. LuTheryn, G.; Hind, C.; Campbell, C.; Crowther, A.; Wu, Q.; Keller, S.B.; Glynne-Jones, P.; Sutton, J.M.; Webb, J.S.; Gray, M.; et al. Bactericidal and Anti-Biofilm Effects of Uncharged and Cationic Ultrasound-Responsive Nitric Oxide Microbubbles on Pseudomonas Aeruginosa Biofilms. *Front. Cell Infect. Microbiol.* **2022**, *12*, 1130. [CrossRef]
37. Plazonic, F.; LuTheryn, G.; Hind, C.; Clifford, M.; Gray, M.; Stride, E.; Glynne-Jones, P.; Hill, M.; Sutton, J.M.; Carugo, D. Bactericidal Effect of Ultrasound-Responsive Microbubbles and Sub-Inhibitory Gentamicin against Pseudomonas Aeruginosa Biofilms on Substrates With Differing Acoustic Impedance. *Ultrasound Med. Biol.* **2022**, *48*, 1888–1898. [CrossRef]
38. Gu, Y.; Chen, C.C. Eliminating the Interference of Oxygen for Sensing Hydrogen Peroxide with the Polyaniline Modified Electrode. *Sensors* **2008**, *8*, 8237–8247. [CrossRef]
39. Soren, O. *Investigation into Novel Nitric Oxide Based Anti-Biofilm Strategies to Target Pseudomonas Aeruginosa Infection in Cystic Fibrosis*; University of Southampton: Southampton, UK, 2019.
40. Owen, J.; Crake, C.; Lee, J.Y.; Carugo, D.; Beguin, E.; Khrapitchev, A.A.; Browning, R.J.; Sibson, N.; Stride, E. A Versatile Method for the Preparation of Particle-Loaded Microbubbles for Multimodality Imaging and Targeted Drug Delivery. *Drug Deliv. Transl. Res.* **2018**, *8*, 342–356. [CrossRef]
41. Delalande, A.; Bastié, C.; Pigeon, L.; Manta, S.; Lebertre, M.; Mignet, N.; Midoux, P.; Pichon, C. Cationic Gas-Filled Microbubbles for Ultrasound-Based Nucleic Acids Delivery. *Biosci. Rep.* **2017**, *37*, 20160619. [CrossRef] [PubMed]
42. Behrens, S.H.; Grier, D.G. The Charge of Glass and Silica Surfaces. *J. Chem. Phys.* **2001**, *115*, 6716. [CrossRef]
43. Westerhof, N.; Stergiopulos, N.; Noble, M.I.M.; Westerhof, N.; Stergiopulos, N.; Noble, M.I.M. Law of Poiseuille. In *Snapshots of Hemodynamics*; Springer: New York, NY, USA, 2010; pp. 9–14.
44. Lewis, A.M.; Boose, E.R. Estimating Volume Flow Rates Through Xylem Conduits. *Am. J. Bot.* **1995**, *82*, 1112–1116. [CrossRef]
45. Wang, D.S.; Panje, C.; Pysz, M.A.; Paulmurugan, R.; Rosenberg, J.; Gambhir, S.S.; Schneider, M.; Willmann, J.K. Cationic versus Neutral Microbubbles for Ultrasound-Mediated Gene Delivery in Cancer. *Radiology* **2012**, *264*, 721–732. [CrossRef] [PubMed]

46. Ginés, L.; Mandal, S.; Ashek-I-Ahmed; Cheng, C.L.; Sow, M.; Williams, O.A. Positive Zeta Potential of Nanodiamonds. *Nanoscale* **2017**, *9*, 12549–12555. [CrossRef]
47. Abaka-Wood, G.B.; Addai-Mensah, J.; Skinner, W. A Study of Flotation Characteristics of Monazite, Hematite, and Quartz Using Anionic Collectors. *Int. J. Min. Process.* **2017**, *158*, 55–62. [CrossRef]
48. Chacko, S.A.; Wenthold, P.G. The Negative Ion Chemistry of Nitric Oxide in the Gas Phase. *Mass. Spectrom. Rev.* **2006**, *25*, 112–126. [CrossRef]
49. Weitzberg, E.; Hezel, M.; Lundberg, J.O. Nitrate-Nitrite-Nitric Oxide Pathway: Implications for Anesthesiology and Intensive Care. *Anesthesiology* **2010**, *113*, 1460–1475. [CrossRef] [PubMed]
50. John, Y.; David, V.E.; Mmereki, D. A Comparative Study on Removal of Hazardous Anions from Water by Adsorption: A Review. *Int. J. Chem. Eng.* **2018**, *2018*, 1–21. [CrossRef]
51. Mautner, A.; Kobkeatthawin, T.; Bismarck, A. Efficient Continuous Removal of Nitrates from Water with Cationic Cellulose Nanopaper Membranes. *Resour. -Effic. Technol.* **2017**, *3*, 22–28. [CrossRef]
52. Choi, S.; Yun, Z.; Hong, S.; Ahn, K. The Effect of Co-Existing Ions and Surface Characteristics of Nanomembranes of the Removal of Nitrate and Fluoride. *Desalination* **2001**, *133*, 53–64. [CrossRef]
53. Lundberg, J.O.; Weitzberg, E.; Gladwin, M.T. The Nitrate-Nitrite-Nitric Oxide Pathway in Physiology and Therapeutics. *Nat. Rev. Drug Discov.* **2008**, *7*, 156–167. [CrossRef] [PubMed]
54. Sang, S.Y.; Coakley, R.; Lau, G.W.; Lymar, S.V.; Gaston, B.; Karabulut, A.C.; Hennigan, R.F.; Hwang, S.H.; Buettner, G.; Schurr, M.J.; et al. Anaerobic Killing of Mucoid Pseudomonas Aeruginosa by Acidified Nitrite Derivatives under Cystic Fibrosis Airway Conditions. *J. Clin. Investig.* **2006**, *116*, 436–446. [CrossRef]
55. Carlsson, S.; Govoni, M.; Wiklund, N.P.; Weitzberg, E.; Lundberg, J.O. In Vitro Evaluation of a New Treatment for Urinary Tract Infections Caused by Nitrate-Reducing Bacteria. *Antimicrob. Agents Chemother.* **2003**, *47*, 3713–3718. [CrossRef]
56. Bremus, C.; Herrmann, U.; Bringer-Meyer, S.; Sahm, H. The Use of Microorganisms in L-Ascorbic Acid Production. *J. Biotechnol.* **2006**, *124*, 196–205. [CrossRef]
57. Barraud, N.; Hassett, D.J.; Hwang, S.-H.; Rice, S.A.; Kjelleberg, S.; Webb, J.S. Involvement of Nitric Oxide in Biofilm Dispersal of Pseudomonas Aeruginosa. *J. Bacteriol.* **2006**, *188*, 7344–7353. [CrossRef]
58. Ballermann, B.J.; Dardik, A.; Eng, E.; Liu, A. Shear Stress and the Endothelium. *Kidney Int.* **1998**, *54*, S100–S108. [CrossRef]
59. Langeveld, S.A.G.; Meijlink, B.; Beekers, I.; Olthof, M.; Van Der Steen, A.F.W.; de Jong, N.; Kooiman, K. Theranostic Microbubbles with Homogeneous Ligand Distribution for Higher Binding Efficacy. *Pharmaceutics* **2022**, *14*, 311. [CrossRef] [PubMed]
60. Rodgers, S.D.; Camphausen, R.T.; Hammer, D.A. Sialyl Lewis(x)-Mediated, PSGL-1-Independent Rolling Adhesion on P-Selectin. *Biophys. J.* **2000**, *79*, 694–706. [CrossRef] [PubMed]
61. Edgeworth, A.; Ross, J.A.; Anderson, T.; Butler, M.; McDicken, W.N.; Moran, C. Novel Flow Chamber to Investigate Binding Strength of a Lipid-Based, High-Frequency, Ultrasonic Contrast Agent. *Ultrasound* **2010**, *18*, 130–139. [CrossRef]
62. Stride, E.P.; Coussios, C.C. Cavitation and Contrast: The Use of Bubbles in Ultrasound Imaging and Therapy. *Proc. Inst. Mech. Eng. H* **2009**, *224*, 171–191. [CrossRef]
63. Ferrara, K.; Pollard, R.; Borden, M. Ultrasound Microbubble Contrast Agents: Fundamentals and Application to Gene and Drug Delivery. *Annu. Rev. Biomed. Eng.* **2007**, *9*, 415–447. [CrossRef]
64. Leighton, T.G. What is ultrasound? *Prog. Biophys. Mol. Biol.* **2007**, *93*, 3–83. [CrossRef]
65. Pereno, V.; Lei, J.; Carugo, D.; Stride, E. Microstreaming inside Model Cells Induced by Ultrasound and Microbubbles. *Langmuir* **2020**, *36*, 6388–6398. [CrossRef]

pharmaceutics

Article

Stable Cavitation-Mediated Delivery of miR-126 to Endothelial Cells

Stephanie He [1], Davindra Singh [1], Hossein Yusefi [2] and Brandon Helfield [1,2,*]

1 Department of Biology, Concordia University, Montreal, QC H4B 1R6, Canada
2 Department of Physics, Concordia University, Montreal, QC H4B 1R6, Canada
* Correspondence: brandon.helfield@concordia.ca

Abstract: In endothelial cells, microRNA-126 (miR-126) promotes angiogenesis, and modulating the intracellular levels of this gene could suggest a method to treat cardiovascular diseases such as ischemia. Novel ultrasound-stimulated microbubbles offer a means to deliver therapeutic payloads to target cells and sites of disease. The purpose of this study was to investigate the feasibility of gene delivery by stimulating miR-126-decorated microbubbles using gentle acoustic conditions (stable cavitation). A cationic DSTAP microbubble was formulated and characterized to carry 6 μg of a miR-126 payload per 10^9 microbubbles. Human umbilical vein endothelial cells (HUVECs) were treated at 20–40% duty cycle with miR-126-conjugated microbubbles in a custom ultrasound setup coupled with a passive cavitation detection system. Transfection efficiency was assessed by RT-qPCR, Western blotting, and endothelial tube formation assay, while HUVEC viability was monitored by MTT assay. With increasing duty cycle, the trend observed was an increase in intracellular miR-126 levels, up to a 2.3-fold increase, as well as a decrease in SPRED1 (by 33%) and PIK3R2 (by 46%) expression, two salient miR-126 targets. Under these ultrasound parameters, HUVECs maintained >95% viability after 96 h. The present work describes the delivery of a proangiogenic miR-126 using an ultrasound-responsive cationic microbubble with potential to stimulate therapeutic angiogenesis while minimizing endothelial damage.

Keywords: gene therapy; miRNA delivery; microRNA-126; microbubbles; sonoporation; passive cavitation detection; ultrasound; endothelial cells; HUVEC; therapeutic angiogenesis

Citation: He, S.; Singh, D.; Yusefi, H.; Helfield, B. Stable Cavitation-Mediated Delivery of miR-126 to Endothelial Cells. *Pharmaceutics* **2022**, *14*, 2656. https://doi.org/10.3390/pharmaceutics14122656

Academic Editor: Udo Bakowsky

Received: 21 October 2022
Accepted: 27 November 2022
Published: 30 November 2022

1. Introduction

Gene therapy is a treatment regime in which a specific cell-function-altering piece of genetic material is introduced into target diseased cells. Typically, DNA, mRNA, siRNA, and anti-sense oligonucleotides are the genetic materials used for this approach, either to restore a specific gene function or to turn off a gene involved in pathogenesis. The first authorized clinical trial for gene therapy was for Gaucher's disease in 1988 (NCT00001234), a lysosomal dysfunction. Mostly due to the fact that it commonly employs viruses to deliver the genetic material, gene therapy was met with mixed success early on in its history [1] owing to complications with the immune response, off-target side-effects, and unwanted or neutral clinical outcomes [2]. Indeed, the critical challenge toward the advancement of this approach is the delivery method itself, and key advances in this field have brought gene therapy back into the spotlight. Nonviral vector approaches, such as systemically injected liposomal constructs, are generally less immunogenic than their viral counterparts, and they can gain target cell specificity through chemical/biological design. Generally, these vectors are considered less efficient than viral vectors, likely due to the fact that they face the endoluminal border and must escape early endosomes to deliver their payload.

More recently, microRNAs (miRs) have been employed in molecular therapeutics. MiRs are noncoding RNA strands that may bind to multiple mRNA targets and modulate their expressions. Nearly three decades ago, the discovery of dysregulated miR in nematodes [3] allowed insight into a better understanding of disease development in a wide array of pathologies [4–6]. The pleiotropic nature of miRs makes for a particularly attractive choice for gene therapy, e.g., in applications of diseases with a multifactorial origin [7]. Indeed, miR therapy is being explored preclinically in many areas, including cancer (e.g., pancreatic [8], breast [9], lung [10], and leukemia [11]) and cardiovascular disease (e.g., cardiac hypertrophy [12], arrhythmia [13], cardiac fibrosis [14], and ischemia [15]).

With regard to the application of therapeutic angiogenesis for ischemia, miR-126 is a potentially potent target. miR-126 is highly expressed in endothelial cells, and feasibility studies have shown its role in modulating angiogenesis by repressing angiogenic repressors, such as PIK3R2 from the Akt pathway and SPRED1 from the ERK pathway, to name a few [16,17]. Indeed, in selecting a potential candidate technique for miR-126 delivery or any other miR that has shown to be a major molecular regulator in cardiovascular disease, there is particular interest in exploring ultrasound-based techniques, as echocardiography is commonly clinically used to diagnose many cardiovascular diseases [18].

Ultrasound-sensitive agents, including clinically employed microbubble contrast agents, provide an exciting alternative to more traditional nonviral vectors. Typically between 1 and 8 μm in diameter, microbubbles remain intravascular and are composed of a thin, flexible lipid monolayer shell and vibrate when exposed to ultrasound [19]. Recent studies have demonstrated that, under specific acoustic conditions, ultrasound-stimulated microbubbles can temporarily alter vascular and cellular permeability, thereby providing an endocytosis-independent pathway for exogenous drug delivery [20–23]. Indeed, as these microbubbles only vibrate as they pass through an ultrasound beam, their activity is spatially and temporally targeted, thus having the potential to limit off-target therapeutic deposition.

In microbubble-mediated gene therapy [24], the design of the microbubble is critical to ensure sufficient loading capacity. One such approach is to synthesize microbubbles with a surface charge to offer a means to attach genetic payload on the microbubble surface, shown to deliver genetic material more efficiently than co-injected with neutral microbubbles [25,26]. Such efficiency is conferred by protection against circulating RNases in the bloodstream to increase the half-life of the short RNA strand, as well as increasing the available genetic material around the tissue when it is coupled to a vehicle, as opposed to free floating in the systematic circulation [27]. Additionally, the noninvasive nature of the methodology allows for repeated treatment to further increase gene delivery efficiency, as well as the ability to spatially target tissues of interest with focused ultrasound limits off target responses [28].

Ultrasound-mediated microbubble behavior is commonly separated into two distinct physical regimes [29]. If bubbles are exposed to acoustic pressures above a specific threshold, they can rapidly expand and violently collapse during the compression of the transmit pulse. This behavior, whereby microbubble disruption occurs, is dominated by the inertia of the surrounding fluid and is typically termed inertial cavitation, accompanied by broadband spectral emissions. Indeed, miR-based delivery using ultrasound and microbubbles has been shown via this acoustic regime, which does result in physical membrane perforation and enhanced cellular and vascular uptake of material [30–32]. In fact, numerous studies have demonstrated the advantage of delivering miR or miR inhibitors through inertial cavitation (typically referred to as ultrasound-targeted microbubble destruction; UTMD) for specific applications, such as cancer cell ablation [30,33,34]. Other applications of gene delivery through UTMD have shown success in preventing organ rejection [35], treating cardiac hypertrophy [31], and even in promoting angiogenesis [32].

While shown to be an effective strategy, microbubble disruption (e.g., inertial cavitation) may trigger severe, undesired bioeffects, including loss of cell viability [36], hemorrhage, or increased inflammation.

Another approach to microbubble-mediated ultrasound gene delivery, which has been less explored, is to ensure a more controlled microbubble oscillation regimen. Generally, bubbles driven by low-pressure ultrasound elicit repeated periodic, volumetric vibrations about their equilibrium size, whereby they may oscillate spherically or non-spherically.

Under specific acoustic conditions, these vibrations can enhance cell permeability due to local fluid microstreaming and prolonged shear stress [37]. This regime is called stable cavitation and is characterized by distinct harmonic scattered pressure emissions [29]. In addition to limiting potential unwanted bioeffects, the prolonged fluid streaming and physical presence of the microbubble may aid in intracellular therapeutic deposition.

Indeed, as the majority of investigations of ultrasound-mediated gene delivery rely on bubble destruction, the novelty of this work presented here is via our intended goal of demonstrating that modest levels of gene therapy can be achieved using gentle bubble vibrations (stable cavitation without bubble disruption), with potential significance toward the design of repeat treatment paradigms. This entails the synthesis of high-capacity gene-loaded microbubble constructs that are clinically feasible in terms of stability, concentration, nonlinear echo, and gene-loading concentration, followed by confirmation of small but significant gene delivery so as to preserve cell viability and ensure successful modulation of salient downstream protein expression and physiological endpoints.

In this present study, we aim to investigate ultrasound-mediated gene delivery using stable cavitation, specifically in the context of therapeutic angiogenesis. First, we synthesized and characterized a cationic lipid microbubble formulation decorated with an miR-126 mimic. Next, we demonstrate that these microbubbles can be used to deliver their miR-126 cargo within endothelial cells in a viable manner while undergoing stable cavitation, monitored with passive cavitation techniques. Lastly, we assess the functional and physiological endpoints of the miR-126 delivery.

2. Materials and Methods

2.1. Contrast Agent Microbubble Preparation

Cationic phospholipid-encapsulated microbubbles were synthesized via a modification of an existing in-house formulation [38]. Briefly, the microbubbles were prepared from a lipid aqueous dispersion composed of 1,2-distearoyl-sn-glycero-3-phosphocholine (DSPC; Avanti Polar Lipids, Alabaster, AL, USA), polyoxyethylene (40) stearate (PEG40S; Sigma-Aldrich, St. Louis, MO, USA), and 1,2-stearoyl-3-trimethylammonium-propane (DSTAP; Avanti Polar Lipids, Alabaster, AL, USA). DSPC, PEG40S, and DSTAP (0.41:0.50:0.09) were dissolved in a solution buffer consisting of PBS-EDTA (1 mM), propylene glycol and glycerol (0.80:0.15:0.05) at a concentration of 1 mg/mL. The lipid solution was then dissolved in a 20 kHz ultrasonic bath (Branson Ultrasonics, Danbury, CT, USA) at 65 °C until a homogenous clear solution was observed. Samples of microbubbles were formed by aliquoting 1.5 mL of this solution in glass vials and sealed with a rubber stopper. The air from the vial was removed via vacuum and replaced with perfluorobutane (C_4F_{10}; FluoroMed, Round Rock, TX, USA). To synthesize the microbubbles, vials were allowed to reach room temperature to ensure consistent size distribution and scattering activity [39], subjected to mechanical agitation using the VialMix™ (Lantheus, Billerica, MA, USA) for the standard 45 s period, and then were set to cool down to room temperature for 15 min. Vials were then decanted for 8 min to ensure the removal of large microbubbles, and bubbles were withdrawn using a 1 inch 19G needle, along with another 19G needle for venting purposes. Agent was consistently extracted from just below the rubber stopper to reduce population variability. After decantation, 0.7 mL was extracted from the glass vial and the microRNA-126 was added to the microtube for incubation. Microbubbles were washed as per described in Wang et al. [25]. Briefly, the mixture was diluted to 1 mL with DEPC-treated diH_2O and centrifuged at 400× g for 3 min in a 3 mL syringe. The bottom 0.5 mL was discarded to remove the smaller microbubbles, and the remainder of the solution was diluted in DEPC-treated diH_2O to fix the concentration to 10^9 microbubbles/mL for the experiments.

2.2. Microbubble Characterization

The size distribution, concentration, and stability of these in-house microbubbles were quantified using a Coulter Counter (Multisizer 4e; Beckman Coulter, CA, USA). Microbubbles were diluted 1:1000 in ISOTON II (Beckman Coulter) and sampled using the 30 μm aperture, which enables an effective measurable size range of 0.6–18 μm. For a given vial, measurements were repeated every 20 min for up to 80 min post activation to assess microbubble stability. For a subset of experiments, the zeta potential was measured using a Zetasizer Nano ZS (Malvern Panalytical, Malvern, UK). Microbubbles were diluted to a 0.2% solution within diH_2O, and measurements were taken within 10 min of having extracted the agent from the vial. For each of these measurements, at least $n = 3$ vials were used for statistical purposes.

To confirm that our agent is acoustically active, the echogenicity of the microbubbles was measured with a clinical ultrasound system (model iU22 Philips Healthcare, Andover, MA, USA) using a C5-2 probe. The microbubbles were diluted 10,000-fold in gas-equilibrated diH_2O and fed into a wall-less 0.8 cm diameter cardiac Doppler flow phantom (model 523A, ATS Laboratories, Norfolk, VA, USA) at a velocity of ~14–16 mm/s. The depth was fixed at 8 mm, and three videos were recorded in B mode and contrast mode for 2 min. The signal-to-noise ratio (SNR) was plotted $\pm$ the standard deviation at different timepoints following microbubble extraction from its vial.

2.3. miR-126 Loading Protocol and Characterization

To confirm gene coupling to the outer shell of our cationic microbubbles, a solution of microbubbles was incubated at room temperature for 15 min with a red fluorescent siRNA (Alexa Fluor 555 BLOCK-iT, Thermo Fisher Scientific, Waltham, Brea, MA, USA) in a 1:1 proportion. These microbubbles were visualized using an RFP light cube from the EVOS M7000 imaging system (Thermo Fisher Scientific). After confirming the cationic nature of our microbubbles, we aimed to load our gene of interest miR-126 (hsa-miR-126-3p; sequence: UCGUACCGUGAGUAAUAAUGCG; Thermo Fisher Scientific).

To quantify miR-126 loading capacity, the microbubbles were incubated with increasing miR-126 concentrations (miR-126 input) at room temperature for 15 min on a tube revolver rotator. These concentrations were fixed from 1 µg miR-126/10^9 microbubbles to 20 µg/10^9 microbubbles, where 20 µL of a RNA loading dye [40] was added to equal volume of the miR-126 and microbubbles solution, and loaded in a 12.5% (19:1 acrylamide/bis-acrylamide) (Thermo Fisher Scientific) urea PAGE [40]. A control lane where equal amounts of miR-126 was diluted in DEPC-treated diH_2O was also loaded next to each miR-126 concentration tested. We ran the gel at 65 V until separation of bromophenol blue (Sigma) and xylene cyanol (Thermo Fisher Scientific). The gel was stained with a 1:10 000 solution of SybrGreen II RNA gel stain (Thermo Fisher Scientific) in TBE buffer for 15 min. The gel was imaged with on a G:BOX F3 (Syngene, Cambridge, UK), and the bands were analyzed on ImageJ (U.S. National Institutes of Health, Bethesda, MD, USA). We assumed that the bands observed corresponded to the unbound miR-126 (when loaded with microbubbles) and were compared to the control bands (miR-126 with diH_2O), which resulted in the percentage of free miR-126 in the gel.

$$\%free = \frac{\textbf{\textit{band intensity value of miR}} - \textbf{126 \textit{with microbubbles}}}{\textbf{\textit{band intensity value of miR}} - \textbf{126 \textit{in diH}}_2\textbf{\textit{O}}}. \tag{1}$$

Lastly, the percentage of miR-126 bound was obtained with the following equation:

$$\textbf{\% bound} = \textbf{\%100} - \textbf{\% free} \tag{2}$$

where we report here the mass of loaded miR-126 as

$$\textbf{mass of bound miR} - \textbf{126} = \textbf{\% bound} \times \textbf{mass of miR} - \textbf{126 input}. \tag{3}$$

These data points were plotted to find the plateau to determine the maximum achievable mass of bound miR-126 per 10^9 microbubbles.

2.4. Cell Culture

Primary human umbilical vein endothelial cells (HUVECs; No. C2519A, Lonza, Verviers, Belgium) were cultured in medium (EGM-2, No. C3162, Lonza, Belgium) and incubated at 37 °C and 5% CO_2. For experiments, the cells were harvested with a 0.05% solution of trypsin-EDTA (Wisent, QC, Canada) at 90% confluency and placed in a suspension within cell culture medium at a concentration of 500 k cells per mL. All experiments were performed on HUVECs characterized by a passage number between 3 and 10.

2.5. Ultrasound Apparatus and Experimental Procedure

We employed a custom-designed ultrasound treatment tank in order to assess microbubble-mediated miR-126 delivery to endothelial cells (Figure 1). The treatment tank was a 30 L plastic container. It was filled with 15 L of gas-equilibrated diH_2O, enough volume to submerge both the transducers, but kept below the opening of the sample chamber. The water temperature was maintained at 37 °C using an immersion heater circulator (VWR model 1120, Radnor, PA, USA). The tank consisted of a sample chamber and two co-aligned single-element transducers. A magnetic stir bar was placed inside the sample chamber, which is made of acrylic and sealed with mylar windows (25 μm thickness) to allow for ultrasound transmission. The unit was placed on a magnetic stir plate. Samples were prepared by incubating miR-126-loaded microbubbles with HUVECs at a fixed ratio of 50 bubbles per cell. Placed within our warmed sample chamber, this cocktail was constantly mixed using a magnetic stir bar to ensure homogeneous distribution of the microbubbles and cells. After a 1 min wait to ensure equilibrium, the sample was treated with ultrasound. The treatment transducer (1 MHz, f = 25.4 mm, f# = 1.33) was driven at 1 MHz with 1000 cycles, varying pulse repetition intervals from 2.5–5 ms (duty cycles of 20–40%) at a peak negative pressure of 52 kPa generated from an arbitrary function generator (AFG31000, Tektronix, Beaverton, OR, USA) and amplified using an RF power amp (model 105A100B, Amplifier Research, Souderton, AR, USA) for a treatment duration of 2 min. The acoustic pressure was measured in free space within a separate water tank using a 'bullet' hydrophone (HGL-0200, ONDA, Sunnyvale, CA, USA). The second transducer (3.5 MHz flat transducer, Olympus) was used as a passive cavitation detector to record microbubble scattering, specifically, to assess the presence of stable and inertial cavitation. Receive echoes were amplified (AU-1579, 0.7–200 MHz, MITEQ, Hauppauge, NY, USA), bandpass-filtered, and then digitized (Gage Razor Express CompuScope, Lockport, IL, USA) for off-line analysis using custom MATLAB software (Mathworks, Natick, MA, USA). The frequency-dependent transfer function of this receive circuit was not determined; thus, all measurements are relative. Joint time–frequency analysis was performed with a window size of 50 ms and a 90% overlap. A Hamming window was applied to the RF data prior to obtaining the fast Fourier transform (FFT).

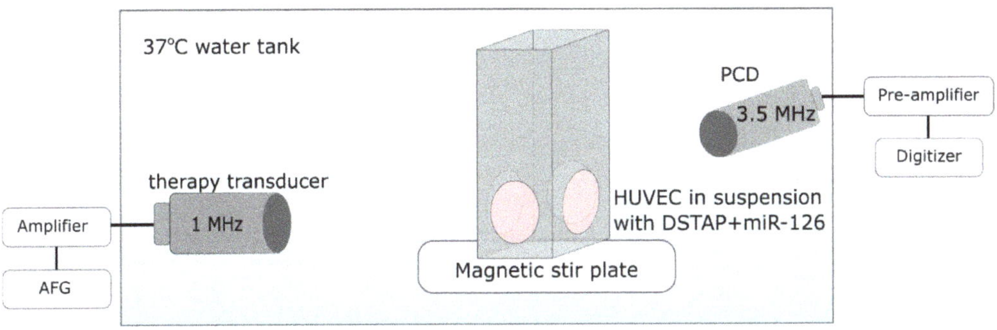

Figure 1. Experimental setup equipped with two co-aligned transducers for ultrasound-mediated gene delivery to endothelial cells. The therapy transducer (1 MHz) focuses within a cell suspension chamber, while the second transducer (3.5 MHz) is used for passive detection of bubble echoes.

Following ultrasound treatment, the cells were transferred from the custom-designed suspension chamber to a conical tube left at room temperature for 5 min to allow for sonoporation recovery. These cells were then placed on ice until all the samples were treated before washing three times at 220× *g* for 5 min at 4 °C per wash to remove the remaining bubbles. The sham controls were handled similarly without ultrasound.

2.6. RT-qPCR

Immediately following the washes, the total RNA content of the cells was extracted using a mirVana™ miRNA isolation kit (Invitrogen), and 10 ng RNA samples were used for PCR. The primers were all purchased from Thermo Fisher Scientific, and, although their sequences are proprietary information, their product numbers (PN) are listed. The PCR primers used to generate cDNA were U6 (PN: 4440887 RT 001973) and has-miR-126 (PN 4427975 39G01 RT 002228) on a thermal cycler (Applied Biosystems, Waltham, MA, USA) set at 16 °C for 30 min, 42 °C for 30 min, and 85 °C for 5 min. RT-qPCR was performed using TaqMan™ Fast Advanced Master Mix no UNG (Thermo Fisher Scientific) on a QuantStudio™ 3 (Thermo Fisher) set at 95 °C for 20 s, then 40 cycles of 95 °C (1 s) to 60 °C (20 s). The primers used for RT-qPCR were U6 (PN: 4440887 TM 001973) and miR-126 (PN: 4440887 74A03 TM 002228). Relative miR-126 levels were calculated using the $2(-\Delta\Delta Ct)$ method using U6 as the housekeeping control.

2.7. Viability Assay

Endothelial cell viability was measured after 2 days and 4 days, whereby 20,000 and 10,000 cells were seeded on 96-well dishes in triplicate. After washing the cells with PBS three times, MTT reagent (Sigma) was added as per the manufacturer's protocol. The absorbance was measured on a Varioskan™ LUX microplate reader (Thermo Fisher Scientific) at 540 nm after a 4 h incubation in 37 °C in 5% CO_2 and solubilizing the formazan with DMSO (BioRad, Hercules, CA, USA) for 10 min. Viability was assessed relative to sham controls.

2.8. Western Blotting

Following the washes after ultrasound treatments, cells were seeded on six-well dishes and harvested 2 days later. Total protein was extracted using RIPA lysis buffer (Alfa Aesar, Haverhill, MA, USA), and 20–30 μg of protein was loaded per lane in a 7.5% acrylamide (37.5:1 acrylamide/bis-acrylamide) (BioRad) SDS-PAGE, 1.5 mm thick. We let the gel run at 70 V for 30 min and 100 V until complete separation of the PageRuler™ Plus prestained protein ladder (Thermo Fisher Scientific). The proteins were transferred on a PVDF membrane overnight in Towbin buffer at 4 °C. The membrane was stained with 0.5% Ponceau S (Sigma) in 1% glacial acetic acid for 5 min to ensure proper protein transfer and destained with TBS. Following blocking in TBS with 5% BSA (Wisent) for 1 h at room temperature, the PVDF membrane was cut in three sections horizontally to separate the target proteins according to their molecular weights.

The antibodies used for blotting were monoclonal anti-SPRED1 (Santa Cruz Biotechnology, Dallas, TX, USA), monoclonal anti-PI-3-kinase p85β (Santa Cruz Biotechnology), and anti-GAPDH (Invitrogen), diluted in TBST-2% BSA for 1 h at room temperature. The first section was cut at the 70 kDa band to probe the PIK3R2. The second section was cut around the 40 kDa band (between the 35 kDa and 55 kDa bands) to probe the SPRED1 antibody. The last section was used to probe GAPDH. When we could not obtain a satisfactory separation from ladder to cut the membrane between the 55 kDa and 35 kDa bands, the membrane was only cut at the 70 kDa band. GAPDH was then probed first, stripped with a mild stripping buffer (Abcam) for 5 min, and washed with PBS and TBS for 5–10 min twice each. Blocking could then be repeated on this half membrane to probe SPRED1. The membrane was then incubated in with goat anti-mouse IgG (H + L) HRP (Invitrogen) secondary antibody diluted in TBST-2% BSA for 1 h at room temperature. PVDF blotted membranes were labeled with Pierce ECL Western Blotting Substrate (Thermo Fisher Scientific) prior to imaging on an Amersham Imager 600 (GE Healthcare, Chicago, IL, USA). Bands were analyzed in ImageJ.

2.9. Endothelial Tube Formation Assay

A growth factor-reduced Matrigel® matrix basement membrane (Corning, NY, USA) was prepared on a 96-well dish according to the manufacturer's protocol. Following the washes, 2000 cells were seeded atop the Matrigel for 16 h. HUVECs were then stained with Calcein-AM (Thermo Fisher) at 1 µg/mL and visualized on an epi-fluorescence microscope at 488 nm (EVOS system, Life Technologies). Angiogenesis quantification was assessed by AutoTube [41–43] on MATLAB.

3. Results

3.1. Characterization of DSTAP Microbubbles and miR-126 Loading

The microbubble size distribution is shown in Figure 2A, depicting a polydisperse population with a peak volume-weighted diameter of 4.56 ± 0.32 µm. Stability tests, where measurements were taken every 20 min, suggest nonsignificant (p-value = 0.313, n = 3) changes in microbubble diameter (Figure 2B) and concentration ($1.01 \times 10^9 \pm 0.31 \times 10^9$ microbubbles/mL) over 80 min. The size and concentration information reported here is consistent with clinically employed agents [44–46].

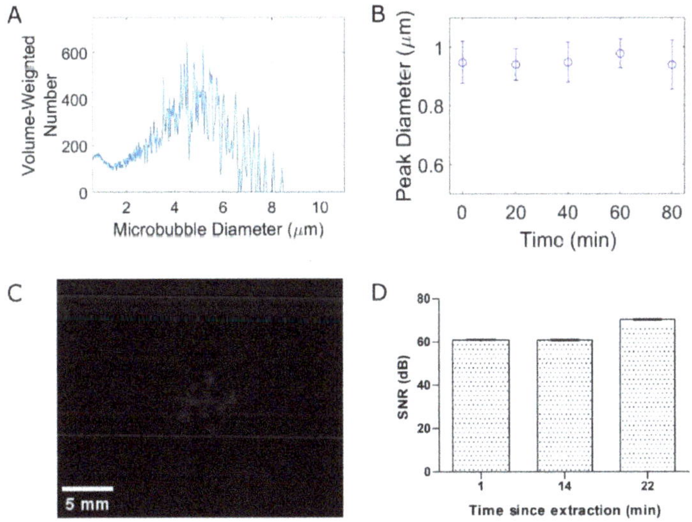

Figure 2. Our stable microbubble formulation can be used for clinical contrast imaging. (**A**) Volume-weighted size distribution with a concentration of $(1.01 \pm 0.31) \times 10^9$ bubbles/mL. (**B**) Stability over 80 min. (**C**) Nonlinear contrast imaging of microbubbles in wall-less tissue phantom. (**D**) SNR analysis of contrast signal over 22 min post-agent extraction.

Clinical contrast echo from the agent—which is indicative of nonlinear scattering of the microbubbles—is shown qualitatively in Figure 2C and quantified in Figure 2D. The large contrast signal-to-noise ratio of the agent remains relatively constant up to 22 min following agent extraction from the sealed vial (62.95 ± 2.56 to 70.24 ± 2.76 dB). These data, along with the concentration and stability properties, suggest its viability as a clinical agent.

In order to test the cationic nature of this in-house agent, as well as to provide evidence for gene coupling along the surface of the microbubbles, Figure 3A highlights the surface distribution of a surrogate fluorescence siRNA. It can be seen from this micrograph that the fluorescent nucleic acid conforms to the outer surface (i.e., the shell) of the microbubbles. Furthermore, the cationic nature of our agent was confirmed via assessment of the surface charge, which was determined to be +38.04 mV pre miR loading, decreasing to +27.60 mV post miR loading (Figure 3B; $p < 0.0002$). To quantify this miR-126 carrying capacity, gel electrophoresis was performed to compare the amount of free-miR-126 in diH_2O compared to unbound miR-126 from microbubbles at varying quantities of miR-126 (Figure 3C). Binding capacity was determined from plotting a saturation curve (red) and the value at which the curve plateaus is defined as the binding capacity. For our formulation, we determined a loading capacity of 6 µg of miR-126 per 10^9 microbubbles (Figure 3D), which is what we used for all subsequent experiments.

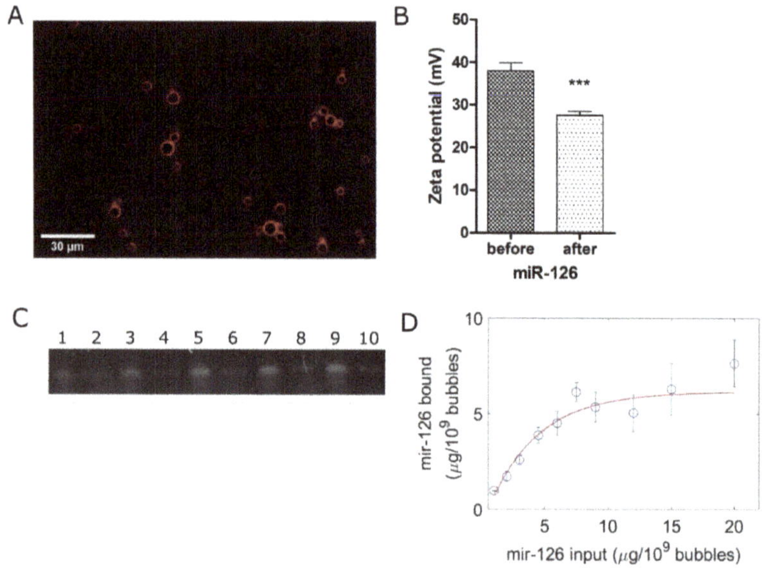

Figure 3. Gene coupling characterization results in 6 µg of miR-126 per 10^9 microbubbles. (**A**) Fluorescent siRNA localization on microbubble surface. Scale bar: 30 µm. (**B**) Surface charge before and after coupling of 6 µg of miR-126 per 10^9 microbubbles. (**C**) Acrylamide/urea PAGE of miR-126 binding capacity on DSTAP: odd-numbered lanes show free miR-126 in DEPC H_2O and even-numbered lanes show unbound miR-126 after incubation with DSTAP. (**D**) Quantification of miR-126 binding capacity on DSTAP. (***: $p < 0.001$).

3.2. *Ultrasound-Mediated miR Delivery Using miR-126-Conjugated DSTAP Microbubbles*

Ultrasound-assisted viable miR-126 delivery to endothelial cells via stable cavitation was demonstrated in Figure 4. Figure 4A indicates a gradual increase in miR-126 levels, from 1.482 to 2.326 relative to the sham control, with increasing duty cycle ($n = 3$ to 8, p-values = 0.004, 0.02, 0.08, respectively). Our data indicate that, under this ultrasound treatment regimen, endothelial cell viability is maintained (>95%) (Figure 4B; $n = 6$ to 24, p-value nonsignificant). Further, as confirmation of functional delivery, we assayed the protein levels of miR-126 target proteins SPRED1 and PIK3R2. These were downregulated 2 days following gene delivery; we observed a 4% to 33% decrease in SPRED1 with increasing duty cycle ($n = 3$ to 4; p-value = 0.0345 at duty-cycle of 40%), while PIK3R2 decreased from 36% to 46% with increasing duty cycle (Figure 4D; $n = 3$–4; p-values = 0.02, 0.008, and 0.003, respectively).

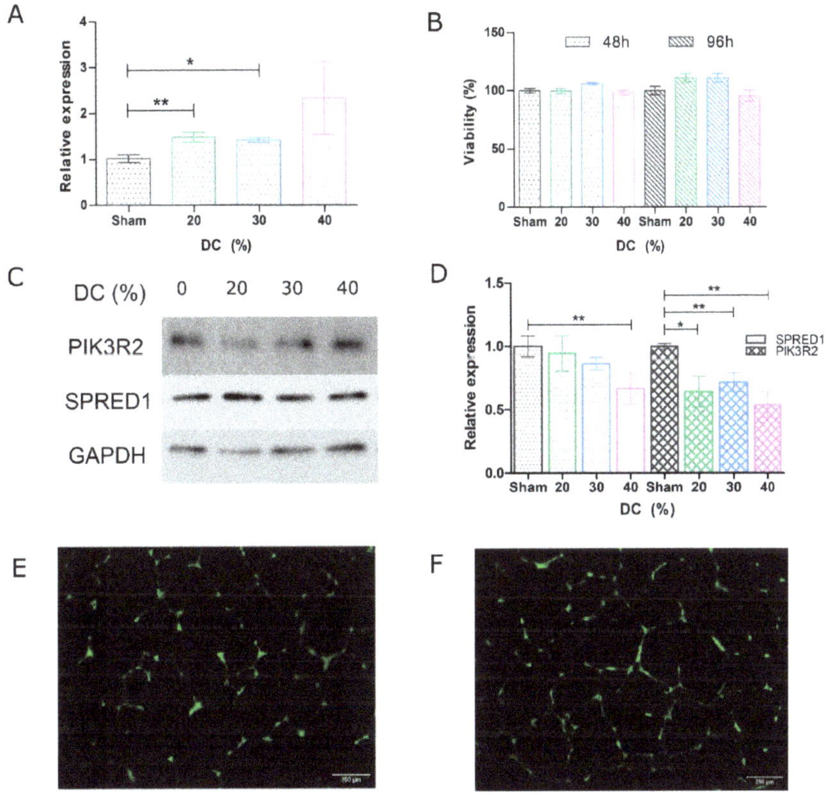

Figure 4. Ultrasound-mediated miR-126 delivery results in viable physiologically relevant treatment of endothelial cells. (**A**) RT-qPCR analysis of the relative miR-126 expression from treated HUVECs. (**B**) Viability assessment of the cells 48 h and 96 h following ultrasound treatment by MTT assay. (**C**) PIK3R2 and SPRED1 protein expression in treated HUVECs determined by Western blotting. (**D**) Aggregate relative PIK3R2 and SPRED1 expression 48 h following HUVEC treatment. (**E**) HUVEC tube formation assay, untreated and (**F**) treated at 20% DC. In aggregate, the tube network increased by 22–26% and exhibited a 5–17% increase in branching nodes. Scale bar: 200 µm. (*: $p < 0.05$, **: $p < 0.01$). Grey: sham, green: 20% DC, blue: 30% DC, magenta: 40% DC.

A second subset of treated cells were used to grow on Matrigel basement membrane matrix, where endothelial tube formation was assessed 16 h after seeding. As can be seen in the representative images shown in Figure 4E,F, tubule networks were significantly more complex in miR-126-treated endothelial cells compared to sham control (Figure 4E). Specifically, tube network increased by 22–26% and exhibited a 5–17% increase in branching nodes.

3.3. Passive Cavitation Detection

To confirm stable microbubble activity during treatment, the gene delivery setup (Figure 1) was concurrently equipped with a passive cavitation detection transducer. Figure 5A shows a representative example of the frequency spectrum at the beginning of the 2 min treatment with microbubbles (30% duty cycle; black) overlaid with pure PBS (blue). It can be clearly seen that, under this acoustic regime, microbubbles underwent stable cavitation, as confirmed by narrowband signal power peaking at second, third, and fourth harmonics (2, 3, and 4 MHz) and an absence of broadband emissions. Furthermore, time–frequency analysis (Figure 5B,C) confirmed that microbubbles were undergoing stable cavitation throughout the entire duration of the ultrasound treatment. Figure 5D depicts the cumulative spectral power from the stable (blue, solid line) versus inertial (black, dashed line) cavitation frequency bands for a given sample. Figure 5E quantifies the total integrated power for microbubble samples and PBS control, highlighting the virtual absence of inertial cavitation and a statistically significant 13.5 dB in stable cavitation activity ($p < 0.02$).

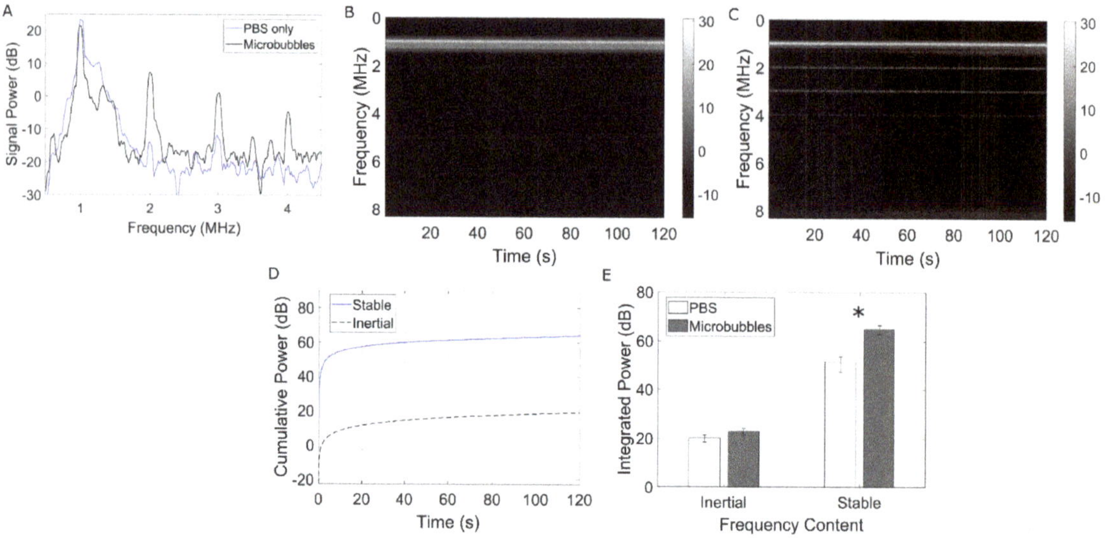

Figure 5. Microbubbles undergo stable cavitation during gene delivery treatment. (**A**) The microbubble echo spectrum at the beginning of treatment (black line) as compared to a PBS only control (blue line). (**B**) Time-frequency plots of the resulting cavitation echoes in PBS (control) and (**C**) miR-126 loaded microbubbles. This time-frequency depiction clearly shows prolonged harmonic content (stable cavitation) throughout the 2 min treatment, and the absence of any broadband spectral emissions (inertial cavitation). (**D**) Quantification of the cumulative spectral power from the stable (blue, solid line) and inertial (black, dashed-line) cavitation frequency bands over the total treatment time of a given sample. (**E**) Quantification over all samples compared to PBS only controls. Asterisks denote statistical significance ($p < 0.02$).

4. Discussion

In this study, we developed a methodology to deliver modest amounts of miR-126 to endothelial cells for the purpose of therapeutic angiogenesis. Our results demonstrate that miR-126 can be successfully delivered intercellularly via ultrasound and microbubbles in sufficient quantities to elicit downstream regulation of target proteins and a viable physiological response.

It is perhaps worth noting that one of the limitations of the present study is that the endothelial cells were treated in suspension, as opposed to within the more biologically relevant monolayer configuration. This could have negatively affected transfection efficiency because the cell morphology of these cells in suspension exhibited less surface area as opposed to their adherent morphology and, incidentally, were less likely to interact with a miR-126-decorated microbubble in the cell chamber. To address this in future work, another nonadherent cell line could be investigated, such a endothelial progenitor cells [47], or cells could be cultivated and treated under fluidic conditions more accurate mimic a vessel [48]. This arrangement, however, was chosen in order to position two co-aligned transducers in such a way as to be able to passively record the acoustic emissions during treatment (Figure 5). This is essential as it provides evidence of microbubble stable cavitation. Due to the specific ultrasound parameters employed (e.g., long pulses), this would have been challenging to conduct using a single transducer with active listening (pulse-echo).

Indeed, while designing the ultrasound regimen, we focused on maintaining cellular viability albeit at the expense of lower transfection efficiency in order to substantiate the feasibility of microbubble-mediated gene delivery as opposed to other gene delivery methods. For instance, in peptide-mediated miR-126 delivery to vascular endothelial cells, Zhou and colleagues reached a 3.5-fold increase in miR-126 but reported cell death up to 20% at the highest peptide dose [49]. Indeed, by devising a microbubble-mediated miR-126 delivery treatment scheme that minimizes cell death, this enables the potential for repeat treatments to further increase the transfection efficiency and enhance the effects of the genetic payload [18,50].

Microbubble-mediated gene therapy, whether employing miRs or other nucleic acid payloads, has been demonstrated in cancer and cardiovascular applications through cavitation regimes in which the bubbles undergo disruption (e.g., ultrasound-targeted microbubble disruption; UTMD [27]). Violent microbubble collapse initiated this way requires strong acoustic forcing conditions, typically large peak-negative pressures [29]. While this technique has shown promising results in gene deposition within the target tissue [51–53], the bubble involution process can initiate high-speed liquid jets along with other cavitation-based phenomena that may lead to localized endothelial bio-effects. Endothelial cell damage and denudation have been reported in both microvessels [54,55] and larger vasculature, events which have been correlated to excessive inertial cavitation doses [54] (a surrogate measure of inertial cavitation and bubble disruption [29]). Further, microvascular hemorrhage and red blood cell extravasation associated with violent microbubble collapse have also been observed [56–59], and recent reports have demonstrated the potential for inertial cavitation to trigger endothelial apoptosis pathways [60]. As a consequence, our treatment paradigm consists of a very low peak negative pressure and large number of cycles, so as to encourage miR payload release from the cationic microbubble surface by stable cavitation—similar in principle to other drug-loaded microbubbles [61]. Indeed, as gene delivery was achieved in the absence of broadband emissions and via these long duration pulses, our results suggest that sustained gentle microbubble vibrations, radiation forces, and agitation lead to delivery of miR-126, a combination that was persistent throughout the 2 min treatment time (see Figure 5). The hypothesized drug delivery mechanism here is two-stage. Firstly, long pulse-driven microbubbles can initiate prolonged lipid and cargo release as compared to those driven by shorter pulses at a given pressure amplitude [62]. Secondly, microbubble translation and constant agitation decrease the average microbubble–cell distance, which both aids in microbubble-mediated cell membrane perforation [63] and lowers the distance over which the free miR mimic is required to diffuse to enter the neighboring cell. It is also of interest to note that ultrasound has been shown to modulate endocytosis activity [64,65]; however given the timescales of the miR-delivery performed here, it is perhaps not likely to be the dominant mechanism [23].

5. Conclusions

We synthesized a miR-126-bearing microbubble agent with a gene loading capacity of 6 μg per 10^9 bubbles, characterized by a bubble concentration, size distribution, and stability similar to that of currently used clinical agents. Using a low-pressure, long-pulse acoustic regime, we were able to show delivery of up to 2.3× in miR-126 to endothelial cells compared to sham controls while maintaining cell viability, resulting in the expected physiological behavior, including downregulation of angiogenic suppressor proteins SPRED1 and PIK3R2. Furthermore, simultaneous passive cavitation detection confirms that this was a stable cavitation treatment, thus minimizing potential damage caused by more violent, inertial cavitation approaches.

Author Contributions: Conceptualization, S.H. and B.H.; writing—original draft preparation, S.H. and B.H.; writing—review and editing, S.H., D.S., H.Y., and B.H.; funding acquisition, B.H. All authors have read and agreed to the published version of the manuscript.

Funding: This research was funded by the Heart and Stroke Foundation of Canada, grant number G-18-0022133, the Burroughs Wellcome Fund, grant number 1018212.03, the NSERC, grant number RGPIN-2019-06969, and the Canada Research Chair program, grant number CRC-2018-00045.

Institutional Review Board Statement: Not applicable.

Informed Consent Statement: Not applicable.

Data Availability Statement: Not applicable.

Acknowledgments: All authors consented to this version of the manuscript. This work was partially supported by Concordia University, and the authors thank Elahe Memari for thoughtful discussions and John Capobianco for access to the zeta sizer.

Conflicts of Interest: The authors declare no conflict of interest.

References

1. Cannatà, A.; Ali, H.; Sinagra, G.; Giacca, M. Gene Therapy for the Heart Lessons Learned and Future Perspectives. *Circ. Res.* **2020**, *126*, 1394–1414. [CrossRef]
2. Goswami, R.; Subramanian, G.; Silayeva, L.; Newkirk, I.; Doctor, D.; Chawla, K.; Chattopadhyay, S.; Chandra, D.; Chilukuri, N.; Betapudi, V. Gene therapy leaves a vicious cycle. *Front. Oncol.* **2019**, *9*, 1–25. [CrossRef] [PubMed]
3. Lee, R.C.; Feinbaum, R.L.; Ambros, V. The C. elegans Heterochronic Gene lin-4 Encodes Small RNAs with Antisense Complementarity to lin-14. *Cell* **1993**, *75*, 843–854. [CrossRef]
4. Chen, W.; Hu, Y.; Ju, D. Gene therapy for neurodegenerative disorders: Advances, insights and prospects. *Acta Pharm. Sin. B* **2020**, *10*, 1347–1359. [CrossRef] [PubMed]
5. Forterre, A.; Komuro, H.; Aminova, S.; Harada, M. A comprehensive review of cancer microRNA therapeutic delivery strategies. *Cancers* **2020**, *12*, 1852. [CrossRef]
6. Kalayinia, S.; Arjmand, F.; Maleki, M.; Malakootian, M.; Singh, C.P. MicroRNAs: Roles in cardiovascular development and disease. *Cardiovasc. Pathol.* **2021**, *50*, 107296. [CrossRef]
7. Rupaimoole, R.; Slack, F.J. MicroRNA therapeutics: Towards a new era for the management of cancer and other diseases. *Nat. Rev. Drug Discov.* **2017**, *16*, 203–221. [CrossRef] [PubMed]
8. Rawat, M.; Kadian, K.; Gupta, Y.; Kumar, A.; Chain, P.S.G.; Kovbasnjuk, O.; Kumar, S.; Parasher, G. MicroRNA in pancreatic cancer: From biology to therapeutic potential. *Genes* **2019**, *10*, 752. [CrossRef]
9. Imani, S.; Wu, R.C.; Fu, J. MicroRNA-34 family in breast cancer: From research to therapeutic potential. *J. Cancer* **2018**, *9*, 3765–3775. [CrossRef]
10. Iqbal, M.A.; Arora, S.; Prakasam, G.; Calin, G.A.; Syed, M.A. MicroRNA in lung cancer: Role, mechanisms, pathways and therapeutic relevance. *Mol. Aspects Med.* **2019**, *70*, 3–20. [CrossRef]
11. Balatti, V.; Croce, C.M. MicroRNA dysregulation and multi-targeted therapy for cancer treatment. *Adv. Biol. Regul.* **2020**, *75*, 100669. [CrossRef] [PubMed]
12. Karakikes, I.; Chaanine, A.H.; Kang, S.; Mukete, B.N.; Jeong, D.; Zhang, S.; Hajjar, R.J.; Lebeche, D. Therapeutic cardiac-targeted delivery of miR-1 reverses pressure overload-induced cardiac hypertrophy and attenuates pathological remodeling. *J. Am. Heart Assoc.* **2013**, *2*, 17–19. [CrossRef] [PubMed]
13. Cheng, W.L.; Kao, Y.H.; Chao, T.F.; Lin, Y.K.; Chen, S.A.; Chen, Y.J. MicroRNA-133 suppresses ZFHX3-dependent atrial remodelling and arrhythmia. *Acta Physiol.* **2019**, *227*, e13322. [CrossRef] [PubMed]
14. Li, P.F.; He, R.H.; Shi, S.B.; Li, R.; Wang, Q.T.; Rao, G.T.; Yang, B. Modulation of miR-10a-mediated TGF-β1/Smads signaling affects atrial fibrillation-induced cardiac fibrosis and cardiac fibroblast proliferation. *Biosci. Rep.* **2019**, *39*, BSR20181931. [CrossRef]
15. Staruschenko, A.; Lan, H.Y.; Jia, Z.; Teng, J.; Song, N.; Zhang, T.; Xu, X.; Lu, Z.; Yu, X.; Fang, Y.; et al. miR-21 Protects Against Ischemia/Reperfusion-Induced Acute Kidney Injury by Preventing Epithelial Cell Apoptosis and Inhibiting Dendritic Cell Maturation. *Front. Physiol.* **2018**, *1*, 790. [CrossRef]
16. Xiao, Z.H.; Wang, L.; Gan, P.; He, J.; Yan, B.C.; Ding, L.D. Dynamic Changes in miR-126 Expression in the Hippocampus and Penumbra Following Experimental Transient Global and Focal Cerebral Ischemia–Reperfusion. *Neurochem. Res.* **2020**, *45*, 1107–1119. [CrossRef]
17. Qu, M.; Pan, J.; Wang, L.; Zhou, P.; Song, Y.; Wang, S.; Jiang, L.; Geng, J.; Zhang, Z.; Wang, Y.; et al. MicroRNA-126 Regulates Angiogenesis and Neurogenesis in a Mouse Model of Focal Cerebral Ischemia. *Mol. Ther.-Nucleic Acids* **2019**, *16*, 15–25. [CrossRef]
18. Chen, H.H.; Matkar, P.N.; Afrasiabi, K.; Kuliszewski, M.A.; Leong-Poi, H. Prospect of ultrasound-mediated gene delivery in cardiovascular applications. *Expert Opin. Biol. Ther.* **2017**, *16*, 815–826. [CrossRef]
19. Yusefi, H.; Helfield, B. Ultrasound Contrast Imaging: Fundamentals and Emerging Technology. *Front. Phys.* **2022**, *10*, 1–16. [CrossRef]
20. He, S.; Singh, D.; Helfield, B. An Overview of Cell Membrane Perforation and Resealing Mechanisms for Localized Drug Delivery. *Pharmaceutics* **2022**, *14*, 886. [CrossRef]

21. Kooiman, K.; Roovers, S.; Langeveld, S.A.G.; Kleven, R.T.; Dewitte, H.; O'Reilly, M.A.; Escoffre, J.M.; Bouakaz, A.; Verweij, M.D.; Hynynen, K.; et al. Ultrasound-Responsive Cavitation Nuclei for Therapy and Drug Delivery. *Ultrasound Med. Biol.* **2020**, *46*, 1296–1325. [CrossRef] [PubMed]
22. Helfield, B.; Chen, X.; Watkins, S.C.; Villanueva, F.S. Transendothelial Perforations and the Sphere of Influence of Single-Site Sonoporation. *Ultrasound Med. Biol.* **2020**, *46*, 1686–1697. [CrossRef]
23. Qin, P.; Han, T.; Yu, A.C.H.; Xu, L. Mechanistic understanding the bioeffects of ultrasound-driven microbubbles to enhance macromolecule delivery. *J. Control Release* **2018**, *272*, 169–181. [CrossRef] [PubMed]
24. Hernot, S.; Klibanov, A.L. Microbubbles in ultrasound-triggered drug and gene delivery. *Adv. Drug Deliv. Rev.* **2008**, *60*, 1153–1166. [CrossRef] [PubMed]
25. Wang, D.S.; Panje, C.; Pysz, M.A.; Paulmurugan, R.; Rosenberg, J.; Gambhir, S.S.; Schneider, M.; Willmann, J.K. Cationic versus neutral microbubbles for ultrasound-mediated gene delivery in cancer. *Radiology* **2012**, *264*, 721–732. [CrossRef]
26. Zhou, Q.; Deng, Q.; Hu, B.O.; Wang, Y.J.; Chen, J.L.; Cui, J.J.; Cao, S.; Song, H.N. Ultrasound combined with targeted cationic microbubble-mediated angiogenesis gene transfection improves ischemic heart function. *Exp. Ther. Med.* **2017**, *13*, 2293–2303. [CrossRef]
27. Carson, A.R.; McTiernan, C.F.; Lavery, L.; Grata, M.; Leng, X.; Wang, J.; Chen, X.; Villanueva, F.S. Ultrasound-targeted microbubble destruction to deliver siRNA cancer therapy. *Cancer Res.* **2012**, *72*, 6191–6199. [CrossRef]
28. Schwartz, M.R.; Debski, A.C.; Price, R.J. Ultrasound-targeted nucleic acid delivery for solid tumor therapy. *J. Control Release* **2021**, *339*, 531–546. [CrossRef]
29. Leighton, T.G. *The Acoustic Bubble*; Academic Press: London, UK, 1994.
30. Wischhusen, J.C.; Chowdhury, S.M.; Lee, T.; Wang, H.; Bachawal, S.; Devulapally, R.; Afjei, R.; Sukumar, U.K.; Paulmurugan, R. Ultrasound-mediated delivery of miRNA-122 and anti-miRNA-21 therapeutically immunomodulates murine hepatocellular carcinoma in vivo. *J. Control Release* **2020**, *321*, 272–284. [CrossRef]
31. Kopechek, J.A.; McTiernan, C.F.; Chen, X.; Zhu, J.; Mburu, M.; Feroze, R.; Whitehurst, D.A.; Lavery, L.; Cyriac, J.; Villanueva, F.S. Ultrasound and microbubble-targeted delivery of a microRNA inhibitor to the heart suppresses cardiac hypertrophy and preserves cardiac function. *Theranostics* **2019**, *9*, 7088–7098. [CrossRef]
32. Cao, W.J.; Rosenblat, J.D.; Roth, N.C.; Kuliszewski, M.A.; Matkar, P.N.; Rudenko, D.; Liao, C.; Lee, P.J.H.; Leong-Poi, H. Therapeutic Angiogenesis by Ultrasound-Mediated MicroRNA-126-3p Delivery. *Arterioscler. Thromb. Vasc. Biol.* **2015**, *37*, 2401–2411. [CrossRef] [PubMed]
33. Zhou, X.; Liu, H.; Pang, Y.; Wang, M.; Liu, S. UTMD-mediated delivery of miR-21-5p inhibitor suppresses the development of lung cancer. *Tissue Cell* **2022**, *74*, 101719. [CrossRef]
34. Li, Y.; Du, M.; Fang, J.; Zhou, J.; Chen, Z. UTMD promoted local delivery of miR-34a-mimic for ovarian cancer therapy. *Drug Deliv.* **2021**, *28*, 1616–1625. [CrossRef] [PubMed]
35. Yi, L.; Chen, Y.; Jin, Q.; Deng, C.; Wu, Y.; Li, H.; Liu, T.; Li, Y.; Yang, Y.; Wang, J.; et al. Antagomir-155 Attenuates Acute Cardiac Rejection Using Ultrasound Targeted Microbubbles Destruction. *Adv. Healthc. Mater.* **2020**, *9*, 2000189. [CrossRef]
36. van Rooij, T.; Skachkov, I.; Beekers, I.; Lattwein, K.R.; Voorneveld, J.D.; Kokhuis, T.J.A.; Bera, D.; Luan, Y.; van der Steen, A.F.W.; de Jong, N.; et al. Viability of endothelial cells after ultrasound-mediated sonoporation: Influence of targeting, oscillation, and displacement of microbubbles. *J. Control Release* **2016**, *238*, 197–211. [CrossRef] [PubMed]
37. Lentacker, I.; De Cock, I.; Deckers, R.; De Smedt, S.C.; Moonen, C.T.W. Understanding ultrasound induced sonoporation: Definitions and underlying mechanisms. *Adv. Drug Deliv. Rev.* **2014**, *72*, 49–64. [CrossRef]
38. Christiansen, J.P.; French, B.A.; Klibanov, A.L.; Kaul, S.; Lindner, J.R. Targeted tissue transfection with ultrasound destruction of plasmid-bearing cationic microbubbles. *Ultrasound Med. Biol.* **2003**, *29*, 1759–1767. [CrossRef]
39. Helfield, B.L.; Huo, X.; Williams, R.; Goertz, D.E. The effect of preactivation vial temperature on the acoustic properties of Definity^{TM}. *Ultrasound Med. Biol.* **2012**, *38*, 1298–1305. [CrossRef]
40. Petrov, A.; Tsa, A.; Puglisi, J.D. *Analysis of RNA by Analytical Polyacrylamide Gel Electrophoresis*, 1st ed.; Elsevier Inc.: Amsterdam, The Netherlands, 2013; Volume 530, ISBN 9780124200371.
41. Montoya-Zegarra, J.A.; Russo, E.; Runge, P.; Jadhav, M.; Willrodt, A.H.; Stoma, S.; Nørrelykke, S.F.; Detmar, M.; Halin, C. AutoTube: A novel software for the automated morphometric analysis of vascular networks in tissues. *Angiogenesis* **2019**, *22*, 223–236. [CrossRef]
42. Zitnick, C.L.; Dollár, P. Edge boxes: Locating object proposals from edges. In Proceedings of the European Conference on Computer Vision, Zurich, Switzerland, 6–12 September 2014; Volume 8693, pp. 391–405. [CrossRef]
43. Dollar, P.; Zitnick, C.L. Structured forests for fast edge detection. In Proceedings of the IEEE International Conference on Computer Vision, Sydney, Australia, 1–8 December 2013; pp. 1841–1848. [CrossRef]
44. Hyvelin, J.M.; Gaud, E.; Costa, M.; Helbert, A.; Bussat, P.; Bettinger, T.; Frinking, P. Characteristics and Echogenicity of Clinical Ultrasound Contrast Agents: An in Vitro and in Vivo Comparison Study. *J. Ultrasound Med.* **2017**, *36*, 941–953. [CrossRef]
45. Shi, W.T.; Forsberg, F. Ultrasonic characterization of the nonlinear properties of contrast microbubbles. *Ultrasound Med. Biol.* **2000**, *26*, 93–104. [CrossRef] [PubMed]
46. Frinking, P.; Segers, T.; Luan, Y.; Tranquart, F. Three Decades of Ultrasound Contrast Agents: A Review of the Past, Present and Future Improvements. *Ultrasound Med. Biol.* **2020**, *46*, 892–908. [CrossRef] [PubMed]

47. Kong, Z.; Wang, Y.; Zhang, Y.; Shan, W.; Wu, J.; Wang, Q. MicroRNA-126 promotes endothelial progenitor cell proliferation and migration ability via the Notch pathway. *Cardiovasc. Diagn. Ther.* **2020**, *10*, 490–499. [CrossRef] [PubMed]
48. Schober, A.; Nazari-Jahantigh, M.; Wei, Y.; Bidzhekov, K.; Gremse, F.; Grommes, J.; Megens, R.T.A.; Heyll, K.; Noels, H.; Hristov, M.; et al. MicroRNA-126-5p promotes endothelial proliferation and limits atherosclerosis by suppressing Dlk1. *Nat. Med.* **2014**, *20*, 368–376. [CrossRef]
49. Zhou, F.; Jia, X.; Yang, Q.; Yang, Y.; Zhao, Y.; Fan, Y.; Yuan, X. Targeted delivery of microRNA-126 to vascular endothelial cells: Via REDV peptide modified PEG-trimethyl chitosan. *Biomater. Sci.* **2016**, *4*, 849–856. [CrossRef]
50. Fujii, H.; Li, S.H.; Wu, J.; Miyagi, Y.; Yau, T.M.; Rakowski, H.; Egashira, K.; Guo, J.; Weisel, R.D.; Li, R.K. Repeated and targeted transfer of angiogenic plasmids into the infarcted rat heart via ultrasound targeted microbubble destruction enhances cardiac repair. *Eur. Heart J.* **2011**, *32*, 2075–2084. [CrossRef]
51. Kaul, S. Myocardial contrast echocardiography—A 25-year retrospective. *Circulation* **2008**, *118*, 291–308. [CrossRef]
52. Su, J.; Wang, J.; Luo, J.; Li, H. Ultrasound-mediated destruction of vascular endothelial growth factor (VEGF) targeted and paclitaxel loaded microbubbles for inhibition of human breast cancer cell MCF-7 proliferation. *Mol. Cell. Probes* **2019**, *46*, 101415. [CrossRef]
53. Jing, Y.; Xiu-Juan, Z.; Hong-Jiao, C.; Zhi-Kui, C.; Qing-Fu, Q.; En-Sheng, X.; Li-Wu, L. Ultrasound-targeted microbubble destruction improved the antiangiogenic effect of Endostar in triple-negative breast carcinoma xenografts. *J. Cancer Res. Clin. Oncol.* **2019**, *145*, 1191–1200. [CrossRef]
54. Hwang, J.H.; Tu, J.; Brayman, A.A.; Matula, T.J.; Crum, L.A. Correlation between inertial cavitation dose and endothelial cell damage in vivo. *Ultrasound Med. Biol.* **2006**, *32*, 1611–1619. [CrossRef]
55. Hwang, J.H.; Brayman, A.A.; Reidy, M.A.; Matula, T.J.; Kimmey, M.B.; Crum, L.A. Vascular effects induced by combined 1-MHz ultrasound and microbubble contrast agent treatments in vivo. *Ultrasound Med. Biol.* **2005**, *31*, 553–564. [CrossRef] [PubMed]
56. Skyba, D.M.; Price, R.J.; Linka, A.Z.; Skalak, T.C.; Kaul, S. Direct in vivo visualization of intravascular destruction of microbubbles by ultrasound and its local effects on tissue. *Circulation* **1998**, *98*, 290–293. [CrossRef] [PubMed]
57. Miller, D.L.; Quddus, J. Diagnostic ultrasound activation of contrast agent gas bodies induces capillary rupture in mice. *Proc. Natl. Acad. Sci. USA* **2000**, *97*, 10179–10184. [CrossRef] [PubMed]
58. Li, P.; Cao, L.Q.; Dou, C.Y.; Armstrong, W.F.; Miller, D. Impact of myocardial contrast echocardiography on vascular permeability: An in vivo dose response study of delivery mode, pressure amplitude and contrast dose. *Ultrasound Med. Biol.* **2003**, *29*, 1341–1349. [CrossRef]
59. Price, R.J.; Skyba, D.M.; Kaul, S.; Skalak, T.C. Delivery of colloidal particles and red blood cells to tissue through microvessel ruptures created by targeted microbubble destruction with ultrasound. *Circulation* **1998**, *98*, 1264–1267. [CrossRef]
60. El Kaffas, A.; Czarnota, G.J. Biomechanical effects of microbubbles: From radiosensitization to cell death. *Futur. Oncol.* **2015**, *11*, 1093–1108. [CrossRef]
61. Yu, F.T.H.; Chen, X.; Wang, J.; Qin, B.; Villanueva, F.S. Low Intensity Ultrasound Mediated Liposomal Doxorubicin Delivery Using Polymer Microbubbles. *Mol. Pharm.* **2016**, *13*, 55–64. [CrossRef]
62. Luan, Y.; Lajoinie, G.; Gelderblom, E.; Skachkov, I.; van der Steen, A.F.W.; Vos, H.J.; Versluis, M.; De Jong, N. Lipid shedding from single oscillating microbubbles. *Ultrasound Med. Biol.* **2014**, *40*, 1834–1846. [CrossRef]
63. Qin, P.; Xu, L.; Han, T.; Du, L.; Yu, A.C.H. Effect of non-acoustic parameters on heterogeneous sonoporation mediated by single-pulse ultrasound and microbubbles. *Ultrason. Sonochem.* **2016**, *31*, 107–115. [CrossRef]
64. Fekri, F.; Delos Santos, R.C.; Karshafian, R.; Antonescu, C.N. Ultrasound microbubble treatment enhances clathrin-mediated endocytosis and fluid-phase uptake through distinct mechanisms. *PLoS ONE* **2016**, *11*, e0156754. [CrossRef]
65. Meijering, B.D.M.; Juffermans, L.J.M.; Van Wamel, A.; Henning, R.H.; Zuhorn, I.S.; Emmer, M.; Versteilen, A.M.G.; Paulus, W.J.; Van Gilst, W.H.; Kooiman, K.; et al. Ultrasound and microbubble-targeted delivery of macromolecules is regulated by induction of endocytosis and pore formation. *Circ. Res.* **2009**, *104*, 679–687. [CrossRef] [PubMed]

pharmaceutics

MDPI

Article

Sonoporation of the Round Window Membrane on a Sheep Model: A Safety Study

Sandrine Kerneis [1,*], Jean-Michel Escoffre [2,*], John J. Galvin III [3,4], Ayache Bouakaz [2], Antoine Presset [2], Corentin Alix [2], Edward Oujagir [2], Antoine Lefèvre [2], Patrick Emond [2,3,5], Hélène Blasco [2,3,5] and David Bakhos [1,2,3,4,*]

1 ENT and Cervico-Facial Surgery Department, University Hospital Center of Tours, 2 Boulevard Tonnellé, 37044 Tours, France
2 UMR 1253, iBrain, Inserm, Université de Tours, 10 Boulevard Tonnellé, 37044 Tours, France
3 Faculty of Medicine, Université de Tours, 10 Boulevard Tonnellé, 37000 Tours, France
4 House Institute Foundation, 2100 W 3rd Street, Suite 111, Los Angeles, CA 90057, USA
5 Department of Biochemistry and Molecular Biology, University Hospital Center of Tours, 2 Boulevard Tonnellé, 37044 Tours, France
* Correspondence: sandrine.kerneis@gmail.com (S.K.); jean-michel.escoffre@univ-tours.fr (J.-M.E.); david.bakhos@univ-tours.fr (D.B.); Tel.: +33-247-474-785 (S.K. & D.B.); Fax: +33-247-473-600 (S.K. & D.B.)

Abstract: Sonoporation using microbubble-assisted ultrasound increases the permeability of a biological barrier to therapeutic molecules. Application of this method to the round window membrane could improve the delivery of therapeutics to the inner ear. The aim of this study was to assess the safety of sonoporation of the round window membrane in a sheep model. To achieve this objective, we assessed auditory function and cochlear heating, and analysed the metabolomics profiles of perilymph collected after sonoporation, comparing them with those of the control ear in the same animal. Six normal-hearing ewes were studied, with one sonoporation ear and one control ear for each. A mastoidectomy was performed on both ears. On the sonoporation side, Vevo MicroMarker® microbubbles (MBs; VisualSonics—Fujifilm, Amsterdam, The Netherlands) at a concentration of 2×10^8 MB/mL were locally injected into the middle ear and exposed to 1.1 MHz sinusoidal ultrasonic waves at 0.3 MPa negative peak pressure with 40% duty cycle and 100 µs interpulse period for 1 min; this was repeated three times with 1 min between applications. The sonoporation protocol did not induce any hearing impairment or toxic overheating compared with the control condition. The metabolomic analysis did not reveal any significant metabolic difference between perilymph samples from the sonoporation and control ears. The results suggest that sonoporation of the round window membrane does not cause damage to the inner ear in a sheep model.

Keywords: sonoporation; round window membrane; metabolomics; perilymph; inner ear

Citation: Kerneis, S.; Escoffre, J.-M.; Galvin, J.J., III; Bouakaz, A.; Presset, A.; Alix, C.; Oujagir, E.; Lefèvre, A.; Emond, P.; Blasco, H.; et al. Sonoporation of the Round Window Membrane on a Sheep Model: A Safety Study. *Pharmaceutics* **2023**, *15*, 442. https://doi.org/10.3390/pharmaceutics15020442

Academic Editor: Kwang-Hyeon Liu

Received: 26 November 2022
Revised: 23 January 2023
Accepted: 26 January 2023
Published: 29 January 2023

1. Introduction

Inner ear disorders are responsible for hearing loss, tinnitus, or vertigo. These pathologies are common and can severely impact patients' health. Hearing loss is the second leading cause of disability worldwide after depression [1] and is the leading modifiable risk factor for dementia [2]. The delivery of therapeutic drugs to the inner ear to treat these pathologies is challenging. The inner ear is a fragile organ, encircled by a dense bone, which offers limited access. Systemic delivery of therapeutic drugs is not an optimal administration route [3]. Indeed, its effectiveness is limited by the presence of the blood-cochlear barrier, the high density of the peri-cochlear bone, and poor peri-cochlear vascularization. Moreover, this administration route of therapeutics induces side effects, including, for example, systemic administration of corticosteroids for diabetic patients.

Intratympanic injection allows for better delivery of therapeutic drugs to the inner ear [4,5] and results in fewer side effects compared with systemic delivery. Molecules injected in the middle ear diffuse to the inner ear, mainly through the round window membrane (RWM), and secondarily through the annular ligament, around the stapes [6,7]. The intratympanic approach is commonly used to treat several diseases. For example, intratympanic injections of gentamicin [8] and corticosteroids [9,10] are used to treat Meniere's disease. Intratympanic injection of corticosteroids is also used to treat sudden sensorineural hearing loss [11,12], acoustic shock, tinnitus, and auto-immune deafness [13,14]. However, the effectiveness of this approach is variable because the intratympanic delivery of drugs is not well controlled. More efficient and targeted delivery methods are required to increase the local concentration of antibiotics and corticosteroids in the endo-cochlear fluids while reducing off-target effects.

Alternative methods have been designed and validated to improve the delivery of therapeutic drugs through the RWM to the hair cells of the inner ear and the spiral ganglion, such as drug-loaded nanoparticles [15–17], liposomal gels [18], and extracellular vesicles [19]. Drug-eluting coating on cochlear implant electrodes [20], as well as perfusion systems with micro-pumps or reservoirs connected directly into cochlear implants, are also under investigation [21,22]. Other therapeutic strategies rely on the transduction of therapeutic genes into target tissues using viral vectors [23]. To date, these approaches are being validated in preclinical models or even undergoing clinical validation but are not yet recommended to treat patients.

Recently, sonoporation-mediated drug delivery has shown great promise in improving the efficacy of therapeutic drugs (e.g., chemotherapeutic drugs, therapeutic antibodies, antibiotics, nucleic acids, etc.) by enhancing their local deposition and reducing off-target effects [24]. Sonoporation uses a combination of high-frequency ultrasound (1–10 MHz) and ultrasound contrast agents (i.e., gas microbubbles (MBs)). The acoustically mediated volumetric oscillations of the MBs induce several local acoustic events near the biological barriers (e.g., cell plasma membrane, endothelial barriers, etc.), thereby enhancing their permeability for therapeutic molecules in a transient, localised, and non-invasive manner. As such, sonoporation improves intra-tissue bioavailability, thereby increasing therapeutic efficacy [25].

Recently, the permeabilization of RWM to fluorescent markers (Biotin-FITC, Texas-red-labelled gentamicin) or therapeutic molecules (plasmid DNA, dexamethasone, and insulin growth factor 1) using sonoporation has been successfully reported in vitro and in guinea pig models [26–32]. These studies described acoustic protection when applying dexamethasone [28,31] and a better uptake by vestibular hair cells when applying gentamicin [29]. The efficiency of acoustically mediated drug delivery depends on the application route of ultrasound (e.g., transcanal or transcranial) and ultrasound parameters, including the central frequency, the duty cycle, the acoustic intensities, and the exposure time [29,30]. No auditory damage or irreversible histological changes to the RWM related to sonoporation have been reported [26,28–32]. In addition, no acoustically induced thermal toxicity (threshold of 1.5 °C [33]) has been observed when the ultrasound parameters were perfectly controlled [29,30]. However, metabolomics profiles of perilymph were not used in these previous studies to confirm the absence of ototoxicity. Furthermore, while these studies demonstrated that sonoporation could enhance the passage of therapeutic molecules to the inner ear, guinea pig models were used. The guinea pig has substantial limitations as a model of the human ear and RWM, as the RWM thickness is 10–15 μm [34], compared with 70 μm in humans [35]. Sheep may be a more relevant large animal model for studies of the human auditory system. The anatomical structures of the sheep's inner, middle, and outer ears are comparable with those of humans [36–39]. Similarities between the sheep ear and the human ear have also been found in terms of the histological composition of temporal bone structures [40], the anatomy and histology of the RWM [41], the auditory spectrum [41], and the recording of auditory brainstem responses (ABRs) [42].

The objective of the present study was to assess the safety of sonoporation of the inner ear in a sheep model. To achieve this objective, we assessed auditory function and cochlear heating and analysed metabolomics profiles of perilymph and plasma with and without sonoporation of RWM.

2. Material and Methods

2.1. Animals and Housing

All six sheep were of the Ile de France breed (PIXANIM platform, INRAE, Nouzilly, Centre Val-de-Loire, France). The mean age was 28 ± 4 months, and the mean weight was 60 ± 5 kg. Only female sheep (ewes) were used because their temporal bone is thinner than that of rams, which facilitates drilling of the temporal bone. The auditory cortex in sheep is mature by 12 months. For the reason that we expected large inter-individual variability in the perilymph used for metabolomic analysis [43], each sheep served as its own control, with one sonoporation ear and one control ear; sonoporation and control ears were randomised across sheep. The number of sheep (n = 6) allowed for sufficient statistical power for the metabolomic study. Indeed, this number has been shown to be sufficient to demonstrate metabolomic changes in the perilymph following acoustic trauma in sheep [44].

Twenty-four hours before the experiment, the sheep were isolated and fasted. The animal study was reviewed and approved by the Animal Care and Regional Committee for Ethics in Animal Experiments, Centre Val-de-Loire (APAFiS code #33161-2021091518241644) in accordance with European Directive 2010/63/EU for animal experiments.

2.2. Sonoporation Protocol

The general procedure is presented in Figure 1. In the protocol, the six sheep were put under general anesthesia, and plasma samples were collected. Furthermore, verification of normal-hearing (NH) status was confirmed using ABRs. For the sonoporation ear, a mastoidectomy was performed, and the promontory temperature was measured before sonoporation. Furthermore, sonoporation of ultrasound-activated MBs was performed in the middle ear. Promontory temperature and hearing thresholds were re-measured after sonoporation. Finally, perilymph and plasma were collected after sonoporation. For the control ear, a mastoidectomy was performed and perilymph was collected. At the end of the experiment, all the sheep were sacrificed. The perilymph and plasma samples were then subjected to metabolomics analysis.

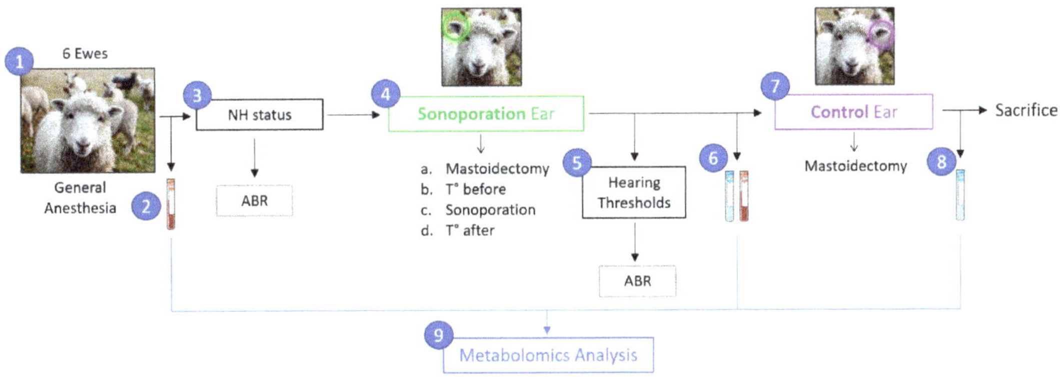

Figure 1. Study protocol. (1) 6 sheep under general anesthesia. (2) Plasma sample after anesthesia (before sonoporation). (3) Verification of NH status using ABR. (4) Sonoporation ear: a. Mastoidectomy, b. Measure of promontory temperature before sonoporation, c. Sonoporation of ultrasound activated MBs injected into the middle ear, and d. Verification of promontory temperature after sonoporation. (5) Verification of hearing thresholds. (6) Perilymph and plasma sampling. (7) Control ear: mastoidectomy. (8) Perilymph sampling before sacrifice of the animal. (9) Metabolomic analysis of the samples.

2.2.1. Surgery

Under general anesthesia (inhalation of isoflurane at 3% and intraveinous ketamine injection at 10 mg/kg), the sheep were intubated and placed in the lateral decubitus position. A surgery similar to a human mastoidectomy was performed. The mastoid of the ewe is unventilated and does not contain mastoid cells. The mastoid was milled following the external auditory canal, taking care not to damage the lateral sinus posteriorly or the temporal meninge superiorly. Once the middle ear was reached, the ossicular chain was respected, and the milling continued posteroinferiorly to open the hypotympanic bulla and to reveal the RWM. Direct visualisation was possible by sacrificing the facial nerve and the pyramid (Figure 2).

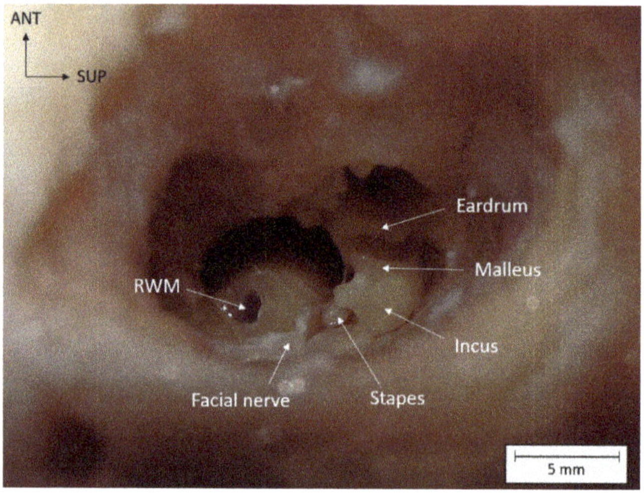

Figure 2. Visualization of the round window membrane (RWM) and anatomical structures of the middle ear after mastoidectomy on a left ewe ear. Spatial orientation is shown with anterior (ANT) and superior (SUP) localizations. mm: millimeters.

2.2.2. Ultrasound Treatment

As previously described [45], Vevo MicroMarker® MBs (VisualSonics—Fujifilm, Amsterdam, the Netherlands) are composed of a mixture of nitrogen and perfluorobutane gases encapsulated in a thin and flexible monolayer of phospholipids. MBs were prepared according to the manufacturer's instructions (with 1 min of hand stirring and 10 min of resting at room temperature) and then diluted 10 times to reach a concentration of 2×10^8 MB/mL. This MB concentration is similar to that exploited in previous sonoporation studies of the inner ear, which used phospholipid MBs (SonoVue®, Bracco Group, Milan, Italy) [28,30,32]. A solution of MBs with an average volume of 886 ± 421 µL (range: 420–1680 µL, depending on the size of the mastoidectomy) was injected into the middle ear. No leakage of MBs through the eustachian tube was observed under an optical microscope before or during sonoporation.

Ultrasound waves were generated using a single-element lab-made transducer with a center frequency of 1.1 MHz, which was introduced into the mastoidectomy cavity. The transducer had a diameter of 20 mm and was naturally focused at 30 mm. The transducer was connected to a degassed, water-filled cone to ensure coupling with the RWM (Figure 3).

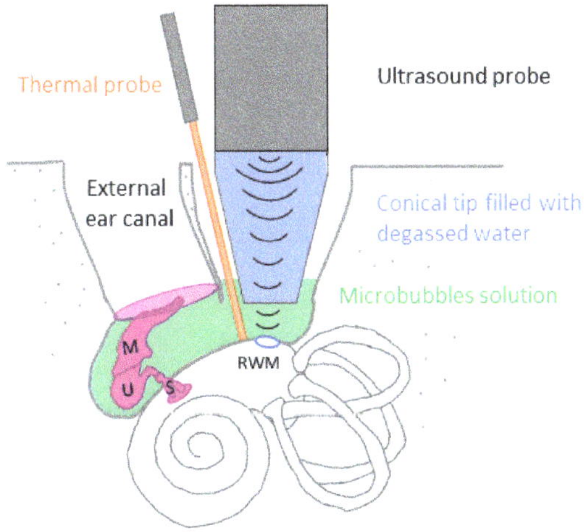

Figure 3. Diagram of the sonoporation of RWM. After surgical mastoidectomy, MBs solution (in green) was injected in the middle ear and ultrasound were applied using an ultrasound probe connected to a degassed water-filled cone. Promontory temperature was measured before and after sonoporation using a thermal probe (in orange). M: Malleus, U: Uncus, S: Stapes, RWM: Round window membrane.

The RWM was exposed to 40 cycles of 1.1 MHz sinusoidal ultrasound waves repeated every 100 μs (i.e., 40% duty cycle) at a peak-negative pressure of 300 kPa for 1 min. This ultrasound sequence was repeated three times, with 1 min between each application to prevent tissue heating. No stirring method was used during sonoporation. For the control ear, MBs were not injected, and ultrasound waves were not applied. At the end of the procedure, the animals were euthanised by an intravenous injection of a lethal dose of Doléthal® (15–20 mL; Laboratoire Vetoquinol, Lure, France).

2.3. Assessment of Auditory Function

Auditory thresholds were assessed using ABRs (NavPRO ONE Bio-logic®, Otometrics, Natus Medical Inc., Middleton, WI, USA), with stimulation via a B71W bone transducer and responses recorded via subcutaneous needle electrodes. The stimuli were clicks at a modulation frequency of 30.7 Hz. Auditory thresholds were measured from 50 dB (transducer's limit) to 20 dB, with 40 dB of contralateral air-conduction masking. Response waves I–IV were obtained. The threshold of disappearance of wave IV was measured, and wave IV latencies were calculated. Note that no wave V is found in sheep. ABRs were recorded bilaterally at the beginning of the procedure and after sonoporation in the sonoporation ear.

2.4. Temperature Measurement

A Dotsmann® P655-LOG digital thermometer with a Pt100 thermocouple was used to measure cochlear heating. The device was calibrated according to the manufacturer's instructions and has an accuracy of ±0.05 °C from −200 °C to +200 °C. The thermocouple was placed in contact with the promontory to measure cochlear heating. The thermal measurement was performed twice on the sonoporation ear: (1) after mastoidectomy but before sonoporation and (2) immediately after sonoporation and aspiration of the MB solution (i.e., 5–10 s after sonoporation ended). With this method, the ultrasound beam did not interact with the thermocouple, inducing no artifact upon the temperature measurement. The light of the surgical microscope was turned off at the end of the mastoidectomy and after MB aspiration to limit artificial overheating.

2.5. Metabolomic Analysis

2.5.1. Sample Collection

Perilymph sampling was performed for the sonoporation and control ears through the RWM. After incision of the RWM with a 22G needle, the perilymph was collected by capillary action (Microcap 15 μL, length 54 mm, Drummond®) to obtain at least a volume of 1 μL. A 5 mL blood sample was collected after induction of anesthesia (before sonoporation) and a second sample was collected immediately after the sonoporation procedure. The samples were collected in heparinised tubes and immediately centrifuged (1509× g at 4 °C for 13 min), and the plasma was collected. All samples were collected in 2 mL polypropylene tubes and placed on dry ice before being stored at −80 °C. The procedures were all performed in the morning to avoid a possible bias related to the nychthemeral cycle of the perilymph composition.

2.5.2. Sample Analysis

The samples were prepared as described by Mavel et al. [46]. Briefly, a volume of 400 μL of methanol is added to 50 μL of perilymph or plasma in order to extract metabolites. The samples were stirred for 5 sec and then incubated at −20 °C for 30 min to deproteinise them. After centrifugation (4193× g at 4 °C for 25 min), the supernatant (350 μL) was collected and evaporated with a SpeedVac concentrator (Labconco Corporation, Kansas City, MO, USA) at 40 °C. The dry residue was resuspended in 100 μL of methanol/water (75/25) and then 5 μL of the sample was injected into the liquid chromatography column for analysis.

As described by Diémé et al. [47], the analyses were performed on an UPLC Ultimate WPS-3000 system (Dionex, Germering, Germany) coupled to a Q-Exactive mass spectrometer (Thermo Fisher Scientific, Bremen, Germany). Liquid chromatography (LC) was performed at 40 °C using two modalities: HILIC (Hydrophilic Interaction Liquid Chromatography) and C18, using complementary stationary phases. Ionisation was performed in the positive electrospray mode (ESI+) for C18 and HILIC and the negative electrospray mode (ESI-) for C18, thus giving results in three modalities. Two mobile phases were used, and the chromatographic gradients were run at a flow rate of 0.3 mL/min. During acquisition, the instrument was operated at a resolution of 70,000 (m/z = 200). The successive use of LC and High-Resolution Mass Spectrometry (HRMS) allowed the separation of compounds from each sample according to their retention time in the LC column for each of the three modalities and according to their molecular weight, allowing for the identification of a peak intensity for each compound.

Identification of metabolites from each sample was performed using an MSML® standard compound library (Mass Spectrometry Metabolite Library of Standards: IROA Technologies™, Sea Girt, NJ, USA). To identify the metabolites, several criteria were used: the retention time of the metabolite had to be within 20 s of the reference, and the molecular weight had to be within 10 ppm of the reference. The signal intensity value was calculated using Xcalibur® 2.2 software (ThermoFisher Scientific, San Jose, CA, USA) by integrating the chromatography peaks corresponding to the selected metabolites.

The stability of the LC and mass spectrometry instruments was assessed by multiple quality control (QC) samples obtained from a 10 mL pool of an equal volume mixture of all samples analysed. QCs were analysed at the start of the assay, then every 10 samples, and at the end of the assay. Coefficients of variation (CVs, defined as standard deviation/mean × 100) were calculated for each metabolite found in the test and control samples. Metabolites with a CV QC > 30% or >the sample CV were considered false positives and excluded from further analysis.

2.6. Statistical Analysis

2.6.1. Statistical Analysis of Auditory Function

Statistical analyses were performed using Prism 9® version 9.4.1 (GraphPad Software ©, Inc., Boston, MA, USA). A two-sided non-parametric Wilcoxon matched pairs signed rank test with a 95% confidence interval was used to compare the wave IV thresholds and latencies of the sonoporation ear before and after sonoporation. Latencies were analysed using a one-way analysis of variance (ANOVA) with three factors: the sonoporation ear before sonoporation, the sonoporation ear after sonoporation, and the control ear. For all tests, a p-value < 0.05 indicated a significant difference. A significantly higher threshold or longer latency after sonoporation indicated a deterioration in hearing.

2.6.2. Statistical Analysis of Temperature Measurements

Statistical analyses were performed using Prism 9® version 9.4.1 (GraphPad Software©, Inc., Boston, MA, USA). Temperature measurements were compared using a two-tailed non-parametric Wilcoxon matched-pairs signed rank test with a 95% confidence interval. A p-value < 0.05 indicated a significant difference.

2.6.3. Statistical Analysis of Metabolomics Samples

Statistical analyses were performed using R® software version 4.2.1. (Bell Laboratories, Inc., Windsor, WI, USA) The data collected in the three modalities were pooled. In cases of redundancy in the identification of metabolites, the modality with the lowest CV QC was chosen for the analysis.

Analyses of the metabolites were conducted using multivariate and univariate analysis. Unsupervised multivariate analysis consisted of a principal component analysis (PCA) to (1) assess the distribution of the samples, (2) group similar samples, and (3) identify outliers. The PCA represents the main directions in which the samples vary. The main dimensions (p1, p2, and p3) explained the maximum variance of the samples. Conventionally, p1 and p2 are used for graphical representations and can be used to summarise a large amount of data into a two-dimensional space.

Supervised multivariate analysis in Partial Least-Squares Discriminant Analysis (PLS-DA) was used to identify a predictive model to differentiate groups by constraining the samples to their own group. The significance of the model established was assessed by two p-values: pR2Y and pQ2; the model was significant if these p-values were <0.05. The predictivity of the model was evaluated by the value of Q2, which must be >0.5. This supervised analysis determined a VIP (Variable Importance in Projection) for each metabolite, which summarised the importance of each metabolite in the predictivity of the model. Metabolites with a VIP > 1 were retained, and metabolites significantly impacting metabolic pathways were sought using MetaboAnalyst V5.0 (www.metaboanalyst.ca, accessed on 22 January 2023).

Univariate analysis was performed using paired Student's t-tests for each metabolite, assuming a normal distribution. A False Discovery Rate (FDR) method was used to adjust p-values for multiple comparisons. Metabolites with adjusted p-values < 0.05 were considered to be significantly altered between groups. If metabolites were significantly altered, MetaboAnalyst V5.0 was used to search for significantly impacted metabolic pathways.

Each significant pathway found after multivariate or univariate analysis was examined for relevance and consistency with pathophysiological hypotheses.

3. Results

3.1. Evaluation of Auditory Thresholds

ABR was used to assess the hearing function of all sheep. Figure 4 shows the results of the recordings for sheep #5, as an example.

The mean threshold for wave IV was 23.3 ± 5.16 dB before sonoporation, 25.0 ± 5.47 dB after sonoporation, and 23.3 ± 5.16 dB for the control ear. ABR thresholds are presented in Table 1. For all sheep, thresholds were unchanged, except for sheep #3, where the wave IV threshold increased from 20 dB before sonoporation to 30 dB after sonoporation. A Wilcoxon matched-pairs signed-rank test did not show any significant changes in thresholds after sonoporation compared with before sonoporation ($p > 0.99$; W = −1.00) or to the control ear ($p > 0.99$; W = 1. 00).

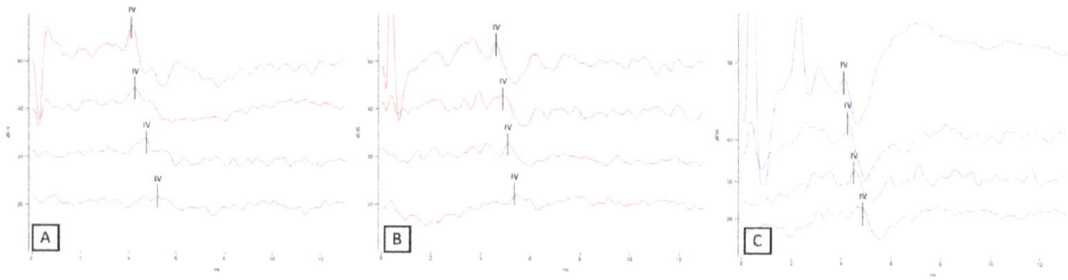

Figure 4. Representative measurement of bone conduction ABR for sonoporation side (right ear) before (**A**) and after sonoporation (**B**) and control side (left ear) (**C**) for sheep #5. Wave IV can be clearly observed and was not modified by sonoporation.

Table 1. ABR thresholds of wave IV detection are presented for each sheep before and after sonoporation session and for control side.

Sheep	Thresholds (dB)		
	Before Sonoporation	After Sonoporation	Control
#1	30	30	30
#2	30	30	30
#3	20	30	20
#4	20	20	20
#5	20	20	20
#6	20	20	20

Latency measurements at 30 and 20 dB are presented in Table 2. Wilcoxon matched pairs signed rank tests showed no significant differences in latency before or after sonoporation at 30 dB ($p = 0.34$; W = −10.00) or 20 dB ($p = 0.25$; W = −6. 00). A one-way ANOVA of the 30 dB latency data showed no significant difference in latency before sonoporation, after sonoporation, or to the control ear [F(2, 17) = 0.8; $p = 0.457$]. A one-way ANOVA of the 20 dB latency data showed no significant difference in latency before sonoporation, after sonoporation, or to the control ear [F(2, 10) = 0.8; $p = 0.479$].

Table 2. Latency measurements for Wave IV in milliseconds (ms) at 30 and 20 dB for each sheep before and after sonoporation, and for control side. # indicates missing values.

Sheep	Measurement	30 dB	20 dB
#1	Before sonoporation	4.49	#
	After sonoporation	5.22	#
	Control side	5.24	#
#2	Before sonoporation	4.9	#
	After sonoporation	4.75	#
	Control side	4.82	#
#3	Before sonoporation	4.74	5.10
	After sonoporation	4.75	#
	Control side	4.61	5.10
#4	Before sonoporation	4.36	4.59
	After sonoporation	4.35	4.85
	Control side	4.77	4.88
#5	Before sonoporation	4.69	5.15
	After sonoporation	5.10	5.40
	Control side	4.49	4.84
#6	Before sonoporation	4.83	5.08
	After sonoporation	4.93	5.50
	Control side	5.03	5.39

3.2. Temperature Measures

The mean temperature measured at the promontory was 33.8 ± 3.65 °C and 34.0 ± 3.57 °C before and after sonoporation, respectively. A Wilcoxon matched pairs signed rank tests showed no significant increase in temperature after sonoporation (mean difference = 0.21 ± 0.47 °C; p = 0.438; W= −9.00).

3.3. Metabolomics Analysis of Perilymph

3.3.1. Perilymph Composition

Metabolomic analysis with LC-HRMS found 201 metabolites, of which 84 were detected in C18 Negative Mode, 63 in C18 Positive Mode, and 107 in HILIC Positive Mode. After the selection of duplicates and the exclusion of metabolites with QC CVs > 30% or >sample CVs, the relative concentrations of 198 metabolites in the 12 samples were retained for statistical analysis. The list of these metabolites is presented in Supplementary Table S1.

3.3.2. Perilymph Comparison by Multivariate Analysis

We conducted the multivariate analysis with and without adjustment for the data from samples S1 (sonoporation side of sheep #1) and C6 (control side of sheep #6). These samples were contaminated by blood, and therefore, some metabolite concentrations were artificially elevated. To avoid bias due to the extreme values found in these samples, we calculated for each of their metabolites the percentage difference between the sample and the average across the other samples of the same group (sonoporation or control). Only the metabolites showing less than a 50% difference relative to the average of the other samples were retained.

The PCA analysis showed that p1 explained 32.2% of the variance and p2 explained 17.9% of the variance. The PCA score plot representing the distribution of the samples according to p1 and p2 is shown in Figure 5. The ellipses are overlapped, and the samples grouped toward the center of the graph show a common metabolomic profile. The PCA according to the other principal components (p2 and p3, p1 and p3) confirmed this profile.

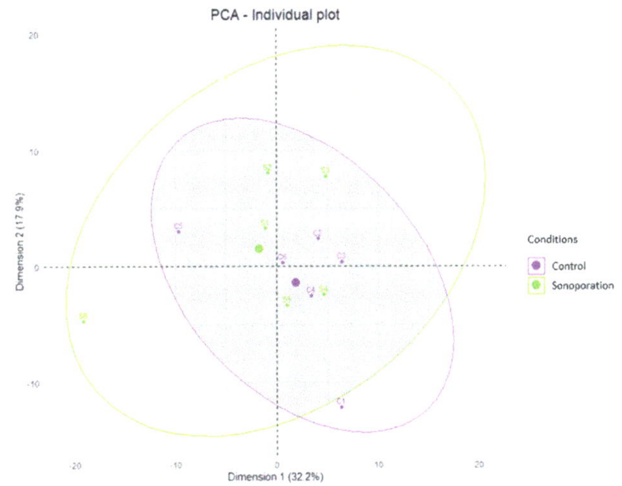

Figure 5. Unsupervised multivariate analysis of perilymph comparing data from sonoporation and control groups. Score-plot of the PCA is shown with adjustment of S1 and C6 data constructed from the metabolites found according to the first two components p1 and p2. The control samples are in purple (C1 to C6) and the sonoporation samples are in green (S1 to S6). Each group is associated with an ellipse regrouping all its samples. The thicker dots correspond to the center of each ellipse. Two samples with the same metabolomic profile appear grouped on this graph (i.e., samples S4 and C4).

We conducted the PLS-DA analysis using p1 and p2 with 100 permutations. The PLS-DA metrics revealed a non-significant predictive model with pR2Y = 0.89, pQ2 = 0.91 and Q2 = −0.89. Thus, there were no predictive metabolites to classify the samples into the post-sonoporation or control groups. Even when the data were not adjusted for the outliers from S1 and C6, PLS-DA analysis did not result in a significant predictive model, with pR2Y = 0.35, pQ2 = 0.87, and Q2 = −0.59. The PCA score plot representing the distribution of all samples according to p1 and p2 is presented in Supplementary Figure S1.

3.3.3. Perilymph Comparison by Univariate Analysis

The univariate analysis was conducted using student *t*-tests on the relative concentrations of the 198 metabolites found in the 12 samples. The relative concentration of each metabolite was compared between the control group and the sonoporation group. Analysis (before or after adjustment for multiple comparisons by the FDR test) did not find any metabolite significantly modified between groups (Figure 6).

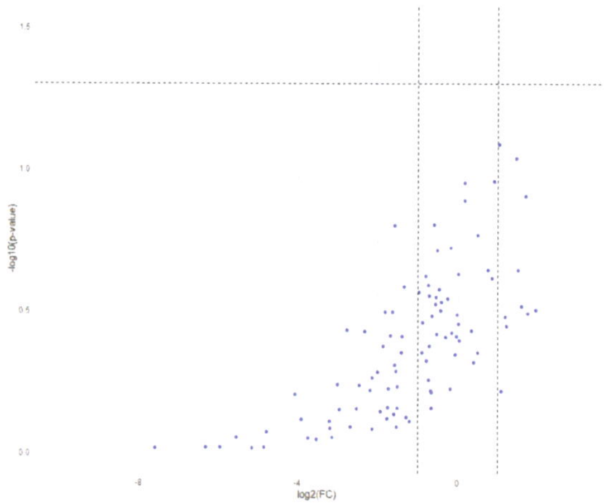

Figure 6. Volcano plot representing metabolite dysregulations between control and sonoporation groups, with thresholds set at log2FC < −1 or >1, and −log10 (*p*-value) > 1.30 ($p < 0.05$). The vertical dotted bars represent the log2FC thresholds at −1 and 1. As all metabolites were under the horizontal dotted line, none of the expression variations were significant. The graph was similar using the FDR adjusted *p*-value.

3.4. *Metabolomics Analysis of Plasma*

3.4.1. Plasma Composition

Metabolomic analysis with LC-HRMS found 269 metabolites, of which 99 were detected in C18 Negative Mode, 83 in C18 Positive Mode and 167 in HILIC Positive Mode. After the selection of duplicates and the exclusion of metabolites with QC CVs > 30% or >sample CVs, we retained for analysis the relative concentrations of 267 metabolites in the 12 samples.

3.4.2. Plasma Comparison before and after Sonoporation by Multivariate Analysis

The PCA analysis showed that p1 explained 26% of the variance and p2 explained 22% of the variance. The score-plot of the PCA (Figure 7) according to the principal components p1 and p2 did not find any outlier samples and separated the plasma samples taken before and after sonoporation into two distinct groups. PCA according to the other principal components (p2 and p3, p1 and p3) confirmed these divergent metabolomic profiles.

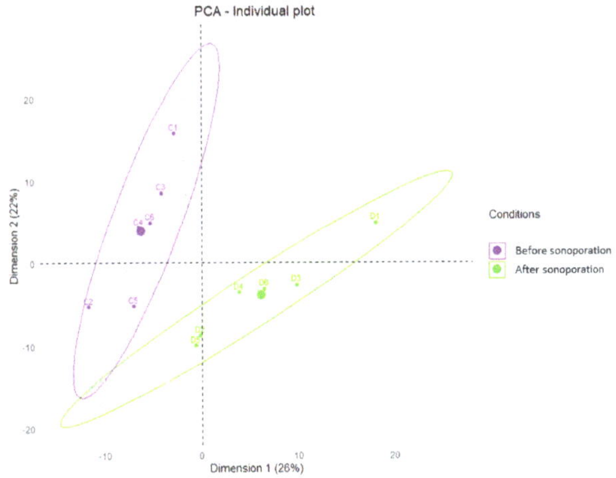

Figure 7. Unsupervised multivariate metabolomic analysis of plasma collected before and after sonoporation. The plot score of the PCA was constructed from the metabolites found according to components p1 and p2. Pre-sonoporation samples are shown in purple (C1 to C6) and post-sonoporation samples are shown in green (D1 to D6). Each group is associated with an ellipse grouping all its samples. The thicker dots correspond to the center of each ellipse.

We conducted a PLS-DA analysis using p1 and p2 with 100 permutations. The metrics revealed a significantly predictive model with pR2Y = 0.02, pQ2 = 0.02, and Q2 = 0.761 (Figure S2). Therefore, there are predictive metabolites for classifying samples into one of the two groups. The VIP analysis allowed us to identify the metabolites that could explain the statistical difference between the two groups. This analysis found 102 metabolites with VIPs > 1. Metabolic pathway analysis of these 102 metabolites using MetaboAnalyst V5.0 (www.metaboanalyst.ca, accessed on 22 January 2023) and the KEGG database identified that phenylalanine metabolism had a significant impact.

3.4.3. Plasma Comparison by Univariate Analysis

A student's *t*-test revealed 108 metabolites with significantly different expression before or after sonoporation. After adjustment for multiple comparisons by the FDR test, 62 metabolites were retained. A volcano plot is shown in Supplementary Figure S3 to illustrate the dysregulation of the expression of these metabolites. 25 metabolites were dysregulated, of which 20 were up-regulated and 5 down-regulated. All the up- and down-regulated metabolites are presented in Supplementary Table S2. Using MetaboAnalyst 5.0, with the 108 significantly modified metabolites, two significantly impacted metabolism pathways were found: (1) D-glutamine and D-glutamate, and (2) arginine biosynthesis.

4. Discussion

In the present study, we validated the safety of sonoporation of the inner ear in a sheep model. No ototoxicity of sonoporation was observed on ABR thresholds, consistent with previous sonoporation studies of the inner ear in guinea pigs [26,28–32]. For sheep #3, the threshold for wave IV detection increased from 20 to 30 dB after sonoporation, but this value remained within normal-hearing levels. This increase may have been due to sonoporation, surgery, or measurement error. The ABR curves were sometimes difficult to interpret despite the use of subcutaneous detection electrodes. Our ABR recordings only assessed auditory function at high frequencies. Another approach to objectively assess would be distortion-product otoacoustic emissions (DPOAEs), which is a multi-frequency method. However, this method may not work because of the morphology of the sheep's tortuous external ear canal, the milling-induced damage during mastoidectomy, and the presence of fluid (MBs, blood) in the middle ear, which may cause conductive hearing loss.

The ultrasound parameters used were chosen given their effectiveness in delivering therapeutic molecules in various contexts. For example, these parameters have been used to increase the permeability of the blood-brain barrier [45] or to increase the extravasation of local drug delivery to tumor tissue [48], as well as in previous studies regarding sonoporation of the RWM [26–32]. Some previous studies demonstrated increased steroid delivery to the inner ear using 1-MHz ultrasound waves [28] and increased safety (according to histology) with 3 rather than 5 one-minute consecutive exposures for the RWM [30]. In a previous study [49], these parameters induced inertial cavitation of Vevo MicroMarker® MBs to stimulate adjacent cells and thus increase the delivery of therapeutic drugs. Note that, after activation by hand stirring, no further mixing method was used during sonoporation, as access to the middle ear was blocked by the transducer and head shaking risked MBs leakage. This could lead to the activation of only a fraction of the MBs by the ultrasound beam. In addition, the Vevo MicroMarker® MBs used in our study are an ultrasound contrast agent indicated for preclinical ultrasound imaging. In previous studies, we reported that these MBs are more efficient than SonoVue® MBs for drug and gene delivery in vitro using sonoporation [49,50]. If they are effective for enhanced delivery of drugs to the inner ear, and human use is intended, more studies will be necessary for regulatory approval.

Consistent with previous studies [29,30], sonoporation of the inner ear did not significantly increase the temperature in the middle ear. It is possible that the temperature could have increased because of interaction between the ultrasound waves and the bone structures of the middle ear, but this was not observed. The increase in temperature must not exceed the permissible toxicity threshold for ultrasound (1.5 °C [33]). To limit this effect, the sonoporation was applied for 1 min and repeated three times with 1 min between applications rather than 3 min of continuous stimulation.

This is the first study to investigate the safety of sonoporation of the inner ear by analysing the metabolome of perilymph from control and sonoporated ears. Several metabolites and metabolic pathways have been identified in the perilymph after inner ear trauma. Trinh et al. [43] reported that N-acetyl neuraminate increased in the metabolomic profiles of perilymph samples from cochlear implant patients with long durations of hearing loss, indicating the destruction of cell membranes. This hearing loss also induced a significant increase in the concentrations of several other metabolites, including glutaric acid, cystine, 2-propanoate, butanoate, and xanthine. In addition, Fujita et al. [51] investigated the metabolic consequences of auditory trauma in the perilymph of guinea pigs using metabolomics. They identified 12 metabolites whose concentrations were significantly altered: 3-hydroxy-butyrate, glycerol, fumaric acid, phosphate, and pyruvate + oxalacetic acid were significantly increased, while citric + isocitric acid, mannose, meso-erythritol, galactosamine, and inositol were significantly decreased. In a mouse model of auditory trauma, Ji et al. [52] reported that the glutamate, aspartate, purine, and alanine pathways were up-regulated in the perilymph, while the phenylalanine, tyrosine, and tryptophan pathways were down-regulated. In the present study, none of these metabolites or pathways were significantly changed in the perilymph of the sonoporated ear, suggesting that sonoporation did not cause damage to the inner ear under our experimental conditions. Theoretically, sonoporation could cause damage to RWM cells, especially when inertial cavitation of the MBs is reached, as it can cause intercellular ruptures, membranous pores, and cell destruction.

Additional adverse events, such as facial palsy, could be induced by ultrasound application in the middle ear. We could not evaluate this possibility here, as sacrifice of the facial nerve was surgically necessary to gain access to the RWM for perilymph sampling. A long-term study of auditory responses and metabolomic perilymph profiles is needed to assess the complete innocuity of this technique. Our perilymph samples were collected approximately 30 min after the sonoporation procedure, which allowed for only short-term metabolomic effects.

Complementary to the metabolomic analysis of the perilymph, a metabolomic study of the plasma was undertaken by taking blood samples before and after sonoporation. This was done to further investigate metabolomic changes if differences were observed in the perilymph. The metabolomic analysis of serum after sonoporation revealed significant changes in the metabolic pathways of phenylalanine metabolism, D-glutamine and D-glutamate metabolism, and arginine biosynthesis. Phenylalanine is an essential amino acid that is converted to tyrosine and used in the biosynthesis of dopamine and noradrenaline. In our study, phenylalanine metabolism was activated, increasing these neuromodulators, which are notably secreted in response to exhaustion or stress. Glutamine is a non-essential amino acid, a substrate for the production of excitatory and inhibitory neurotransmitters (glutamate and GABA), but also a major source of energy for the nervous system. In our study, the D-glutamine and D-glutamate metabolism pathways were activated towards the production of 2-oxo-glutarate, which is used in the Krebs cycle for energy production. L-citrulline is the most dysregulated metabolite of the arginine pathway. It is also a precursor of the Krebs cycle and therefore contributes to energy production. The data used for metabolomic analysis of plasma should be interpreted with caution, as the blood sample taken before sonoporation was taken right after anesthetic induction. The anesthesia and surgical procedure may have contributed to the stress and energy consumption observed here. A second limitation is that a human KEGG database was used for the metabolomic analysis and pathway research. To our best knowledge, an ovine metabolomic database is not available.

Finally, the surgical technique used to access the inner ear here was a mastoidectomy, which was necessary to easily gain access to the round window, to deliver the ultrasound stimulation, and to collect perilymph. Other approaches to the round window could be developed with adapted ultrasound probes, as done for guinea pigs [29,31].

5. Conclusions

The present study established the safety of sonoporation of the inner ear in a sheep model. This procedure did not cause deafness, harmful heating of the inner ear, or metabolomic changes in the perilymph. Further experiments are required to investigate the efficacy of sonoporation for delivering therapeutic drugs to the inner ear in a sheep model.

Supplementary Materials: The following supporting information can be downloaded at: https://www.mdpi.com/article/10.3390/pharmaceutics15020442/s1, Table S1. List of metabolites found in the perilymph on sonoporation and control sides. For each metabolite, the mean relative concentration and standard deviation are presented for the sonoporation side and control side; Table S2. List of down- and up-regulated metabolites in plasma after sonoporation with mean concentration and standard deviation before and after sonoporation, as well as *p*-values from Student' *t*-test and after FDR adjustment and Fold-Change; Figure S1. Unsupervised multivariate analysis of perilymph comparing sonoporation and control groups. Score-plot of the PCA constructed from the metabolites found according to components p1 and p2. The control samples are represented in purple (C1 to C6) and the sonoporation samples are in green (S1 to S6). Each group is associated with an ellipse regrouping all its samples. The thicker dots correspond to the center of each ellipse; Figure S2. Supervised multivariate analysis of plasma comparing before and after sonoporation groups. PLS-DA constructed from the metabolites found according to components p1 and p2 with 100 permutations. The samples before sonoporation (Before) are represented in red (C1 to C6) and the samples after sonoporation (After) are represented in blue (D1 to D6). Each group is associated with an ellipse regrouping all its samples; Figure S3. Volcano-plot comparing the dysregulation of metabolites before and after sonoporation, with thresholds set at log2FC < -1 or >1 and adjusted *p*-value $-10\log > 1.3$ ($p < 0.05$). The blue symbols represent down-regulated metabolites, the red symbols represent up-regulated metabolites, and the gray symbols represent no significant dysregulation.

Author Contributions: Conceptualization, D.B., J.-M.E., A.B., H.B. and S.K.; methodology, D.B., J.-M.E., S.K., A.B. and H.B.; software, S.K. and A.P.; validation, S.K., D.B., J.-M.E., A.B. and H.B.; formal analysis, S.K., A.P., D.B. and H.B.; investigation, S.K., D.B., J.-M.E., E.O., A.L. and C.A.; resources, S.K., D.B., J.-M.E., A.L., C.A., H.B. and P.E.; data curation, S.K.; writing—original draft preparation, S.K., D.B., J.-M.E. and J.J.G.III; writing—review and editing, S.K., D.B., J.-M.E., J.J.G.III, H.B., A.B., A.P., C.A., A.L, E.O. and P.E.; visualization, S.K. and A.P.; supervision, D.B. and J.-M.E.; project administration, S.K., D.B. and J.-M.E.; funding acquisition, D.B. All authors have read and agreed to the published version of the manuscript.

Funding: This research was supported by the House Institute Foundation (HIF-121-5).

Institutional Review Board Statement: The animal study protocol was approved by the French Ethics Committee of Animal Experimentation of Centre—Val de Loire (protocol code APAFiS #33161-2021091518241644, approved in February 2022).

Informed Consent Statement: Not applicable.

Data Availability Statement: The datasets generated during and/or analyzed during the current study are available in supplementary materials. Other data are available on reasonable request from the corresponding author.

Acknowledgments: We thank the House Institute Foundation, Los Angeles, CA, USA for their financial support. We thank the technical staff from PIXANIM platform (INRAE, Université de Tours, Tours, France) and Jean-Yves Tartu (UMR 1253, iBrain, Université de Tours, Inserm, Tours, France) for their help in this project.

Conflicts of Interest: The authors declare no conflict of interest.

References

1. Mathers, C.D.; Stein, C.; Ma Fat, D.; Rao, C.; Inoue, M.; Tomijima, N.; Bernard, C.; Lopez, A.D.; Murray, C.J.L. *Global Burden of Disease 2000: Version 2 Methods and Results*; World Health Organization: Geneva, Switzerland, 2002.
2. Livingston, G.; Sommerlad, A.; Orgeta, V.; Costafreda, S.G.; Huntley, J.; Ames, D.; Ballard, C.; Banerjee, S.; Burns, A.; Cohen-Mansfield, J.; et al. Dementia prevention, intervention, and care. *Lancet* **2017**, *390*, 2673–2734. [CrossRef] [PubMed]
3. Li, Y.; Kanzaki, S.; Shibata, S.; Nakamura, M.; Ozaki, M.; Okano, H.; Ogawa, K. Comparison of inner ear drug availability of combined treatment with systemic or local drug injections alone. *Neurosci. Res.* **2020**, *155*, 27–33. [CrossRef] [PubMed]
4. Yang, J.; Wu, H.; Zhang, P.; Hou, D.M.; Chen, J.; Zhang, S.G. The pharmacokinetic profiles of dexamethasone and methylprednisolone concentration in perilymph and plasma following systemic and local administration. *Acta Otolaryngol.* **2008**, *128*, 496–504. [CrossRef]
5. Li, Y.; Kanzaki, S.; Shibata, S.; Nakamura, M.; Ozaki, M.; Okano, H.; Ogawa, K. Comparison of Drug Availability in the Inner Ear After Oral, Transtympanic, and Combined Administration. *Front. Neurol.* **2021**, *12*, 641593. [CrossRef] [PubMed]
6. King, E.B.; Salt, A.N.; Eastwood, H.T.; O'Leary, S.J. Direct entry of gadolinium into the vestibule following intratympanic applications in Guinea pigs and the influence of cochlear implantation. *J. Assoc. Res. Otolaryngol.* **2011**, *12*, 741–751. [CrossRef]
7. Salt, A.N.; King, E.B.; Hartsock, J.J.; Gill, R.M.; O'Leary, S.J. Marker entry into vestibular perilymph via the stapes following applications to the round window niche of guinea pigs. *Hear. Res.* **2012**, *283*, 14–23. [CrossRef]
8. Lange, G. Gentamicin and other ototoxic antibiotics for the transtympanic treatment of Menière's disease. *Arch. Otorhinolaryngol.* **1989**, *246*, 269–270. [CrossRef]
9. Patel, M.; Agarwal, K.; Arshad, Q.; Hariri, M.; Rea, P.; Seemungal, B.M.; Golding, J.F.; Harcourt, J.P.; Bronstein, A.M. Intratympanic methylprednisolone versus gentamicin in patients with unilateral Ménière;s disease: A randomised, double-blind, comparative effectiveness trial. *Lancet* **2016**, *388*, 2753–2762. [CrossRef]
10. Ahmadzai, N.; Cheng, W.; Kilty, S.; Esmaeilisaraji, L.; Wolfe, D.; Bonaparte, J.; Schramm, D.; Fitzpatrick, E.; Lin, V.; Skidmore, B.; et al. Pharmacologic and surgical therapies for patients with Meniere's disease: A systematic review and network meta-analysis. *PLoS ONE* **2020**, *15*, e0237523. [CrossRef]
11. Lim, H.J.; Kim, Y.T.; Choi, S.J.; Lee, J.B.; Park, H.Y.; Park, K.; Choung, Y.H. Efficacy of 3 different steroid treatments for sudden sensorineural hearing loss: A prospective, randomized trial. *Otolaryngol. Head Neck Surg.* **2013**, *148*, 121–127. [CrossRef]
12. Ahmadzai, N.; Kilty, S.; Cheng, W.; Esmaeilisaraji, L.; Wolfe, D.; Bonaparte, J.P.; Schramm, D.; Fitzpatrick, E.; Lin, V.; Skidmore, B.; et al. A systematic review and network meta-analysis of existing pharmacologic therapies in patients with idiopathic sudden sensorineural hearing loss. *PLoS ONE* **2019**, *14*, e0221713. [CrossRef] [PubMed]
13. Parnes, L.S.; Sun, A.H.; Freeman, D.J. Corticosteroid pharmacokinetics in the inner ear fluids: An animal study followed by clinical application. *Laryngoscope* **1999**, *109 Pt 2*, 1–17. [CrossRef] [PubMed]
14. Silverstein, H.; Choo, D.; Rosenberg, S.I.; Kuhn, J.; Seidman, M.; Stein, I. Intratympanic steroid treatment of inner ear disease and tinnitus (preliminary report). *Ear. Nose Throat J.* **1996**, *75*, 468–488. [CrossRef] [PubMed]

15. Xu, X.; Wang, S.; Shu, Y.; Zhang, H. Nanoparticle-based inner ear delivery systems for the treatment of hearing loss. *Smart Mater. Med.* **2021**, *2*, 350–353. [CrossRef]
16. Wise, A.K.; Tan, J.; Wang, Y.; Caruso, F.; Shepherd, R.K. Improved Auditory Nerve Survival with Nanoengineered Supraparticles for Neurotrophin Delivery into the Deafened Cochlea. *PLoS ONE* **2016**, *11*, e0164867. [CrossRef]
17. Tan, J.; Wang, Y.; Yip, X.; Glynn, F.; Shepherd, R.K.; Caruso, F. Nanoporous peptide particles for encapsulating and releasing neurotrophic factors in an animal model of neurodegeneration. *Adv. Mater.* **2012**, *24*, 3362–3366. [CrossRef]
18. Jaudoin, C.; Carré, F.; Gehrke, M.; Sogaldi, A.; Steinmetz, V.; Hue, N.; Cailleau, C.; Tourrel, G.; Nguyen, Y.; Ferrary, E.; et al. Transtympanic injection of a liposomal gel loaded with N-acetyl-L-cysteine: A relevant strategy to prevent damage induced by cochlear implantation in guinea pigs? *Int. J. Pharm.* **2021**, *604*, 120757. [CrossRef]
19. Warnecke, A.; Prenzler, N.; Harre, J.; Köhl, U.; Gärtner, L.; Lenarz, T.; Laner-Plamberger, S.; Wietzorrek, G.; Staecker, H.; Lassacher, T.; et al. First-in-human intracochlear application of human stromal cell-derived extracellular vesicles. *J. Extracell. Vesicles* **2021**, *10*, e12094. [CrossRef]
20. Dinh, C.; Hoang, K.; Haake, S.; Chen, S.; Angeli, S.; Nong, E.; Eshraghi, A.A.; Balkany, T.J.; Van De Water, T.R. Biopolymer-released dexamethasone prevents tumor necrosis factor alpha-induced loss of auditory hair cells in vitro: Implications toward the development of a drug-eluting cochlear implant electrode array. *Otol. Neurotol.* **2008**, *29*, 1012–1019. [CrossRef]
21. Paasche, G.; Bögel, L.; Leinung, M.; Lenarz, T.; Stöver, T. Substance distribution in a cochlea model using different pump rates for cochlear implant drug delivery electrode prototypes. *Hear. Res.* **2006**, *212*, 74–82. [CrossRef]
22. Scheper, V.; Hessler, R.; Hütten, M.; Wilk, M.; Jolly, C.; Lenarz, T.; Paasche, G. Local inner ear application of dexamethasone in cochlear implant models is safe for auditory neurons and increases the neuroprotective effect of chronic electrical stimulation. *PLoS ONE* **2017**, *12*, e0183820. [CrossRef]
23. Hakuba, N.; Watabe, K.; Hyodo, J.; Ohashi, T.; Eto, Y.; Taniguchi, M.; Yang, L.; Tanaka, J.; Hata, R.; Gyo, K. Adenovirus-mediated overexpression of a gene prevents hearing loss and progressive inner hair cell loss after transient cochlear ischemia in gerbils. *Gene Ther.* **2003**, *10*, 426–433. [CrossRef]
24. Escoffre, J.M.; Bouakaz, A. Minireview: Biophysical Mechanisms of Cell Membrane Sonopermeabilization. Knowns and Unknowns. *Langmuir* **2019**, *35*, 10151–10165. [CrossRef]
25. Presset, A.; Bonneau, C.; Kazuyoshi, S.; Nadal-Desbarats, L.; Mitsuyoshi, T.; Bouakaz, A.; Kudo, N.; Escoffre, J.M.; Sasaki, N. Endothelial Cells, First Target of Drug Delivery Using Microbubble-Assisted Ultrasound. *Ultrasound Med. Biol.* **2020**, *46*, 1565–1583. [CrossRef]
26. Shih, C.P.; Chen, H.C.; Chen, H.K.; Chiang, M.C.; Sytwu, H.K.; Lin, Y.C.; Li, S.L.; Shih, Y.F.; Liao, A.H.; Wang, C.H. Ultrasound-aided microbubbles facilitate the delivery of drugs to the inner ear via the round window membrane. *J. Control. Release* **2013**, *167*, 167–174. [CrossRef]
27. Liao, A.H.; Hsieh, Y.L.; Ho, H.C.; Chen, H.K.; Lin, Y.C.; Shih, C.P.; Chen, H.C.; Kuo, C.Y.; Lu, Y.J.; Wang, C.H. Effects of microbubble size on ultrasound-mediated gene transfection in auditory cells. *Biomed. Res. Int.* **2014**, *2014*, 840852. [CrossRef]
28. Shih, C.P.; Chen, H.C.; Lin, Y.C.; Chen, H.K.; Wang, H.; Kuo, C.Y.; Lin, Y.Y.; Wang, C.H. Middle-ear dexamethasone delivery via ultrasound microbubbles attenuates noise-induced hearing loss. *Laryngoscope* **2019**, *129*, 1907–1914. [CrossRef]
29. Liao, A.H.; Wang, C.H.; Weng, P.Y.; Lin, Y.C.; Wang, H.; Chen, H.K.; Liu, H.L.; Chuang, H.C.; Shih, C.P. Ultrasound-induced microbubble cavitation via a transcanal or transcranial approach facilitates inner ear drug delivery. *JCI Insight* **2020**, *5*, e132880. [CrossRef]
30. Lin, Y.C.; Chen, H.C.; Chen, H.K.; Lin, Y.Y.; Kuo, C.Y.; Wang, H.; Hung, C.L.; Shih, C.P.; Wang, C.H. Ultrastructural Changes Associated with the Enhanced Permeability of the Round Window Membrane Mediated by Ultrasound Microbubbles. *Front. Pharmacol.* **2020**, *10*, 1580. [CrossRef]
31. Liao, A.H.; Shih, C.P.; Li, M.W.; Lin, Y.C.; Chuang, H.C.; Wang, C.H. Development of thermosensitive poloxamer 407-based microbubble gel with ultrasound mediation for inner ear drug delivery. *Drug Deliv.* **2021**, *28*, 1256–1271. [CrossRef]
32. Lin, Y.C.; Lin, Y.Y.; Chen, H.C.; Kuo, C.Y.; Liao, A.H.; Chou, Y.L.; Hung, C.L.; Shih, C.P.; Wang, C.H. Ultrasound Microbubbles Enhance the Efficacy of Insulin-Like Growth Factor-1 Therapy for the Treatment of Noise-Induced Hearing Loss. *Molecules* **2021**, *26*, 3626. [CrossRef] [PubMed]
33. Ter Haar, G. Ultrasonic imaging: Safety considerations. *Interface Focus* **2011**, *1*, 686–697. [CrossRef]
34. Nordang, L.; Linder, B.; Anniko, M. Morphologic changes in round window membrane after topical hydrocortisone and dexamethasone treatment. *Otol. Neurotol.* **2003**, *24*, 339–343. [CrossRef]
35. Schachern, P.A.; Paparella, M.M.; Duvall, A.J., 3rd; Choo, Y.B. The human round window membrane. An electron microscopic study. *Arch. Otolaryngol.* **1984**, *110*, 15–21. [CrossRef]
36. Trinh, T.T.; Cohen, C.; Boullaud, L.; Cottier, J.P.; Bakhos, D. Sheep as a large animal model for cochlear implantation. *Braz. J. Otorhinolaryngol.* **2023**, *88*, 24–32. [CrossRef]
37. Seibel, V.A.; Lavinsky, L.; Irion, K. CT-Scan sheep and human inner ear morphometric comparison. *Braz. J. Otorhinolaryngol.* **2006**, *72*, 370–376. [CrossRef]
38. Seibel, V.A.; Lavinsky, L.; De Oliveira, J.A. Morphometric study of the external and middle ear anatomy in sheep: A possible model for ear experiments. *Clin. Anat.* **2006**, *19*, 503–509. [CrossRef] [PubMed]
39. Péus, D.; Dobrev, I.; Pfiffner, F.; Sim, J.H. Comparison of sheep and human middle-ear ossicles: Anatomy and inertial properties. *J. Comp. Physiol. A Neuroethol. Sens. Neural Behav. Physiol.* **2020**, *206*, 683–700. [CrossRef]

40. Soares, H.B.; Lavinsky, L. Histology of sheep temporal bone. *Braz. J. Otorhinolaryngol.* **2011**, *77*, 285–292. [CrossRef]
41. Han, S.; Suzuki-Kerr, H.; Suwantika, M.; Telang, R.S.; Gerneke, D.A.; Anekal, P.V.; Bird, P.; Vlajkovic, S.M.; Thorne, P.R. Characterization of the Sheep Round Window Membrane. *J. Assoc. Res. Otolaryngol.* **2021**, *22*, 1–17. [CrossRef]
42. Hill, M.W.; Heavens, R.P.; Baldwin, B.A. Auditory evoked potentials recorded from conscious sheep. *Brain Res. Bull.* **1985**, *15*, 453–458. [CrossRef] [PubMed]
43. Trinh, T.T.; Blasco, H.; Emond, P.; Andres, C.; Lefevre, A.; Lescanne, E.; Bakhos, D. Relationship between Metabolomics Profile of Perilymph in Cochlear-Implanted Patients and Duration of Hearing Loss. *Metabolites* **2019**, *9*, 262. [CrossRef] [PubMed]
44. Boullaud, L.; Blasco, H.; Caillaud, E.; Emond, P.; Bakhos, D. Immediate-Early Modifications to the Metabolomic Profile of the Perilymph Following an Acoustic Trauma in a Sheep Model. *J. Clin. Med.* **2022**, *11*, 4668. [CrossRef] [PubMed]
45. Presset, A.; Bodard, S.; Lefèvre, A.; Millet, A.; Oujagir, E.; Dupuy, C.; Iazourène, T.; Bouakaz, A.; Emond, P.; Escoffre, J.M.; et al. First Metabolomic Signature of Blood-Brain Barrier Opening Induced by Microbubble-Assisted Ultrasound. *Front. Mol. Neurosci.* **2022**, *15*, 888318. [CrossRef]
46. Mavel, S.; Lefèvre, A.; Bakhos, D.; Dufour-Rainfray, D.; Blasco, H.; Emond, P. Validation of metabolomics analysis of human perilymph fluid using liquid chromatography-mass spectroscopy. *Hear. Res.* **2018**, *367*, 129–136. [CrossRef]
47. Diémé, B.; Mavel, S.; Blasco, H.; Tripi, G.; Bonnet-Brilhault, F.; Malvy, J.; Bocca, C.; Andres, C.R.; Nadal-Desbarats, L.; Emond, P. Metabolomics Study of Urine in Autism Spectrum Disorders Using a Multiplatform Analytical Methodology. *J. Proteome Res.* **2015**, *4*, 5273–5282. [CrossRef]
48. Bressand, D.; Novell, A.; Girault, A.; Raoul, W.; Fromont-Hankard, G.; Escoffre, J.M.; Lecomte, T.; Bouakaz, A. Enhancing Nab-Paclitaxel Delivery Using Microbubble-Assisted Ultrasound in a Pancreatic Cancer Model. *Mol. Pharm.* **2019**, *16*, 3814–3822. [CrossRef]
49. Escoffre, J.M.; Novell, A.; Piron, J.; Zeghimi, A.; Doinikov, A.; Bouakaz, A. Microbubble attenuation and destruction: Are they involved in sonoporation efficiency? *IEEE Trans. Ultrason. Ferroelectr. Freq. Control* **2013**, *60*, 46–52. [CrossRef]
50. Escoffre, J.M.; Piron, J.; Novell, A.; Bouakaz, A. Doxorubicin delivery into tumor cells with ultrasound and microbubbles. *Mol. Pharm.* **2011**, *8*, 799–806. [CrossRef]
51. Fujita, T.; Yamashita, D.; Irino, Y.; Kitamoto, J.; Fukuda, Y.; Inokuchi, G.; Hasegawa, S.; Otsuki, N.; Yoshida, M.; Nibu, K. Metabolomic profiling in inner ear fluid by gas chromatography/mass spectrometry in guinea pig cochlea. *Neurosci. Lett.* **2015**, *606*, 188–193. [CrossRef]
52. Ji, L.; Lee, H.J.; Wan, G.; Wang, G.P.; Zhang, L.; Sajjakulnukit, P.; Schacht, J.; Lyssiotis, C.A.; Corfas, G. Auditory metabolomics, an approach to identify acute molecular effects of noise trauma. *Sci. Rep.* **2019**, *9*, 9273. [CrossRef] [PubMed]

pharmaceutics

MDPI

Article

Impact of Perfluoropentane Microdroplets Diameter and Concentration on Acoustic Droplet Vaporization Transition Efficiency and Oxygen Scavenging

Rachel P. Benton [1], Nour Al Rifai [1], Kateryna Stone [1], Abigail Clark [1], Bin Zhang [2,3] and Kevin J. Haworth [1,3,*]

1 Department of Internal Medicine, Division of Cardiovascular Health and Disease, University of Cincinnati, Cincinnati, OH 45267, USA
2 Cincinnati Children's Hospital Medical Center, Cincinnati, OH 45229, USA
3 Department of Pediatrics, University of Cincinnati, Cincinnati, OH 45229, USA
* Correspondence: hawortkn@ucmail.uc.edu; Tel.: +1-513-558-3536

Abstract: Acoustic droplet vaporization is the ultrasound-mediated phase change of liquid droplets into gas microbubbles. Following the phase change, oxygen diffuses from the surrounding fluid into the microbubble. An in vitro model was used to study the effects of droplet diameter, the presence of an ultrasound contrast agent, ultrasound duty cycle, and droplet concentration on the magnitude of oxygen scavenging in oxygenated deionized water. Perfluoropentane droplets were manufactured through a microfluidic approach at nominal diameters of 1, 3, 5, 7, 9, and 12 μm and studied at concentrations varying from 5.1×10^{-5} to 6.3×10^{-3} mL/mL. Droplets were exposed to an ultrasound transduced by an EkoSonic™ catheter (2.35 MHz, 47 W, and duty cycles of 1.70%, 2.34%, or 3.79%). Oxygen scavenging and the total volume of perfluoropentane that phase-transitioned increased with droplet concentration. The ADV transition efficiency decreased with increasing droplet concentration. The increasing duty cycle resulted in statistically significant increases in oxygen scavenging for 1, 3, 5, and 7 μm droplets, although the increase was smaller than when the droplet diameter or concentration were increased. Under the ultrasound conditions tested, droplet diameter and concentration had the greatest impact on the amount of ADV and subsequent oxygen scavenging occurred, which should be considered when using ADV-mediated oxygen scavenging in therapeutic ultrasounds.

Keywords: perfluoropentane microdroplets; cavitation; microfluidic emulsion manufacturing; polydispersity; ultrasound contrast agent; ultrasound duty cycle; intravascular ultrasound

Citation: Benton, R.P.; Al Rifai, N.; Stone, K.; Clark, A.; Zhang, B.; Haworth, K.J. Impact of Perfluoropentane Microdroplets Diameter and Concentration on Acoustic Droplet Vaporization Transition Efficiency and Oxygen Scavenging. *Pharmaceutics* **2022**, *14*, 2392. https://doi.org/10.3390/pharmaceutics14112392

Academic Editors: Brandon Helfield, Shashank Sirsi, James Kwan and Michael Gray

Received: 16 September 2022
Accepted: 2 November 2022
Published: 5 November 2022

1. Introduction

The application of ultrasound to convert liquid perfluorocarbon droplets into gas microbubbles is known as acoustic droplet vaporization (ADV). ADV is being actively investigated for therapeutic applications, such as targeted drug delivery [1–6], thermal ablation [7,8], mechanical ablation [9–11], tissue regeneration with acoustically responsive scaffolds (ARS) [12–14], and reperfusion injury [15,16]. The later approach relies on gas scavenging that occurs following ADV [17–20]. The effectiveness of most ADV-based therapies is dependent on the fraction of perfluorocarbon droplets that undergo the phase transition to gas microbubbles, known as the ADV transition efficiency [20].

Fabiilli et al. demonstrated that the ADV transition efficiency improved with increasing droplet diameter when using double-phase perfluorocarbon droplets [21]. These findings were subsequently confirmed by Mercado et al. with the use of relatively polydisperse single-phase perfluorocarbon droplets [22]. It has also been shown that the threshold pressure amplitude necessary to nucleate ADV decreases with the increasing droplet diameter [18,21–23].

The relationship between ADV transition efficiency and droplet concentration does not have a single clear trend in the literature. Aliabouzar et al. demonstrated that the ADV transition efficiency was higher for larger droplet volume fractions near the ADV threshold at 2.25 MHz; however, well above the ADV threshold, 1% (v/v) droplet fractions had a saturating efficiency, whereas 0.05% (v/v) droplet fractions did not [13]. The authors postulated that acoustic shadowing may cause the reduced ADV transition efficiency at higher concentrations [13]. Dong et al. demonstrated that as the volume fraction of droplets increased, so did the release of the growth factor from double droplets, suggesting a higher ADV transition efficiency with increased droplet concentration [12]. Furthermore, Kang et al. observed that as the droplet concentration increased, the final ADV microbubble diameter decreased, implying a difference in ingassing [19].

The presence of microbubbles in proximity to droplets can impact ADV. Lo et al. found that the ADV threshold could be reduced by co-administering droplets with the ultrasound contrast agent Definity® [24]. The use of microbubbles coupled to microdroplets to nucleate ADV is key to acoustic cluster therapy (ACT), which is being investigated for its potential in ultrasound-mediated drug delivery [1–3,25,26]. Studies involving ACT have found that the presence of these clusters enhanced drug delivery and uptake in tumors [2,3]. The increased drug delivery is suggestive of higher ADV transition efficiency as the drug is carried in the microdroplet portion of the cluster and must be released.

Ultrasound parameters can also impact the number of ADV microbubbles formed. Fabiilli et al. observed that as a fluid was exposed to ultrasound multiple times, the ADV transition efficiency increased [21]. The conditions under which this happens can be complex. Kang et al. observed that as the pulse duration or pulse repetition frequency were increased, fewer microbubbles were observed, which they attributed to ADV occurring and the microbubbles subsequently being destroyed by additional ultrasound insonation [19]. This observation is consistent with the fact that they also observed that when fewer microbubbles were present, each microbubble scavenged a larger number of moles of dissolved gas, resulting in larger ADV microbubbles.

To further investigate how the presence of a contrast agent, droplet size, droplet concentration, and ultrasound duty cycle impact ADV, several studies are reported herein. For each of the aforementioned parameters, the ADV transition efficiency was measured by comparing the volume-weighted concentration of perfluoropentane droplets without and with exposure to ultrasound. Additionally, the magnitude of oxygen scavenged was measured in the flow phantom before and during the ultrasound exposure of the fluid containing perfluoropentane droplets. Per our prior work, the magnitude of oxygen scavenged depends on the total volume of perfluoropentane that undergoes ADV per volume of total fluid [20]. As noted above, the impact of droplet diameter and concentration on ADV transition efficiency was investigated in prior studies. In this work, droplets with monodisperse size distributions and modal diameters between 1 and 12 μm were used to minimize the impact of droplet–droplet interactions for droplets of different diameters and to provide additional information that could guide future therapeutic applications of ADV. Similarly, the data reported herein on the effect of the presence of an ultrasound contrast agent are carried out with relatively monodisperse droplet distributions, whereas prior studies used polydisperse droplet distributions [24–26].

2. Materials and Methods

2.1. Droplet Preparation and Characterization

Kolliphor® P188-coated perfluoropentane droplets of 1, 3, 5, 7, 9, and 12 μm nominal diameters were manufactured with an in-house microfluidic system using a flow-focusing chip with either 5 or 14 μm junctions (model numbers: 300152 and 3200146, Dolomite-Blacktrace, Royston, UK) [27,28]. The microfluidic system was placed at room temperature (23.5 ± 1.5 °C), which is lower than the boiling point temperature of PFP. Table 1 details the manufacturing parameters for the droplets. Kolliphor® P188 was purchased from Sigma Aldrich (St. Louis, MO, USA) and perfluoropentane was purchased from FluoroMed, L.P.

(Round Rock, TX, USA). Kolliphor® P188 was dissolved in filtered deionized water from a Nanopure system (model number: 04751, Barnstead, Dubuque, IA, USA). Red food dye (Kroger, Inc., Cincinnati, OH, USA) was used for visualizing fluids during manufacturing. The kinematic viscosity of each polymer with dye solution was measured 5 times using a Cannon-Fenske routine viscometer (Cannon Instrument Company, State College, PA, USA). Droplets were collected in a 50 mL conical vial.

Table 1. Droplet manufacturing parameters for polymer flow rate, PFP flow rate, and polymer concentration are listed. For droplets of 1 and 3 μm nominal diameters, a chip with a 5 μm junction was used. For droplets of 5, 7, 9, and 12 μm nominal diameters, a chip with a 14 μm junction was used. Measured polymer viscosity, measured production rate, and anticipated production rate (based on nominal droplet diameter and PFP flow rate) are listed.

Nominal Droplet Diameter (μm)	Polymer Flow Rate (μL/min)	PFP Flow Rate (μL/min)	Polymer Concentration (mg/mL)	Polymer Kinematic Viscosity (cp)	Measured Production Rate (droplets/sec)	Anticipated Production Rate (droplets/sec)
1	8.33	0.1	45	2.23 ± 0.02	1.69×10^6	2.53×10^6
3	5	0.3	45	2.23 ± 0.02	6.15×10^5	4.69×10^5
5	11.67	0.8	105	4.79 ± 0.01	1.21×10^5	1.94×10^5
7	11.67	1.2	105	4.79 ± 0.01	1.01×10^5	1.00×10^5
9	8.33	1.7	125	5.05 ± 0.05	5.88×10^4	6.63×10^4
12	6.67	3.4	125	5.05 ± 0.05	1.22×10^5	6.92×10^4

2.2. Droplet Size and Stability Measurements

The modal diameter and concentration of the droplets were measured using a Coulter Counter (Multisizer 4, Beckman Coulter Inc., Brea, CA, USA) equipped with a 30 μm aperture. The system measured droplets between 0.6 to 18 μm in diameter. To reduce the amount of noise in the measurements originating from particulate in the flow phantom, only a subset of diameters was used when calculating the pre and peri ADV droplet concentrations in the effluent. Data were analyzed in bins from 0.6 to 3 μm, 0.6 to 4 μm, 3 to 7 μm, 5 to 9 μm, 7 to 11 μm, and 11 to 13 μm for samples containing 1, 3, 5, 7, 9, and 12 μm, respectively. The measurements were performed by diluting 6 μL of stock solution into 10 mL of phosphate-buffered saline (PBS) for droplets of nominal diameters 3, 5, 7, and 9 μm and 3 μL of stock solution in 10 mL of PBS for droplets of nominal diameters 1 and 12 μm. Volume-weighted size distributions were used to calculate the droplet modal diameter and concentration (mL/mL). The size distribution was used to compute the polydispersity index (PDI, the number-weighted droplet size distribution standard deviation divided by the mean volume-weighted particle diameter, quantity squared) for each droplet. After the measurement, droplets were stored in a freezer at −20 °C, given their instability observed in pilot experiments at 4 °C, and used within 45 days.

Due to Ostwald ripening [29,30] and potential spontaneous droplet vaporization during storage, a time-dependent stability study of the 1, 3, 5, 7, 9, and 12 μm nominal diameter droplets was conducted. On the day of manufacturing (D0), droplets from a stock solution were aliquoted into multiple Eppendorf tubes and stored at −20 °C. For each measurement, droplets in one Eppendorf tube were thawed and the size distribution was measured every day for the first week and every four days up to day 33.

2.3. In Vitro Experimental Setup

An in vitro flow phantom was used to perform experiments testing the ADV transition efficiency and oxygen scavenging (Figure 1). The flow phantom was constructed using 3.18 mm of inner diameter EVA tubing (McMaster-Carr, Aurora, OH, USA). The tubing was water-jacketed, such that all ADV was performed at a temperature of 37 ± 1 °C, which was monitored by an internal thermocouple immediately upstream to the site of the oxygen partial pressure (PO_2) measurement. A peristaltic pump was used to induce 16 mL/min flow of oxygenated deionized water (Section 2.1) through the flow phantom. The oxygenated deionized water was prepared in a covered 4 L reservoir by constantly bubbling a 95%/5% O_2/CO_2 gas mixture through it as is typically used for ex vivo systems. The oxygenated water's temperature was maintained in the range of 36.5 to 38.0 °C with a water bath [31]. The corresponding PO_2 of the oxygenated water was 560.8 ± 7.6 mmHg.

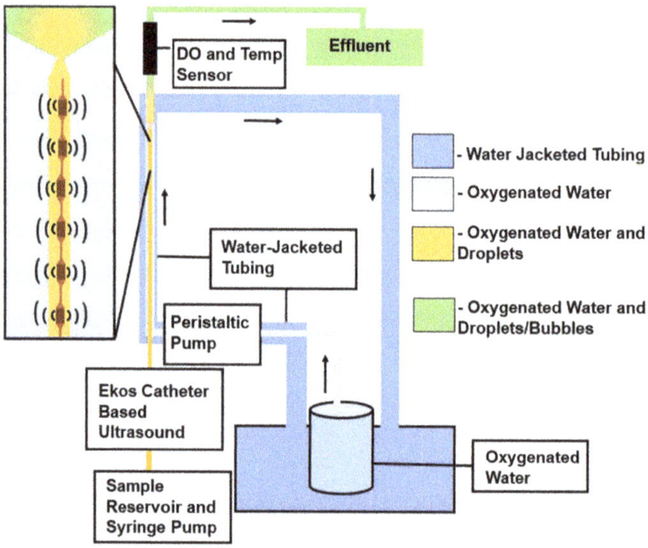

Figure 1. Schematic of the in vitro flow phantom with black arrows showing direction of flow. Droplets were pumped from the sample reservoir and through the flow system maintained at 37 °C. The oxygenated water was saturated to a PO_2 of 560.8 ± 7.6 mmHg. Droplets were exposed to a 2.35 MHz-pulsed ultrasound using a 6 cm treatment zone EkoSonic® catheter. The PO_2 in the fluid distal to the catheter was measured by a flow-through oxygen sensor. The effluent was collected to measure the size distribution and volume concentration of droplets that did not undergo ADV.

An EkoSonic® catheter (Boston Scientific, Model 500-56112, Natick, MA, USA) was inserted into the flow phantom using a hemostasis valve. The EkoSonic® catheter is composed of a MicroSonic™ Device (MSD) that contained 6 pairs of ultrasound transducer elements (used to nucleate ADV) inserted into an Intelligent™ Drug Delivery Catheter (IDDC). The EkoSonic® catheter is designed to be disposable in the clinical environment. However, due to cost, the catheters were reused over multiple experiments, with the catheters flushed with deionized water between trials. The health of the MSD was assessed each day via an impedance measurement (Aim4170D, Array Solutions, Sunnyvale, TX, USA). Based on pilot data, MSDs were only used if the phase of the impedance at 2.35 MHz (the ultrasound insonation frequency) was between –15 and 10 degrees. Droplets were diluted (5×10^{-4} mL/mL, unless noted otherwise) in deionized water taken from the reservoir and infused through the coolant port of the IDDC via a syringe pump operating at 5 mL/min. This flow rate matched prior ADV studies [16]. A 1-inch stir bar was placed within the syringe and manually agitated to prevent droplets from settling. Unless as noted below, the EkoSonic® catheter was driven with a 40-cycle tone burst with a burst period of 1.000 ms (AFG3500B, Keysight Technologies, Inc., Santa Rosa, CA, USA) through a power amplifier (2200L, Electronics and Innovation, Ltd., Rochester, NY, USA). A 23B RF Wattmeter (Sonic Concepts, Inc., Bothell, WA, USA) was used to monitor the electrical drive power to the catheter and the applied voltage was adjusted to ensure a pulse-average power of 47 W, the maximum output power used in FDA-cleared protocols. The acoustic field pattern from a transducer pair in the catheter is complex [32].

The droplet stock concentration was measured and appropriately diluted to the target concentration for the experiment. The droplets were then infused through the coolant port and into the oxygenated water. The PO_2 was measured downstream using an inline dissolved oxygen and temperature flow-through sensor (TOFTC2, PyroScience GmbH, Aachen, Germany). The flow time from the tip of the EkoSonic® catheter to the flow-through sensor was 3 s. Measurements were made over 40 s periods with the droplet syringe pump off (pre-droplets), droplet syringe pump on but without driving the EkoSonic® catheter (peri-droplets), and droplet syringe pump on with the EkoSonic® catheter driven at 47 W (peri-ADV). All experiments measuring the PO_2 and ADV transition efficiency were repeated five times. The size distribution of droplets in the effluent was measured (Section 2.2). Care was taken to remove any bubbles that floated to the top of the effluent. The ADV transition efficiency was calculated by taking one minus the ratio of the total volume concentration of the peri-ADV effluent to the peri-droplet effluent.

2.4. Effect of Lumason on ADV-Mediated Oxygen Scavenging

The effect of the ultrasound contrast agent Lumason® (Bracco Diagnostics, Inc., Milan, Italy) on the ADV transition efficiency and oxygen scavenging was studied by comparing measurements with and without Lumason® co-administered with droplets. Lumason® was activated according to the manufacturer's package insert and used within 48 h of activation. A 19 G needle (Hamilton, Reno, NV, USA) connected to a 25 µL gas-tight syringe (Becton Dickinson, Franklin Lakes, NJ, USA) was used to withdraw 24 µL of Lumason®. Lumason® was transferred to a 60 mL syringe (Becton Dickinson, Franklin Lakes, NJ, USA) containing 60 mL of 95% DI water and perfluorocarbon droplets ($4.8 \times 10^{-4} \pm 0.6 \times 10^{-4}$ mL/mL final concentration). The Lumason® dose was based on the package insert dose per weight (0.03 mL/kg). Co-diluted Lumason® and droplets were infused through the coolant delivery port of an EkoSonic® catheter with a Harvard Apparatus Elite syringe pump (Harvard Apparatus, Holliston, MA, USA) and exposed to ultrasound, as described in Section 2.3.

2.5. Effect of Ultrasound Duty Cycle on ADV-Mediated Oxygen Scavenging

The effect of ultrasound duty cycle was determined by fixing the pulse duration at 40 cycles (17.0 µs) and varying the burst period (0.450 ms, 0.725 ms, and 1.000 ms) or by fixing the burst period at 1.000 ms and varying the pulse duration (17.0 µs, 23.4 µs, and 37.9 µs). These combinations were selected so that varying the burst period or the pulse duration corresponded to the duty cycles of 1.70%, 2.34%, or 3.79%. The time-average electrical drive power used in these studies was near the maximum operating range for the single-use MSD, and thus, higher duty cycles could not be tested without damaging the catheter. Because the transition efficiency and oxygen scavenging did not change substantially between droplets of nominal diameters of 7, 9, and 12 µm, the effect of ultrasound duty cycle was only investigated for droplets with nominal diameters between 1 and 7 µm. The initial volume-weighted droplet concentrations were $5.0 \times 10^{-4} \pm 0.4 \times 10^{-4}$ mL/mL and $4.8 \times 10^{-4} \pm 0.6 \times 10^{-4}$ mL/mL when investigating the effect of burst period and pulse duration, respectively.

2.6. Effect of Droplet Concentration on ADV-Mediated Oxygen Scavenging

The effect of droplet concentration on the ADV transition efficiency and oxygen scavenging was measured for droplets with nominal diameters of 3, 5, and 7 µm. Droplets of nominal diameter 9 and 12 µm were not investigated as described in Section 2.5. The 1 µm nominal diameter droplets were not investigated because the droplet production rates were too slow. The volume-weighted droplet concentrations used were between 5.1×10^{-5} and 6.5×10^{-3} mL/mL.

2.7. Comparison of Measured Oxygen Scavenging to the Transition Efficiency-Based Model

For each of the experiments described in Sections 2.4–2.6, a previously reported physics-based model was used to estimate the magnitude of oxygen scavenging based on the volume of perfluoropentane transitioned from a liquid to a gas (i.e., the transition efficiency times the initial volume-weighted droplet concentration) [20]. Within the model, the relative partial pressures in the peri-droplet fluids were 166 mmHg, 560 mmHg, 26.7 mmHg, 2.0 mmHg for N_2, O_2, CO_2, and Ar, respectively. The vapor pressure of water was 47 mmHg. These values were based on the measured PO_2 values and then assuming the remaining partial pressures of gases were proportionate to their composition in standard air. An ambient hydrostatic pressure of 1 atm and a temperature of 37 °C were assumed. Perfluoropentane was assumed to be insoluble in all fluids and the solubility of oxygen was assumed to be 964 L atm/mol in water [33,34]. The expansion factor was 125, based on the change in density of liquid and gaseous perfluoropentane [20]. Surface tension and viscosity were not taken into account for the numerical model.

2.8. Statistical Analysis

The amount of oxygen scavenging and transition efficiency for the absence and presence of Lumason® was compared using a 2-tailed Mann–Whitney analysis. A Kruskal–Wallis analysis for each droplet diameter with Dunn's multiple testing correction was used to compare the amount of oxygen scavenging and transition efficiency for studies of the ultrasound duty cycle. The agreement between the measured and modeled oxygen scavenging was evaluated using an intraclass correlation coefficient with 95% confidence intervals and the correlation was assessed using Pearson correlation coefficients. We selected the consistency and a two-way mixed-effects ICC model of a single score per observed measurement type (C,1) [35]. The scores were defined as measurements predicted by the numerical oxygen scavenging model. All statistical analyses were performed using Prism 9 (GraphPad Software Inc., La Jolla, CA, USA) and SAS, Version 9.4 (SAS Institute, Cary, NC, USA). A p-value of less than 0.05 was used to determine the significant differences for all statistical tests.

3. Results

3.1. Droplet Manufacturing and Characterization

In general, polymer solution viscosity increased with polymer concentration (Table 1). The normalized volume-weighted size distribution of the droplets is shown in Figure 2. As the microfluidic chip junction decreased from 14 µm to 5 µm, the droplet size decreased. Generally, for both chips used, as the ratio of polymer flow rate to the perfluoropentane flow rate increased, the modal diameter decreased. However, this trend was not observed with the 9 µm droplets. The calculated droplet production rate (volume of PFP per droplet divided by the PFP flow rate) agreed within a factor of two with the measured droplet production rate (Table 1).

3.2. Droplet Stability at −20 °C

Table 2 lists the nominal diameter, modal diameter, and polydispersity index (PDI) at day 0 (D_0) and at day 33 (D_{33}). The modal diameter remained constant over 33 days, except for the 5 µm nominal diameter droplets, which size-shifted 80 nm when stored at −20 °C. The PDI increased during storage primarily due to an increase in measured particulate less than 1 µm in diameter. The average PDI for each nominal droplet size across all time points measured was 0.102 ± 0.0216, 0.165 ± 0.0561, 0.2292 ± 0.13, 0.285 ± 0.137, 0.536 ±0.003, and 0.09 ± 0.068 for 1, 3, 5, 7, 9, and 12 µm, respectively. The concentration stability was measured for 7 µm droplets and varied by less than 6% over 33 days. The modal diameter of droplets measured before and after use within the in vitro system (~3 min at 37 °C) showed no change.

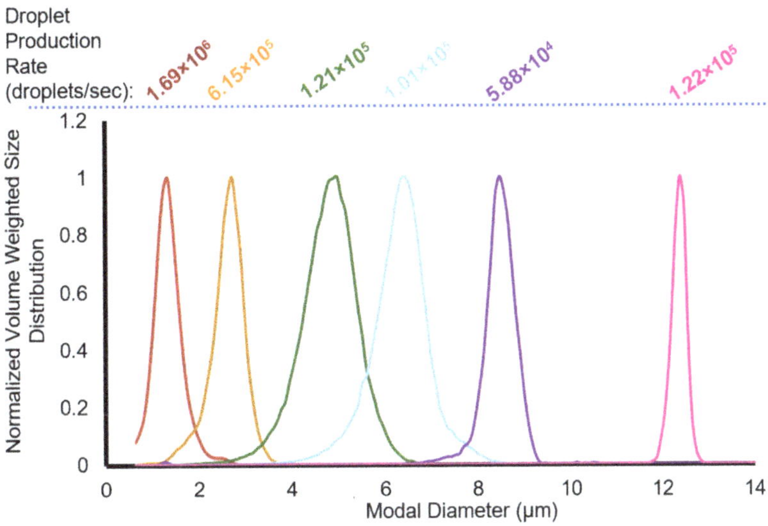

Figure 2. Normalized volume-weighted distribution for microfluidic droplets of nominal diameters 1, 3, 5, 7, 9, and 12 μm. The droplet production rate (droplets/s) is listed at the top of the figure for each droplet diameter. The colors (red, yellow, green, blue, violet, and magenta) used to plot the data correspond to nominal droplet diameters for each microdroplet (1, 3, 5, 7, 9, and 12 μm, respectively) and are used for all subsequent figures.

Table 2. Nominal size and modal diameter of droplets from the initial day of manufacturing (D_0) are compared with the modal diameter of droplets from 33 days (D_{33}) after manufacturing. PDI D_0 is compared with the PDI for D_{33}.

Nominal Size (μm)	D_0 Modal Diameter (μm)	D_{33} Modal Diameter (μm)	D_0 PDI	D_{33} PDI
1	1.34	1.34	0.068	0.0812
3	2.73	2.73	0.081	0.188
5	4.99	4.91	0.109	0.291
7	6.47	6.47	0.185	0.342
9	8.47	8.47	0.5324	0.5581
12	12.39	12.39	0.0967	0.2224

3.3. Effect of Lumason® on ADV-Mediated Oxygen Scavenging

ADV transition efficiency and oxygen scavenging were compared each with and without Lumason® for droplet diameters between 1 and 12 μm (Figure 3). The total oxygen scavenging was composed of the effect of the addition of droplets without ultrasound (i.e., peri-droplet and pre-ADV, solid color Figure 3b) and the effect of ADV (peri-ADV, white or hatched component Figure 3b). Oxygen scavenging due to droplets without ADV was expected due to the solubility of oxygen in liquid perfluoropentane [20]. Experiments performed without droplets measured no statistically significant change in the oxygen partial pressure with and without Lumason (-0.2 ± 1.7 mmHg). Both ADV transition efficiency and oxygen scavenging had an increasing trend with increasing droplet diameter between 1 and 7 μm, with minimal differences between 7, 9, and 12 μm droplets. The magnitude of the difference between oxygen scavenging with or without Lumason® was relatively small compared to the total amount of oxygen scavenged. The average difference between the amount of oxygen scavenging with and without Lumason® across all droplet diameters was 2.67 ± 19.03 mmHg. The addition of Lumason® resulted in increased oxygen scavenging when using 5 μm droplets (14.85 mmHg, p-value < 0.05) and decreased oxygen scavenging with 1 μm droplets (32.3 mmHg, p-value < 0.05). The transition efficiency with and without Lumason® was different only for droplets of nominal diameter 7 μm (12 percentage points, p-value < 0.05). There was significantly less (p-value < 0.05) oxygen scavenging observed with 1 μm droplets than with 7 μm, 9 μm, and 12 μm droplets when Lumason® was added (15 comparisons). Additionally, the ADV of 3 μm droplets with Lumason® resulted in less oxygen scavenging than using 12 μm droplets. The same pairs of significant differences in oxygen scavenging between droplet diameters without the addition of Lumason® were the same as with Lumason®, except no difference was observed between 1 and 9 μm droplets. The transition efficiency was only different with 12 μm and 9 μm droplets when compared to 1 μm droplets. Without Lumason®, the transition efficiency was only different with 7 μm and 12 μm droplets in comparison to 1 μm droplets.

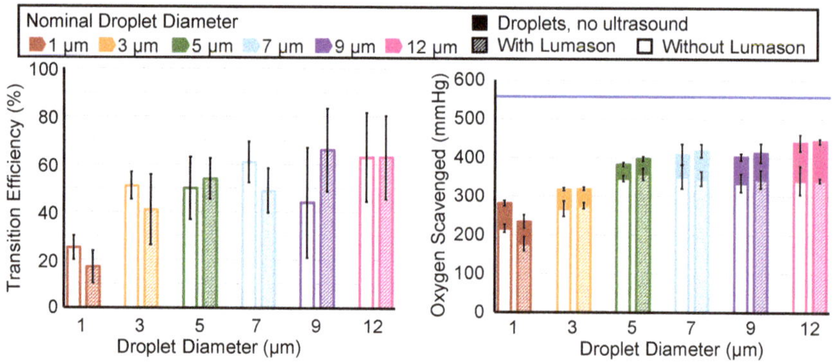

(**a**) (**b**)

Figure 3. (**a**) The measured ADV transition efficiency without (white fill) and with (hatched fill) Lumason® as a percentage of the change in the volume-weighted size distribution of the droplets with and without ADV. (**b**) The amount of oxygen scavenging with and without Lumason®. The solid filled top of the stacked bar graph is the amount of oxygen scavenging observed peri-droplet (i.e., no ultrasound exposure, and thus, no ADV). The bottom portion of each bar describes the amount of oxygen scavenging observed peri-ADV without (white fill) and with (hatched fill) Lumason®. The horizontal blue bar represents the initial PO_2 before droplets were infused, and thus, the maximum amount of PO_2 that could be scavenged. The error bars denote the standard deviation for five samples. The legend applies to both panels.

3.4. *Effect of Ultrasound Duty Cycle on ADV-Mediated Oxygen Scavenging*

For each nominal droplet diameter, the ADV transition efficiency did not show a consistent trend as a function of burst period (Figure 4a) or pulse duration (Figure 5a). There were no statistically significant differences in ADV transition efficiency for the varying pulse duration. The only statistically significant difference in transition efficiency for the varying burst period was observed between 0.450 ms and 0.725 ms (17 percentage points, p-value < 0.05). For all burst periods and all pulse durations (except at 1 μm), a trend existed for increasing oxygen scavenging with increasing duty cycle. More oxygen was scavenged with a 0.450 ms burst period than with a 1.000 ms burst period with 1, 3, 5, and 7 μm droplets (Figure 4a, p-value < 0.05). However, there was not a statistically significant difference in oxygen scavenging between 0.750 ms and the other two burst periods. More oxygen was scavenged with a 37.9 μs pulse duration than with 17 μs for 5 and 7 μm droplets, and there was a statistically significant difference between 17.0 μs and 23.4 μs for 3 μm droplets (Figure 5a, p-value < 0.05). Regardless of the burst period or pulse duration, there was significantly less oxygen scavenging observed with 1 μm droplets than with either 5 or 7 μm droplets (6 comparisons). Moreover, the transition efficiency was also significantly different between 1 μm and 7 μm droplets for all ultrasound parameters tested except for the 17.0 μs pulse duration. A significant difference in transition efficiency between 3 and 7 μm droplets was also observed for 0.725 ms burst period and 37.9 μs pulse duration; and a difference between 1 and 5 μm for 1.000 ms burst period.

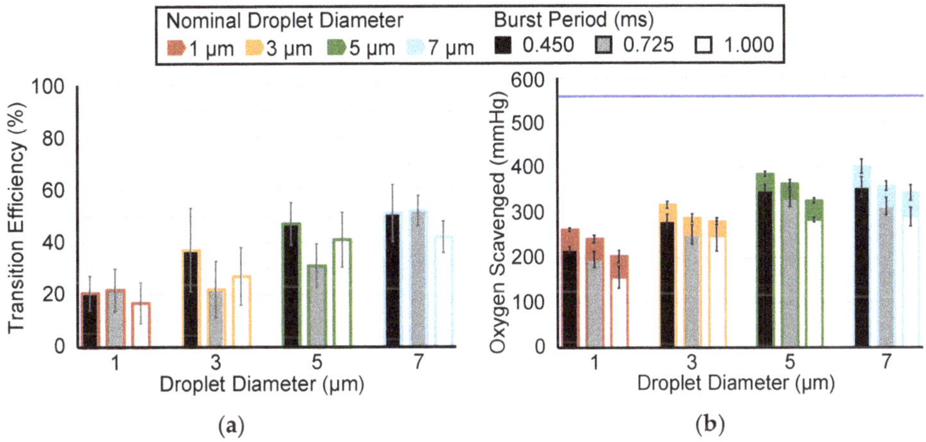

Figure 4. (**a**) ADV transition efficiency with burst periods of 0.450 ms (3.79% duty cycle), 0.725 ms (2.34% duty cycle), and 1.000 ms (1.70% duty cycle). The transition efficiency is calculated as a percent change based on the volume-weighted size distributions of the droplets. (**b**) The measured amount of oxygen scavenging peri-droplet (colored fill) or peri-ADV with burst periods of 0.450 ms, 0.725 ms, and 1.000 ms for droplets of 1, 3, 5, and 7 μm diameters. The horizontal blue bar represents the initial PO_2 before droplets were infused, and thus, the maximum amount of oxygen that could be scavenged. The error bars denote the standard deviation for five samples. The legend applies to both panels.

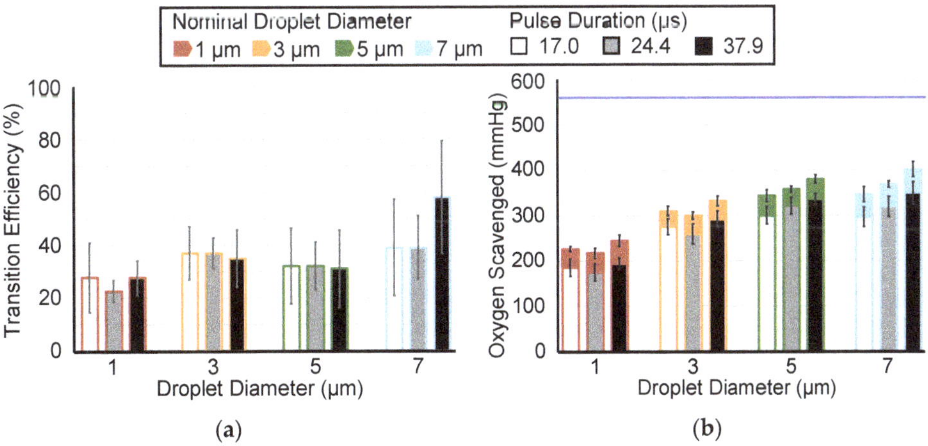

Figure 5. (**a**) ADV transition efficiency with pulse durations of 17.0 μs (1.70% duty cycle), 23.4 μs (2.34% duty cycle), and 37.9 μs (3.79% duty cycle). The transition efficiency is calculated as a percent change based on the volume-weighted size distributions of the droplets. (**b**) The measured amount of oxygen scavenging peri-droplet (colored fill) or peri-ADV with pulse durations of 17.0 μs, 23.4 μs, and 37.9 μs for droplets of 1, 3, 5, and 7 μm diameters. The horizontal blue bar represents the initial PO_2 before droplets were infused, and thus, the maximum amount of PO_2 that could be scavenged. The error bars denote the standard deviation for five samples. The legend applies to both panels.

3.5. Effect of Droplet Concentration on ADV-Mediated Oxygen Scavenging

The ADV transition efficiency and magnitude of oxygen scavenging for droplets with nominal diameters of 3, 5, and 7 μm are shown in Figure 6. A moving average line with a period of 4 was calculated and plotted to facilitate the visualization of the trend that the ADV transition efficiency decreased (Figure 6a) and oxygen scavenging increased (Figure 6b) with the increasing droplet volume-weighted concentration. The largest changes in transition efficiency and oxygen scavenging occurred for volume concentrations up to 0.001 mL/mL. Note that although the ADV transition efficiency decreased (i.e., the percentage of droplets that phase-transitioned) with increasing droplet volume concentration, the total volume of droplets that phase-transitioned was observed to increase, which is consistent with the observations in Figure 6b.

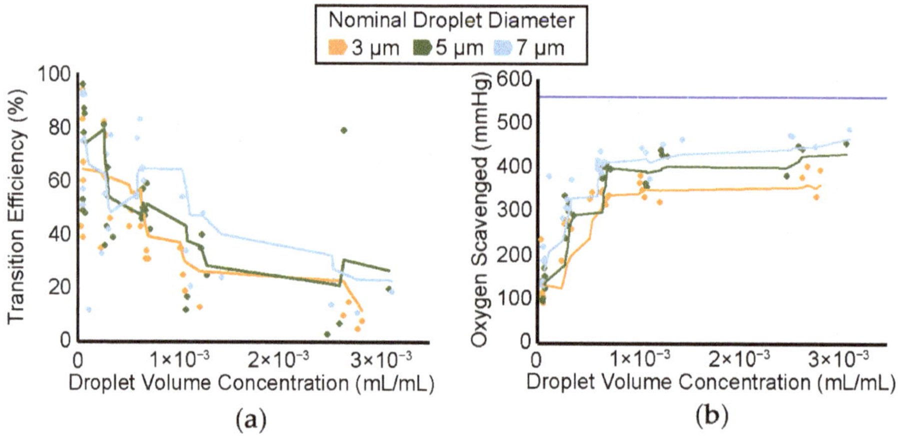

Figure 6. (**a**) ADV transition efficiency for droplets of nominal diameters 3 (yellow), 5 (green), and 7 (blue) μm for initial droplet concentrations between 5.10×10^{-5} to 6.30×10^{-3} mL/mL without Lumason. The transition efficiency is calculated as a percent change based on the volume-weighted size distributions of the droplets. Individual measurements are denoted with diamonds. Lines are the moving average of the measurements with a period of 6. (**b**) The measured amount of oxygen scavenging for droplets of nominal diameters 3 (blue diamonds), 5 (green diamonds), and 7 (yellow diamonds) μm with corresponding lines representing moving averages. The horizontal blue bar represents the initial PO_2 before droplets were infused, and thus, the maximum amount of PO_2 that could be scavenged. The legend applies to both panels.

3.6. Comparison of Measured Oxygen Scavenging to the Transition Efficiency-Based Model

The ADV-mediated measured oxygen scavenging was compared to the predicted oxygen scavenging based on the Radhakrishnan model that uses measured transition efficiency as an input [20]. Figure 7 plots the measured versus modeled oxygen scavenging values for each trial. For readability, the data are separated by whether the data were used to compare the effect of Lumason® (Figure 7a), droplet concentration (Figure 7b), burst period (Figure 7c), and pulse duration (Figure 7d). The ICC and 95% confidence intervals were 0.62 (0.47, 0.73), 0.55 (0.38, 0.68), 0.45 (0.26, 0.60), and 0.84 (0.78, 0.88) for the Lumason® trials, burst period trials, pulse duration trials, and droplet concentration trials, respectively. Collectively, the ICC and 95% confidence interval for all trials was 0.72 (0.67, 0.76). The Pearson correlation coefficient and 95% confidence intervals were 0.67 (0.50, 0.79), 0.57 (0.37, 0.72), 0.50 (0.28, 0.67), 0.88 (0.82, 0.91), and 0.76 (0.71, 0.81), for Lumason® trials, burst period trials, pulse duration trials, droplet concentration trials, and all trials, respectively.

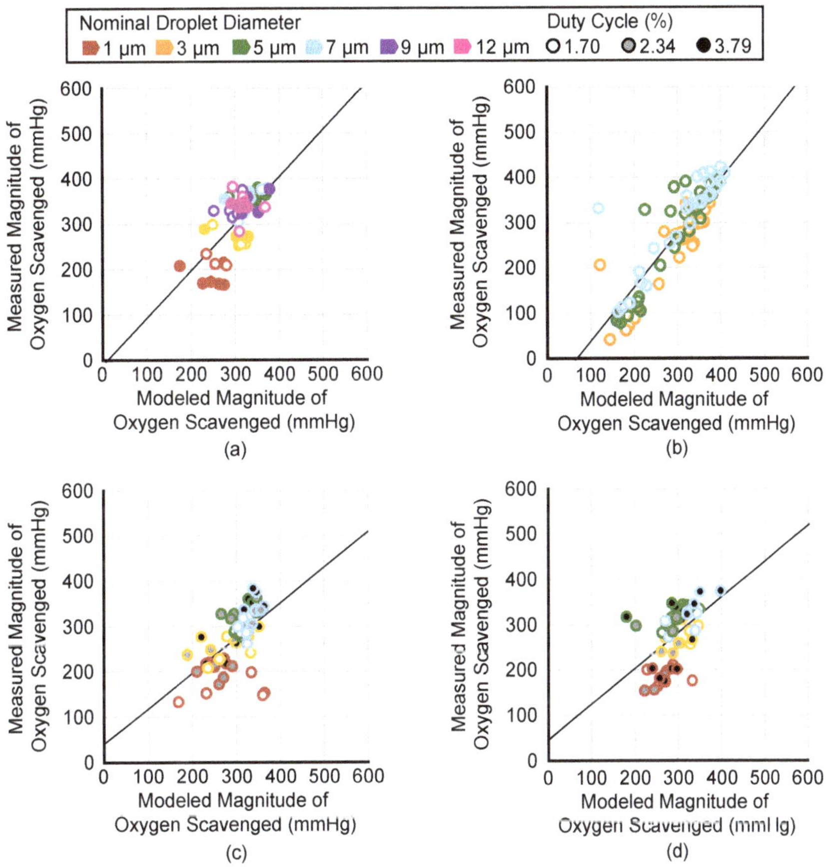

Figure 7. The measured amount of oxygen scavenged versus the modeled amount of oxygen scavenged for (**a**) trials investigating the impact of Lumason®, (**b**) droplet concentration, (**c**) burst period, and (**d**) pulse duration. The legend at the top applies to all panels. Data points represent individual trials with the outer marker color indicating the nominal droplet diameter and the fill indicating the duty cycle, except in panel (**a**) where filled data points are from trials that used Lumason® and open data points did not. The solid black line is a best-fit line based on the Pearson correlation analysis.

4. Discussion

4.1. Droplet Manufacturing and Characterization

The current study showed that the droplet diameter would decrease as the ratio between the polymer surfactant flow rate and PFP flow rate increased. However, it was observed that smaller microfluidic junctions were necessary to achieve the smallest diameter droplets. Although manufacturing droplets of specific diameters was repeatable, the production rates decreased for smaller droplets. Running the chip overnight for 1 μm droplets resulted in sufficient droplet stock to run approximately 4 trials with dilutions of 5×10^{-4} mL/mL. These production rates may be limiting in certain applications. Storage stability is an important parameter for low-boiling point PFC-in-water droplets to be of practical use, which was a feature of the droplets in this study. The modal diameter did not change substantially over 33 days. However, the polydispersity increased, primarily due to increased particle counts at diameters of 1 μm and smaller.

4.2. Effect of Lumason® on ADV-Mediated Oxygen Scavenging

In the present study, the effect of using the ultrasound contrast agent Lumason® showed no statistically significant effect in most cases, and where it did, the effect was inconsistent (8.5% increase and 2.6% decrease for 1 and 5 μm droplet, respectively). Prior studies have demonstrated that ultrasound contrast agents decrease the ultrasound pressure amplitude necessary to nucleate ADV (nominally 6 MPa to 2 MPa peak rarefactional pressure) using 20 μs pulses at a burst period of 2 ms at 1.44 MHz (1% duty cycle) [24]. In that study, Definity® was used at concentrations of 10^3, 10^4, and 10^5 microsphere per mL with highly polydisperse droplets at a concentration of 5.3×10^7 droplets/mL. Based on the Lumason® package insert concentration (1.5 to 5.6×10^8 microspheres/mL), our studies were performed at a concentration of 8.4×10^6 microspheres/mL in the IDDC coolant port, assuming no loss of Lumason® during infusion. The droplet volume fraction for our study (5×10^{-4} mL of droplet volume per mL of diluent) corresponded to droplet concentrations of 9.6×10^8 and 5.5×10^5 droplets/mL for 1 and 12 μm droplets, respectively. Thus, the ultrasound insonation parameters, droplet concentrations, and ultrasound contrast agent concentrations are similar between the studies. Given the differing impact of ultrasound contrast agents between the work of Lo et al. and those reported herein, it may be inferred that, as long as the ultrasound insonation pressure amplitude is greater than the ADV threshold pressures, reducing the threshold may not correspond to increased ADV transition efficiencies. There are notable differences between the work of Lo et al. and the results of the present study that prevent a definitive conclusion. Key differences include the droplet size distributions (polydisperse in Lo et al. versus monodisperse in the present study) and the ultrasound beam profile (relatively uniform in Lo et al. and complex from an EkoSonic® catheter [32]). Additionally, a broader range of ultrasound contrast agent concentrations in our study may have produced more pronounced differences. However, decreasing the ultrasound contrast agent concentration to be more similar to Lo et al. would likely decrease any observed effect. A final key difference between the two studies is that Lo et al. observed the greatest effect of pulse duration when millisecond-long pulses were used. These pulse durations were not possible at the pulse repetition frequencies used in this study without causing damage to the catheters. When studying acoustic cluster therapy, where microbubbles and microdroplets are electrostatically bound, Healey et al. observed that the nucleation efficiency decreased when albumin was added to a solution, which can cause an uncoupling of the microbubbles and microdroplets [25]. Their work suggests that increasing the distances between droplets and microbubbles decreases the impact of the microbubble on transitioning the droplet. Neither the work of Lo et al. nor that reported herein studied whether any microbubble and microdroplets were clustered together.

4.3. Ultrasound Insonation Parameters

Both the pulse duration and burst period affect the amount of oxygen scavenging during ADV. The burst period had the greater impact, although the effect was a modest increase in oxygen scavenging (10.8% ± 0.2%). Our observation that increasing the ultrasound duty cycle increased the ADV transition efficiency (and oxygen scavenging) is in concert with the observation from Lo et al. [24] that the ADV threshold pressure decreased with increasing duty cycle. However, as discussed in Section 4.2, there was not a coherence between ADV transition efficiency and ADV threshold pressure amplitude in the presence of an ultrasound contrast agent. Additional studies without ultrasound contrast agents would be needed to further elucidate the relationship between the ADV transition efficiency and threshold pressure amplitude. A limitation of this work is that hardware limitations restricted the range of ultrasound duty cycles from 1.70% to 3.79%.

Varying ultrasound insonation parameters did not substantially affect oxygen scavenging and, for the most part, transition efficiency when the droplet diameter was varied. In agreement with our observations made in experiments with and without the ultrasound contrast agent, using 1 μm and, in part, 3 μm droplets, result in significantly less oxygen scavenged than with 7, 9, or 12 μm droplets when varying duty cycles. However, differerices in transition efficiency were not as consistent as those in oxygen scavenging within and between groups.

4.4. Effect of Droplet Concentration on Ultrasound Mediated Oxygen Scavenging

The current study observed that as droplet concentration increased, the amount of ultrasound-mediated oxygen scavenging increased. The PO_2 reduction observed within the surrounding fluid with increasing droplet concentrations showed a saturating effect at around 400 mmHg (approximately 70% of the total PO_2). Kang et al. observed with increasing droplet concentrations that the ADV microbubbles were smaller [19]. The ADV microbubbles measured by Kang et al. included ingassing, and the smaller microbubble sizes may indicate that at higher concentrations, there is less ingassing per microbubble, which would be consistent with the saturation effect observed in Figure 6 and the theoretical model of oxygen scavenging plotted by Mercado-Shekhar et al. [16]. At concentrations below approximately 5×10^{-4} mL/mL, the magnitude of oxygen scavenging has a strong dependence on concentration. For nanodroplets, Yang et al. observed a decrease in ADV threshold pressure amplitude (8.5 MPa to 5.6 MPa peak rarefactional), as determined by changes in B-mode echogenicity, for increasing droplet concentrations (10^4 to 10^8 droplets/mL) [36]. Notably, these pressure amplitudes are significantly greater than the pressure amplitudes used in this study (1.5 MPa), though this could be consistent with the decreasing transition efficiency with a decreasing droplet diameter.

The total volume of PFP that phase-transitioned increased with increasing droplet concentrations, consistent with the increased oxygen scavenging (Figure 7). However, even though more oxygen scavenging was observed with increasing droplet concentrations, the transition efficiency as a percent of infused droplets decreased at higher droplet con centrations. The results of this study are not able to indicate the reason for the decrease, but it is possible that acoustic shadowing or droplet–droplet interactions are a reason. Supplemental Table S1 provides the average distance between droplets for each nominal diameter at the lowest and highest droplet concentration in these studies. Aliabouzar et al. observed a complex relationship between droplet concentration and ADV transition efficiency for double emulsion microdroplets. At higher droplet concentrations (1%), the ADV transition efficiency saturated with an increasing insonation pressure amplitude, while lower concentrations (0.05%) showed a linear increase up to 6.5 MPa [13]. Similarly, Dong et al. observed that, at 3.3 MPa insonation, the amount of growth factor released from an ADV double droplet was similar for 0.25% and 1% droplet concentrations [12]. However, at 8.8 MPa insonation, the higher droplet concentration released more growth factors, suggesting a higher ADV transition efficiency [12]. Together, these results highlight that the absolute trends observed in our studies may vary depending on the ultrasound pressure amplitude, which was not studied herein.

4.5. Comparison of Measured Oxygen Scavenging to Transition Efficiency-Based Model

Reasonable agreement was observed between the Radhakrishnan analytic model of oxygen scavenging [20], which is based on the measured amounts of PFP that phase-transitioned per milliliter of fluid, and the measured amount of oxygen scavenging droplet concentrations of 5×10^{-4} mL/mL (Figure 7a). The Pearson correlation coefficient quantifies that the data in Figure 7b demonstrate a strong linear relationship between the measured and modeled magnitude of oxygen scavenging as the droplet concentration was varied; however, there was an offset in the slope and intercept. This effect is observed with the intraclass correlation coefficient being slightly lower than the Pearson correlation coefficient. Potential reasons for the disagreement between measured and modeled values may include errors in the estimated volume expansion of the PFP, particularly because of surface tension [18,37] or the measured transition efficiency. In all experiments, the coefficient of variance was larger for the measured transition efficiency than the measured oxygen scavenging. Correspondingly, significant differences between experimental groups were more often seen in the magnitude of oxygen scavenging than the transition efficiency due to the larger standard deviation in the transition efficiency measurements (Figures 3–5). Fabiilli et al. [21] and Radhakrishnan et al. [20] also observed relatively large transition efficiency standard deviations, similar to this study, though their large standard deviations may be due to the use of polydisperse droplet distributions. Despite potential errors in the model, measuring the oxygen scavenging and reversing the model to compute an estimated transition efficiency measurement could be a more accurate way to obtain transition efficiency values. Additional studies designed to estimate the precision and accuracy of this approach would be needed to validate it. Alternative particle sizing techniques, such as nanoparticle tracking analysis [38] that can measure particles below the limits of a Coulter counter, could provide more accurate and less variable estimates of the transition efficiency. An additional source of error is that the ADV microbubbles are assumed to remain microbubbles. However, depending on the temperature, size, and surface tension, it is possible for droplets to recondense [37,39].

4.6. Study Limitations

The large variability observed with transition efficiency for pulse duration, burst period, and Lumason® trials could be attributed to challenges in measuring effluent samples. Our measurements assume that only droplets and not bubbles are measured. When measuring peri-ADV samples, bubbles floating to the top of the sample were removed. Even though care was taken to only remove bubbles, approximately 10–100 μL of the sample was also removed in the peri-ADV samples. Another limitation with the method used to measure samples showed that over time, droplets within a sample will settle to the bottom. Larger droplets settle to the bottom faster than smaller droplets. Although samples were mixed well before each measurement, the droplets would settle during the measurement (note that the terminal velocity is between 4 and 14 mm/s for perfluoropentane droplets between 1 and 12 μm in diameter in water).

A limitation of the current study was the re-use of the single-use EkoSonic® catheters due to their cost. At the beginning of each experimental day, an electrical impedance measurement was taken from the EkoSonic® device. If the phase was between –15 to 10 degrees, the EkoSonic® device was used. In pilot experiments, it was observed that the magnitude of ADV varied outside of this range. When the phase was greater than 10 degrees, the amount of ADV was higher and below −15 degrees it was lower. The pilot experiments demonstrated that the phase of the impedance changed relatively quickly the first several times a previously unused catheter was used. After that, the amount of ADV stabilized and the catheter could be used for approximately 70 to 120 trials. Therefore, the first six trials of a new catheter were not used in the final data analysis.

5. Conclusions

The presented results indicate that the ADV transition efficiency and magnitude of oxygen scavenging when infusing perfluoropentane droplets through the coolant port of an EkoSonic® device driven at 47 W does not change substantially in the presence of Lumason® at a concentration based on clinical dosing. A modest effect was observed where more ADV occurs as the ultrasound duty cycle was increased from 1.70% to 3.79%. More significant changes in transition efficiency and oxygen scavenging were observed as the droplet concentration was increased from 5×10^{-5} mL/mL to 5×10^{-4} mL/mL, but then saturated for higher concentrations. These results indicate the oxygen scavenging effect can be modulated most readily by varying the droplet concentration. Furthermore, alternate approaches to increasing the transition efficiency at higher droplet concentrations may provide a route to overcoming the saturation effect.

Supplementary Materials: The following supporting information can be downloaded at: https://www.mdpi.com/article/10.3390/pharmaceutics14112392/s1, Table S1: Average droplet separation for each nominal droplet diameter at the lowest and highest droplet concentrations used.

Author Contributions: R.P.B.: conceptualization, methodology, investigation, validation, formal analysis, visualization, writing—original draft, writing—review and editing; N.A.R.: methodology, investigation, formal analysis, writing—original draft, writing—review and editing; K.S.: validation, formal analysis, writing—original draft, writing—review and editing; A.C.: methodology, visualization, writing—original draft preparation, writing—review and editing; B.Z.: methodology, formal analysis, writing—review and editing; K.J.H.: conceptualization, methodology, software, validation, formal analysis, visualization, writing—review and editing, funding acquisition, resources, project administration, supervision. All authors have read and agreed to the published version of the manuscript.

Funding: This research was funded, in part, by the National Heart, Lung, and Blood Institute of the National Institutes of Health under award number R01HL148451.

Institutional Review Board Statement: Not applicable.

Informed Consent Statement: Not applicable.

Data Availability Statement: The data presented in this study are available from the corresponding author upon reasonable request.

Acknowledgments: The authors gratefully acknowledge Curtis Genstler of Boston Scientific for assistance in using the EkoSonic® catheters.

Conflicts of Interest: The authors declare no conflict of interest. The funders had no role in the design of the study; in the collection, analyses, or interpretation of data; in the writing of the manuscript; or in the decision to publish the results.

References

1. Van Wamel, A.; Sontum, P.C.; Healey, A.; Kvåle, S.; Bush, N.; Bamber, J.; de Lange Davies, C. Acoustic Cluster Therapy (ACT) Enhances the Therapeutic Efficacy of Paclitaxel and Abraxane® for Treatment of Human Prostate Adenocarcinoma in Mice. *J. Control. Release* **2016**, *236*, 15–21. [CrossRef] [PubMed]
2. Andersen, K.K.; Healey, A.J.; Bush, N.L.; Frijlink, M.E.; Hoff, L. Design, Fabrication, and Testing of a Dual-Frequency Transducer for Acoustic Cluster Therapy Activation. In Proceedings of the 2018 IEEE International Ultrasonics Symposium (IUS), Kobe, Japan, 22–25 October 2018; pp. 1–4.
3. Kotopoulis, S.; Stigen, E.; Popa, M.; Safont, M.M.; Healey, A.; Kvåle, S.; Sontum, P.; Gjertsen, B.T.; Gilja, O.H.; McCormack, E. Sonoporation with Acoustic Cluster Therapy (ACT®) Induces Transient Tumour Volume Reduction in a Subcutaneous Xenograft Model of Pancreatic Ductal Adenocarcinoma. *J. Control. Release* **2017**, *245*, 70–80. [CrossRef] [PubMed]
4. Burgess, M.T.; Porter, T.M. Control of Acoustic Cavitation for Efficient Sonoporation with Phase-Shift Nanoemulsions. *Ultrasound Med. Biol.* **2019**, *45*, 846–858. [CrossRef]
5. Lea-Banks, H.; O'Reilly, M.A.; Hamani, C.; Hynynen, K. Localized Anesthesia of a Specific Brain Region Using Ultrasound-Responsive Barbiturate Nanodroplets. *Theranostics* **2020**, *10*, 2849–2858. [CrossRef] [PubMed]
6. Fan, C.-H.; Lin, Y.-T.; Ho, Y.-J.; Yeh, C.-K. Spatial-Temporal Cellular Bioeffects from Acoustic Droplet Vaporization. *Theranostics* **2018**, *8*, 5731–5743. [CrossRef]
7. Zhang, M.; Fabiilli, M.L.; Haworth, K.J.; Padilla, F.; Swanson, S.D.; Kripfgans, O.D.; Carson, P.L.; Fowlkes, J.B. Acoustic Droplet Vaporization for Enhancement of Thermal Ablation by High Intensity Focused Ultrasound. *Acad. Radiol.* **2011**, *18*, 1123–1132. [CrossRef]
8. Zhang, P.; Porter, T. An in Vitro Study of a Phase-Shift Nanoemulsion: A Potential Nucleation Agent for Bubble-Enhanced HIFU Tumor Ablation. *Ultrasound Med. Biol.* **2010**, *36*, 1856–1866. [CrossRef]
9. Peng, C.; Sun, T.; Vykhodtseva, N.; Power, C.; Zhang, Y.; Mcdannold, N.; Porter, T. Intracranial Nonthermal Ablation Mediated by Transcranial Focused Ultrasound and Phase-Shift Nanoemulsions. *Ultrasound Med. Biol.* **2019**, *45*, 2104–2117. [CrossRef]
10. Pajek, D.; Burgess, A.; Huang, Y.; Hynynen, K. High Intensity Focused Ultrasound Sonothrombolysis: The Use of Perfluorocarbon Droplets to Achieve Clot Lysis at Reduced Acoustic Powers. *Ultrasound Med. Biol.* **2014**, *40*, 2151–2161. [CrossRef]
11. Aydin, O.; Vlaisavljevich, E.; Yuksel Durmaz, Y.; Xu, Z.; ElSayed, M.E.H. Noninvasive Ablation of Prostate Cancer Spheroids Using Acoustically-Activated Nanodroplets. *Mol. Pharm.* **2016**, *13*, 4054–4065. [CrossRef]
12. Dong, X.; Lu, X.; Kingston, K.; Brewer, E.; Juliar, B.A.; Kripfgans, O.D.; Fowlkes, J.B.; Franceschi, R.T.; Putnam, A.J.; Liu, Z.; et al. Controlled Delivery of Basic Fibroblast Growth Factor (BFGF) Using Acoustic Droplet Vaporization Stimulates Endothelial Network Formation. *Acta Biomater.* **2019**, *97*, 409–419. [CrossRef] [PubMed]
13. Aliabouzar, M.; Lu, X.; Kripfgans, O.D.; Fowlkes, J.B.; Fabiilli, M.L. Acoustic Droplet Vaporization in Acoustically Responsive Scaffolds: Effects of Frequency of Excitation, Volume Fraction and Threshold Determination Method. *Ultrasound Med. Biol.* **2019**, *45*, 3246–3260. [CrossRef] [PubMed]

14. Lu, X.; Dong, X.; Natla, S.; Kripfgans, O.D.; Fowlkes, J.B.; Wang, X.; Franceschi, R.; Putnam, A.J.; Fabiilli, M.L. Parametric Study of Acoustic Droplet Vaporization Thresholds and Payload Release from Acoustically-Responsive Scaffolds. *Ultrasound Med. Biol.* **2019**, *45*, 2471–2484. [CrossRef]
15. Fischesser, D.M.; Bo, B.; Benton, R.P.; Su, H.; Jahanpanah, N.; Haworth, K.J. Controlling Reperfusion Injury With Controlled Reperfusion: Historical Perspectives and New Paradigms. *J. Cardiovasc. Pharmacol. Ther.* **2021**, *26*, 504–523. [CrossRef]
16. Mercado-Shekhar, K.P.; Su, H.; Kalaikadal, D.S.; Lorenz, J.N.; Manglik, R.M.; Holland, C.K.; Redington, A.N.; Haworth, K.J. Acoustic Droplet Vaporization-Mediated Dissolved Oxygen Scavenging in Blood-Mimicking Fluids, Plasma, and Blood. *Ultrason. Sonochem.* **2019**, *56*, 114–124. [CrossRef]
17. Kripfgans, O.D.; Fowlkes, J.B.; Miller, D.L.; Eldevik, O.P.; Carson, P.L. Acoustic Droplet Vaporization for Therapeutic and Diagnostic Applications. *Ultrasound Med. Biol.* **2000**, *26*, 1177–1189. [CrossRef]
18. Sheeran, P.S.; Wong, V.P.; Luois, S.; McFarland, R.J.; Ross, W.D.; Feingold, S.; Matsunaga, T.O.; Dayton, P.A. Decafluorobutane as a Phase-Change Contrast Agent for Low-Energy Extravascular Ultrasonic Imaging. *Ultrasound Med. Biol.* **2011**, *37*, 1518–1530. [CrossRef]
19. Kang, S.-T.; Huang, Y.-L.; Yeh, C.-K. Characterization of Acoustic Droplet Vaporization for Control of Bubble Generation under Flow Conditions. *Ultrasound Med. Biol.* **2014**, *40*, 551–561. [CrossRef]
20. Radhakrishnan, K.; Holland, C.K.; Haworth, K.J. Scavenging Dissolved Oxygen via Acoustic Droplet Vaporization. *Ultrason. Sonochem.* **2016**, *31*, 394–403. [CrossRef]
21. Fabiilli, M.L.; Haworth, K.J.; Sebastian, I.E.; Kripfgans, O.D.; Carson, P.L.; Fowlkes, J.B. Delivery of Chlorambucil Using an Acoustically-Triggered Perfluoropentane Emulsion. *Ultrasound Med. Biol.* **2010**, *36*, 1364–1375. [CrossRef]
22. Mercado, K.P.; Radhakrishnan, K.; Stewart, K.; Snider, L.; Ryan, D.; Haworth, K.J. Size-Isolation of Ultrasound-Mediated Phase Change Perfluorocarbon Droplets Using Differential Centrifugation. *J. Acoust. Soc. Am.* **2016**, *139*, EL142. [CrossRef] [PubMed]
23. Schad, K.C.; Hynynen, K. In Vitro Characterization of Perfluorocarbon Droplets for Focused Ultrasound Therapy. *Phys. Med. Biol.* **2010**, *55*, 4933–4947. [CrossRef] [PubMed]
24. Lo, A.H.; Kripfgans, O.D.; Carson, P.L.; Rothman, E.D.; Fowlkes, J.B. Acoustic Droplet Vaporization Threshold: Effects of Pulse Duration and Contrast Agent. *IEEE Trans. Ultrason. Ferroelectr. Freq. Control* **2007**, *54*, 933–946. [CrossRef] [PubMed]
25. Healey, A.J.; Sontum, P.C.; Kvåle, S.; Eriksen, M.; Bendiksen, R.; Tornes, A.; Østensen, J. Acoustic Cluster Therapy: In Vitro and Ex Vivo Measurement of Activated Bubble Size Distribution and Temporal Dynamics. *Ultrasound Med. Biol.* **2016**, *42*, 1145–1166. [CrossRef] [PubMed]
26. Sontum, P.; Kvåle, S.; Healey, A.J.; Skurtveit, R.; Watanabe, R.; Matsumura, M.; Østensen, J. Acoustic Cluster Therapy (ACT)—A Novel Concept for Ultrasound Mediated, Targeted Drug Delivery. *Int. J. Pharm.* **2015**, *495*, 1019–1027. [CrossRef]
27. Haworth, K.J.; Goldstein, B.H.; Mercado-Shekhar, K.P.; Srivastava, R.; Arunkumar, P.; Su, H.; Privitera, E.M.; Holland, C.K.; Redington, A.N. Dissolved Oxygen Scavenging by Acoustic Droplet Vaporization Using Intravascular Ultrasound. In Proceedings of the 2017 IEEE International Ultrasonics Symposium (IUS), Washington, DC, USA, 6–9 September 2017; Volume 2017. [CrossRef]
28. Moncion, A.; Lin, M.; O'Neill, E.G.; Franceschi, R.T.; Kripfgans, O.D.; Putnam, A.J.; Fabiilli, M.L. Controlled Release of Basic Fibroblast Growth Factor for Angiogenesis Using Acoustically-Responsive Scaffolds. *Biomaterials* **2017**, *140*, 26–36. [CrossRef]
29. Houk, L.R.; Challa, S.R.; Grayson, B.; Fanson, P.; Datye, A.K. The Definition of "Critical Radius" for a Collection of Nanoparticles Undergoing Ostwald Ripening. *Langmuir* **2009**, *25*, 11225–11227. [CrossRef]
30. Taylor, P. Ostwald Ripening in Emulsions: Estimation of Solution Thermodynamics of the Disperse Phase. *Adv. Colloid Interface Sci.* **2003**, *106*, 261–285. [CrossRef]
31. Bell, R.M.; Mocanu, M.M.; Yellon, D.M. Retrograde Heart Perfusion: The Langendorff Technique of Isolated Heart Perfusion. *J. Mol. Cell. Cardiol.* **2011**, *50*, 940–950. [CrossRef]
32. Lafond, M.; Salido, N.G.; Haworth, K.J.; Hannah, A.S.; Macke, G.P.; Genstler, C.; Holland, C.K. Cavitation Emissions Nucleated by Definity Infused through an EkoSonic Catheter in a Flow Phantom. *Ultrasound Med. Biol.* **2021**, *47*, 693–709. [CrossRef]
33. Fasciolo, J.C.; Chiodi, H. Arterial Oxygen Pressure during Pure O_2 Breathing. *Am. J. Physiol.* **1946**, *147*, 54–65. [CrossRef] [PubMed]
34. Christoforides, C.; Laasberg, L.H.; Hedley-Whyte, J. Effect of Temperature on Solubility of O_2 in Human Plasma. *J. Appl. Physiol.* **1969**, *26*, 56–60. [CrossRef] [PubMed]
35. McGraw, K.O. "Forming Inferences about Some Intraclass Correlations Coefficients": Correction. *Psychol. Methods* **1996**, *1*, 390. [CrossRef]
36. Yang, Y.; Yang, D.; Zhang, Q.; Guo, X.; Raymond, J.L.; Roy, R.A.; Zhang, D.; Tu, J. The Influence of Droplet Concentration on Phase Change and Inertial Cavitation Thresholds Associated with Acoustic Droplet Vaporization. *J. Acoust. Soc. Am.* **2020**, *148*, EL375. [CrossRef]
37. Burgess, M.T.; Aliabouzar, M.; Aguilar, C.; Fabiilli, M.L.; Ketterling, J.A. Slow-Flow Ultrasound Localization Microscopy Using Recondensation of Perfluoropentane Nanodroplets. *Ultrasound Med. Biol.* **2022**, *48*, 743–759. [CrossRef]
38. Filipe, V.; Hawe, A.; Jiskoot, W. Critical Evaluation of Nanoparticle Tracking Analysis (NTA) by NanoSight for the Measurement of Nanoparticles and Protein Aggregates. *Pharm. Res.* **2010**, *27*, 796–810. [CrossRef]
39. Aliabouzar, M.; Kripfgans, O.D.; Wang, W.Y.; Baker, B.M.; Brian Fowlkes, J.; Fabiilli, M.L. Stable and Transient Bubble Formation in Acoustically-Responsive Scaffolds by Acoustic Droplet Vaporization: Theory and Application in Sequential Release. *Ultrason. Sonochem.* **2021**, *72*, 105430. [CrossRef]

pharmaceutics

Article

Cavitation Characterization of Size-Isolated Microbubbles in a Vessel Phantom Using Focused Ultrasound

Payton Martinez [1,2], Nick Bottenus [1,3] and Mark Borden [1,3,*]

1 Biomedical Engineering Program, University of Colorado, Boulder, CO 80309, USA
2 IQ Biology Program, University of Colorado, Boulder, CO 80309, USA
3 Mechanical Engineering Department, University of Colorado, Boulder, CO 80309, USA
* Correspondence: mark.borden@colorado.edu

Abstract: Pharmaceutical delivery can be noninvasively targeted on-demand by microbubble (MB) assisted focused ultrasound (FUS). Passive cavitation detection (PCD) has become a useful method to obtain real-time feedback on MB activity due to a FUS pulse. Previous work has demonstrated the acoustic PCD response of MBs at a variety of acoustic parameters, but few have explored variations in microbubble parameters. The goal of this study was to determine the acoustic response of different MB size populations and concentrations. Four MB size distributions were prepared (2, 3, 5 μm diameter and polydisperse) and pulled through a 2% agar wall-less vessel phantom. FUS was applied by a 1.515 MHz geometrically focused transducer for 1 ms pulses at 1 Hz PRF and seven distinct mechanical indices (MI) ranging from 0.01 to 1.0 (0.0123 to 1.23 MPa PNP). We found that the onset of harmonic (HCD) and broadband cavitation dose (BCD) depends on the mechanical index, MB size and MB concentration. When matched for MI, the HCD and BCD rise, plateau, and decline as microbubble concentration is increased. Importantly, when microbubble size and concentration are combined into gas volume fraction, all four microbubble size distributions align to similar onset and peak; these results may help guide the planning and control of MB + FUS therapeutic procedures.

Keywords: passive cavitation detection; size isolated microbubbles; focused ultrasound; stable cavitation; inertial cavitation

Citation: Martinez, P.; Bottenus, N.; Borden, M. Cavitation Characterization of Size-Isolated Microbubbles in a Vessel Phantom Using Focused Ultrasound. *Pharmaceutics* **2022**, *14*, 1925. https://doi.org/10.3390/pharmaceutics14091925

Academic Editors: Shashank Sirsi, James Kwan and Michael Gray

Received: 11 August 2022
Accepted: 6 September 2022
Published: 12 September 2022
Corrected: 21 October 2022

1. Introduction

Microbubble ultrasound contrast agents (MBs) typically comprise a phospholipid shell and a low-solubility weight gas (e.g., perfluorocarbons) [1] and range from 0.1 to 10 um in diameter [2,3]. Several compositions are clinically approved for ultrasound contrast imaging due to their echogenicity [3]. The stabilizing shell can be modified to provide ligand targeting or drug attachment [4]; this versatility has allowed the emergence of microbubbles as an agent for more precise imaging applications, including super-resolution and molecular imaging [5–7].

A unique property of microbubbles is their ability to cavitate (oscillate) strongly under ultrasound. The MB oscillations in response to ultrasound induce localized forces that can provide a means of cell membrane permeabilization and drug extravasation [8–10]. The combination of these effects with the noninvasive targeting precision of focused ultrasound (FUS) has become an area of interest for novel pharmaceutical therapeutics; it has been shown that FUS and MBs can disrupt the blood–brain barrier (BBB) at a lower mechanical index (MI) than for FUS alone [11–13]. At the lower MI, there is minimal damage while creating a pathway for macromolecules to enter the parenchyma [8,14]. To create safer and more effective therapies, efforts are underway to better understand and control the mechanisms behind MB cavitation.

Many acoustic parameters have been investigated with respect to the resulting cavitation. For example, the pressure threshold of blood–brain barrier disruption (BBBD) was shown to decrease with increasing pulse length, where an increasing number of cycles increased the extent of BBBD [11,15]. However, the pulse repetition frequency has not been shown to affect the threshold [11,12]. The mechanical index is one of the most important factors for achieving BBBD [11–13]. While these studies have varied the ultrasound parameters, few have discussed the microbubble parameters.

Microbubble size has been of interest to the effect on therapy, including BBBD; it was previously shown that dye-delivery by BBBD depends on microbubble diameter [16,17]. Larger microbubbles had a higher BBBD when concentrations were matched by the number of bubbles; these studies demonstrated that larger microbubbles had a lower threshold to observe significant BBBD [16,17]. Wang et al. Furthermore showed that microbubble composition (e.g., choice of gas core, shell) affects the intensity of BBBD, although differences were not as significant as size [18].

Clinically approved microbubbles come in a variety of concentrations and size distributions that can even vary vial-to-vial for the same product [19]. Studies have shown that this variety may result in different acoustic responses in both imaging [20–23] and therapeutics, which can result in different biological effects [16,24,25]. MB size distribution and concentration can be combined into a single parameter, called the gas volume fraction (GVF, μL/mL), or microbubble volume dose (MVD) when prescribed to the subject's body weight (μL/kg) [26]. Interestingly, effects of different MB size distributions at matched MVD were shown to be similar for both pharmacokinetics [25,27] and BBBD drug extravasation [26]; this begs the question as to whether cavitation activity of different MB size distributions and concentrations also collapses to a master curve when plotted against MVD. Such a result would help to bridge the knowledge gap between MB pharmacokinetics and MB + FUS bioeffects.

A passive cavitation detector (PCD) allows real-time analysis of the acoustic emissions of the microbubbles [28–32]. Previous reports have shown two defining forms of MB cavitation, "stable" and "inertial" cavitation [33]. Stable cavitation occurs at lower acoustic pressures and is defined by relatively small oscillations and minimal effect on the resting size of the microbubble [33–35]. Stable cavitation has been identified by harmonic emissions during ultrasonic excitation [4,36]. However, harmonic signals can be produced by tissue or coupling media and therefore are not strictly indicative of microbubbles [37]. On the contrary, sub/ultra-harmonic emissions only arise from microbubbles and are, therefore better indicators of bubble activity [38]. As the acoustic driving pressure increases, the MBs can move into the inertial cavitation regime. Inertial cavitation emissions have been identified by a broadband PCD response [28,30,36], where they collapse violently to produce a broadband echo. At this point, jetting, shockwaves and other inertial effects can occur [33,35,39].

However, microbubbles observed under high-speed video microscopy have been shown to display relatively large oscillations and may change the size by rectified diffusion, coalescence, fragmentation or dissolution at the same acoustic forcing in which they would be characterized as undergoing "stable" cavitation with PCD data [40–42]. In fact, the MBs may be dynamic (changing size and number), rather than stable, while they are emitting sub/ultra-harmonic echoes. MBs can also move into intermittent inertial cavitation and still produce sub and ultra-harmonic echoes [43]. We, therefore, use the terms "harmonic" or "broadband" cavitation, rather than "stable" or "inertial" cavitation, to describe the behavior.

The purpose of this study was to investigate the effects of microbubble size and concentration on the harmonic and broadband cavitation response in a vessel phantom. To achieve this, microbubbles were pulled through a wall-less agar vessel phantom while FUS was applied. The resulting acoustic response was recorded with a PCD. Four size distributions were analyzed at mechanical indices ranging from 0.01 to 1.0. Each size distribution was also examined at a range of concentrations from 1.5×10^4 to 1.5×10^8 MBs/mL. The harmonic cavitation dose (HCD) and broadband cavitation dose (BCD) were calculated and compared at different MB diameters, concentrations and GVFs.

2. Materials and Methods

2.1. Materials

All solutions were made using filtered, sterile, deionized water (Direct-Q 3 Millipore, Billerica, MA, USA). All glassware was cleaned with 70 vol% isopropyl alcohol (Supelco, Burlington, MA, USA). 1,2-distearoyl-sn-glycero-3-phosphocholine (DSPC) was purchased from Avanti Polar Lipids (Alabaster, AL, USA). Perfluorobutane gas (n-C_4F_{10}, PFB) was purchased from FluroMed (Round Rock, TX, USA). Polyoxyethylene-40 stearate (PEG40S) and chloroform were purchased from Sigma-Aldrich (St. Louis, MO, USA). Sterile saline solution and phosphate-buffered saline solution (PBS) were purchased from Fisher Scientific (Pittsburg, PA, USA). The purity of all the reagents was $\geq$99%, and they were used as received without further purification.

2.2. Experimental Set-up

To target focused ultrasound, we designed and built a custom wall-less vessel phantom (Figure 1). The RK-50 system (Stereotactic-guided Focused Ultrasound, FUS Instruments, Toronto, ON, Canada) was used for our experiments. The RK-50 system has a single element, 1.515 MHz, focused transducer with a 20-mm diameter. To receive returning signals from the microbubbles, a built-in passive cavitation detector (PCD) positionedcoaxially within the focused transducer was used. The PCD had a wide bandwidth and center frequency of 0.7575 MHz, with an 8 mm diameter. PCD voltage vs. time data were collected by the RK-50 software. The phantom was placed under the transducer. The phantom shell was filled with filtered and degassed water. On either end of the vessel, female luers were used to attach 0.9 mm outer diameter tubing. One end of the tubing was placed in a beaker of microbubbles on a stir-plate with medium stirring (150 RPM). The other end of the tubing was attached to a 12 mL syringe (Monoject, Coviden, Mansfield, MA, USA) on a withdrawing syringe pump (Model 1000, New Era Pump Systems, Farmingdale, NY, USA). Microbubbles were diluted down to the desired concentration just before the flow was started.

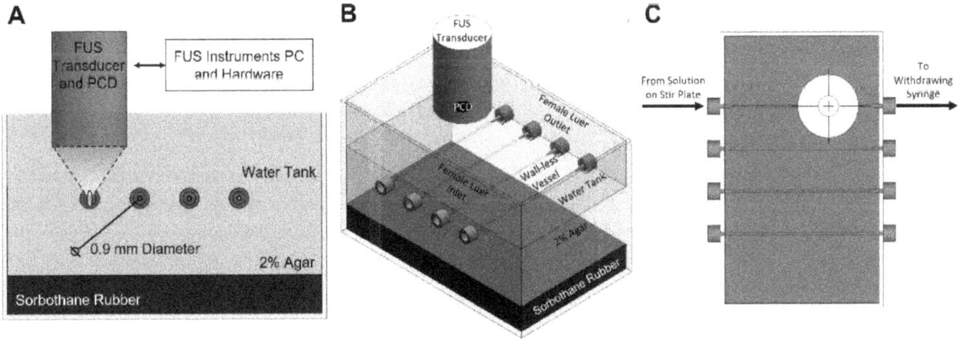

Figure 1. Diagram of experimental set up. A passive cavitation detector (PCD) was positioned perpendicular to the wall-less vessel and coaxially to the focused ultrasound (FUS) beam. The body of the phantom was made of 2% agar. The space between the agar and FUS transducer was degassed water. Sorbothane rubber was used beneath the agar to minimize reflections. A block diagram and side view of the phantom (**A**) shows a connection to the FUS Instruments RK50 system. An isometric view of the phantom (**B**) illustrates the four wall-less vessels with luer inlets and outlets. The microbubble flow path from solution on the stir plate through the phantom to a withdrawing syringe can be seen in the top view (**C**).

2.3. Fabrication of Agar Phantom

The agar phantom was created within an acrylic box with dimensions shown in Supplementary Figure S1. Barbed female luers were added to four evenly spaced points along the length of the box with barbed pieces facing inwards. A sorbothane vibration isolating rubber (Isolate It! Burlington, NC, USA) layer was added to the bottom of the box as an acoustic absorber. Prior to preparing the agar, polyethylene tubing (Warner Instruments, Holliston, MA, USA) with an outer diameter of 0.9 mm was pulled through both ends of the luers. The 0.9 mm diameter was chosen to be small enough to be clearly within the focal zone of the FUS and PCD. Agar (2%) was prepared using 11 g of Agar powder (Millipore, Burlington, MA, USA) mixed with 550 mL of filtered and distilled water at room temperature for 20 min or until the powder had fully dissolved. The solution was then placed in an 1100-Watt microwave and heated for 30 s, mixed, then reheated until it was transparent. The solution was then placed in a vacuum chamber to degas. The pressure was slowly reduced by 80 kPa over a five-minute period. The solution was then removed and slowly poured into the acrylic box. Polyethylene tubing was adjusted as needed to confirm a perpendicular angle to the side of the phantom. All, if any, air bubbles were removed by popping or scooping them out. The phantom was allowed to cool for one and a half hours at room temperature.

2.4. Microbubble Preparation

Lipid-coated MBs with a PFB gas core was prepared by sonication, as described previously [44]. Briefly, under sterile conditions, a dried lipid film comprising DSPC:PEG40S (90:10) was hydrated with filtered and sterile PBS (1X) to a final lipid/emulsifier concentration of 2 mg/mL at 65 °C for 40 min. Once fully hydrated, the solution was allowed to cool to room temperature. The lipid solution was sonicated with a 20 kHz probe (model 250A, Branson Ultrasonics; Danbury, CT, USA) at low power (3/10; 3 W) for four minutes. After cooling to room temperature, PFB was delivered to the surface of the lipid suspension for 10 s. Then, the solution was sonicated at high power (10/10; 33 W) for 30 s to produce polydisperse MBs. Polydisperse MBs were then collected into 30 mL sterile syringes and isolated by differential centrifugation into three diameters: 2, 3 and 5 μm diameter. The size isolation process, including time and speeds of centrifugation, can be found in Supplemental Figure S2. MB concentration and number- and volume-weighted size distributions were measured using a Multisizer 3 Coulter Counter (Beckman Coulter). MB concentration (c, MBs/μL) versus MB volume (v, μL/MB) was plotted, and MB gas volume fraction (ΦMB) was estimated as follows:

$$\Phi MB = \sum_{i=1}^{n} c_i \times v_i \quad (1)$$

where i is the index of the sizing bin, and there were 300 logarithmically spaced bins ranging from 0.7 to 18 μm in diameter. Three independent size-isolated *MB* preparations were evaluated after synthesis and three hours prior to sonication to confirm the size distributions and concentration. Finally, after isolation, *MB* cakes were stored in the refrigerator at 4 °C for subsequent use.

2.5. Focused Ultrasound Stimulation of Microbubbles

A sinusoidal burst signal with a center frequency of 1.515 MHz was produced using the RK-50 system. The signal had a pulse repetition frequency (PRF) of 1 Hz, pulse duration (PD) of 1 ms, the peak negative pressures (*PNP*) varied from 12.3 kPa to 1.23 MPa. Mechanical Index (*MI*) is a unifying ultrasound parameter that combines the ultrasound center frequency (f in units of MHz) and peak negative pressure (*PNP* in units of MPa). The relationship is:

$$MI = \frac{PNP}{\sqrt{f}} \quad (2)$$

Previous studies have shown that *MI* correlates well with biological effects including BBBD over a range of peak negative pressures and fundamental frequencies [12]. Therefore, the *MI*s used during sonication were between 0.01 and 1.0. Microbubbles were kept at 4 °C just prior to dilution into room temperature water and subsequent PCD measurement.

MB stability during stirring was confirmed by measuring the concentration and mean diameter periodically after 15 min of stirring (Supplemental Figure S3). The syringe pump withdrew at 34 mL/h to allow new microbubbles in the focal area of ultrasound between each pulse. The flowrate was slow enough to avoid a large pressure gradient (~150 Pa), yet high enough to avoid significant bubble flotation on the top of the tube due to buoyancy.

2.6. PCD Data Collection and Analysis

To analyze the acoustic cavitation of the microbubbles, returning pressure was converted into voltage via a PCD device built into the FUS transducer. The resulting voltage over time signals were stored on the RK-50 desktop. Data were collected at a 37 MHz sampling frequency. The collection started just prior to initial sonication. All data were analyzed using MATLAB (MathWorks, Natick, MA, USA). Voltage data were first cropped to remove any signal prior to the first expected return signal (pre-signal sensitivity plus travel time of the first pulse). Tukey windowing was applied to the cropped signal to prevent spectral leakage on the fast Fourier transform (FFT). The FFT was then taken for every pulse (26). The FFT magnitudes were averaged together within each treatment.

2.7. Harmonic Cavitation Dose

Sub-harmonic and ultra-harmonic ($F \times n/2$; $n = 1, 3, 5 \ldots$) frequencies are associated with the harmonic (non-broadband) cavitation of microbubbles. Once voltage data were converted to the frequency domain, the power spectrum was used to analyze these frequencies. A lowpass filter was applied at the end of the fourth harmonic (6.06 MHz). The sub-harmonic and the first three ultra-harmonic (e.g., 0.7575, 2.2725, 3.7875, 5.3025 MHz) components were extracted using a minimum order bandpass filter, with a bandwidth of 0.1 MHz, at each frequency. Fast Fourier transform was used to obtain the frequency spectrum:

$$X(k) = \sum_{i=0}^{N-1} x(i)e^{-j\frac{2\pi}{N}ik}(k = 0, 1, \ldots, N-1) \tag{3}$$

where k is the frequency point in the frequency spectrum. To determine the intensity of the harmonic frequency components, the power spectrum of the signal was then calculated by:

$$P(k) = X(k)^2(k = 0, 1, \ldots, N-1) \tag{4}$$

The resulting power spectrums were summed together and multiplied by the total sampling time:

$$HCDm = Ts \sum_{1}^{a} \sum_{1}^{b} P(k) \tag{5}$$

where a and b are the number of sub/ultra-harmonic components and the number of frequency points in every sub/ultra-harmonic component, respectively. Note that the four sub/ultra-harmonic components ranged from 0.7075 MHz to 5.3525 MHz, ($a = 1, 2, \cdots, 4$), and the total number of frequency points in every component can be calculated by dividing the bandwidth of analysis (0.1 MHz) by the frequency resolution (1587 Hz), giving 63 points per component. In Equation (5), the coefficient was the sampling time (Ts) for harmonic cavitation signals. The sampling time used was 850×10^{-6} s. The real harmonic cavitation dose (HCD) from the microbubbles can be calculated by subtracting the measurement from the control degassed saline solution ($HCDc$, Figure S2):

$$HCD = HCDm - HCDc \tag{6}$$

2.8. Broadband Cavitation Dose

The broadband signal is associated with the inertial cavitation of microbubbles. Once voltage data were converted to the frequency domain, the power spectrum was used to analyze these frequencies. A lowpass filter was applied at the end of the fourth harmonic

(6.06 MHz). The subharmonic, the first three ultra-harmonics and the first four harmonics (e.g., 0.7575, 1.515, 2.2725, 3.03... etc. MHz) components were removed using a stopband filter, with a bandwidth of 0.1 MHz, at each frequency. Fast Fourier transform was used to obtain the frequency spectrum:

$$X(k) = \sum_{i=0}^{N-1} x(i)e^{-j\frac{2\pi}{N}ik}(k = 0,1,\ldots,N-1) \tag{7}$$

where again k is the frequency point in the frequency spectrum. To determine the intensity of the broadband frequency components, the power spectrum of the signal was then calculated by:

$$P(k) = X(k)^2(k = 0,1,\ldots,N-1) \tag{8}$$

The resulting power spectrums were summed together and multiplied by the total sampling time:

$$BCDm = Ts \sum_1^a \sum_1^b P(k) \tag{9}$$

where a and b are the number of broadband components and the number of frequency points in every broadband component, respectively. Noted that seven broadband components ranged from 0.8075 MHz to 6.01 MHz, (a = 1, 2, $\cdots$, 7), and the total number (414) of frequency points in every broadband component can be calculated by dividing the bandwidth of analysis (0.6575 MHz) by the frequency resolution (1587 Hz). In Equation (9), the coefficient was the sampling time (*Ts*) for broadband cavitation signals. The sampling time used was 850×10^{-6} s. After broadband cavitation dose control (*BCDc*) from the control degassed saline solution was obtained, the real broadband cavitation dose (*BCD*) from the microbubbles can be calculated by:

$$BCD = BCDm - BCDc \tag{10}$$

2.9. Statistical Analysis

For each treatment a total of 26 bursts were completed. An FFT was taken for each burst individually and the magnitudes were averaged together. A minimum of four replicate treatments were done for each concentration, mechanical index, and microbubble size distribution using a new batch of microbubbles and agar. The saline control solution was also completed after changing microbubble size distribution, new agar, or moving to a new channel. The control values were ensured to be similar in magnitude for each change (Supplemental Figure S4). The standard deviation and mean were calculated for the respective groups. An ANOVA was completed at all plots where *HCD* or *BCD* was expressed. Tukey's multiple comparisons test was used to determine the significance between points. All adjusted p-values obtained are shown in Supplemental Tables S1–S4.

3. Results and Discussion

3.1. MB Characterization

Following size isolation of microbubble populations, each size was visually inspected using microscope brightfield images shown in Figure 2A–D. Coulter Multisizer sizing confirmed an initially broad polydisperse population with uniform sizes following centrifugal isolations. Figure 2E illustrates the number weighted percent of the populations giving a mean diameter of 2.0, 2.9, 4.4 and 1.9 μm for 2, 3, 5 μm and polydisperse respectively. Volume weighted percent had similar narrow peaks with mean diameters of 2.4, 3.5, 5.5 and 8.8 μm (Figure 2F). Figure 2G shows the plot used to determine the gas volume fraction (ΦMB) of each microbubble (MB) population. The mean ΦMB per MB for 10^{10} MBs/mL was shown to be 5.28, 14.8, 54.7, and 18.7 μL/mL for 2, 3, 5 μm and polydisperse respectively (Table 1). Statistical analysis was done for each plot, showing significant differences between all size isolated populations ($p < 0.05$).

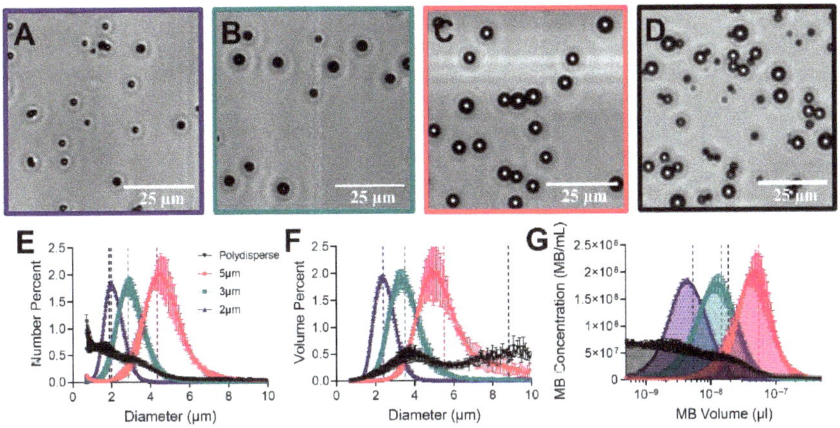

Figure 2. Characterization of size-selected and polydisperse microbubbles. Representative brightfield images of each diameter (2, 3 and 5 µm) and polydisperse microbubbles ((**A–D**), respectively). Number weighted (**E**) and volume-weighted (**F**) size distributions. (**G**) Microbubble concentration against the microbubble volume; the shaded area under the curve represents the gas volume fraction at the given total concentration (10^{10} MBs/mL). Vertical dotted lines of matched color represent mean values for the size distribution. Statistical analysis showed a significant difference ($p < 0.05$) between all isolated diameters (2, 3 and 5 µm) but not the polydisperse population. Data points represent the mean and error bars show the standard deviation ($n = 4$).

Table 1. Microbubble size distribution statistics. Table shows the mean diameter, mode diameter, standard deviation, d_{10} and d_{90} (where 10 or 90 percent of all particles fall below this diameter), in the number weighted graph (Figure 2E). The mean diameter for the volume weighted plot (Figure 2F) is shown next. The average gas volume fraction (ΦMB) per microbubble is given in the rightmost column and found using Figure 2G. All values were averaged over 4 separate measurements.

Size Distribution Characterization							
	Number % Parameters					Volume % Mean	ΦMB @ 10^{10} MBs/mL
	Mean	**Mode**	**SD**	d_{10}	d_{90}		
2 µm	2.0 µm	1.8 µm	0.46 µm	1.3 µm	2.4 µm	2.4 µm	5.28 µL/mL
3 µm	2.9 µm	2.9 µm	0.7 µm	2.2 µm	3.6 µm	3.5 µm	14.8 µL/mL
5 µm	4.4 µm	4.2 µm	1.1 µm	3.2 µm	5.4 µm	5.5 µm	54.7 µL/mL
Polydisperse	1.9 µm	1.0 µm	1.6 µm	0.8 µm	3.6 µm	8.8 µm	18.7 µL/mL

3.2. Signal Analysis

The acoustic response of control PBS is shown in Figure 3. As expected, the saline showed very little acoustic feedback at all mechanical indices. Supplemental Figure S4 shows the difference in harmonic cavitation dose between the control PBS and the microbubbles. Figure 4 shows the voltage and frequency response of 5 µm diameter microbubbles (1.5×10^6 MBs/mL; MI = 1.0) flowing (34 mL/h) through a 900 µm diameter wall-less vessel. After signal processing depicted in Figures 3 and 4, two types of responses could be determined: harmonic cavitation dose (HCD) and broadband cavitation dose (BCD).

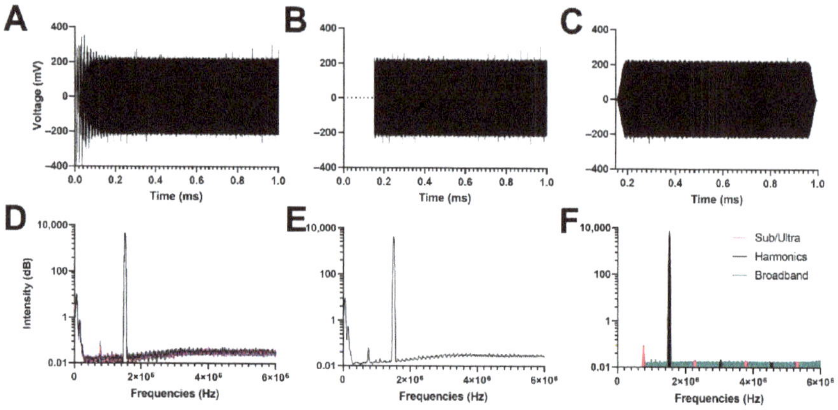

Figure 3. Signal processing of voltage versus time data obtained from the passive cavitation detector for a PBS control (**A**). Initial 150 μs removed to decrease noise prior to first sonication reflection at the vessel (**B**). Tukey window is applied (**C**). The Fast Fourier Transform (FFT) was taken for all 26 sonications (**D**). For all replicates, the average FFT was obtained (**E**). The area under the curve for sub/ultra-harmonics, harmonics, and broadband peaks were calculated (**F**).

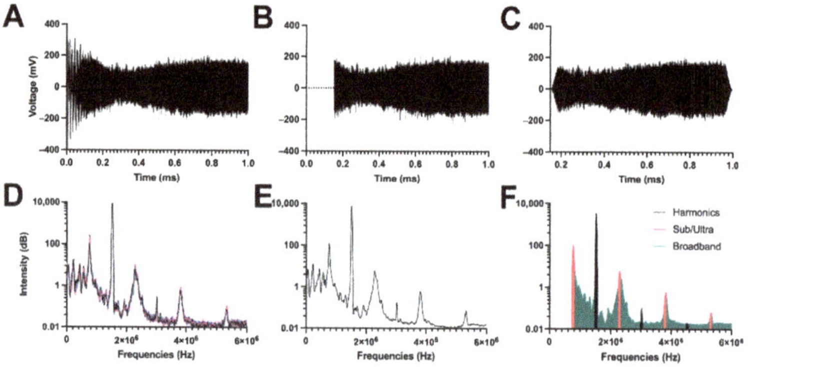

Figure 4. Signal processing of voltage versus time data obtained from the passive cavitation detector for 1.5×10^6 MBs/mL of 5 μm MBs sonicated at 1.0 mechanical index. (**A**–**F**) Similar signal processing as Figure 3.

3.3. Harmonic Cavitation Dose vs. Mechanical Index

The subharmonic and ultraharmonic response of four different size distributions of microbubbles across a range of mechanical indices (MIs) from 0.01 to 1.0 is illustrated in Figure 5. For all diameters at a given concentration, the HCD increased only slightly up to a threshold MI, above which a larger increase was observed. The general trend was a linear increase for both regimes of HCD vs. MI, below and above the threshold MI. Our threshold MI for the onset of harmonic cavitation dose (HCD) matched previously determined MIs both in vivo [37] and in vitro [45]. The threshold MI appeared to increase in magnitude for decreasing bubble diameter (e.g., from 0.4 MI for 2 μm MBs to 0.1 MI for 5 μm MBs at a matched concentration of 1.5×10^5 MBs/mL). Additionally, at a given MI, the HCD increased to a peak and then decreased with MB concentration, presumably due to attenuation of the US signal during both transmit and receive; this phenomenon can also be explained by bubble-bubble interactions during the sonication [46,47]. The peak HCD was at a concentration of ~10^7 MBs/mL for 2 μm MBs and ~10^6 MBs/mL for the other MB sizes.

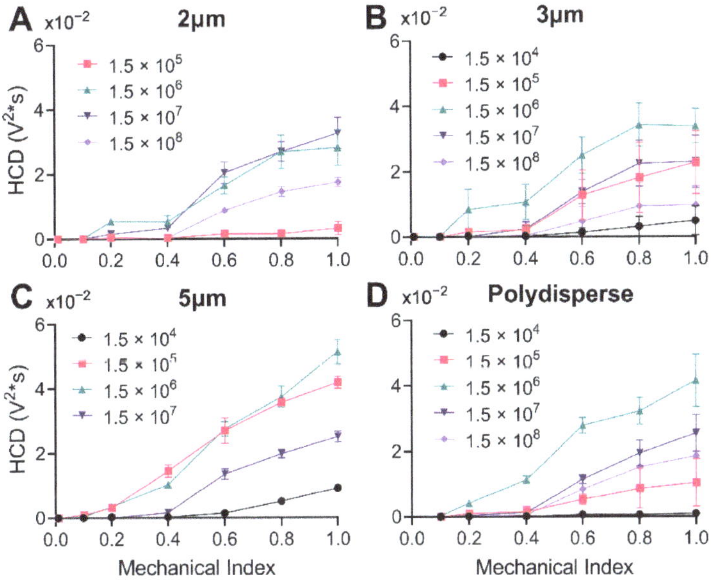

Figure 5. Harmonic Cavitation Dose (HCD) versus mechanical index at varying concentrations. (**A**) 2 µm microbubbles at four concentrations (1.5×10^5, 1.5×10^6, 1.5×10^7 and 1.5×10^8 MBs/mL). (**B**) 3 µm microbubbles at five concentrations (1.5×10^4–1.5×10^8 MBs/mL). (**C**) 5 µm microbubbles at four concentrations (1.5×10^4–1.5×10^7 MBs/mL). (**D**) Polydisperse microbubbles at five different concentrations (1.5×10^4–1.5×10^8 MBs/mL). Statistical significance can be found in Supplemental Table S1. Data represent the mean and errors bars show the standard deviation (n = 4).

3.4. Broadband Cavitation Dose vs. Mechanical Index

The broadband response of four different size distributions of microbubbles across a range of mechanical indices (MIs) from 0.01 to 1.0 is shown in Figure 6. There was a clear onset of BCD after 0.4 MI for all MB sizes, which matches previous literature [28–30]. After 0.4 MI, the BCD increased linearly, although at different slopes depending on the concentration. In some cases (e.g., for 1.5×10^8 MBs/mL for 2 µm MBs), there was a slight reduction from the linear trend at the highest MI of 1.0, presumably due to MB destruction from pulse to pulse.

3.5. Harmonic Cavitation Dose vs. Concentration

Figure 7 illustrates the sub/ultraharmonic response as number concentration was increased. The HCD increased, peaked, and then declined as MB concentration increased. Comparing across size distributions, the larger microbubbles (5 µm) had a peak around 1.5×10^6 MBs/mL whereas the smallest microbubble diameter (2 µm) had a peak around 1.5×10^7 MBs/mL. Both polydisperse and 3 µm diameter distributions had a similar peak around 1.5×10^6 MBs/mL. From a theoretical perspective, increasing HCD with MB concentration follows simply as an increase in the number of scatterers. The decreasing trend is likely due to attenuation (due to scattering and absorption) of the ultrasound signal during both transmit and receive. Previous work using numerical simulations has shown this decrease in sub/ultra-harmonic emissions as the number of microbubbles increases [48]. The simulations showed that bubble-bubble interactions intensified broadband signaling and therefore hid the harmonic emissions [46,48]. Similarly, using randomly distributed microbubbles, it was shown that sub and ultra-harmonic emissions plateau and decrease as the concentration of MBs is increased [47,49,50]; it is also reported that a peak acoustic emission occurs around 10^6 MBs/mL, matching our results (Figure 7) [50]. Additionally, previous work has shown that microbubbles themselves can attenuate acoustic signals through scattering and absorption of the acoustic energy during both transmit and receive [51]. Given the nonlinearity of MB attenuation and the complexity of bubble-bubble interactions, further work is required to determine the exact mechanism for the peak acoustic emission observed here. The maximal concentration corresponds to the optimum between these two opposing trends. Interestingly, these optimal concentrations roughly correspond to the physiologically relevant dose for commercial ultrasound contrast agent MBs [52].

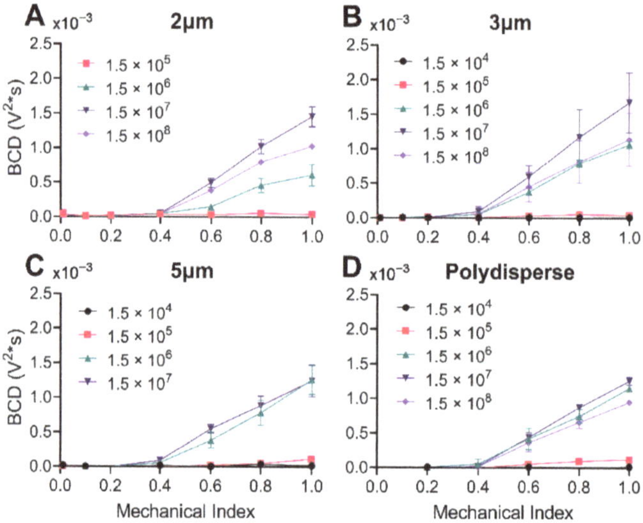

Figure 6. Broadband Cavitation Dose (BCD) versus the mechanical index. (**A**) 2 µm microbubbles at four concentrations (1.5×10^5–1.5×10^8 MBs/mL). (**B**) 3 µm microbubbles at five concentrations (1.5×10^4–1.5×10^8 MBs/mL). (**C**) 5 µm microbubbles at four concentrations (1.5×10^4–1.5×10^7 MBs/mL). (**D**) Polydisperse microbubbles at five different concentrations (1.5×10^4–1.5×10^8 MBs/mL). Statistical analyses can be found in Supplemental Table S2. Data represent the mean and error bars show the standard deviation ($n = 4$).

3.6. Harmonic Cavitation Dose vs. Gas Volume Fraction

In the spirit of using the MVD metric from prior pharmacological and therapeutic studies [26,27], we converted the MB number concentration to gas volume fraction (GVF), essentially matching total gas volume rather than the number of MBs. Figure 8 shows that as GVF increased, the sub/ultraharmonic response (HCD) rose to a peak and then declined. Importantly, there was a clear overlap between all size distributions.

At matching MI (Figure 8 left column), all MB size distributions had a similar onset, peak and decline. For all MB sizes, HCD peaked at a GVF ~0.02 µL/mL; this optimal GVF fell slightly above the estimated initial blood GVF at the recommended dose of clinically approved UCAs [53] (black dotted lines) and on the lower end of the current doses used for therapeutic MB + FUS [26] (pink dotted lines).

At matching MB size (Figure 8 right column), the trend between HCD and GVF is elucidated as a family of curves that increase with MI. Plotting as GVF aligns the onsets, peaks, and declines for all four size distributions. Interestingly, the peak HCD is approximately the same magnitude (0.0325–0.05 $V^2 \cdot s$) for all sizes and occurs at a similar MI (0.4–1.0). The family of curves for 3 µm diameter MBs is remarkable in that a clear triangular shape (owing to the onset, liner increase, peak, and linear decline) is observed for HCD vs. GVF for all MIs; this should greatly simplify the application of feedback control strategies.

3.7. Broadband Cavitation Dose vs. Gas Volume Fraction

Figure 9 shows the broadband response as the gas volume fraction was increased. Similarly, to the sub/ultra-harmonic response, there is a gas volume fraction where the onset occurs. The onset is consistently higher than the respective onset of sub/ultra-harmonic emissions. As gas volume fraction is increased, the broadband response increases to a peak and begins to plateau or decrease. In Figure 9A,C,E,G, all microbubble size distributions have similar onset, peak and declines relative to gas volume fraction. The magnitudes of BCD are also very similar for each MB population at matching MI and gas volume fractions.

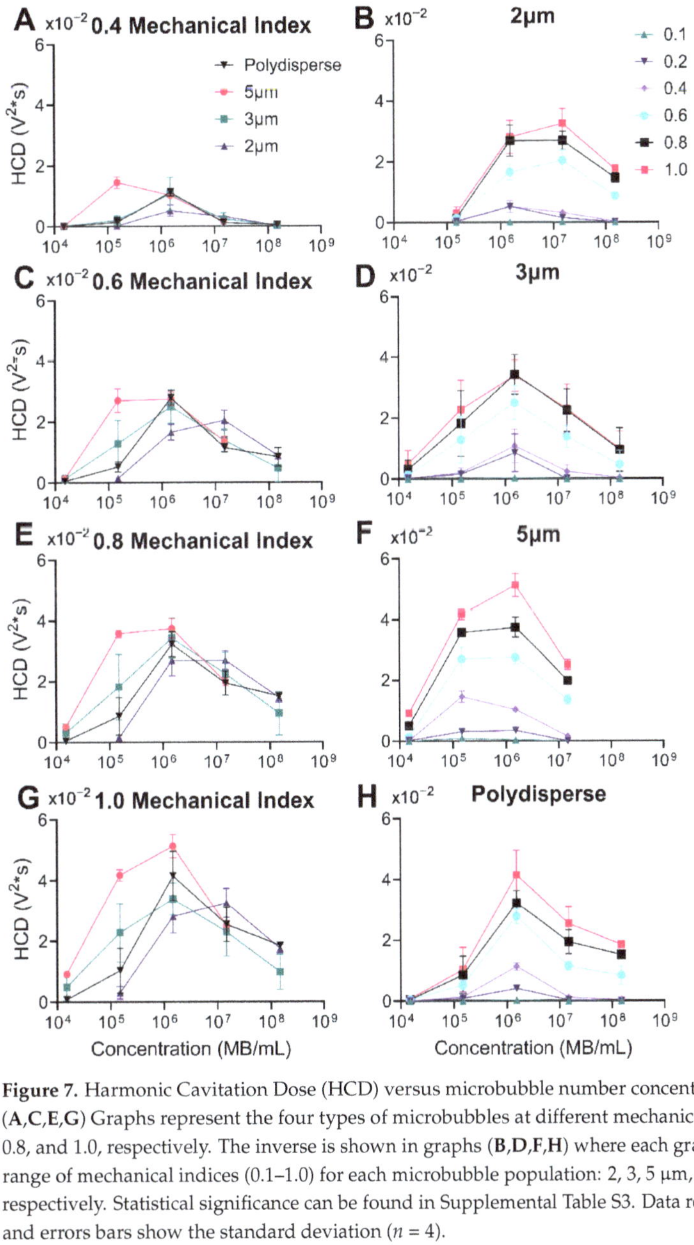

Figure 7. Harmonic Cavitation Dose (HCD) versus microbubble number concentration (MBs/mL). (**A**,**C**,**E**,**G**) Graphs represent the four types of microbubbles at different mechanical indices: 0.4, 0.6, 0.8, and 1.0, respectively. The inverse is shown in graphs (**B**,**D**,**F**,**H**) where each graph represents the range of mechanical indices (0.1–1.0) for each microbubble population: 2, 3, 5 μm, and polydisperse, respectively. Statistical significance can be found in Supplemental Table S3. Data represent the mean and errors bars show the standard deviation (n = 4).

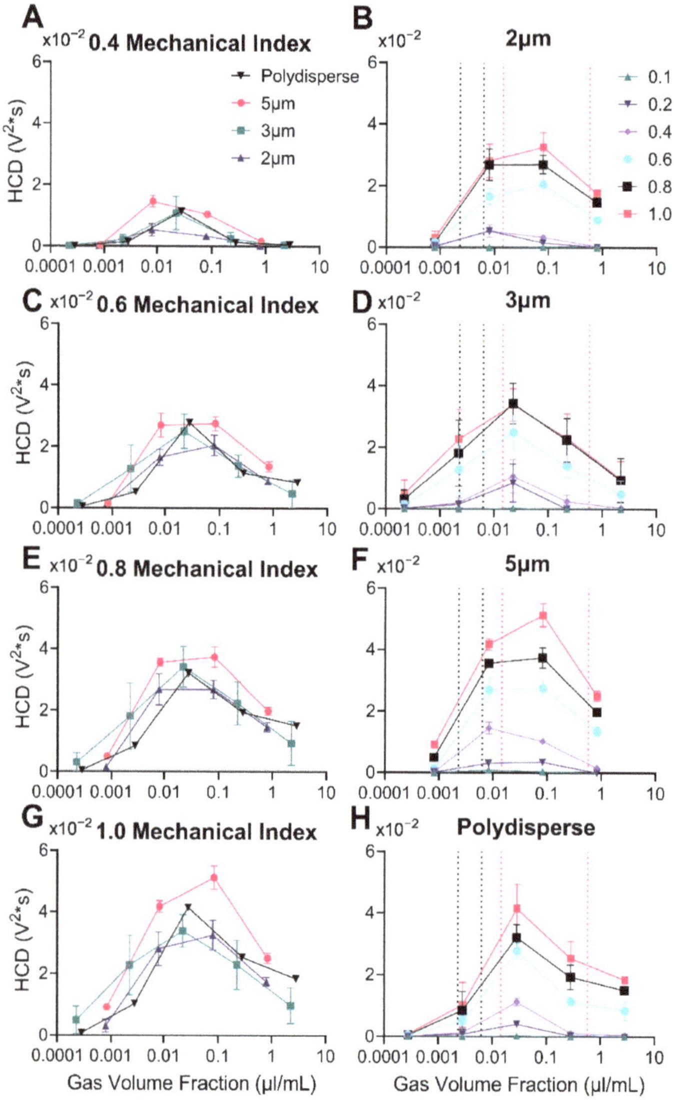

Figure 8. Harmonic Cavitation Dose (HCD) versus gas volume fraction (μL/mL). (**A**,**C**,**E**,**G**) Graphs represent the four types of microbubbles at different mechanical indices: 0.4, 0.6, 0.8, and 1.0, respectively. The inverse is shown in graphs (**B**,**D**,**F**,**H**) where each graph represents the range of mechanical indices (0.1–1.0) for each microbubble population: 2, 3, 5 μm diameter, and polydisperse respectively. Vertical dotted lines represent regions of gas volume fraction used in ultrasound imaging [53] (black) and blood–brain barrier disruption [26] (pink). Data represent the mean and error bars show the standard deviation (n = 4).

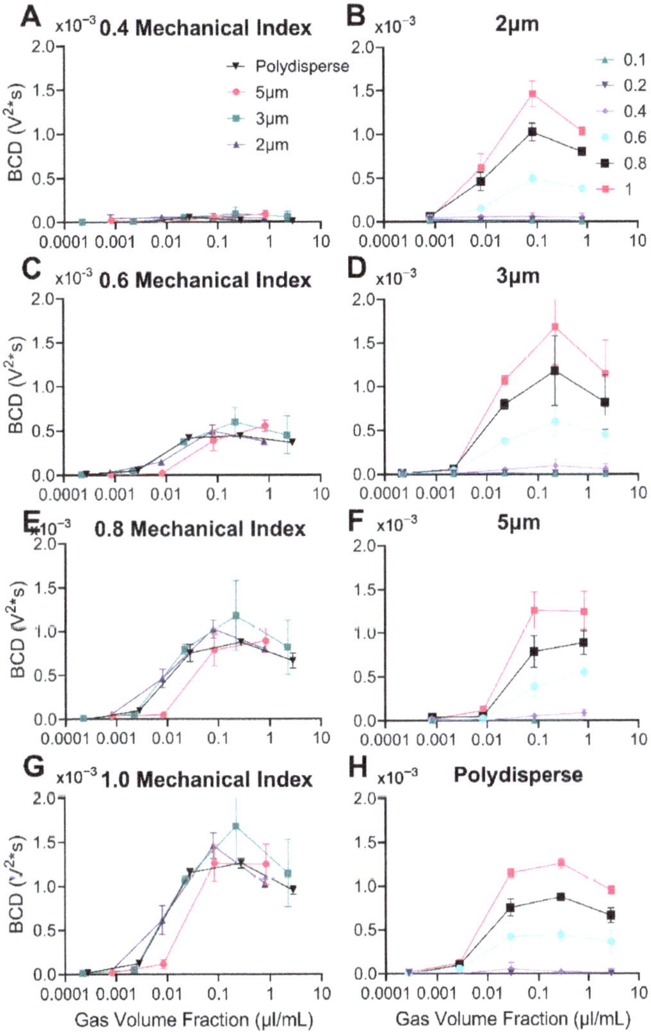

Figure 9. Broadband Cavitation Dose (BCD) versus gas volume fraction (μL/mL). (**A**,**C**,**E**,**G**) Graphs represent the four types of microbubbles at different mechanical indices: 0.4, 0.6, 0.8, and 1.0, respectively. The inverse is shown in graphs (**B**,**D**,**F**,**H**) where each graph represents the range of mechanical indices (0.1–1.0) for each microbubble type: 2, 3, 5 μm diameter, and polydisperse, respectively. Statistical significance can be found in Supplemental Table S4. Data represent the mean and error bars show the standard deviation (n = 4).

3.8. Master Surface Plot of Harmonic and Broadband Caviation Dose vs. Gas Volume Fraction vs. Mechanical Index

In Figure 10, all four microbubble types (2, 3, 5 μm diameter and polydisperse) were combined into a single 3D mesh to illustrate the master curve alignment for both HCD and BCD. Given how both HCD and BCD (Figure 10) show similar reductions to a single curve, resonance comes to question; it is well known that microbubble diameter influences the resonance frequency of the MB; although we do not see significant differences between the larger 5 μm MBs and smaller 2 μm MBs, even though at 1.515 MHz driving frequency we would expect the larger MBs to be closer to resonance [2,54–57]. The resonance effect could be lost due to the dynamic behavior of the MBs from pulse to pulse; it has been determined that the resonance curve and subsequent peak correspond to how much pressure is applied to the microbubble [20]. During the pulse, the MBs may fragment and/or dissolve [3,58,59], coalesce [60], displace by primary radiation force [61–63] and/or aggregate by secondary radiation force [63–65]. Thomas et al. demonstrated as MBs decayed due to ultrasound the MBs would move through their resonance peak [66]. The amount of substance that each shell lost was dependent on how high of pressure the microbubbles were being driven at [66]; these phenomena are not mutually exclusive. Additionally, depending on the driving frequency and peak negative pressure, MB oscillations can extend beyond the ultrasound pulse [58]. The change in MB number and/or diameter during the subsequent pulses may therefore change.

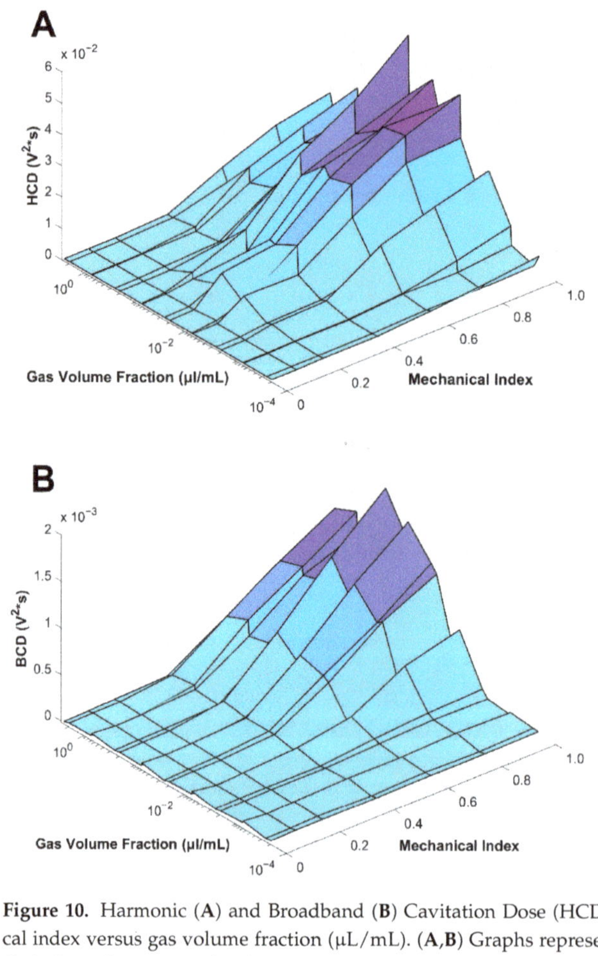

Figure 10. Harmonic (**A**) and Broadband (**B**) Cavitation Dose (HCD and BCD) versus mechanical index versus gas volume fraction (µL/mL). (**A**,**B**) Graphs represent all four size distributions (2, 3, 5 µm diameter and polydisperse). Error bars are not shown but can be found in previous Figures 5, 6, 8 and 9). Statistical significance can be found in Supplemental Tables S1–S4.

This study had a few limitations. First, we only investigated a single microbubble composition. As discussed previously, MB composition can greatly influence the acoustic response and colloidal stability. Although we have significant differences in size, we only apply a single frequency, 1.515 MHz; this prevents the illustration of how moving towards or away from resonance would influence the cavitation echoes for given size distribution.

Similarly, our size isolation process was completed using differential centrifugation, which can result in wider size distributions than other methods such as microfluidic flow focusing. Although we observed some overlap between the intermediate 3 μm diameter MB size distribution and the others, the 2 and 5 μm distributions were not overlapping. However, when matched to the same GVF, these latter two distributions gave similar results. Thus, we do not expect that more monodisperse distributions will yield a different conclusion. However, this should be tested experimentally in future work.

4. Conclusions

In this article, we discovered that all four microbubble size distributions (2, 3, 5 μm diameter and polydisperse) align to similar onset and peak in both harmonic and broadband cavitation doses when matched for microbubble gas volume fraction, rather than number concentrations. Thus, microbubble gas volume fraction, a precursor of the microbubble volume dose (MVD), appears to be sufficient to capture variations in the microbubble size distribution. MVD may therefore be used as a single microbubble input, along with the derated mechanical index to represent the ultrasound input, to predict the resulting bubble activity as determined by passive cavitation detection; this result provides a missing link between prior literature showing that pharmacokinetics and blood-brain barrier disruption also scale with MVD. Future work should determine if this effect holds in vivo, and if other parameters held constant in this study (e.g., bubble chemistry and ultrasound pulsing scheme) affect the harmonic and broadband cavitation doses.

Supplementary Materials: The following supporting information can be downloaded at: https://www.mdpi.com/article/10.3390/pharmaceutics14091925/s1, Figure S1: Phantom Box Model and Dimensions; Figure S2: Size Isolation of Monodisperse Microbubbles; Figure S3: Microbubble Stability During Stirring; Figure S4: Harmonic Cavitation Dose Controls; Table S1: Statistical Analysis of HCD vs. MI; Table S2: Statistical Analysis of BCD vs. MI; Table S3: Statistical Analysis of HCD vs. Concentration; Table S4: Statistical Analysis of BCD vs. Concentration.

Author Contributions: Conceptualization, P.M. and M.B.; methodology, P.M. and N.B.; software, P.M. and N.B.; validation, P.M.; formal analysis, P.M.; investigation, P.M.; resources, M.B.; data curation, P.M.; writing—original draft preparation, P.M.; writing—review and editing, M.B. and N.B.; visualization, P.M.; supervision, M.B. and N.B.; project administration, M.B.; funding acquisition, M.B. All authors have read and agreed to the published version of the manuscript.

Funding: This research was funded by the US National Institutes of Health, award R01CA239465.

Institutional Review Board Statement: Not applicable.

Informed Consent Statement: Not applicable.

Data Availability Statement: The data presented in this study are available on request from the corresponding author. The data are not publicly available due to the large size of raw data collected.

Conflicts of Interest: The authors declare no conflict of interest.

References

1. Borden, M.A.; Song, K.-H. Reverse Engineering the Ultrasound Contrast Agent. *Adv. Colloid Interface Sci.* **2018**, *262*, 39–49. [CrossRef] [PubMed]
2. Frinking, P.; Segers, T.; Luan, Y.; Tranquart, F. Three Decades of Ultrasound Contrast Agents: A Review of the Past, Present and Future Improvements. *Ultrasound Med. Biol.* **2020**, *46*, 892–908. [CrossRef]
3. Stride, E.; Segers, T.; Lajoinie, G.; Cherkaoui, S.; Bettinger, T.; Versluis, M.; Borden, M. Microbubble Agents: New Directions. *Ultrasound Med. Biol.* **2020**, *46*, 1326–1343. [CrossRef] [PubMed]
4. Roovers, S.; Segers, T.; Lajoinie, G.; Deprez, J.; Versluis, M.; De Smedt, S.C.; Lentacker, I. The Role of Ultrasound-Driven Microbubble Dynamics in Drug Delivery: From Microbubble Fundamentals to Clinical Translation. *Langmuir* **2019**, *35*, 10173–10191. [CrossRef]
5. Chen, Q.; Song, H.; Yu, J.; Kim, K. Current Development and Applications of Super-Resolution Ultrasound Imaging. *Sensors* **2021**, *21*, 2417. [CrossRef] [PubMed]

6. Abou-Elkacem, L.; Bachawal, S.V.; Willmann, J.K. Ultrasound Molecular Imaging: Moving toward Clinical Translation. *Eur. J. Radiol.* **2015**, *84*, 1685–1693. [CrossRef]
7. Kosareva, A.; Abou-Elkacem, L.; Chowdhury, S.; Lindner, J.R.; Kaufmann, B.A. Seeing the Invisible—Ultrasound Molecular Imaging. *Ultrasound Med. Biol.* **2020**, *46*, 479–497. [CrossRef]
8. Kooiman, K.; Vos, H.J.; Versluis, M.; de Jong, N. Acoustic Behavior of Microbubbles and Implications for Drug Delivery. *Adv. Drug Deliv. Rev.* **2014**, *72*, 28–48. [CrossRef]
9. Song, K.-H.; Fan, A.C.; Brlansky, J.T.; Trudeau, T.; Gutierrez-Hartmann, A.; Calvisi, M.L.; Borden, M.A. High Efficiency Molecular Delivery with Sequential Low-Energy Sonoporation Bursts. *Theranostics* **2015**, *5*, 1419–1427. [CrossRef]
10. Leighton, T. *The Acoustic Bubble*; Academic Press: Cambridge, MA, USA, 2012; ISBN 0-323-14413-6.
11. McDannold, N.; Vykhodtseva, N.; Hynynen, K. Effects of Acoustic Parameters and Ultrasound Contrast Agent Dose on Focused-Ultrasound Induced Blood-Brain Barrier Disruption. *Ultrasound Med. Biol.* **2008**, *34*, 930–937. [CrossRef]
12. Chu, P.-C.; Chai, W.-Y.; Tsai, C.-H.; Kang, S.-T.; Yeh, C.-K.; Liu, H.-L. Focused Ultrasound-Induced Blood-Brain Barrier Opening: Association with Mechanical Index and Cavitation Index Analyzed by Dynamic Contrast-Enhanced Magnetic-Resonance Imaging. *Sci. Rep.* **2016**, *6*, 33264. [CrossRef] [PubMed]
13. McDannold, N.; Arvanitis, C.D.; Vykhodtseva, N.; Livingstone, M.S. Temporary Disruption of the Blood–Brain Barrier by Use of Ultrasound and Microbubbles: Safety and Efficacy Evaluation in Rhesus MacaquesBlood–Brain Barrier Disruption via Focused Ultrasound. *Cancer Res.* **2012**, *72*, 3652–3663. [CrossRef] [PubMed]
14. Liu, H.-L.; Wai, Y.-Y.; Chen, W.-S.; Chen, J.-C.; Hsu, P.-H.; Wu, X.-Y.; Huang, W.-C.; Yen, T.-C.; Wang, J.-J. Hemorrhage Detection during Focused-Ultrasound Induced Blood-Brain-Barrier Opening by Using Susceptibility-Weighted Magnetic Resonance Imaging. *Ultrasound Med. Biol.* **2008**, *34*, 598–606. [CrossRef] [PubMed]
15. Tung, Y.-S.; Marquet, F.; Teichert, T.; Ferrera, V.; Konofagou, E.E. Feasibility of Noninvasive Cavitation-Guided Blood-Brain Barrier Opening Using Focused Ultrasound and Microbubbles in Nonhuman Primates. *Appl. Phys. Lett.* **2011**, *98*, 163704. [PubMed]
16. Choi, J.J.; Feshitan, J.A.; Baseri, B.; Wang, S.; Tung, Y.-S.; Borden, M.; Konofagou, E.E. Microbubble-Size Dependence of Focused Ultrasound-Induced Blood–Brain Barrier Opening in Mice In Vivo. *IEEE Trans. Biomed. Eng.* **2010**, *57*, 145–154. [CrossRef]
17. Tung, Y.-S.; Vlachos, F.; Feshitan, J.A.; Borden, M.A.; Konofagou, E.E. The Mechanism of Interaction between Focused Ultrasound and Microbubbles in Blood-Brain Barrier Opening in Mice. *J. Acoust. Soc. Am.* **2011**, *130*, 3059–3067. [CrossRef]
18. Wang, S.; Samiotaki, G.; Olumolade, O.; Feshitan, J.A.; Konofagou, E.E. Microbubble Type and Distribution Dependence of Focused Ultrasound-Induced Blood–Brain Barrier Opening. *Ultrasound Med. Biol.* **2014**, *40*, 130–137. [CrossRef]
19. Goertz, D.E.; de Jong, N.; van der Steen, A.F.W. Attenuation and Size Distribution Measurements of Definity™ and Manipulated Definity™ Populations. *Ultrasound Med. Biol.* **2007**, *33*, 1376–1388. [CrossRef]
20. Segers, T.; de Jong, N.; Versluis, M. Uniform Scattering and Attenuation of Acoustically Sorted Ultrasound Contrast Agents: Modeling and Experiments. *J. Acoust. Soc. Am.* **2016**, *140*, 2506–2517. [CrossRef]
21. Segers, T.; Kruizinga, P.; Kok, M.P.; Lajoinie, G.; De Jong, N.; Versluis, M. Monodisperse versus Polydisperse Ultrasound Contrast Agents: Non-Linear Response, Sensitivity, and Deep Tissue Imaging Potential. *Ultrasound Med. Biol.* **2018**, *44*, 1482–1492. [CrossRef]
22. Jafari Sojahrood, A.; de Leon, A.C.; Lee, R.; Cooley, M.; Abenojar, E.C.; Kolios, M.C.; Exner, A.A. Toward Precisely Controllable Acoustic Response of Shell-Stabilized Nanobubbles: High Yield and Narrow Dispersity. *ACS Nano* **2021**, *15*, 4901–4915. [CrossRef]
23. Helbert, A.; Gaud, E.; Segers, T.; Botteron, C.; Frinking, P.; Jeannot, V. Monodisperse versus Polydisperse Ultrasound Contrast Agents: In Vivo Sensitivity and Safety in Rat and Pig. *Ultrasound Med. Biol.* **2020**, *46*, 3339–3352. [CrossRef]
24. Samiotaki, G.; Vlachos, F.; Tung, Y.-S.; Konofagou, E.E. A Quantitative Pressure and Microbubble-Size Dependence Study of Focused Ultrasound-Induced Blood-Brain Barrier Opening Reversibility in Vivo Using MRI: FUS-Induced BBB Opening Reversibility. *Magn. Reson. Med.* **2012**, *67*, 769–777. [CrossRef] [PubMed]
25. Sirsi, S.; Feshitan, J.; Kwan, J.; Homma, S.; Borden, M. Effect of Microbubble Size on Fundamental Mode High Frequency Ultrasound Imaging in Mice. *Ultrasound Med. Biol.* **2010**, *36*, 935–948. [CrossRef]
26. Song, K.-H.; Harvey, B.; Borden, M. Microbubble Gas Volume: A Unifying Dose Parameter in Blood-Brain Barrier Disruption by Focused Ultrasound. *J. Acoust. Soc. Am.* **2018**, *144*, 1823. [CrossRef]
27. Navarro-Becerra, J.A.; Song, K.-H.; Martinez, P.; Borden, M.A. Microbubble Size and Dose Effects on Pharmacokinetics. *ACS Biomater. Sci. Eng.* **2022**, *8*, 1686–1695. [CrossRef] [PubMed]
28. Tung, Y.-S.; Choi, J.J.; Baseri, B.; Konofagou, E.E. Identifying the Inertial Cavitation Threshold and Skull Effects in a Vessel Phantom Using Focused Ultrasound and Microbubbles. *Ultrasound Med. Biol.* **2010**, *36*, 840–852. [CrossRef]
29. Sassaroli, E.; Hynynen, K. Cavitation Threshold of Microbubbles in Gel Tunnels by Focused Ultrasound. *Ultrasound Med. Biol.* **2007**, *33*, 1651–1660. [CrossRef]
30. Chen, W.-S.; Matula, T.J.; Brayman, A.A.; Crum, L.A. A Comparison of the Fragmentation Thresholds and Inertial Cavitation Doses of Different Ultrasound Contrast Agents. *J. Acoust. Soc. Am.* **2003**, *113*, 643–651. [CrossRef]
31. Li, T.; Chen, H.; Khokhlova, T.; Wang, Y.-N.; Kreider, W.; He, X.; Hwang, J.H. Passive Cavitation Detection during Pulsed HIFU Exposures of Ex Vivo Tissues and In Vivo Mouse Pancreatic Tumors. *Ultrasound Med. Biol.* **2014**, *40*, 1523–1534. [CrossRef]
32. Xu, S.; Ye, D.; Wan, L.; Shentu, Y.; Yue, Y.; Wan, M.; Chen, H. Correlation Between Brain Tissue Damage and Inertial Cavitation Dose Quantified Using Passive Cavitation Imaging. *Ultrasound Med. Biol.* **2019**, *45*, 2758–2766. [CrossRef] [PubMed]

33. Lentacker, I.; De Cock, I.; Deckers, R.; De Smedt, S.C.; Moonen, C.T.W. Understanding Ultrasound Induced Sonoporation: Definitions and Underlying Mechanisms. *Adv. Drug Deliv. Rev.* **2014**, *72*, 49–64. [CrossRef] [PubMed]
34. Stride, E. Physical Principles of Microbubbles for Ultrasound Imaging and Therapy. *Cereb. Dis* **2009**, *27*, 1–13. [CrossRef] [PubMed]
35. Hernot, S.; Klibanov, A.L. Microbubbles in Ultrasound-Triggered Drug and Gene Delivery. *Adv. Drug Deliv. Rev.* **2008**, *60*, 1153–1166. [CrossRef]
36. Tung, Y.-S.; Vlachos, F.; Choi, J.J.; Deffieux, T.; Selert, K.; Konofagou, E.E. In Vivo Transcranial Cavitation Threshold Detection during Ultrasound-Induced Blood–Brain Barrier Opening in Mice. *Phys. Med. Biol.* **2010**, *55*, 6141. [CrossRef]
37. O'Reilly, M.A.; Hynynen, K. Blood-Brain Barrier: Real-Time Feedback-Controlled Focused Ultrasound Disruption by Using an Acoustic Emissions-Based Controller. *Radiology* **2012**, *263*, 96–106. [CrossRef]
38. Neppiras, E.A. Acoustic Cavitation. *Phys. Rep.* **1980**, *61*, 159–251. [CrossRef]
39. Miller, M.W.; Miller, D.L.; Brayman, A.A. A Review of in Vitro Bioeffects of Inertial Ultrasonic Cavitation from a Mechanistic Perspective. *Ultrasound Med. Biol.* **1996**, *22*, 1131–1154. [CrossRef]
40. Yamamoto, T.; Hatanaka, S.; Komarov, S.V. Fragmentation of Cavitation Bubble in Ultrasound Field under Small Pressure Amplitude. *Ultrason. Sonochem.* **2019**, *58*, 104684. [CrossRef]
41. Wu, Q.; Gray, M.; Smith, C.; Bau, L.; Coussios, C.; Stride, E.P. Correlating High-Speed Optical Imaging and Passive Acoustic Mapping of Cavitation Dynamics. *J. Acoust. Soc. Am.* **2022**, *151*, A174. [CrossRef]
42. Pereno, V.; Aron, M.; Vince, O.; Mannaris, C.; Seth, A.; de Saint Victor, M.; Lajoinie, G.; Versluis, M.; Coussios, C.; Carugo, D.; et al. Layered Acoustofluidic Resonators for the Simultaneous Optical and Acoustic Characterisation of Cavitation Dynamics, Microstreaming, and Biological Effects. *Biomicrofluidics* **2018**, *12*, 034109. [CrossRef] [PubMed]
43. Johnston, K.; Tapia-Siles, C.; Gerold, B.; Postema, M.; Cochran, S.; Cuschieri, A.; Prentice, P. Periodic Shock-Emission from Acoustically Driven Cavitation Clouds: A Source of the Subharmonic Signal. *Ultrasonics* **2014**, *54*, 2151–2158. [CrossRef] [PubMed]
44. Feshitan, J.A.; Chen, C.C.; Kwan, J.J.; Borden, M.A. Microbubble Size Isolation by Differential Centrifugation. *J. Colloid Interface Sci.* **2009**, *329*, 316–324. [CrossRef] [PubMed]
45. Lin, Y.; Lin, L.; Cheng, M.; Jin, L.; Du, L.; Han, T.; Xu, L.; Yu, A.C.H.; Qin, P. Effect of Acoustic Parameters on the Cavitation Behavior of SonoVue Microbubbles Induced by Pulsed Ultrasound. *Ultrason. Sonochem.* **2017**, *35*, 176–184. [CrossRef]
46. Yasui, K.; Lee, J.; Tuziuti, T.; Towata, A.; Kozuka, T.; Iida, Y. Influence of the Bubble-Bubble Interaction on Destruction of Encapsulated Microbubbles under Ultrasound. *J. Acoust. Soc. Am.* **2009**, *126*, 973–982. [CrossRef]
47. Sojahrood, A.J.; Earl, R.; Haghi, H.; Li, Q.; Porter, T.M.; Kolios, M.C.; Karshafian, R. Nonlinear Dynamics of Acoustic Bubbles Excited by Their Pressure-Dependent Subharmonic Resonance Frequency: Influence of the Pressure Amplitude, Frequency, Encapsulation and Multiple Bubble Interactions on Oversaturation and Enhancement of the Subharmonic Signal. *Nonlinear Dyn.* **2021**, *103*, 429–466. [CrossRef]
48. Yasui, K.; Tuziuti, T.; Lee, J.; Kozuka, T.; Towata, A.; Iida, Y. Numerical Simulations of Acoustic Cavitation Noise with the Temporal Fluctuation in the Number of Bubbles. *Ultrason. Sonochem.* **2010**, *17*, 460–472. [CrossRef]
49. Haghi, H.; Sojahrood, A.J.; Kolios, M.C. Collective Nonlinear Behavior of Interacting Polydisperse Microbubble Clusters. *Ultrason. Sonochem.* **2019**, *58*, 104708. [CrossRef]
50. Haghi, H.; Jafari Sojahrood, A.; De Leon, A.C.; Exner, A.; Kolios, M.C. Experimental and Numerical Investigation of Backscattered Signal Strength from Different Concentrations of Nanobubble and Microbubble Clusters. *J. Acoust. Soc. Am.* **2018**, *144*, 1888. [CrossRef]
51. Keller, S.; Bruce, M.; Averkiou, M.A. Ultrasound Imaging of Microbubble Activity during Sonoporation Pulse Sequences. *Ultrasound Med. Biol.* **2019**, *45*, 833–845. [CrossRef]
52. *Definity [Package Insert]*; DEFINITY: Waterloo, ON, Canada, 2011. Available online: https://www.accessdata.fda.gov/drugsatfda_docs/label/2011/020664s011lbl.pdf (accessed on 8 September 2021).
53. Hyvelin, J.-M.; Gaud, E.; Costa, M.; Helbert, A.; Bussat, P.; Bettinger, T.; Frinking, P. Characteristics and Echogenicity of Clinical Ultrasound Contrast Agents: An in Vitro and in Vivo Comparison Study. *J. Ultrasound Med.* **2017**, *36*, 941–953. [CrossRef] [PubMed]
54. Doinikov, A.A.; Haac, J.F.; Dayton, P.A. Resonance Frequencies of Lipid-Shelled Microbubbles in the Regime of Nonlinear Oscillations. *Ultrasonics* **2009**, *49*, 263–268. [CrossRef] [PubMed]
55. Paul, S.; Katiyar, A.; Sarkar, K.; Chatterjee, D.; Shi, W.T.; Forsberg, F. Material Characterization of the Encapsulation of an Ultrasound Contrast Microbubble and Its Subharmonic Response: Strain-Softening Interfacial Elasticity Model. *J. Acoust. Soc. Am.* **2010**, *127*, 3846–3857. [CrossRef] [PubMed]
56. Khismatullin, D.B. Resonance Frequency of Microbubbles: Effect of Viscosity. *J. Acoust. Soc. Am.* **2004**, *116*, 1463–1473. [CrossRef]
57. Butler, M.B.; Thomas, D.H.; Pye, S.D.; Moran, C.M.; McDicken, W.N.; Sboros, V. The Acoustic Response from Individual Attached and Unattached Rigid Shelled Microbubbles. *Appl. Phys. Lett.* **2008**, *93*, 223906. [CrossRef]
58. Ferrara, K.; Pollard, R.; Borden, M. Ultrasound Microbubble Contrast Agents: Fundamentals and Application to Gene and Drug Delivery. *Annu. Rev. Biomed. Eng.* **2007**, *9*, 415–447. [CrossRef]
59. Chomas, J.E.; Dayton, P.; May, D.; Ferrara, K. Threshold of Fragmentation for Ultrasonic Contrast Agents. *J. Biomed. Opt.* **2001**, *6*, 141. [CrossRef]

60. Postema, M.; Marmottant, P.; Lancée, C.T.; Hilgenfeldt, S.; Jong, N. de Ultrasound-Induced Microbubble Coalescence. *Ultrasound Med. Biol.* **2004**, *30*, 1337–1344. [CrossRef]
61. Lum, A.F.H.; Borden, M.A.; Dayton, P.A.; Kruse, D.E.; Simon, S.I.; Ferrara, K.W. Ultrasound Radiation Force Enables Targeted Deposition of Model Drug Carriers Loaded on Microbubbles. *J. Control. Release* **2006**, *111*, 128–134. [CrossRef]
62. Acconcia, C.N.; Wright, A.; Goertz, D.E. Translational Dynamics of Individual Microbubbles with Millisecond Scale Ultrasound Pulses. *J. Acoust. Soc. Am.* **2018**, *144*, 2859–2870. [CrossRef]
63. Dayton, P.; Klibanov, A.; Brandenburger, G.; Ferrara, K. Acoustic Radiation Force in Vivo: A Mechanism to Assist Targeting of Microbubbles. *Ultrasound Med. Biol.* **1999**, *25*, 1195–1201. [CrossRef]
64. Wang, S.; Hossack, J.A.; Klibanov, A.L.; Mauldin, F.W. Binding Dynamics of Targeted Microbubbles in Response to Modulated Acoustic Radiation Force. *Phys. Med. Biol.* **2014**, *59*, 465–484. [CrossRef] [PubMed]
65. Wang, S.; Wang, C.Y.; Unnikrishnan, S.; Klibanov, A.L.; Hossack, J.A.; Mauldin, F.W. Optical Verification of Microbubble Response to Acoustic Radiation Force in Large Vessels With In Vivo Results. *Investig. Radiol.* **2015**, *50*, 772–784. [CrossRef] [PubMed]
66. Thomas, D.H.; Butler, M.; Anderson, T.; Emmer, M.; Vos, H.; Borden, M.; Stride, E.; de Jong, N.; Sboros, V. The "Quasi-Stable" Lipid Shelled Microbubble in Response to Consecutive Ultrasound Pulses. *Appl. Phys. Lett.* **2012**, *101*, 071601. [CrossRef]

pharmaceutics

Review

Targeted Microbubbles for Drug, Gene, and Cell Delivery in Therapy and Immunotherapy

J. Angel Navarro-Becerra [1] and Mark A. Borden [1,2,*]

[1] Mechanical Engineering Department, University of Colorado Boulder, Boulder, CO 80309, USA; jose.navarro@colorado.edu

[2] Biomedical Engineering Program, University of Colorado Boulder, Boulder, CO 80309, USA

* Correspondence: mark.borden@colorado.edu; Tel.: +1-303-492-7750

Abstract: Microbubbles are 1–10 μm diameter gas-filled acoustically-active particles, typically stabilized by a phospholipid monolayer shell. Microbubbles can be engineered through bioconjugation of a ligand, drug and/or cell. Since their inception a few decades ago, several targeted microbubble (tMB) formulations have been developed as ultrasound imaging probes and ultrasound-responsive carriers to promote the local delivery and uptake of a wide variety of drugs, genes, and cells in different therapeutic applications. The aim of this review is to summarize the state-of-the-art of current tMB formulations and their ultrasound-targeted delivery applications. We provide an overview of different carriers used to increase drug loading capacity and different targeting strategies that can be used to enhance local delivery, potentiate therapeutic efficacy, and minimize side effects. Additionally, future directions are proposed to improve the tMB performance in diagnostic and therapeutic applications.

Keywords: targeted microbubbles; ultrasound imaging probes; ultrasound-targeted delivery; therapy; immunotherapy

Citation: Navarro-Becerra, J.A.; Borden, M.A. Targeted Microbubbles for Drug, Gene, and Cell Delivery in Therapy and Immunotherapy. *Pharmaceutics* **2023**, *15*, 1625. https://doi.org/10.3390/pharmaceutics15061625

Academic Editors: Shashank Sirsi, James Kwan, Michael Gray and Brandon Helfield

Received: 31 March 2023
Revised: 18 May 2023
Accepted: 26 May 2023
Published: 30 May 2023

1. Introduction

Microbubbles (MBs) are gas-filled particles of 1–10 μm in diameter suspended in an aqueous medium. The gas core is highly compressible, making the MB an ideal acoustically responsive agent for ultrasound. However, uncoated MBs are highly unstable owing to their Laplace pressure [1] and tendency to coalesce [2]. Therefore, shells comprising lipid, surfactant, protein, polymer or other materials have been developed to stabilize the MBs for in vitro and in vivo use (Figure 1) [3,4]. Current clinically approved MBs comprise sulfur hexafluoride, perfluorobutane or perfluoropropane gas, and a phospholipid or protein shell (Table 1) [4–6]. These commercial MBs have been approved as ultrasound contrast agents for different diagnostic purposes including echocardiography, radiology, and other diagnostic imaging purposes [5–7].

Following their clinical translation as ultrasound imaging contrast agents, MB formulations have been engineered as ultrasound-responsive carriers to promote and enhance the local delivery and uptake of a wide variety of drugs [8–11], genes [12–14], and cells [15–20] for various therapeutic applications. Many of these applications include the delivery of therapeutic agents to treat the brain [19–23], heart [15,24–26], and cancer [27–29]. For immunotherapy, MBs can facilitate the delivery of immune cells, cytokines, antigens, and antibodies to promote the activation and infiltration of immune cells from the level of single cells to tissues, organs, and even physiological systems [28,30–32]. In cancer, for example, the concept is to induce the modulation and modification of the tumor microenvironment with a subsequent enhanced adaptive immune-cell activation to destroy the primary tumor and its metastases.

At the cellular level, increased uptake of therapeutic molecules into the cytosol has been attributed to sonoporation, i.e., the creation of a transient pore in the cell plasma membrane that increases permeability to molecules into the cytosol [33,34]. This effect is elicited by MB cavitation at the cell surface. MB cavitation refers to volumetric oscillations of the gas core and depends on the frequency, pressure, and waveform of the acoustic pulse. Fortuitously, MBs resonate at clinical ultrasound scanner frequencies, in the 1–10 MHz range. At low acoustic pressures, MBs exhibit stable cavitation, whereby repeated harmonic oscillations induce pushing-and-pulling effects and microstreaming that impart contact and shear forces against the cell membrane (Figure 2A) [35,36]. This mode of operation can be detected by passive acoustic cavitation and cavitation mapping as harmonics of the fundamental diving frequency [37]. At high acoustic pressures, MBs oscillate more violently, leading to rapid collapse and other inertial effects, such as microjets, shockwaves, and fragmentation (Figure 2B) [35,36]. This mode of operation can be detected by passive acoustic cavitation and cavitation mapping as a broadband response, although inertial cavitation can also enhance harmonics [37]. Stable and inertial cavitation can trigger various cellular effects, such as plasma membrane pore formation, cytoskeleton reorganization, transmembrane calcium (Ca^{2+}) influx, potassium (K^+) efflux (hyperpolarization), and reactive oxygen species production (Figure 2C). These biological effects lead to enhanced cellular delivery through mechanisms such as drug convection and diffusion through the pores (4–70 kDa) and endocytosis (70–500 kDa) [33–35,38].

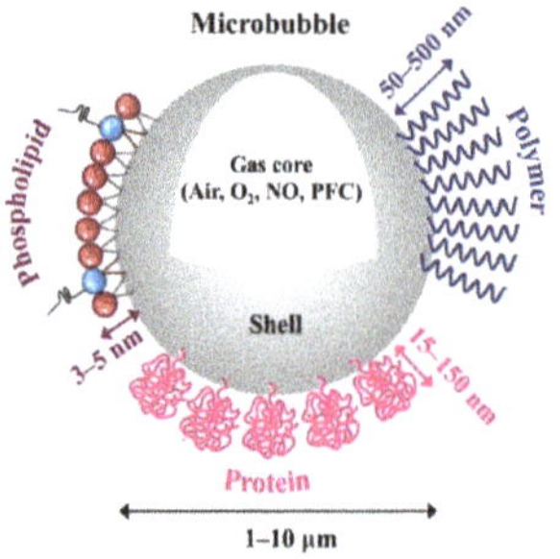

Figure 1. Illustration of a typical engineered microbubble used in biomedical applications. Microbubbles are particles filled with gas, such as air, oxygen (O_2), nitric oxide (NO) or perfluorocarbon (PFC), with a diameter of 1–10 µm. They are stabilized by shells made of phospholipid (3–5 nm thick), protein (15–150 nm thick) or polymer (50–500 nm thick).

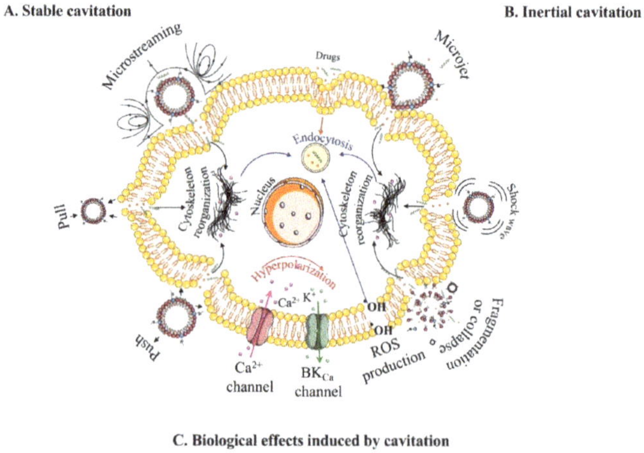

Figure 2. Schematic overview of the underlying physical and biological effects produced during microbubble cavitation on the cell membrane. (**A**,**B**) Mechanical effects associated with sonoporation, including stable and inertial cavitation of microbubbles. (**C**) Biological effects induced by microbubble cavitation, which enhance the internalization of drugs.

Table 1. Current clinically approved MB ultrasound contrast agents.

Contrast Agent	Manufacturer	Indications	Shell	Gas	Concentration (MBs/mL)	Size (Diameter)	Half-Life (min)	Volume Dose (μL/kg) a	Mechanical Properties	Reference
Optison	GE Healthcare	LVO/EBD	Protein: HSA $\psi_0 = -9.5$ to -25.3 mV	C_3F_8	$5–8 \times 10^8$	3.0–4.5 μm (max. 32 μm) 95% < 10 μm	0.5 ± 0.3	6	f = 2–4 MHz χ = 0.9 N/m	[39–43]
Definity	Lantheus	LVO/EBD, breast, liver, vascular.	Phospholipid: DPPC, DPPA, DPPE-mPEG5000 $\psi_0 = -1.1$ to -4.2 mV	C_3F_8	1.2×10^{10}	1.1–3.3 (max. 20 μm) 98% < 10 μm	2.0 ± 0.3	10	f = 2–6 MHz χ = 0.5–2.5 N/m	[40,42–46]
SonoVue	Bracco	LVO/EBD, breast, liver, vascular, urinary tract.	Phospholipid: DSPC, DPPG, PA $\psi_0 = -28.3$ mV	SF_6	$1.5–2.5 \times 10^8$	1.5–2.5 (max 20 μm) 99% < 10 μm	1.04 ± 0.15	25	f = 1.5–2 MHz χ = 0.2–0.3 N/m	[40,44,47–49]
Sonazoid	GE Healthcare	Myocardial perfusion, liver, breast.	Phospholipid: H-EPS $\psi_0 = -76$ to -82 mV	C_4F_{10}	1.2×10^9	1.0–5.0 (max 10 μm) 99.9% < 7 μm	2.6 ± 0.2	15	f = 4–6 MHz χ = 0.6 N/m	[50–52]

EBD = endocardial border definition; LVO = left ventricular opacification; HSA = human serum albumin; DPPC = 1,2-dipalmitoyl-sn-glycero-3-phosphocholine; DPPA = 1,2-dipalmitoyl-sn-glycero-3-phosphate; DPPE-mPEG5000 = 1,2- dipalmitoyl-sn-glycero-3-phosphoethanolamine-N-[methoxy(polyethylene glycol)-5000]; DSPC = 1,2-distearoyl-sn-glycero-3-phosphocholine; DPPG = 1,2-dipalmitoyl-sn-glycero-3-phospho-(1′-rac-glycerol); PA = palmitic acid; H-EPS = hydrogenated egg phosphatidylserine sodium; ψ_0 = zeta potential measured; C_3F_8 = perfluoropropane; SF_6 = sulfur hexafluoride, C_4F_{10} = perfluorobutane; MBs = microbubbles; [a] Bolus intravenous injection; f = resonance frequency; χ = shell elasticity. Adapted with permission from references [4,5,21].

At the vascular level, increased transport of therapeutic molecules from the vessel lumen into the surrounding tissue parenchyma also occurs by sonoporation, including the mechanisms described above for uptake into the endothelium and then transcytosis or disruption of the endothelium (e.g., rupture of tight junctions to create transport paths between endothelial cells). This effect can be captured by an increase in the Biot number (Bi), which is defined as

$$Bi = \frac{K_m L_{tiss}}{D_{tiss}} \quad (1)$$

where K_m is the mass transfer coefficient, equivalent to the endothelial cell permeability (m/s), and is proportional to the pharmacokinetic first-order rate constant ($K = K_m A$, where A is the endothelial surface area) between the central and tissue compartments; D_{tiss} is the drug diffusivity in the tissue; and L_{tiss} is the diffusion length in tissue, typically the half-length between capillaries (~100 μm). Microbubble sonoporation increases K_m in a dose-dependent manner, making diffusion into tissue the limiting transport barrier [53]. Here, the dose includes not only the drug dose, but also the microbubble dose (microbubble volume dose) and ultrasound dose (mechanical index), which can be estimated by the received echo intensity by passive cavitation detection.

Therapeutic agents can be co-administered with MBs or loaded onto the MB shell. This review focuses on shell modification to bioconjugate molecular ligands, drugs (including nucleic acids), nanoparticles and/or cells for therapeutic applications. Articles utilizing so-called "nanobubble" formulations (<1 μm diameter) were excluded from this review. By incorporating specific targeting moieties into the shell, MBs can be engineered to create targeted formulations (tMBs) that exhibit improved pharmacokinetic and acoustic performance, ligand-receptor targeting, biological activation, immunomodulation, and many other functions. With so many options for MB composition, processing, size, microstructure and properties, there is a wealth of opportunity to synthesize innovative targeted MB designs for enhanced performance. Thus, the aim of this review is to summarize the state-of-the-art of current engineered MB formulations and their ultrasound-targeted delivery in different biomedical applications.

2. Microbubble Formulations

2.1. Shell Composition

2.1.1. Phospholipid-Coated Microbubbles

Phospholipid-coated MBs have a thin and soft shell (3–5 nm) (Figure 1) that provides excellent acoustic response for US-contrast imaging. However, for prolonged imaging and drug delivery, the MBs have limited circulation persistence and drug loading capacity [3,54,55]. To optimize MB performance by controlling their structure, properties, targeting, drug loading, and acoustic response, the shell is made up of saturated diacyl phosphatidylcholine lipids (PC) (80–90 mol%) and PEG-lipid emulsifiers (10–20 mol%) [56]. Longer lipid chains increase intermolecular forces between the phospholipids [57], increase shell rigidity [58,59], decrease shell permeability [60,61], and improve MB stability [62–64]. This results in superior acoustic stability, stiffer shell elasticity, and reduced fragmentation propensity [62,65]. They also prolong the in vivo half-life [63] and increase the response at the second harmonic frequency [62,65]. Moreover, MBs with long lipid chains can achieve high delivery efficiency for larger molecules [66].

MBs can be quickly eliminated by opsonization, leading to decreased circulation time. To address this issue, buried-ligand architecture has been developed to extend lipid-coated MB circulation time [67–69], preserve targeted specificity, and reduce the immunogenic response by blocking complement protein C3b fixation [67,70,71].

2.1.2. Protein-Coated Microbubbles

Protein-coated MBs have a thicker and stiffer layer (15–150 nm) composed of natural proteins (Figure 1). These MBs exhibit moderate rigidity and drug-loading capabilities, as well as good stability and moderate acoustic response [3,72,73]. At present, only Optison is clinically approved and commercially available as a protein-coated MB for UCAs (Table 1). However, the development of novel albumin-tMBs is gaining interest due to their potential for rapid approval and clinical translation as a drug delivery system.

To maximize albumin-MBs' loading efficiency, the physicochemical properties of drugs play a critical role. Drug loading can be increased by pre-loading the drugs prior to MB synthesis or using drugs with high albumin binding percentages at higher feed ratios [74]. Additionally, chemical modifications such as disulfide bonds and glutaraldehyde cross-linking [75] can link therapeutic molecules or other loaded carriers to albumin-MBs. Physical properties such as size, shape, storage stability, and acoustic response can be tuned by incorporating dextrose [74,76,77], glycerol, and propylene glycol [77], amphipathic molecules [78], polymers [79], PEGylation [80] or increasing the number of thiols groups [81] into the albumin shell. Several in vitro and in vivo studies have employed these methods to propose using of albumin-MBs as targeted carriers [82–87] and multimodal contrast agents [84,86,88–91]. These strategies offer promising options to develop targeted albumin-MB formulations that can enhance their performance in US-image-guided drug delivery applications. Additionally, proteins such as lysozyme and oleosin [92] are alternative candidates that could be used to create new protein-tMB constructs.

2.1.3. Polymer-Coated Microbubbles

Polymer-coated MBs are composed of a thick and stiff layer (50–500 nm) of biodegradable, natural or synthetic polymers (Figure 1). These MBs can encapsulate either hydrophilic and hydrophobic molecules, carry higher drug doses and enhance MB stability during circulation, more so than typical protein and phospholipid MBs. Nevertheless, their incompressibility and rigidity lead to suboptimal US imaging capability and limited acoustic response [3,72,93]. To enhance polymer-MB drug loading capacity and acoustic properties, their chemical composition or polymer molecular weight (MW) is often modified to adjust the shell's elasticity and thickness. Increasing the drug's hydrophobicity and molecular weight also boosts loading capacity, stability, and drug release [94]. Proteins [95] or high MW surfactants [96] can also be used to fine-tune MB loading capacity and acoustic response by regulating their stability, size, and shell thickness.

2.2. Gas Composition

The composition of the gas core plays a significant role in the stability of MBs. MBs composed of PFC and SF_6 gases, which have lower solubility in blood and higher molecular weight, diffuse more slowly across the MB shell, leading to longer circulation persistence. A recent study has shown that MBs filled with C_3F_8 or C_4F_{10} were highly stable, exhibited sustained in vivo circulation, and demonstrated more efficient delivery of Evans blue into the brain without causing side effects as compared to SF_6-MBs [97].

MBs are now being used as carriers for therapeutic gases, such as NO and O_2, to potentially regulate the tumor microenvironment or to enhance immune responses. O_2-MBs, for example, have been used to locally release O_2 into tumors, thereby improving therapeutic efficacy by reducing hypoxic treatment resistance [98,99]. Additionally, O_2-MBs can normalize dysfunctional vessels, enhancing vascular maturity, blood perfusion, and drug penetration [100]. Meanwhile, NO-MBs can control the release of NO to mitigate oxidative stress and apoptosis during ischemia–reperfusion injury [101], enhance the targeted delivery of mesenchymal cells into the infarcted myocardium, and induce regional angiogenic response [15]. In addition, NO-MBs accelerate deep vein thrombosis resolution by reducing platelets and inflammatory cells aggregation, enhancing collagen turnover and stimulating an anticoagulant condition of endothelium [102].

The use of O_2- and NO-MBs is still limited due to the properties of the diffusible gases, resulting in lower storage stability and significantly shorter circulation persistence than PFC-MBs. To solve this problem, our group has demonstrated that the lipid shell can be engineered with longer acyl chains (C22:0) to increase O_2 payload and enhance local delivery [103]. Additionally, O_2-MBs with larger diameters (2–10 µm) have demonstrated an 8-fold increase in half-life compared to smaller diameters (0.5–2 μm) [104]. Recent approaches propose incorporating folate ligands or SPIONs into drug-loaded-O_2MBs to accumulate them in a target region by molecular targeting or using an external magnetic field. This incorporation enhances stability, improves drug delivery, and suppresses tumor growth [105–107]. Additionally, C4-d-tMBs composed with a gas core combination of PFC/NO have been successfully used to alleviate cardiac allograft rejection by suppressing thrombosis and inflammatory cell infiltrates, while prolonging the survival time two-fold [108].

2.3. Size and Microbubble Volume Dose

The varying components of MBs, including the shell, gas core, size, concentration, and recommended dosage, lead to distinct acoustic responses [109,110], circulation half-life [110,111], and biological effects [111–114] for both clinically approved MBs (Table 1) and experimental tMB formulations (Tables 2–6). The biological impact of MB acoustic behavior is dose-dependent [115], so selecting dosing strategies for each MB formulation that produce comparable bioeffects is crucial. The MB gas volume dose (MVD) is a unified dose metric that integrates MB size distribution and concentration into a single parameter [22,116]. Our group has demonstrated that MVD can be used to: (1) maximize US imaging contrast and circulation persistence (Figure 3A) [112]; (2) achieve similar, consistent, and comparable molecular delivery across the blood–brain barrier (Figure 3B) [116]; (3) predict, compare, and characterize the MB pharmacokinetics behavior (Figure 3C) [117]; and (4) match the harmonic and broadband cavitation doses regardless of MB size (Figure 3D) [37]. Moreover, matching the gas volume fractions of different commercial MBs with varying sizes and distributions results in equivalent permeabilization effects [118].

Table 2. Microbubbles targeted with a therapeutic ligand used for therapy.

Ligand Type	Ligand Species	Conjugation Chemistry	Microbubble Composition	Ultrasound Parameters	Animal/Disease Model	Reference
Protein	SDF-1α	EDC/sulfo-NHS	Shell: DSPC, DPPG, PEG4000-COOH Gas: C_3F_8 Diameter: 1–5 µm Dose: unknown	f = 4 MHz MI = 1.5 Time = 1.15 min	Rat/Diabetic nephropathy	[119]
	GPVI-Fc	Avidin-biotin	MicroMarker™ Shell Phospholipid Gas: C_4F_{10}/N_2 Diameter: 2.3–2.9 µm Dose: 2.8 × 10^8 MBs/kg or 5.3 µL/kg MVD	f = 24 MHz Time = 2 min	Mouse/ Atherosclerosis	[120]
Antibody	Aβ 1-42	Avidin-biotin	Shell: DSPC, DPPA, PEG4000 Gas: C_3F_8 Diameter: 2.13 µm Dose: 5.8 × 10^8 MBs/kg or 2.9 µL/kg MVD	f = 3 MHz MI = 0.8 PRF = 50 Hz Time = 5 min	Mouse/Alzheimer's disease	[20]
	C4d mAb	Avidin-biotin	Shell: DPPC, DSPE-PEG-2000, -Biotin Gas: C_3F_8/NO Diameter: 0.96 ± 0.07 µm Dose: 5.8 × 10^6 MBs/kg or 0.003 µL/kg MVD	f = 13 MHz MI = 0.33 Time = 5 min	Rat/Heterotopic Heart transplant	[108]
	IL-8 mAb	Avidin-biotin	USphere™ Labeler Shell: Phospholipid Diameter: 1.8 µm Dose: 1.2 × 10^9 MBs/kg or 3.6 µL/kg MVD	f = 1 MHz MI = 1.5 Power = 5 W Time = 1 min	Rabbit/ Atherosclerosis	[121]

DSPC = 1,2-distearoyl-sn-glycero-3-phosphocholine; DPPG = 1,2-dipalmitoyl-sn-glycero-3-phospho-(1′-rac-glycerol); DPPA = 1,2-dipalmitoyl-sn-glycero-3-phosphate; PEG4000-COOH = polyethylene glycol-4000-carboxy; DSPE-PEG2000 = 1,2-distearoyl-sn-glycero-3-phosphoethanolamine-N-[methoxy(polyethylene glycol)-2000]; C_3F_8 = perfluoropropane; C_4F_{10} = perfluorobutane; NO = nitric oxide; MVD = Microbubble Volume Dose; f = frequency; MI = mechanical index; PRF = pulse repetition frequency.

Table 3. Drug or gene loaded-MBs targeted with a ligand used for therapy.

Ligand Type	Targeting Ligand	Therapeutic Molecule	Conjugation Chemistry	Microbubble Composition	Ultrasound Parameters	Animal/Disease Model	Reference
Antibody	VEGFR2 mAb	pHSV-TK	Avidin-biotin Electrostatic	Shell: DPPC, DPTAP, DSPE-PEG2000, -Biotin Gas: C_3F_8 Diameter: 1.1 ± 0.1 μm Dose: 1.8 × 10^{10} MBs/kg or 12.4 μL/kg MVD	f = 1 MHz PNP = 0.7 MPa PRF = 5 Hz Time = 2 min	Mouse/Glioma tumor	[122]
	VEGFR2 mAb	BCNU	Avidin-biotin Hydrophobic	Shell: DPPC, DSPE-PEG2000, -Biotin Gas: C_3F_8 Diameter: 1.79 ± 0.13 μm Dose: 1.4 × 10^{10} MBs/kg or 41 μL/kg MVD	f = 1 MHz PNP = 0.7 MPa DC = 5% PRF = 5 Hz Time = 1 min	Rat/Glioma tumor	[123]
	CD105	pEZ-M46-ES	Avidin-biotin Electrostatic	Shell: DPPC, Cholesterol, DSPE-PEG2000, -Biotin Gas: C_3F_8 Diameter: 1.44 ± 0.21 μm Dose: 1.3 × 10^{10} MBs/kg or 20.6 μL/kg MVD	f = 1 MHz PNP = 0.7 MPa DC = 50% I = 2 W/cm^2 PRF = 5 Hz Time = 30 s	Mouse/Hind limb tumor	[124]
	RGD	Urokinase	Electrostatic	SonoVue Diameter: 1.5–2.5 μm Dose: 4.4 × 10^7 MBs/kg or 0.18 μL/kg MVD	f = 1.6 MHz MI = 1.1 PRF = 24 kHz Time = 10 min	Pig/Thrombosis	[125]
	VCAM-1	miR-126	Avidin-biotin	VisualSonics Target-Ready MBs Shell: Phospholipid Gas: C_4F_{10}/N_2 Diameter: 1.5 μm Dose: 3.7 × 10^9 MBs/kg or 6.6 μL/kg MVD	f = 10 MHz MI = 0.66 I = 0.076 W/cm^2 Time = 2.5 min	Mouse/Aortic Aneurysm	[126]
Peptides	LHRa	Paclitaxel	Avidin-biotin Hydrophobic	Shell: DPPC, DSPE-PEG2000-Biotin Gas: C_3F_8 Diameter: 1.8 ± 0.2 μm Dose: 1.5 × 10^9 MBs/kg or 4.5 μL/kg MVD	f = 0.3 MHz I = 1 W/cm^2 DC = 50% Time = 3 min	Mouse/Breast cancer	[127]
	RGD	tPA	Amine Hydrophobic	Shell: DPPC, DSPC PEG-Amine Gas: C_3F_8 Diameter: 2.08 ± 0.93 μm Dose: 8.0 × 10^8 MBs/kg or 3.8 μL/kg MVD	f = 2 MHz MI = 1.4 I = 1.8 W/cm^2 DC = 95% PRF = 15 Hz Time = 30 min	Rabbit/ Thrombolysis	[128]

Table 3. *Cont.*

Ligand Type	Targeting Ligand	Therapeutic Molecule	Conjugation Chemistry	Microbubble Composition	Ultrasound Parameters	Animal/Disease Model	Reference
Vitamins	Folate	Paclitaxel	Hydrophobic	Shell: DPPC. DSPE-PEG2000-Folate Gas: O_2 Diameter: 1.81 ± 0.04 µm Dose: unknown	f = 0.3 MHz I = 1 W/cm^2 Time = 3 min	Mouse/Ovarian cancer	[105]
	Folate	p-FLuc	Amide Electrostatic	Shell: DPPC, DPTAP, DSPE-PEG2000 Gas: C_3F_8 Diameter: 3.2 ± 0.1 µm Dose: 1.8 × 10^8 MBs/kg or 3.1 µL/kg MVD	f = 1 MHz PRF = 5 Hz PNP = 0.7 MPa Time = 1 min	Rat/ Glioma tumor	[129]

DPPC = 1,2-dipalmitoyl-sn-glycero-3-phosphocholine; DSPE-PEG2000 = 1,2-distearoyl-sn-glycero-3-phosphoethanolamine-N-[methoxy(polyethylene glycol)-2000]; DPPG = 1,2-dipalmitoyl-sn-glycero-3-phospho-(1′-rac-glycerol); PA = palmitic acid; DSPC = 1,2-distearoyl-sn-glycero-3-phosphocholine; DPTAP = 1,2-dipalmitoyl-3-trimethylammonium-propane; C_3F_8 = perfluoropropane; SF_6 = sulfur hexafluoride, C_4F_{10} = perfluorobutane; N_2 = nitrogen; O_2 = oxygen; MVD = Microbubble Volume Dose; f = frequency; MI = mechanical index; PRF = pulse repetition frequency; I = intensity; DC = duty cycle; PNP = peak negative pressure.

Table 4. Drug, gene, or cell loaded-MBs targeted with two different ligands.

Targeting Ligands	Therapeutic Molecule	Conjugation Chemistry	Microbubble Composition	Ultrasound Parameters	Animal/Disease Model	Reference
CD90 Ab ICAM-1 Ab	Adipose-derived stem cells	Avidin-biotin	Shell: DSPC, PEG40S, DSPE-PEG2000, -Biotin Gas: C_4F_{10} Diameter: 3.5 µm Dose: 9.3 × 10^7 MBs/kg or 2.1 µL/kg MVD	f = 1 MHz PNP: 0.1 MPa PRF = 1 kHz DC = 50% Time = 10 min	Rat/Myocardial Infarction	[130]
CD90 Ab ICAM-1 Ab	Adipose-derived stem cells	Avidin-biotin	Shell: DSPC, PEG40S, DSPE-PEG2000, -Biotin Gas: C_4F_{10} Diameter: 3.5 µm Dose: 4.6 × 10^7 MBs/kg or 1 µL/kg MVD	f = 1 MHz PNP: 0.1 MPa PRF = 1 kHz DC = 50% Time = 1 min	Mouse/ Atherosclerosis	[131]

Table 4. *Cont.*

Targeting Ligands	Therapeutic Molecule	Conjugation Chemistry	Microbubble Composition	Ultrasound Parameters	Animal/Disease Model	Reference
CCR2 Ab iRGD peptide	shAKT2	Avidin-biotin, Electrostatic	Shell: DSPC, Stearic-PEI600, DSPE-PEG2000-iRGD-Biotin Gas: C_4F_{10} Diameter: 1.32 ± 0.22 µm Dose: 5.3 × 10^9 MBs/kg or 6.3 µL/kg MVD	f = 1 MHz PNP: 1.2 MPa PRF = 1 kHz DC = 50% Time = 10 min	Mouse/Breast cancer	[132]
cRGD peptide Folate	Doxorubicin	Avidin-biotin	Shell: DSPC, DSPE-PEG2000, -Biotin Gas: C_3F_8 Diameter: 5.8 ± 2.1 µm Dose: 1.1 × 10^{10} MBs/kg or 1.1 mL/kg MVD	f = 10 MHz MI: 0.64 Time = 0.5 min	Mouse/Breast cancer	[133]
cRGD and cCLT1 peptide	Paclitaxel	Amide, Hydrophobic	Shell: DSPC, DSPE-PEG2000, -COOH Gas: SF_6 Diameter: 1.59 ± 0.54 µm Dose: 1.0×10^{10} MBs/kg or 21 µL/kg MVD	f = 1 MHz MI = 1.17 DC = 10% I = 2.5 W/cm^2 Time = 3 min	Mouse/Pancreatic cancer	[134]

DSPC = 1,2-distearoyl-sn-glycero-3-phosphocholine; PEG-40S = Polyoxyethylene (40) stearate; DSPE-PEG2000 = 1,2-distearoyl-sn-glycero-3-phosphoethanolamine-N-[methoxy(polyethylene glycol)-2000]; PEI = Branched Polyethylenimine; C_3F_8 = perfluoropropane; SF_6 = sulfur hexafluoride, C_4F_{10} = perfluorobutane; MVD= Microbubble Volume Dose; f = frequency; MI = mechanical index; PRF = pulse repetition frequency; I = intensity; DC = duty cycle; PNP = peak negative pressure.

Table 5. Microbubbles conjugated with targeted drug-loaded carriers used for therapy.

Carrier	Type of Targeting	Therapeutic Molecule	Conjugation Chemistry	Microbubble Composition	Ultrasound Parameters	Animal/Disease Model	Reference
Nanoparticles	Magnetic	5-fluorouracil Rose Bengal	Avidin-biotin	Shell: DBPC, DSPE-PEG2000, -Biotin Gas: O_2 Diameter: 1–2 µm Dose: 7.0 × 10^9 MBs/kg or 12.3 µL/kg MVD	f = 1 MHz PNP = 0.85 MPa DC = 30% I = 3.5 W/cm^2 PRF = 100 Hz Time = 3.5 min	Mouse/Pancreatic tumor	[106]
	Magnetic	Doxorubicin	Amide	Shell: DSPC, DSPG DSPE-PEG2000, -Biotin Gas: C_3F_8 Diameter: 5.4 ± 1.1 µm Dose: 5.7 × 10^8 MBs/kg or 46.7 µL/kg MVD	f = 1 MHz PNP = 0.3 MPa DC = 30% I = 3.5 W/cm^2 PRF = 1 Hz Time = 4 min	Rat/Glioma tumor	[107]
	Magnetic	Gemcitabine Rose Bengal	Avidin-biotin	Shell: DBPC, DSPE-PEG2000, -Biotin Gas: O_2 Diameter: 1.9 ± 0.4 µm Dose: 5.4 × 10^9 MBs/kg or 19.5 µL/kg MVD	f = 1.17 MHz PNP = 0.7 MPa DC = 30% Time = 3.5 min	Mouse/Pancreatic tumor	[135]

Table 5. *Cont.*

Carrier	Type of Targeting	Therapeutic Molecule	Conjugation Chemistry	Microbubble Composition	Ultrasound Parameters	Animal/Disease Model	Reference
	Magnetic	Doxorubicin	Electrostatic and hydrophobic	Shell: DSPC, DSPG DSPE-PEG2000, -Biotin Gas: C_3F_8 Diameter: 1.04 ± 0.01 µm Dose: 9.7 × 10^{10} MBs/kg or 56.8 µL/kg MVD	f = 0.4 MHz PNP = 0.325 MPa Power = 4 W PRF = 1 Hz Time = 1.5 min	Rat/Glioma tumor	[136]
	Magnetic	tPA	Electrostatic	Shell: SDS Gas: Air Diameter: 5.36 ± 1.44 µm Dose: 6.4 × 10^{10} MBs/kg or 0.50 µL/kg MVD	f = 18 MHz DC = 10% Time = 5 min	Mouse/ Thrombolysis	[137]
	Magnetic/ molecular	RGD	Amide	Shell: PVA Gas: Air Diameter: 1.37 µm Dose: 5.4 × 10^{8} MBs/kg or 0.70 µL/kg MVD	f = 30 MHz Time = 10 min	Mouse/Colon cancer	[138]
Liposomes	Magnetic	Doxorubicin	Maleimide	Shell: DPPC, DSPE-PEG2000-SPDP Gas: C_3F_8 Diameter: 4 µm Dose: 7.0 × 10^{9} MBs/kg or 233 µL/kg MVD	f = 1 MHz DC = 30% I = 2 W/cm^2 Time = 2 min	Mouse/Pancreatic tumor	[139]
	Molecular: RGD peptide	Paclitaxel	Avidin-biotin	Shell: DSPC, DSPE-PEG2000, -Biotin Gas: C_3F_8 Diameter: 1.5 µm Dose: 1.1 × 10^{10} MBs/kg or 17.8 µL/kg MVD	f = 1 MHz DC = 1% PRF = 1 Hz Time = 2 min	Mouse/Breast cancer	[140]
	Molecular: NGR peptide	shBirc5	Avidin-biotin	Shell: DPPC, Cholesterol DSPE-PEG2000-Biotin Gas: C_3F_8 Diameter: 2.90 ± 0.38 µm Dose: 2.1 × 10^{10} MBs/kg or 265.6 µL/kg MVD	f = 1 MHz DC = 50% Power = 1.84 W Time = 1–5 min	Rat/Glioma tumor	[141]
	Molecular: VEGFR2 Ab	Irinotecan	Avidin-biotin	MicroMarker™ Shell: Phospholipid Gas: C_4F_{10}/N_2 Diameter: 1.5 µm Dose: 4.7 × 10^{9} MBs/kg or 8.2 µL/kg MVD	f = 2.2 MHz PNP = 0.26 MPa PRF = 1 Hz Time = 4 min	Mouse/Colorectal cancer	[142]
Nanodroplets	Molecular: VEGFR2 Ab	Combretastatin A4	Avidin-biotin	Shell: DPPC, DSPE-PEG2000-Biotin Gas: C_4F_{10} Diameter: 2.6 ± 1.5 µm Dose: unknown	f = 2.2 MHz PNP = 0.26 MPa PRF = 1 Hz Time = 5 s	Mouse/Colorectal cancer	[143]

DBPC:1,2-dibehenoyl-sn-glycero-3-phosphocholine; DSPG: 1,2-Distearoyl-sn-glycero-3-phospho-rac-glycerol; DSPE-PEG2000: 1,2-distearoyl-sn-glycero-3-phosphoethanolamine-N-[methoxy(polyethylene glycol) -2000]; DSPC: 1,2-distearoyl-sn-glycero-3-phosphocholine; SDS: Sodium Dodecyl Sulfate; PVA = Poly(vinyl alcohol); DPPC = 1,2-dipalmitoyl-sn-glycero-3-phosphocholine; C_3F_8 = perfluoropropane; C_4F_{10} = perfluorobutane; O_2 = oxygen; MVD = Microbubble Volume Dose; f = frequency; PRF = pulse repetition frequency; I = intensity; DC = duty cycle; PNP = peak negative pressure.

Table 6. Targeted microbubbles used for cancer immunotherapy.

Immunotherapy Modality	Therapeutic Molecule	Conjugation Chemistry	MicrobubbleComposition	Ultrasound Parameters	Animal/Disease Model	Reference
Monoclonal antibody	EGFR mAb	Avidin-biotin	Targestar™-SA Shell: Phospholipid Gas: C_4F_{10} Diameter: 2.5 μm Dose: 3.6 × 10^9 MBs/kg or 29.7 μL/kg MVD	f = 400 kHz PRF = 1 Hz Power = 5 W Time = 3–4 min	Mouse/Glioma tumor	[144]
	DLL4 mAb	Avidin-biotin	Targestar™-SA Shell: Phospholipids Gas: C_4F_{10} Diameter: 2 μm Dose: 5.3 × 10^8 MBs/kg or 2.2 μL/kg MVD	f = 1 MHz DC= 50% I = 2 W/cm^2 Time = 1.5 min	Mouse/Gastric cancer	[145]
	Trastuzumab mAb	NHS	Shell: DPSC, DSPE-PEG-2000-NHS, Cholesterol and pyropheophorbide Gas: SF_6 Diameter: 1.654 ± 1.07 μm Dose: unknown	f = 1 MHz DC= 50% I = 2 W/cm^2 Time = 5 min	Mouse/Gastric cancer	[146]
Immune checkpoint inhibitors	PDL-1 mAb	NHS	Shell: DSPC, DSPE-PEG-2000-NHS. Gas: SF_6 Diameter: 1.06 ± 0.31 μm Dose: 6.9 × 10^{10} MBs/kg or 43.3 μL/kg MVD	f = 1.1 MHz DC = 5% PRF = 100 Hz Time = 0.5 min	Mouse/Colon cancer	[147]
	PDL-1 mAb and Cisplatin	Avidin-biotin Unbounded	Shell: DPSC, DSPE-PEG-2000, -Biotin Gas: C_3F_8 Diameter: 1.01 ± 0.14 μm Dose: 1 × 10^8 MBs/kg or 0.06 μL/kg MVD	f = 1 MHz DC = 50% I = 1W/cm^2 PRF = 1 kHz Time = 1.5 min	Mouse/Cervical cancer	[148]
	PDL-1 mAb and miR-34a	Avidin-biotin Electrostatic	Shell: DSPC, DSPE-PEG-2000, DSPE-PEG-2000-Biotin, PEI-600 Gas: C_3F_8 Diameter: 0.940 ± 0.080 μm Dose: 4 × 10^9 MBs/kg or 1.7 μL/kg MVD	f = 18 MHz DC = 50% I = 1 W/cm^2 Time = 1.5 min	Mouse/Cervical cancer	[149]
Vaccine	CD11b mAb and CGAMP	Maleimide	Shell: DSPC, DSPE-PEG-2000, DSPE-PEG-5000-Maleimide Gas: C_4F_{10} Diameter: 2.6 μm Dose: 1.4 × 10^9 MBs/kg or 13.1 μL/kg MVD	f = 1 MHz DC = 50% I = 4 W/cm^2 Time = 2 min	Mouse/Breast cancer	[150]

Table 6. *Cont.*

Immunotherapy Modality	Therapeutic Molecule	Conjugation Chemistry	MicrobubbleComposition	Ultrasound Parameters	Animal/Disease Model	Reference
	HSP70-MAGEA1	Electrostatic	Shell: Span 60 and Tween 80 Gas: SF_6 Diameter: 6 µm Dose: 1.3×10^9 MBs/kg or 144.7 µL/kg MVD	MI = 0.75	Mouse/Melanoma tumor	[151]
	Dendritic cell plasma membrane fragments	Hydrophobic	Shell: DPPC, DPPA, DSPE-PEG5000 Gas: C_3F_8 Diameter: 1.21 ± 1.0 µm Dose: 5×10^8 MBs/kg or 0.5 µL/kg MVD	f = 18 MHz MI = 0.75 Time = 5 min	Mouse/Breast cancer	[152]

DSPC = 1,2-distearoyl-sn-glycero-3-phosphocholine; DPPA: 1,2-dipalmitoyl-sn-glycero-3-phosphate, DPPC: (1,2-dipalmitoyl-sn-glycero-3-phosphocholine); DSPE: 1,2-Distearoyl-sn-glycero-3-phosphorylethanolamine; PEI: Branched Polyethylenimine; DSPE-PEG-2000: (1,2-distearoyl-sn-glycero-3-phosphoethanolamine-N-methoxypolyethylene glycol)-2000]; DSPE-PEG-5000: (1,2-distearoyl-sn-glycero-3-phosphoethanolamine-N-methoxypolyethylene glycol)-5000]; C_3F_8 = perfluoropropane; SF_6 = sulfur hexafluoride, C_4F_{10} = perfluorobutane; MVD = Microbubble Volume Dose; f = frequency; DC = duty cycle; I = intensity; PRF = pulse repetition frequency; MI = mechanical index.

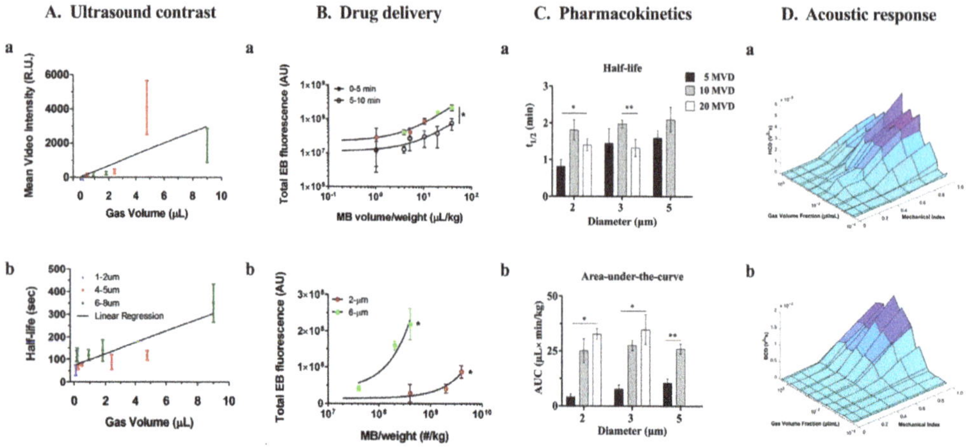

Figure 3. Effect of microbubble volume dose (MVD) on microbubble performance. (**A**) The use of MVD as a dose metric can maximize US imaging contrast (**a**) and circulation persistence (**b**) for isolated microbubbles (MBs) of different diameters. Adapted with permission from reference [112]. Copyright © 2023 World Federation for Ultrasound in Medicine & Biology. Published by Elsevier Inc. All rights reserved. (**B**) The use of MVD results in similar and consistent delivery of Evans blue across the blood–brain barrier (**a**) compared to using MB concentration as dose metric (**b**). The delivery effect of size distribution and concentration of MBs can be collapsed by using MVD. Best fit trend lines, determined by linear regression analysis, are represented by black lines for the 2-μm and 6-μm data sets during the initial 5-minute period of BBB disruption (solid circles; right striatum) and the subsequent 5-minute period (empty circles; left striatum). * $p < 0.0001$. Adapted with permission from reference [116]. (**C**) In vivo pharmacokinetics of size-isolated MBs demonstrated that MVD has greater impact on half-life (**a**) and area-under-the-curve (**b**) than size. * $p < 0.05$ and ** $p \leq 0.01$. Adapted with permission from [117]. Copyright © 2023, American Chemical Society. (**D**) The in vitro harmonic (**a**) and broadband (**b**) cavitation dose (HCD and BCD) versus mechanical index versus gas volume fraction can be aligned by matching the MVD for 2, 3, and 5 μm diameter and polydisperse MBs. Adapted with permission from [37]. © 2023 by the authors.

The acoustic response of MBs is significantly affected by their size. Changes in size can lead to variations in important parameters such as cavitation threshold [153], radiation force experienced [154,155], and resonance frequency [156,157]. Monodisperse tMB size distributions, combined with acoustic radiation force, have been suggested as a new approach to improve MB targeting, promote ligand–receptor interactions and enhance the sensitivity of US detection in various disease models, including inflammation [158–161], prostate cancer [162], and fibrosarcoma tumors [67,163]. Our recent results have shown that monodisperse tMBs, driven near their resonance frequency and at low concentrations, can maximize the adhesion efficiency and specificity of tMB [164]. Thus, the use of monodisperse tMB formulations and MVD as a dose metric could be advantageous in future biomedical applications to enhance US-image contrast sensitivity, increase MB adhesion efficiency, maximize MB interaction with endothelial cells, control cavitation behavior, and increase drug delivery efficacy at the target site.

2.4. Methods to Control Microbubble Size Distributions

Regulating the size of MBs is critical to control their acoustic response, generate high contrast and maximize therapeutic efficiency. Therefore, ensuring a narrow size distribution is desirable for future formulations. Although mechanical agitation is currently employed to produce clinically approved MB suspensions, the resulting size distributions are highly polydisperse, which is unsuitable for therapeutic applications. Hence, to engineer monodisperse MB suspensions, research has mainly focused on employing differential centrifugation and microfluidic technologies.

2.4.1. Differential Centrifugation

To obtain stable size-selected MBs with narrow size distributions, our group developed the differential centrifugation method [165]. This method involves two steps: First, highly concentrated, polydisperse MB suspensions (10^9 to 10^{10} MBs/mL) are rapidly produced by the sonication method. Second, the MBs are isolated based on their migration in a centrifugal field. The sorting process relies on the application of Stokes' equation (Equation (2)) which describes the velocity at which a buoyant particle raises relative to the surrounding fluid under conditions of creeping flow:

$$u_i = \frac{2(\rho_2 - \rho_{1i})}{9\eta_2} r_i^2 g \tag{2}$$

where i refers to the MB size index desired for isolation, r_i is the MB radius, g is the relative centrifugal force (RCF) required for the MB rise through a column of length L, and η_2 is the effective viscosity.

Size-isolated MB populations obtained by this method can be biochemically targeted with specific ligands for ultrasound molecular purposes [67–70,164]. Thus, this method represents a simple way to generate size-isolated tMBs suspensions in high yield that can be adopted for therapeutic applications.

2.4.2. Microfluidic Devices

Microfluidic devices, such as flow focusing [2,166–168] and T-junctions [169], have emerged as promising one-step methods to generate stable monodisperse MB suspensions in the clinical application range. In these geometries, a gas phase is concentrated between two liquid flows through an orifice. Due to unstable capillary instability, the gas becomes unstable and pinches off, resulting in the release monodisperse MBs. Moreover, these devices, offer precise control over the flow rate, viscosity, and interfacial tension, allowing for the production of MBs with uniform and narrow size distributions at the desired size [170–173]. However, one must account for a transient Ostwald ripening process when using microfluidic microbubbles [2,167].

The use of microfluidic devices has been proposed for the in situ production of monodisperse drug-loaded MBs for delivery applications. The use of US and MB enhances the drug delivery to ex vivo rat aortic smooth muscle cells, with over 70% of cells internalization under physiological flow and shear stress conditions [174]. Recently, the feasibility of coating monodisperse MBs with liposomes containing quantum dots and a drug model via a biotin-streptavidin linkage using microfluidic devices has been demonstrated [175]. This technology allows the production of the liposome-MB complex at clinically relevant concentrations with high reproducibility. These results suggest the possibility of adapting this method to simultaneously produce future tMB formulations with homogenous ligand and therapeutic molecule distribution, high payloads, and deliver therapeutic agents in real time.

3. Targeted Microbubbles for Therapeutic Applications

MBs can be biochemically targeted by loading their shell with ligands that avidly bind to specific cell receptors, e.g., those overexpressed during disease states. The ligand can be linked to the MB shell by covalent or noncovalent coupling [6,176,177]. The covalent coupling uses PEG-functionalized lipids with thiol-maleimide, disulfide, DBCO-azide, and folate bonds, whereas noncovalent coupling uses electrostatic interactions and biotin-avidin bridge interactions (Figure 4A).

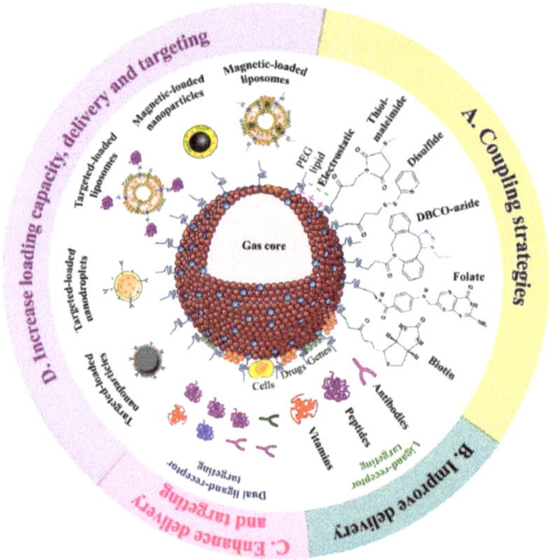

Figure 4. Schematic illustration of current strategies to improve the performance of targeted microbubbles (tMBs) in therapeutic applications. (**A**) Microbubble targeting can be achieved through electrostatic interactions or through the use of PEGylated lipids linked with conjugation groups such as thiol-maleimide, disulfide, DBCO-azide, folate or biotin. (**B**) Ligand–receptor targeting of drug, gene or cell-loaded MBs with antibodies, peptides or vitamins can improve delivery at the target site. (**C**) Dual targeting of loaded MBs with antibody-antibody, antibody-peptide, peptide-peptide, or peptide-vitamin conjugation can improve delivery efficiency and spatial targeting. (**D**) MBs conjugated with molecular or magnetic targeted carriers, such as nanoparticles, liposomes, and nanodroplets, can increase drug loading capacity and delivery efficiency at the target site.

To improve the delivery of drugs, genes, and cells in different biomedical applications, MB targeting can be divided into four strategies: (1) MBs targeted with a therapeutic ligand; (2) drug or gene loaded-MBs targeted with a ligand; (3) drug, gene, or cell loaded-MBs targeted with two different ligands; and (4) MBs conjugated with magnetically or molecularly targeted loaded carriers.

3.1. Microbubbles Targeted with a Therapeutic Ligand

MBs can be loaded with a therapeutic ligand (i.e., where the ligand aids microbubble adhesion *and* has a therapeutic effect) to improve targeted binding efficiency, spatial delivery, and facilitate real-time imaging of tissue structure, function, and molecular characteristics in neurodegenerative and cardiovascular diseases (Table 2). Zhu et al. [20] demonstrated that administering anti-Aβ-tMBs with neural stem cells is a safe and effective dual delivery approach to accelerate Aβ plaque clearance, increase BDNF expression, and improve learning and spatial memory function impairment in Alzheimer's disease. Similarly, Gao et al. [119] showed that injecting anti-SDF-1-tMBs with mesenchymal stem cells can simultaneously monitor real-time renal perfusion, promote cell homing, and repair early diabetic neuropathy kidneys. Using GPVI-tMBs [120] and anti-IL8-tMBs [121] can aid in diagnosing atherosclerotic lesions, slowing down atheroprogression, reducing inflammation, and enhancing plaque stability. Additionally, administering nitric oxide-filled MBs that are targeted with anti-Cd4 can help alleviate cardiac allograft rejection by suppressing thrombosis and inflammatory cell infiltrates, while prolonging the survival time two-fold [108].

3.2. Drug or Gene Loaded-MBs Targeted with a Ligand

Drug- or gene-loaded MB can be targeted with a ligand (here, just to aid microbubble adhesion) to enable site-specific delivery, enhance therapeutic efficacy, and enable image-guided treatment (Figure 4B). Antibody/peptide-tMBs loaded with various genes and chemotherapeutic drugs (Table 3) have been proposed in several studies as successful treatments for various tumors, including glioma [122,123], hind limb [124], and breast tumor [127]. Targeted delivery in these therapies acts as both a molecular contrast agent and a delivery system, improving tumor-specific targeting without harming normal tissues, and increasing local agent delivery, circulation times, and median survival times in mouse and rat models.

A recent study has suggested using folate-tMBs loaded with oxygen and paclitaxel as a dual-targeted system for ovarian tumor cells and tumor-associated macrophages, resulting in two-fold increase in drug concentration, apoptosis index, and median survival [105]. Similarly, folate-tMB loaded with pFLuc enhanced gene transfection efficiency in the brain by 1.5-fold higher than MBs without folate conjugation [129].

On the other hand, different studies have demonstrated that RGD-tMBs loaded with fibrinolytic agents, such as urokinase [125] and tPA [128], can improve arterial blood flow and reduce thrombus size locally without causing adverse side effects, making it a promising and safe option for thrombolysis therapy. Furthermore, VCAM-1-tMBs loaded with mir-126 can serve as a non-invasive and risk-free anti-inflammatory therapy to prevent the growth of abdominal aortic aneurysms and assess the inflammatory state of the endothelium [126].

3.3. Drug-, Gene- or Cell-Loaded MBs Targeted with Two Different Ligands

Drug-, gene- or cell-loaded MB can be targeted with two different ligands (dual-tMBs) to achieve higher adhesion efficiency, enhance ultrasound contrast signal, and have more control over drug delivery at the target site (Figure 4C, Table 4). Adipose-derived stem cells conjugated with CD90/ICAM-1-tMBs, known as StemBells, have been proposed as an image guided treatment for myocardial infarction (MI) [130] and to prevent atherosclerosis acceleration post-MI [131]. Studies have shown that StemBells can improve cardiac function, reach the infarct area without complications, increase cap thickness, decrease intra-plaque macrophage density, and promote the presence of anti-inflammatory macrophages and chemokines in both the plaque and infarcted myocardium, as well as in circulating monocytes.

Additionally, dual-tMBs loaded with genes or chemotherapeutic agents have shown superior performance in treating breast and pancreatic cancer compared to those conjugated with only one ligand [132–134]. These studies have shown that dual-tMBs had 2–3 times higher adhesion efficiency, resulting in a 2–4 times greater ultrasound contrast signal for tumor visualization. Additionally, localized release has improved the accumulation of molecules in the tumor region, resulting in tumor suppression 1.5–4 times greater. This has led to reduced side-effects and better drug tolerability.

3.4. MBs Conjugated with Magnetically or Molecularly Targeted Loaded Carriers

MBs can be conjugated with other magnetically [106,107,135–139] or molecularly [140–143] targeted drug/gene-loaded carriers such as nanoparticles, liposomes, and nanodroplets to increase drug loading capacity, stability in circulation, and delivery efficiency (Figure 4D, Table 5). Drug-loaded superparamagnetic iron oxide nanoparticles (SPIONs) conjugated with MBs have been developed to control drug delivery in brain tumors through magnetic targeting and enabling direct delivery visualization using magnetic resonance imaging (MRI) [107,136]. This conjugation provided significant superparamagnetic/acoustic properties for imaging, resulting in a two-fold increase in MB half-life and US contrast signal, as well as 2–4 times more drug and SPION deposition in the tumor region, leading to enhanced MRI signals.

Recently, a magnetically responsive and ultrasound-sensitive delivery system, called doxorubicin-loaded magneto-liposome MBs, has been proposed for enhanced therapy of pancreatic tumors [139]. This combination resulted in a 1.5-fold improvement in drug penetration at the tumor site and a 2-fold increase in tumor suppression compared to the control group without magnetic targeting. Additionally, formulations of drug- or gene-molecularly targeted liposomes conjugated to MBs have shown the potential to enhance the cytotoxic effects in breast, glioma, and colorectal cancer treatment [140–142], by increasing the concentration of encapsulated drug in circulation, promoting drug accumulation in tumors, and reducing toxicity in normal tissues. Moreover, liposomes conjugated to MBs, which are molecularly and magnetically responsive, can increase the targeting efficiency to tumor neovasculature, enhance MRI, enhance US tumor imaging by 2.5-fold, prolong the MB half-life by 5-times, and decrease the tumor growth by 2-fold [138].

4. Targeted Microbubbles for Immunotherapy Applications

The combination of MBs with US has demonstrated promising potential in modulating and modifying the tumor microenvironment. This is achieved by promoting the penetration of immunotherapeutic agents, enhancing blood perfusion, increasing therapeutic delivery, and inducing tumor cell death [28,30–32]. To further enhance the performance of MBs in cancer immunotherapy, the use of loaded-tMBs with drugs, genes or cells has emerged as an attractive strategy. This approach enables precise control of immune stimulation, and it enhances the delivery and pharmacokinetics of immunomodulatory agents at the target site. This strategy has been shown for various cancer immunotherapy modalities, including monoclonal antibodies, immune checkpoint inhibitors, adoptive cell transfer, cytokine therapy, and vaccines (Table 6).

4.1. Monoclonal Antibody Immunotherapy

Monoclonal antibody immunotherapy aims to induce cell death by targeting specific antigens, sequences, or epitopes expressed at the disease target site [178]. Therapeutic monoclonal antibodies (mAb) can be administered unconjugated or conjugated with chemotherapeutic drugs and radioisotopes to target tumors and minimize the toxicity effects of conventional chemotherapy [179]. Although mAb immunotherapy has potential therapeutic benefits, poor penetration and heterogenous distribution can impact the therapeutic effectiveness [178,179]. High concentrations are often required, which can result in adverse side effects due to the rapid metabolism and clearance rate through the kidneys [180]. mAb-tMBs + US have been proposed as an image-guided delivery method to increase targeting, enhance local penetration, and potentiate the therapeutic effect of mAbs in different cancer therapies (Table 6). Liao et al. [144] improved glioma treatment by administrating EGFR-tMB + US, resulting in increased tumor vessel permeability and enhanced tumor-suppressing effect 7 days after treatment, with no tumor regrowth in the following 10 days. Kang et al. [145] found that combining anti-DLL4-tMBs + US with DAPT for gastric tumor therapy was more effective than DAPT alone, showing synergistic antitumor proapoptotic effects. These effects were attributed to the regulation of apoptosis-related proteins Bcl-2 and BAX, as well as the tumor suppressor protein P53. Recently, Sun et al. [146] developed a tMB construct that delivers pyropheophorbide sensitizer and therapeutic trastuzumab mAbs for targeted combination of sonodynamic and antibody therapies in gastric cancer. The therapy resulted in enhanced antibody accumulation at the tumor site, increased tumor cell apoptosis and tumor growth inhibition by suppressing AKT phosphorylation.

4.2. Immune Checkpoint Inhibitor Therapy

Immune checkpoint inhibitors (ICI) are mAbs variants that activate T cells by blocking immune checkpoint receptors [178]. Approved ICIs for therapy include programmed cell death protein-1 (PD-1), programmed death ligand-1 (PD-L1), and cytotoxic T-lymphocyte-associated antigen-4 (CTLA-4) [178,181]. However, systemic ICI therapy is limited by severe side effects associated with dosage, low treatment response, and overactivation of the immune response [182]. To address these limitations, a controlled delivery strategy called PD-L1 mAb–tMBs + US has been proposed (Table 6). Kim et al. [147] reported that PD-L1 mAb–tMBs can improve therapeutic efficacy, increase the therapeutic index, reduce toxicity, and avoid immune responses and fatalities associated with PD-L1 mAb systemic administration in the treatment of colon cancer. Ma et al. [148] and Liu et al. [149] both demonstrated that combining PD-L1 mAb–tMBs with chemotherapeutic drugs or loading PD-L1 mAb–tMBs with genes exhibited strong synergistic effects in inhibiting cervical tumor growth, improving survival rates, and reducing side effects compared to using either drug/gene or PD-L1 mAb–tMBs alone. The combination treatment showed better immunological activity, indicated by increased CD8+ T cell infiltration and cytokine expression, ultimately resulting in an effective antitumor immune killing effect.

4.3. Adoptive Cell Immunotherapy

Adoptive cell-mediated immunotherapy (ACT) involves the intravenous transfer in vitro of resident T cells or genetically modified T cells to target tumor antigens and mediate anti-tumor function. The three types of ACT are tumor-infiltrating lymphocytes (TIL), T cell receptor (TCR) gene therapy, and chimeric antigen receptor-modified T cells (CAR-T) [178,183]. However, ACT is limited by the lack of in vivo persistence of transferred cells, toxicities related to lymphodepletion, immune response, and cytokine release [183]. Preliminary studies using MBs targeted with specific antibodies and retroviruses (CD3, CD8, CD45RA, CD62L, CD3/CD28, and CD-19CAR) have shown that tMBs represent a potential method to stimulate [184,185], activate, transduce, and precisely sort specific phenotypes of CAR-T cells [184]. These stimulations lead to greater in vivo persistence, decreased toxicity, and improved antitumor response of adoptively transferred CAR-T cells when compared to CAR-T cells obtained through conventional methods [185].

4.4. Cytokine Immunotherapy

Cytokine-mediated immunotherapy involves the systemic administration of cytokines to enhance the immune response [178]. Commonly used cytokines for immunotherapy in clinics and research include Interferon-alpha (IFN-α), Interleukins (ILs), and granulocyte-macrophage colony-stimulating factor (GM-CSF) [178,186,187]. However, this therapy has several limitations, including low efficacy, high levels of toxicity, and immune response activation [188]. IL 27-tMBs have shown promising results by enhancing cytokine bioactivity, inhibiting prostate tumor growth, and efficiently improving the recruitment of natural killer cells (NKT) and CD8+ cells to the tumor compared to untargeted delivery [189]. Additionally, IL-16–tMBs [190] have been introduced to evaluate myocardial ischemia–reperfusion, detect atherosclerosis, and detect ovarian tumors, respectively.

4.5. Vaccine Immunotherapy

Vaccine-mediated immunotherapy involves administering specific antigens or protein fragments to stimulate an immune response [178]. Different types of vaccines are used, including peptide-based vaccines [191], DNA-based vaccines [192], and cell-based vaccines such as NK cells, dendritic cells (DC), and CAR-T cells [193]. However, the effectiveness of vaccine therapy is limited by tissue-specific antigens, low humoral responses, and heterogenous immune responses [178,194]. Gene- and protein-loaded MBs targeted with tumor-specific antigens offer promising image-guided vaccine immunotherapy for breast and melanoma tumors, as demonstrated by Li et al. [150] and Gao et al. [151]. This approach can increase delivery efficiency, prolong survival rates, activate systemic anti-tumor immunity, inhibit and delay tumor growth [150,151], reduce systemic toxicity, and inhibit cancer metastasis by bridging the innate and adaptive immune responses [150]. Recently, Jungio et al. [152] introduced the first cell-free vaccine in the form of tMBs with activated DC plasma membranes to enhance breast cancer tumor targeting, reduce tumor growth, and increase survival rates. Studies also showed tumor growth inhibition and/or antigen-specific protection through DCs activated by mRNA-loaded MBs [195] or tMBs conjugated with NK cells to promote the controlled delivery at the target site [196].

5. Challenges of tMBs in Therapy and Immunotherapy

Tables 2–6 highlight the potential of combining drug-, gene- or cell-loaded tMBs with US for treating various cancers and cardiovascular diseases. However, before exploring future clinical applications, it is crucial to address the in vivo safety of tMBs. The diversity of MB compositions, ligands, therapeutic molecules, and US settings applied pose a significant challenge for clinical translation. While the studies report enhanced therapeutic uptake associated with MB cavitation without side effects, evidence of the latter is lacking. Depending on the US and MB parameters, cavitation can induce adverse bioeffects, such as endothelial cell damage, endothelial dysfunction [197,198], vascular rupture [199,200], and petechial hemorrhages [66,201–203]. Moreover, acoustic cavitation can cause long-term side effects on the target tissues [35,197]. Therefore, a deeper understanding of how cavitation events interact with cells and tissues, along with the resulting cellular and molecular responses, can lead to the development of novel design strategies that improve treatment efficacy while minimizing potential safety concerns.

The MB doses and US parameters used in current studies are based on the reference dose of clinically approved MBs, but comparing bioeffects among different MB formulations and animal models is challenging due to the differences in size distribution, dose, shell composition, gas core, and half-life [66,109,111,113,114,204]. Strategies for approximating equivalent MB or cavitation doses are necessary to improve understanding of induced bioeffects. While US and MB doses are critical factors in comparing bioeffects, physiological variations also significantly impact biological outcomes [22]. Therefore, understanding MB pharmacokinetics and accounting for physiological variability is essential for developing safe and consistent treatment protocols.

On the other hand, the reported experimental therapy durations are typically shorter than those in clinic applications and success is limited to reporting tumor growth suppression, tumor cell apoptosis, and survival time. However, MB cavitation can trigger the activation of immune cells, cytokine secretion or protein production, which can modulate and modify the tumor microenvironment to fight against cancer [28,30–32]. These immune cells, such as T cells, macrophages, neutrophils, fibroblasts, and B cells, participate in the tumor suppression [205]. Therefore, to explain the molecular mechanisms associated with tumor suppression in detail, it is desirable to extend the analysis to characterize the activation and production of these immune components during therapy and immunotherapy protocols. Moreover, the evaluation of only one or four doses of tMB formulations is insufficient, and further pharmacological and toxicological experiments are necessary to identify the optimal dose, administration frequency and protocols, as well as the US parameters, to increase the delivery and maximize therapeutic effects in future studies.

The majority of experimental studies have used murine-origin monoclonal antibodies linked to the MB surface via avidin and biotin coupling, which could potentially cause severe immune responses in human patients [206,207]. To eliminate immunogenic effects, future studies should consider using humanized antibodies, antibody fragments or peptides linked by covalent binding.

Finally, the issue of MB dose is oftentimes ambiguous. We have proposed the use of microbubble volume dose (MVD) as a unifying dose metric that combines MB size distribution and concentration dose to provide a useful metric for correlating pharmacokinetic parameters (e.g., half-life), acoustic response, and the amount of drug delivered to target tissue (Figure 3) [116]. For approved, commercially available MB ultrasound contrast agents, the MVD ranges from 6 to 25 μL/kg (Table 1). However, for targeted MBs, the MVD typically ranges from 1 to 50 μL/kg, although some reports have very low MVD (<0.1 μL/kg), while others have very high MVD (>100 μL/kg). Note that these MVDs were calculated by multiplying the dose given to the animal, usually in number of MBs per animal weight (although sometimes animal weights were not reported, so we had to use average for the species and age). Importantly, there is often a significant difference between the number-weighted distribution (Figure 5D) and volume-weighted distribution (Figure 5E). Note the difference in mean diameters, for example (Figure 5G), even for monodisperse size-isolated microbubbles. For example, as a basis of 10^{10} MB/mL, the volume fractions estimated by the number and volume distributions are quite different (208, 350, and 698 μL/kg vs. 406, 699, and 1170 μL/kg for 2, 3, and 5 μm diameter MBs, respectively). Additionally, the size distribution is often skewed from a normal distribution, even for monodisperse MBs, owing to Ostwald ripening and other colloidal mechanisms. Therefore, one should take the area-under-the-curve of the MB volume vs. concentration plot (Figure 5F) to determine the volume fraction ($\varnothing_{MB}$), which can then be used to determine an accurate MVD [117]. Therefore, a major challenge in the field is for researchers to report an accurate MVD for their tMBs, or at least the combination of dose (number of MBs/kg body weight) and volume-weighted mean diameter.

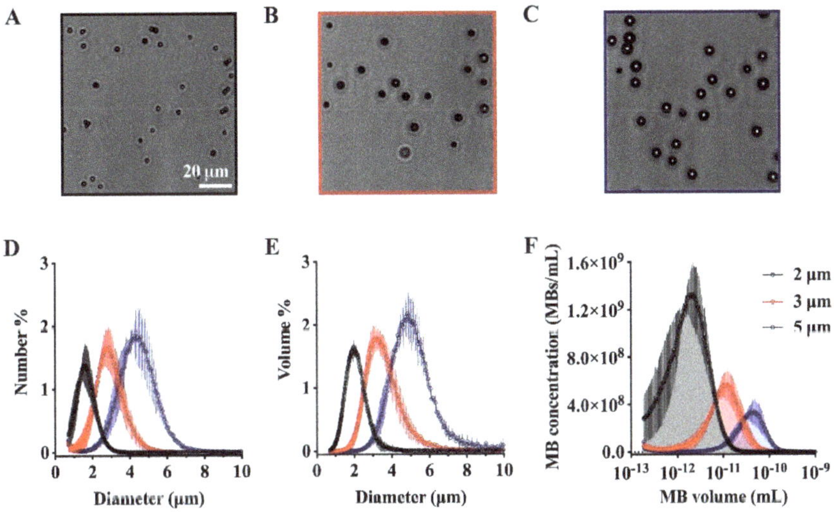

G

MB diameter (μm)	Number % (μm)	Volume % (μm)	φ_{MB} (μL/mL)	Concentration (10^{10} MBs/mL)
2	1.6 ± 0.1	2.0 ± 0.1	7.8 ± 1.4	9.7 ± 0.5
3	2.7 ± 0.3	3.4 ± 0.2	14.6 ± 4.3	3.4 ± 0.2
5	4.2 ± 0.2	5.0 ± 0.2	25.6 ± 5.4	1.8 ± 0.1

Figure 5. Importance of microbubble volume dose (MVD) as a unifying dose metric. Example of size characteristics for three size-isolated lipid-coated microbubbles: (**A**–**C**) microscope images of MBs from the three different sizes; (**D**) number-weighted size distributions; (**E**) volume-weighted size distributions; (**F**) MB volume vs. concentration curve used to determine the gas volume fraction by taking the area-under-the-curve; (**G**) corresponding size metrics. Adapted with permission from [117]. Copyright © 2023, American Chemical Society.

6. Future Directions

This review is focused on tMB formulations, which include direct bioconjugation of a ligand, drug, nanoparticle and/or cell onto the MB shell. One may argue that loading onto the MB is more efficient because the payload is administered at a lower dose and is co-localized with the cavitation event. However, it is currently thought that the regulatory burden of approving a specific MB/drug combination for clinical use makes this approach less financially viable. Additionally, MBs are cleared quickly from circulation (typically 5–10 min) with accumulation primarily in the lung, liver, kidney, and spleen, leading to loss and rapid elimination of the therapeutic cargo before it can become available to the target cells [9,34]. Therefore, current clinical trials favor co-administration of commercial MBs (Table 1) and free drug, either as a single cocktail bolus or a sequence of injections. This co-administration without bioconjugation approach is more amenable as a platform, where a single MB formulation could be used for multiple drug/disease indications. However, this simplistic approach also limits the creativity and potential benefits of MB engineering, several examples of which are reviewed here (Tables 2–6). Therefore, it is advantageous to continue to develop novel targeted microbubble formulations and applications.

As discussed in this review, targeted MBs offer many benefits as innovative agents for ultrasound image-guided drug, gene, and cell delivery in therapy and immunotherapy. While research continues on targeted MBs for research purposes in animal models, more work is necessary to improve their safety, efficacy, and financial viability for clinical translation. Further research and advancements should be made in the following areas:

1. The design and characterization of tMB formulations to achieve high payload capacity, stable drug loading, homogenous and reproducible size distributions and colloidal stability, and strong echogenicity and ultrasound responsiveness.
2. The understanding of tMB pharmacokinetics with the goal of extending circulation time and improving biodistribution, and determination of minimum effective dose and maximum tolerated dose for a given drug and US scheme.
3. The evaluation of tMB interactions with cells and tissues, their correlation to short- and long-term bioeffects in vivo, and molecular description of biological mechanisms induced by tMB cavitation.
4. The optimization of US protocols with consideration of tMB pharmacokinetics and bioeffects to ensure treatment safety and efficacy.
5. The establishment of drug, microbubble, and ultrasound dose metrics to compare therapeutic index between tMB formulations.

Innovations in these areas may help reduce over-reliance on commercially available ultrasound contrast agents and promote clinical translation of engineered tMBs.

However, to justify investment into the clinical translation of tMBs, researchers must show a significant improvement in safety and/or efficacy in comparison to currently approved and commercially available MB formulations (see Table 1). For example, it would be helpful to show an increase in therapeutic index for delivery of the target payload using tMBs vs. commercial MBs (vs. systemic administration without MBs). Additionally, it would be helpful for comparison to report dose optimization in terms of microbubble MVD and pharmacokinetics, along with ultrasound MI, duty cycle, and sonication time. Such characterization and reporting would help the community standardize treatments and optimize passive cavitation monitoring methods.

7. Conclusions

This review highlights recent developments in targeted MB formulations engineered for drug and cell delivery for therapy and immunotherapy. Basic MB formulations are described prior to citing some recent innovative designs for targeted MBs and their applications. The bioconjugation approach has demonstrated promise in enhancing the delivery of various drugs, genes, and cells to target tissues. Issues for future research are discussed, including the need to better define tMB pharmacokinetics and bioeffects, as well as the standardization of MB and US parameters to compare therapeutic index between different MB formulations.

Author Contributions: Conceptualization J.A.N.-B. and M.A.B.; investigation J.A.N.-B.; writing—original draft preparation, J.A.N.-B.; writing—review and editing, M.A.B.; visualization J.A.N.-B.; supervision, M.A.B.; project administration, M.A.B.; funding acquisition, M.A.B. All authors have read and agreed to the published version of the manuscript.

Funding: This work was supported by the US National Institutes of Health award R01CA239465.

Institutional Review Board Statement: Not applicable.

Informed Consent Statement: Not applicable.

Data Availability Statement: Not applicable.

Conflicts of Interest: The authors declare no conflict of interest.

References

1. Epstein, P.S.; Plesset, M.S. On the Stability of Gas Bubbles in Liquid-Gas Solutions. *J. Chem. Phys.* **1950**, *18*, 1505–1509. [CrossRef]
2. Segers, T.; Lohse, D.; Versluis, M.; Frinking, P. Universal Equations for the Coalescence Probability and Long-Term Size Stability of Phospholipid-Coated Monodisperse Microbubbles Formed by Flow Focusing. *Langmuir* **2017**, *33*, 10329–10339. [CrossRef]
3. Sirsi, S.R.; Borden, M.A. Microbubble Compositions, Properties and Biomedical Applications. *Bubble Sci. Eng. Technol.* **2009**, *1*, 3–17. [CrossRef]
4. Borden, M.A.; Song, K.-H. Reverse Engineering the Ultrasound Contrast Agent. *Adv. Colloid Interface Sci.* **2018**, *262*, 39–49. [CrossRef]
5. Frinking, P.; Segers, T.; Luan, Y.; Tranquart, F. Three Decades of Ultrasound Contrast Agents: A Review of the Past, Present and Future Improvements. *Ultrasound Med. Biol.* **2020**, *46*, 892–908. [CrossRef]
6. Borden, M.A.; Dayton, P.A.; Slagle, C.; Walmer, R.W. Ultrasound Contrast Agents. In *Molecular Imaging*; Elsevier: Amsterdam, The Netherlands, 2021; pp. 639–653. ISBN 978-0-12-816386-3.
7. Stride, E.; Segers, T.; Lajoinie, G.; Cherkaoui, S.; Bettinger, T.; Versluis, M.; Borden, M. Microbubble Agents: New Directions. *Ultrasound Med. Biol.* **2020**, *46*, 1326–1343. [CrossRef]
8. Sennoga, C.A.; Kanbar, E.; Auboire, L.; Dujardin, P.-A.; Fouan, D.; Escoffre, J.-M.; Bouakaz, A. Microbubble-Mediated Ultrasound Drug-Delivery and Therapeutic Monitoring. *Expert Opin. Drug Deliv.* **2017**, *14*, 1031–1043. [CrossRef]
9. Lammertink, B.H.A.; Bos, C.; Deckers, R.; Storm, G.; Moonen, C.T.W.; Escoffre, J.-M. Sonochemotherapy: From Bench to Bedside. *Front. Pharmacol.* **2015**, *6*, 138. [CrossRef]
10. He, J.; Liu, Z.; Zhu, X.; Xia, H.; Gao, H.; Lu, J. Ultrasonic Microbubble Cavitation Enhanced Tissue Permeability and Drug Diffusion in Solid Tumor Therapy. *Pharmaceutics* **2022**, *14*, 1642. [CrossRef]
11. Escoffre, J.-M.; Sekkat, N.; Oujagir, E.; Bodard, S.; Mousset, C.; Presset, A.; Chautard, R.; Ayoub, J.; Lecomte, T.; Bouakaz, A. Delivery of Anti-Cancer Drugs Using Microbubble-Assisted Ultrasound in Digestive Oncology: From Preclinical to Clinical Studies. *Expert Opin. Drug Deliv.* **2022**, *19*, 421–433. [CrossRef]
12. Negishi, Y.; Endo-Takahashi, Y.; Maruyama, K. Gene Delivery Systems by the Combination of Lipid Bubbles and Ultrasound. *DDT* **2016**, *10*, 248–255. [CrossRef]
13. Sirsi, S.R.; Borden, M.A. Advances in Ultrasound Mediated Gene Therapy Using Microbubble Contrast Agents. *Theranostics* **2012**, *2*, 1208–1222. [CrossRef]
14. Rychak, J.J.; Klibanov, A.L. Nucleic Acid Delivery with Microbubbles and Ultrasound. *Adv. Drug Deliv. Rev.* **2014**, *72*, 82–93. [CrossRef]
15. Tong, J.; Ding, J.; Shen, X.; Chen, L.; Bian, Y.; Ma, G.; Yao, Y.; Yang, F. Mesenchymal Stem Cell Transplantation Enhancement in Myocardial Infarction Rat Model under Ultrasound Combined with Nitric Oxide Microbubbles. *PLoS ONE* **2013**, *8*, e80186. [CrossRef]
16. Toma, C.; Fisher, A.; Wang, J.; Chen, X.; Grata, M.; Leeman, J.; Winston, B.; Kaya, M.; Fu, H.; Lavery, L.; et al. Vascular Endoluminal Delivery of Mesenchymal Stem Cells Using Acoustic Radiation Force. *Tissue Eng. Part A* **2011**, *17*, 1457–1464. [CrossRef]
17. Sun, T.; Gao, F.; Li, X.; Cai, Y.; Bai, M.; Li, F.; Du, L. A Combination of Ultrasound-Targeted Microbubble Destruction with Transplantation of Bone Marrow Mesenchymal Stem Cells Promotes Recovery of Acute Liver Injury. *Stem Cell Res* **2018**, *9*, 356. [CrossRef]
18. Zhu, T.; Huang, X.; Peng, S.; Ye, Y.; Zhao, J. Ultrasound Targeted Microbubble Destruction Promotes the Therapeutic Effect of HUMSC Transplantation on Glaucoma-Caused Optic Nerve Injury in Rabbits. *Trans. Vis. Sci. Tech.* **2022**, *11*, 12. [CrossRef]
19. Cui, H.; Zhu, Q.; Xie, Q.; Liu, Z.; Gao, Y.; He, Y.; Tan, X.; Xu, Y. Low Intensity Ultrasound Targeted Microbubble Destruction Assists MSCs Delivery and Improves Neural Function in Brain Ischaemic Rats. *J. Drug Target.* **2020**, *28*, 320–329. [CrossRef]
20. Zhu, Q.; Xu, X.; Chen, B.; Liao, Y.; Guan, X.; He, Y.; Cui, H.; Rong, Y.; Liu, Z.; Xu, Y. Ultrasound-targeted Microbubbles Destruction Assists Dual Delivery of Beta-amyloid Antibody and Neural Stem Cells to Restore Neural Function in Transgenic Mice of Alzheimer's Disease. *Med. Phys.* **2022**, *49*, 1357–1367. [CrossRef]
21. Schoen, S.; Kilinc, M.S.; Lee, H.; Guo, Y.; Degertekin, F.L.; Woodworth, G.F.; Arvanitis, C. Towards Controlled Drug Delivery in Brain Tumors with Microbubble-Enhanced Focused Ultrasound. *Adv. Drug Deliv. Rev.* **2022**, *180*, 114043. [CrossRef]
22. Song, K.-H.; Harvey, B.K.; Borden, M.A. State-of-the-Art of Microbubble-Assisted Blood-Brain Barrier Disruption. *Theranostics* **2018**, *8*, 4393–4408. [CrossRef]
23. Wang, J.; Li, Z.; Pan, M.; Fiaz, M.; Hao, Y.; Yan, Y.; Sun, L.; Yan, F. Ultrasound-Mediated Blood–Brain Barrier Opening: An Effective Drug Delivery System for Theranostics of Brain Diseases. *Adv. Drug Deliv. Rev.* **2022**, *190*, 114539. [CrossRef]
24. Unger, E.; Porter, T.; Lindner, J.; Grayburn, P. Cardiovascular Drug Delivery with Ultrasound and Microbubbles. *Adv. Drug Deliv. Rev.* **2014**, *72*, 110–126. [CrossRef]

25. Qian, L.; Thapa, B.; Hong, J.; Zhang, Y.; Zhu, M.; Chu, M.; Yao, J.; Xu, D. The Present and Future Role of Ultrasound Targeted Microbubble Destruction in Preclinical Studies of Cardiac Gene Therapy. *J. Thorac. Dis.* **2018**, *10*, 1099–1111. [CrossRef]
26. Chen, H.H.; Matkar, P.N.; Afrasiabi, K.; Kuliszewski, M.A.; Leong-Poi, H. Prospect of Ultrasound-Mediated Gene Delivery in Cardiovascular Applications. *Expert Opin. Biol. Ther.* **2016**, *16*, 815–826. [CrossRef]
27. Wanigasekara, J.; de Carvalho, A.M.A.; Cullen, P.J.; Tiwari, B.; Curtin, J.F. Converging Technologies: Targeting the Hallmarks of Cancer Using Ultrasound and Microbubbles. *Trends Cancer* **2021**, *7*, 886–890. [CrossRef]
28. Tu, J.; Zhang, H.; Yu, J.; Liufu, C.; Chen, Z. Ultrasound-Mediated Microbubble Destruction: A New Method in Cancer Immunotherapy. *OTT* **2018**, *11*, 5763–5775. [CrossRef]
29. Li, H.; Zhang, Y.; Shu, H.; Lv, W.; Su, C.; Nie, F. Highlights in Ultrasound-Targeted Microbubble Destruction-Mediated Gene/Drug Delivery Strategy for Treatment of Malignancies. *Int. J. Pharm.* **2022**, *613*, 121412. [CrossRef]
30. Han, Y.; Sun, J.; Wei, H.; Hao, J.; Liu, W.; Wang, X. Ultrasound-Targeted Microbubble Destruction: Modulation in the Tumor Microenvironment and Application in Tumor Immunotherapy. *Front. Immunol.* **2022**, *13*, 937344. [CrossRef]
31. Ho, Y.-J.; Li, J.-P.; Fan, C.-H.; Liu, H.-L.; Yeh, C.-K. Ultrasound in Tumor Immunotherapy: Current Status and Future Developments. *J. Control. Release* **2020**, *323*, 12–23. [CrossRef]
32. Omata, D.; Munakata, L.; Maruyama, K.; Suzuki, R. Ultrasound and Microbubble-Mediated Drug Delivery and Immunotherapy. *J. Med. Ultrason.* **2022**, *Online ahead of print*. [CrossRef]
33. Yang, Y.; Li, Q.; Guo, X.; Tu, J.; Zhang, D. Mechanisms Underlying Sonoporation: Interaction between Microbubbles and Cells. *Ultrason. Sonochem.* **2020**, *67*, 105096. [CrossRef]
34. Deprez, J.; Lajoinie, G.; Engelen, Y.; De Smedt, S.C.; Lentacker, I. Opening Doors with Ultrasound and Microbubbles: Beating Biological Barriers to Promote Drug Delivery. *Adv. Drug Deliv. Rev.* **2021**, *172*, 9–36. [CrossRef]
35. Qin, P.; Han, T.; Yu, A.C.H.; Xu, L. Mechanistic Understanding the Bioeffects of Ultrasound-Driven Microbubbles to Enhance Macromolecule Delivery. *J. Control. Release* **2018**, *272*, 169–181. [CrossRef]
36. Kooiman, K.; Vos, H.J.; Versluis, M.; de Jong, N. Acoustic Behavior of Microbubbles and Implications for Drug Delivery. *Adv. Drug Deliv. Rev.* **2014**, *72*, 28–48. [CrossRef]
37. Martinez, P.; Bottenus, N.; Borden, M. Cavitation Characterization of Size-Isolated Microbubbles in a Vessel Phantom Using Focused Ultrasound. *Pharmaceutics* **2022**, *14*, 1925. [CrossRef]
38. Wischhusen, J.; Padilla, F. Ultrasound-Targeted Microbubble Destruction (UTMD) for Localized Drug Delivery into Tumor Tissue. *IRBM* **2019**, *40*, 10–15. [CrossRef]
39. Chatterjee, D.; Sarkar, K. A Newtonian Rheological Model for the Interface of Microbubble Contrast Agents. *Ultrasound Med. Biol.* **2003**, *29*, 1749–1757. [CrossRef]
40. Ja'afar, F.; Leow, C.H.; Garbin, V.; Sennoga, C.A.; Tang, M.-X.; Seddon, J.M. Surface Charge Measurement of SonoVue, Definity and Optison: A Comparison of Laser Doppler Electrophoresis and Micro-Electrophoresis. *Ultrasound Med. Biol.* **2015**, *41*, 2990–3000. [CrossRef]
41. Shi, W.T.; Forsberg, F. Ultrasonic Characterization of the Nonlinear Properties of Contrast Microbubbles. *Ultrasound Med. Biol.* **2000**, *26*, 93–104. [CrossRef]
42. Tan, J.-K.Y.; Pham, B.; Zong, Y.; Perez, C.; Maris, D.O.; Hemphill, A.; Miao, C.H.; Matula, T.J.; Mourad, P.D.; Wei, H.; et al. Microbubbles and Ultrasound Increase Intraventricular Polyplex Gene Transfer to the Brain. *J. Control. Release* **2016**, *231*, 86–93. [CrossRef]
43. Fix, S.M.; Nyankima, A.G.; McSweeney, M.D.; Tsuruta, J.K.; Lai, S.K.; Dayton, P.A. Accelerated Clearance of Ultrasound Contrast Agents Containing Polyethylene Glycol Is Associated with the Generation of Anti-Polyethylene Glycol Antibodies. *Ultrasound Med. Biol.* **2018**, *44*, 1266–1280. [CrossRef]
44. Tu, J.; Swalwell, J.E.; Giraud, D.; Cui, W.; Chen, W.; Matula, T.J. Microbubble Sizing and Shell Characterization Using Flow Cytometry. *IEEE Trans. Ultrason. Ferroelect. Freq. Contr.* **2011**, *58*, 955–963. [CrossRef] [PubMed]
45. Helfield, B.L.; Goertz, D.E. Nonlinear Resonance Behavior and Linear Shell Estimates for Definity™ and MicroMarker™ Assessed with Acoustic Microbubble Spectroscopy. *J. Acoust. Soc. Am.* **2013**, *133*, 1158–1168. [CrossRef]
46. Chatterjee, D.; Sarkar, K.; Jain, P.; Schreppler, N.E. On the Suitability of Broadband Attenuation Measurement for Characterizing Contrast Microbubbles. *Ultrasound Med. Biol.* **2005**, *31*, 781–786. [CrossRef]
47. Tu, J.; Guan, J.; Qiu, Y.; Matula, T.J. Estimating the Shell Parameters of SonoVue ® Microbubbles Using Light Scattering. *J. Acoust. Soc. Am.* **2009**, *126*, 2954–2962. [CrossRef]
48. Gorce, J.-M.; Arditi, M.; Schneider, M. Influence of Bubble Size Distribution on the Echogenicity of Ultrasound Contrast Agents: A Study of SonoVue? *Investig. Radiol.* **2000**, *35*, 661–671. [CrossRef]
49. Wu, S.-K.; Chu, P.-C.; Chai, W.-Y.; Kang, S.-T.; Tsai, C.-H.; Fan, C.-H.; Yeh, C.-K.; Liu, H.-L. Characterization of Different Microbubbles in Assisting Focused Ultrasound-Induced Blood-Brain Barrier Opening. *Sci. Rep.* **2017**, *7*, 46689. [CrossRef]
50. Sarkar, K.; Shi, W.T.; Chatterjee, D.; Forsberg, F. Characterization of Ultrasound Contrast Microbubbles Using in Vitro Experiments and Viscous and Viscoelastic Interface Models for Encapsulation. *J. Acoust. Soc. Am.* **2005**, *118*, 539–550. [CrossRef]
51. Sontum, P.C. Physicochemical Characteristics of Sonazoid™, A New Contrast Agent for Ultrasound Imaging. *Ultrasound Med. Biol.* **2008**, *34*, 824–833. [CrossRef]

52. Landmark, K.E.; Johansen, P.W.; Johnson, J.A.; Johansen, B.; Uran, S.; Skotland, T. Pharmacokinetics of Perfluorobutane Following Intravenous Bolus Injection and Continuous Infusion of Sonazoid™ in Healthy Volunteers and in Patients with Reduced Pulmonary Diffusing Capacity. *Ultrasound Med. Biol.* **2008**, *34*, 494–501. [CrossRef]
53. Valdez, M.A.; Fernandez, E.; Matsunaga, T.; Erickson, R.P.; Trouard, T.P. Distribution and Diffusion of Macromolecule Delivery to the Brain via Focused Ultrasound Using Magnetic Resonance and Multispectral Fluorescence Imaging. *Ultrasound Med. Biol.* **2020**, *46*, 122–136. [CrossRef] [PubMed]
54. Upadhyay, A.; Dalvi, S.V. Microbubble Formulations: Synthesis, Stability, Modeling and Biomedical Applications. *Ultrasound Med. Biol.* **2019**, *45*, 301–343. [CrossRef] [PubMed]
55. Al-Jawadi, S.; Thakur, S.S. Ultrasound-Responsive Lipid Microbubbles for Drug Delivery: A Review of Preparation Techniques to Optimise Formulation Size, Stability and Drug Loading. *Int. J. Pharm.* **2020**, *585*, 119559. [CrossRef]
56. Borden, M.A. Lipid-Coated Nanodrops and Microbubbles. In *Handbook of Ultrasonics and Sonochemistry*; Springer: Singapore, 2016; pp. 1075–1100. ISBN 978-981-287-277-7.
57. Borden, M.A. Intermolecular Forces Model for Lipid Microbubble Shells. *Langmuir* **2019**, *35*, 10042–10051. [CrossRef]
58. Lum, J.S.; Dove, J.D.; Murray, T.W.; Borden, M.A. Single Microbubble Measurements of Lipid Monolayer Viscoelastic Properties for Small-Amplitude Oscillations. *Langmuir* **2016**, *32*, 9410–9417. [CrossRef] [PubMed]
59. Kim, D.H.; Costello, M.J.; Duncan, P.B.; Needham, D. Mechanical Properties and Microstructure of Polycrystalline Phospholipid Monolayer Shells: Novel Solid Microparticles. *Langmuir* **2003**, *19*, 8455–8466. [CrossRef]
60. Borden, M.A.; Longo, M.L. Oxygen Permeability of Fully Condensed Lipid Monolayers. *J. Phys. Chem. B* **2004**, *108*, 6009–6016. [CrossRef]
61. Pu, G.; Longo, M.L.; Borden, M.A. Effect of Microstructure on Molecular Oxygen Permeation through Condensed Phospholipid Monolayers. *J. Am. Chem. Soc.* **2005**, *127*, 6524–6525. [CrossRef]
62. Borden, M.A.; Kruse, D.E.; Caskey, C.F.; Shukui, Z.; Dayton, P.A.; Ferrara, K.W. Influence of Lipid Shell Physicochemical Properties on Ultrasound-Induced Microbubble Destruction. *IEEE Trans. Ultrason. Ferroelect. Freq. Contr.* **2005**, *52*, 1992–2002. [CrossRef]
63. Garg, S.; Thomas, A.A.; Borden, M.A. The Effect of Lipid Monolayer In-Plane Rigidity on in Vivo Microbubble Circulation Persistence. *Biomaterials* **2013**, *34*, 6862–6870. [CrossRef] [PubMed]
64. Borden, M.A.; Longo, M.L. Dissolution Behavior of Lipid Monolayer-Coated, Air-Filled Microbubbles: Effect of Lipid Hydrophobic Chain Length. *Langmuir* **2002**, *18*, 9225–9233. [CrossRef]
65. van Rooij, T.; Luan, Y.; Renaud, G.; van der Steen, A.F.W.; Versluis, M.; de Jong, N.; Kooiman, K. Non-Linear Response and Viscoelastic Properties of Lipid-Coated Microbubbles: DSPC versus DPPC. *Ultrasound Med. Biol.* **2015**, *41*, 1432–1445. [CrossRef] [PubMed]
66. Wu, S.-Y.; Chen, C.C.; Tung, Y.-S.; Olumolade, O.O.; Konofagou, E.E. Effects of the Microbubble Shell Physicochemical Properties on Ultrasound-Mediated Drug Delivery to the Brain. *J. Control. Release* **2015**, *212*, 30–40. [CrossRef]
67. Borden, M.A.; Streeter, J.E.; Sirsi, S.R.; Dayton, P.A. In Vivo Demonstration of Cancer Molecular Imaging with Ultrasound Radiation Force and Buried-Ligand Microbubbles. *Mol. Imaging* **2013**, *12*, 357–363. [CrossRef]
68. Chen, C.C.; Sirsi, S.R.; Homma, S.; Borden, M.A. Effect of Surface Architecture on In Vivo Ultrasound Contrast Persistence of Targeted Size-Selected Microbubbles. *Ultrasound Med. Biol.* **2012**, *38*, 492–503. [CrossRef]
69. Borden, M.A.; Sarantos, M.R.; Stieger, S.M.; Simon, S.I.; Ferrara, K.W.; Dayton, P.A. Ultrasound Radiation Force Modulates Ligand Availability on Targeted Contrast Agents. *Mol. Imaging* **2006**, *5*, 139–147. [CrossRef]
70. Borden, M.A.; Zhang, H.; Gillies, R.J.; Dayton, P.A.; Ferrara, K.W. A Stimulus-Responsive Contrast Agent for Ultrasound Molecular Imaging. *Biomaterials* **2008**, *29*, 597–606. [CrossRef]
71. Chen, C.C.; Borden, M.A. The Role of Poly(Ethylene Glycol) Brush Architecture in Complement Activation on Targeted Microbubble Surfaces. *Biomaterials* **2011**, *32*, 6579–6587. [CrossRef]
72. Kiessling, F.; Fokong, S.; Bzyl, J.; Lederle, W.; Palmowski, M.; Lammers, T. Recent Advances in Molecular, Multimodal and Theranostic Ultrasound Imaging. *Adv. Drug Deliv. Rev.* **2014**, *72*, 15–27. [CrossRef]
73. Rudakovskaya, P.G.; Barmin, R.A.; Kuzmin, P.S.; Fedotkina, E.P.; Sencha, A.N.; Gorin, D.A. Microbubbles Stabilized by Protein Shell: From Pioneering Ultrasound Contrast Agents to Advanced Theranostic Systems. *Pharmaceutics* **2022**, *14*, 1236. [CrossRef] [PubMed]
74. Liu, M.; Dasgupta, A.; Qu, N.; Rama, E.; Kiessling, F.; Lammers, T. Strategies to Maximize Anthracycline Drug Loading in Albumin Microbubbles. *ACS Biomater. Sci. Eng.* **2021**, *Online ahead of print*. [CrossRef] [PubMed]
75. Ji, J.; Ji, J.; He, X. Preparation of Ultrasound Microbubbles Crosslinked to Albumin Nanoparticles Packaged with Tissue-Type Plasminogen Activator Gene Plasmid and Method of in Vivo Transfection. *JEP* **2011**, *24*, 35–41. [CrossRef] [PubMed]
76. Borrelli, M.J.; O'Brien, W.D.; Bernock, L.J.; Williams, H.R.; Hamilton, E.; Wu, J.; Oelze, M.L.; Culp, W.C. Production of Uniformly Sized Serum Albumin and Dextrose Microbubbles. *Ultrason. Sonochem.* **2012**, *19*, 198–208. [CrossRef] [PubMed]
77. Chen, J.L.; Dhanaliwala, A.H.; Dixon, A.J.; Klibanov, A.L.; Hossack, J.A. Synthesis and Characterization of Transiently Stable Albumin-Coated Microbubbles via a Flow-Focusing Microfluidic Device. *Ultrasound Med. Biol.* **2014**, *40*, 400–409. [CrossRef]
78. Upadhyay, A.; Dalvi, S.V. Synthesis, Characterization and Stability of BSA-Encapsulated Microbubbles. *RSC Adv.* **2016**, *6*, 15016–15026. [CrossRef]
79. Lentacker, I.; De Geest, B.G.; Vandenbroucke, R.E.; Peeters, L.; Demeester, J.; De Smedt, S.C.; Sanders, N.N. Ultrasound-Responsive Polymer-Coated Microbubbles That Bind and Protect DNA. *Langmuir* **2006**, *22*, 7273–7278. [CrossRef]

80. Upadhyay, A.; Dalvi, S.V.; Gupta, G.; Khanna, N. Effect of PEGylation on Performance of Protein Microbubbles and Its Comparison with Lipid Microbubbles. *Mater. Sci. Eng. C* **2017**, *71*, 425–430. [CrossRef]
81. Ma, X.; Bussonniere, A.; Liu, Q. A Facile Sonochemical Synthesis of Shell-Stabilized Reactive Microbubbles Using Surface-Thiolated Bovine Serum Albumin with the Traut's Reagent. *Ultrason. Sonochem.* **2017**, *36*, 454–465. [CrossRef]
82. Liu, X.; Gong, P.; Song, P.; Xie, F.; Miller II, A.L.; Chen, S.; Lu, L. Fast Functionalization of Ultrasound Microbubbles Using Strain Promoted Click Chemistry. *Biomater. Sci.* **2018**, *6*, 623–632. [CrossRef]
83. Narihira, K.; Watanabe, A.; Sheng, H.; Endo, H.; Feril, L.B.; Irie, Y.; Ogawa, K.; Moosavi-Nejad, S.; Kondo, S.; Kikuta, T.; et al. Enhanced Cell Killing and Apoptosis of Oral Squamous Cell Carcinoma Cells with Ultrasound in Combination with Cetuximab Coated Albumin Microbubbles. *J. Drug Target.* **2018**, *26*, 278–288. [CrossRef] [PubMed]
84. Wang, Y.-H.; Liao, A.-H.; Chen, J.-H.; Chris Wang, C.-R.; Li, P.-C. Photoacoustic/Ultrasound Dual-Modality Contrast Agent and Its Application to Thermotherapy. *J. Biomed. Opt.* **2012**, *17*, 045001. [CrossRef] [PubMed]
85. Liou, Y.-R.; Wang, Y.-H.; Lee, C.-Y.; Li, P.-C. Buoyancy-Activated Cell Sorting Using Targeted Biotinylated Albumin Microbubbles. *PLoS ONE* **2015**, *10*, e0125036. [CrossRef]
86. Wang, Y.-H.; Chen, S.-P.; Liao, A.-H.; Yang, Y.-C.; Lee, C.-R.; Wu, C.-H.; Wu, P.-C.; Liu, T.-M.; Wang, C.-R.C.; Li, P.-C. Synergistic Delivery of Gold Nanorods Using Multifunctional Microbubbles for Enhanced Plasmonic Photothermal Therapy. *Sci. Rep.* **2014**, *4*, 5685. [CrossRef]
87. Porter, T.R.; Xie, F.; Knapp, D.; Iversen, P.; Marky, L.A.; Tsutsui, J.M.; Maiti, S.; Lof, J.; Radio, S.J.; Kipshidze, N. Targeted Vascular Delivery of Antisense Molecules Using Intravenous Microbubbles. *Cardiovasc. Revascularization Med.* **2006**, *7*, 25–33. [CrossRef]
88. Liao, A.-H.; Wu, S.-Y.; Wang, H.-E.; Weng, C.-H.; Wu, M.-F.; Li, P.-C. Evaluation of 18F-Labeled Targeted Perfluorocarbon-Filled Albumin Microbubbles as a Probe for MicroUS and MicroPET in Tumor-Bearing Mice. *Ultrasonics* **2013**, *53*, 320–327. [CrossRef]
89. Barmin, R.; Rudakovskaya, P.; Gusliakova, O.; Sindeeva, O.; Prikhozhdenko, E.; Maksimova, E.; Obukhova, E.; Chernyshev, V.; Khlebtsov, B.; Solovev, A.; et al. Air-Filled Bubbles Stabilized by Gold Nanoparticle/Photodynamic Dye Hybrid Structures for Theranostics. *Nanomaterials* **2021**, *11*, 415. [CrossRef] [PubMed]
90. Maksimova, E.A.; Barmin, R.A.; Rudakovskaya, P.G.; Sindeeva, O.A.; Prikhozhdenko, E.S.; Yashchenok, A.M.; Khlebtsov, B.N.; Solovev, A.A.; Huang, G.; Mei, Y.; et al. Air-Filled Microbubbles Based on Albumin Functionalized with Gold Nanocages and Zinc Phthalocyanine for Multimodal Imaging. *Micromachines* **2021**, *12*, 1161. [CrossRef]
91. Yoon, Y.I.; Pang, X.; Jung, S.; Zhang, G.; Kong, M.; Liu, G.; Chen, X. Smart Gold Nanoparticle-Stabilized Ultrasound Microbubbles as Cancer Theranostics. *J. Mater. Chem. B* **2018**, *6*, 3235–3239. [CrossRef]
92. Chen, Z.; Chattaraj, R.; Pulsipher, K.W.; Karmacharya, M.B.; Hammer, D.A.; Lee, D.; Sehgal, C.M. Photoacoustic and Ultrasound Dual-Mode Imaging via Functionalization of Recombinant Protein-Stabilized Microbubbles with Methylene Blue. *ACS Appl. Bio Mater.* **2019**, *2*, 4020–4026. [CrossRef]
93. Xiong, X.; Zhao, F.; Shi, M.; Yang, H.; Liu, Y. Polymeric Microbubbles for Ultrasonic Molecular Imaging and Targeted Therapeutics. *J. Biomater. Sci. Polym. Ed.* **2011**, *22*, 417–428. [CrossRef] [PubMed]
94. Liu, M.; Dasgupta, A.; Koczera, P.; Schipper, S.; Rommel, D.; Shi, Y.; Kiessling, F.; Lammers, T. Drug Loading in Poly(Butyl Cyanoacrylate)-Based Polymeric Microbubbles. *Mol. Pharm.* **2020**, *17*, 2840–2848. [CrossRef]
95. Estifeeva, T.M.; Barmin, R.A.; Rudakovskaya, P.G.; Nechaeva, A.M.; Luss, A.L.; Mezhuev, Y.O.; Chernyshev, V.S.; Krivoborodov, E.G.; Klimenko, O.A.; Sindeeva, O.A.; et al. Hybrid (Bovine Serum Albumin)/Poly (*N*-Vinyl-2-Pyrrolidone-*Co*-Acrylic Acid)-Shelled Microbubbles as Advanced Ultrasound Contrast Agents. *ACS Appl. Bio. Mater.* **2022**, *5*, 3338–3348. [CrossRef] [PubMed]
96. Barmin, R.A.; Dasgupta, A.; Bastard, C.; De Laporte, L.; Rütten, S.; Weiler, M.; Kiessling, F.; Lammers, T.; Pallares, R.M. Engineering the Acoustic Response and Drug Loading Capacity of PBCA-Based Polymeric Microbubbles with Surfactants. *Mol. Pharm.* **2022**, *19*, 3256–3266. [CrossRef] [PubMed]
97. Omata, D.; Maruyama, T.; Unga, J.; Hagiwara, F.; Munakata, L.; Kageyama, S.; Shima, T.; Suzuki, Y.; Maruyama, K.; Suzuki, R. Effects of Encapsulated Gas on Stability of Lipid-Based Microbubbles and Ultrasound-Triggered Drug Delivery. *J. Control. Release* **2019**, *311–312*, 65–73. [CrossRef] [PubMed]
98. Sun, J.; Yin, M.; Zhu, S.; Liu, L.; Zhu, Y.; Wang, Z.; Xu, R.X.; Chang, S. Ultrasound-Mediated Destruction of Oxygen and Paclitaxel Loaded Lipid Microbubbles for Combination Therapy in Hypoxic Ovarian Cancer Cells. *Ultrason. Sonochem.* **2016**, *28*, 319–326. [CrossRef] [PubMed]
99. Eisenbrey, J.R.; Shraim, R.; Liu, J.-B.; Li, J.; Stanczak, M.; Oeffinger, B.; Leeper, D.B.; Keith, S.W.; Jablonowski, L.J.; Forsberg, F.; et al. Sensitization of Hypoxic Tumors to Radiation Therapy Using Ultrasound-Sensitive Oxygen Microbubbles. *Int. J. Radiat. Oncol. Biol. Phys.* **2018**, *101*, 88–96. [CrossRef]
100. Ho, Y.-J.; Chu, S.-W.; Liao, E.-C.; Fan, C.-H.; Chan, H.-L.; Wei, K.-C.; Yeh, C.-K. Normalization of Tumor Vasculature by Oxygen Microbubbles with Ultrasound. *Theranostics* **2019**, *9*, 7370–7383. [CrossRef]
101. Liang, Z.; Chen, H.; Gong, X.; Shi, B.; Lin, L.; Tao, F.; Wu, Q.; Fang, M.; Li, H.; Lu, C.; et al. Ultrasound-Induced Destruction of Nitric Oxide–Loaded Microbubbles in the Treatment of Thrombus and Ischemia–Reperfusion Injury. *Front. Pharmacol.* **2022**, *12*, 745693. [CrossRef]
102. Wang, C.; Yang, F.; Xu, Z.; Shi, D.; Chen, D.; Dai, J.; Gu, N.; Jiang, Q. Intravenous Release of NO from Lipidic Microbubbles Accelerates Deep Vein Thrombosis Resolution in a Rat Model. *Thromb. Res.* **2013**, *131*, e31–e38. [CrossRef]
103. Kwan, J.J.; Kaya, M.; Borden, M.A.; Dayton, P.A. Theranostic Oxygen Delivery Using Ultrasound and Microbubbles. *Theranostics* **2012**, *2*, 1174–1184. [CrossRef] [PubMed]

104. Reusser, T.D.; Song, K.-H.; Ramirez, D.; Benninger, R.K.; Papadopoulou, V.; Borden, M.A. Phospholipid Oxygen Microbubbles for Image-Guided Therapy. *Nanotheranostics* **2020**, *4*, 83–90. [CrossRef] [PubMed]
105. Luo, T.; Sun, J.; Zhu, S.; He, J.; Hao, L.; Xiao, L.; Zhu, Y.; Wang, Q.; Pan, X.; Wang, Z.; et al. Ultrasound-Mediated Destruction of Oxygen and Paclitaxel Loaded Dual-Targeting Microbubbles for Intraperitoneal Treatment of Ovarian Cancer Xenografts. *Cancer Lett.* **2017**, *391*, 1–11. [CrossRef] [PubMed]
106. Sheng, Y.; Beguin, E.; Nesbitt, H.; Kamila, S.; Owen, J.; Barnsley, L.C.; Callan, B.; O'Kane, C.; Nomikou, N.; Hamoudi, R.; et al. Magnetically Responsive Microbubbles as Delivery Vehicles for Targeted Sonodynamic and Antimetabolite Therapy of Pancreatic Cancer. *J. Control. Release* **2017**, *262*, 192–200. [CrossRef] [PubMed]
107. Fan, C.-H.; Cheng, Y.-H.; Ting, C.-Y.; Ho, Y.-J.; Hsu, P.-H.; Liu, H.-L.; Yeh, C.-K. Ultrasound/Magnetic Targeting with SPIO-DOX-Microbubble Complex for Image-Guided Drug Delivery in Brain Tumors. *Theranostics* **2016**, *6*, 1542–1556. [CrossRef]
108. Liao, T.; Li, Q.; Zhang, Y.; Yang, Z.; Huang, Z.; Han, F.; Chen, X.; Yin, T.; Ren, J.; Sun, Q. Precise Treatment of Acute Antibody-Mediated Cardiac Allograft Rejection in Rats Using C4d-Targeted Microbubbles Loaded with Nitric Oxide. *J. Heart Lung Transplant.* **2020**, *39*, 481–490. [CrossRef] [PubMed]
109. Sonne, C. Differences in Definity and Optison Microbubble Destruction Rates at a Similar Mechanical Index with Different Real-Time Perfusion Systems. *J. Am. Soc. Echocardiogr.* **2003**, *16*, 1178–1185. [CrossRef]
110. Hyvelin, J.-M.; Gaud, E.; Costa, M.; Helbert, A.; Bussat, P.; Bettinger, T.; Frinking, P. Characteristics and Echogenicity of Clinical Ultrasound Contrast Agents: An In Vitro and In Vivo Comparison Study: Comparison of Clinical Ultrasound Contrast Agents. *J. Ultrasound Med.* **2017**, *36*, 941–953. [CrossRef]
111. McMahon, D.; Lassus, A.; Gaud, E.; Jeannot, V.; Hynynen, K. Microbubble Formulation Influences Inflammatory Response to Focused Ultrasound Exposure in the Brain. *Sci. Rep.* **2020**, *10*, 21534. [CrossRef] [PubMed]
112. Sirsi, S.; Feshitan, J.; Kwan, J.; Homma, S.; Borden, M. Effect of Microbubble Size on Fundamental Mode High Frequency Ultrasound Imaging in Mice. *Ultrasound Med. Biol.* **2010**, *36*, 935–948. [CrossRef]
113. Choi, J.J.; Feshitan, J.A.; Baseri, B.; Shougang, W.; Yao-Sheng, T.; Borden, M.A.; Konofagou, E.E. Microbubble-Size Dependence of Focused Ultrasound-Induced Blood–Brain Barrier Opening in Mice In Vivo. *IEEE Trans. Biomed. Eng.* **2010**, *57*, 145–154. [CrossRef]
114. Samiotaki, G.; Vlachos, F.; Tung, Y.-S.; Konofagou, E.E. A Quantitative Pressure and Microbubble-Size Dependence Study of Focused Ultrasound-Induced Blood-Brain Barrier Opening Reversibility in Vivo Using MRI: FUS-Induced BBB Opening Reversibility. *Magn. Reson. Med.* **2012**, *67*, 769–777. [CrossRef] [PubMed]
115. McMahon, D.; Hynynen, K. Acute Inflammatory Response Following Increased Blood-Brain Barrier Permeability Induced by Focused Ultrasound Is Dependent on Microbubble Dose. *Theranostics* **2017**, *7*, 3989–4000. [CrossRef]
116. Song, K.-H.; Fan, A.C.; Hinkle, J.J.; Newman, J.; Borden, M.A.; Harvey, B.K. Microbubble Gas Volume: A Unifying Dose Parameter in Blood-Brain Barrier Opening by Focused Ultrasound. *Theranostics* **2017**, *7*, 144–152. [CrossRef] [PubMed]
117. Navarro-Becerra, J.A.; Song, K.-H.; Martinez, P.; Borden, M.A. Microbubble Size and Dose Effects on Pharmacokinetics. *ACS Biomater. Sci. Eng.* **2022**, *8*, 1686–1695. [CrossRef] [PubMed]
118. Bing, C.; Hong, Y.; Hernandez, C.; Rich, M.; Cheng, B.; Munaweera, I.; Szczepanski, D.; Xi, Y.; Bolding, M.; Exner, A.; et al. Characterization of Different Bubble Formulations for Blood Brain Barrier Opening Using a Focused Ultrasound System with Acoustic Feedback Control. *Sci. Rep.* **2018**, *8*, 7986. [CrossRef] [PubMed]
119. Gao, Y.; Wu, S.; Li, L.; Wang, G.; Shen, W.; Xu, Y.; Liu, Z.; Zhuo, Z.; Xia, H.; Tan, K. Ultrasound-Targeted Stromal Cell-Derived Factor-1-Loaded Microbubble Destruction Promotes Mesenchymal Stem Cell Homing to Kidneys in Diabetic Nephropathy Rats. *IJN* **2014**, *9*, 5639–5651. [CrossRef]
120. Metzger, K.; Vogel, S.; Chatterjee, M.; Borst, O.; Seizer, P.; Schönberger, T.; Geisler, T.; Lang, F.; Langer, H.; Rheinlaender, J.; et al. High-Frequency Ultrasound-Guided Disruption of Glycoprotein VI-Targeted Microbubbles Targets Atheroprogressison in Mice. *Biomaterials* **2015**, *36*, 80–89. [CrossRef]
121. Yang, H.; Sun, Y.; Wei, J.; Xu, L.; Tang, Y.; Yang, L.; Zhang, X.; Lu, Y. The Effects of Ultrasound-Targeted Microbubble Destruction (UTMD) Carrying IL-8 Monoclonal Antibody on the Inflammatory Responses and Stability of Atherosclerotic Plaques. *Biomed. Pharmacother.* **2019**, *118*, 109161. [CrossRef]
122. Chang, E.-L.; Ting, C.-Y.; Hsu, P.-H.; Lin, Y.-C.; Liao, E.-C.; Huang, C.-Y.; Chang, Y.-C.; Chan, H.-L.; Chiang, C.-S.; Liu, H.-L.; et al. Angiogenesis-Targeting Microbubbles Combined with Ultrasound-Mediated Gene Therapy in Brain Tumors. *J. Control. Release* **2017**, *255*, 164–175. [CrossRef]
123. Fan, C.-H.; Ting, C.-Y.; Liu, H.-L.; Huang, C.-Y.; Hsieh, H.-Y.; Yen, T.-C.; Wei, K.-C.; Yeh, C.-K. Antiangiogenic-Targeting Drug-Loaded Microbubbles Combined with Focused Ultrasound for Glioma Treatment. *Biomaterials* **2013**, *34*, 2142–2155. [CrossRef] [PubMed]
124. Zhou, Y.; Gu, H.; Xu, Y.; Li, F.; Kuang, S.; Wang, Z.; Zhou, X.; Ma, H.; Li, P.; Zheng, Y.; et al. Targeted Antiangiogenesis Gene Therapy Using Targeted Cationic Microbubbles Conjugated with CD105 Antibody Compared with Untargeted Cationic and Neutral Microbubbles. *Theranostics* **2015**, *5*, 399–417. [CrossRef] [PubMed]
125. Nederhoed, J.H.; Ebben, H.P.; Slikkerveer, J.; Hoksbergen, A.W.J.; Kamp, O.; Tangelder, G.-J.; Wisselink, W.; Musters, R.J.P.; Yeung, K.K. Intravenous Targeted Microbubbles Carrying Urokinase versus Urokinase Alone in Acute Peripheral Arterial Thrombosis in a Porcine Model. *Ann. Vasc. Surg.* **2017**, *44*, 400–407. [CrossRef]

126. Wang, X.; Searle, A.K.; Hohmann, J.D.; Liu, A.L.; Abraham, M.-K.; Palasubramaniam, J.; Lim, B.; Yao, Y.; Wallert, M.; Yu, E.; et al. Dual-Targeted Theranostic Delivery of MiRs Arrests Abdominal Aortic Aneurysm Development. *Mol. Ther.* **2018**, *26*, 1056–1065. [CrossRef] [PubMed]
127. Pu, C.; Chang, S.; Sun, J.; Zhu, S.; Liu, H.; Zhu, Y.; Wang, Z.; Xu, R.X. Ultrasound-Mediated Destruction of LHRHa-Targeted and Paclitaxel-Loaded Lipid Microbubbles for the Treatment of Intraperitoneal Ovarian Cancer Xenografts. *Mol. Pharm.* **2014**, *11*, 49–58. [CrossRef] [PubMed]
128. Hua, X.; Zhou, L.; Liu, P.; He, Y.; Tan, K.; Chen, Q.; Gao, Y.; Gao, Y. In Vivo Thrombolysis with Targeted Microbubbles Loading Tissue Plasminogen Activator in a Rabbit Femoral Artery Thrombus Model. *J. Thromb. Thrombolysis.* **2014**, *38*, 57–64. [CrossRef]
129. Fan, C.-H.; Chang, E.-L.; Ting, C.-Y.; Lin, Y.-C.; Liao, E.-C.; Huang, C.-Y.; Chang, Y.-C.; Chan, H.-L.; Wei, K.-C.; Yeh, C.-K. Folate-Conjugated Gene-Carrying Microbubbles with Focused Ultrasound for Concurrent Blood-Brain Barrier Opening and Local Gene Delivery. *Biomaterials* **2016**, *106*, 46–57. [CrossRef]
130. Woudstra, L.; Krijnen, P.A.J.; Bogaards, S.J.P.; Meinster, E.; Emmens, R.W.; Kokhuis, T.J.A.; Bollen, I.A.E.; Baltzer, H.; Baart, S.M.T.; Parbhudayal, R.; et al. Development of a New Therapeutic Technique to Direct Stem Cells to the Infarcted Heart Using Targeted Microbubbles: StemBells. *Stem Cell Res.* **2016**, *17*, 6–15. [CrossRef]
131. Woudstra, L.; Meinster, E.; Van Haren, L.; Kay, A.M.; Koopman, M.; Belien, J.A.M.; Morrison, M.C.; Van Rossum, A.C.; Helder, M.N.; Juffermans, L.J.M.; et al. StemBell Therapy Stabilizes Atherosclerotic Plaques after Myocardial Infarction. *Cytotherapy* **2018**, *20*, 1143–1154. [CrossRef]
132. Liu, Y.; Zhou, Y.; Xu, J.; Luo, H.; Zhu, Y.; Zeng, X.; Dong, F.; Wei, Z.; Yan, F.; Zheng, H. Ultrasound Molecular Imaging-Guided Tumor Gene Therapy through Dual-Targeted Cationic Microbubbles. *Biomater. Sci.* **2021**, *9*, 2454–2466. [CrossRef]
133. Luo, W.; Wen, G.; Yang, L.; Tang, J.; Wang, J.; Wang, J.; Zhang, S.; Zhang, L.; Ma, F.; Xiao, L.; et al. Dual-Targeted and PH-Sensitive Doxorubicin Prodrug-Microbubble Complex with Ultrasound for Tumor Treatment. *Theranostics* **2017**, *7*, 452–465. [CrossRef]
134. Zhang, L.; Sun, L.; Tang, Q.; Sun, S.; Zeng, L.; Ma, J.; Li, X.; Ge, H.; Liang, X. Cascade Drug Delivery through Tumor Barriers of Pancreatic Cancer via Ultrasound in Combination with Functional Microbubbles. *ACS Biomater. Sci. Eng.* **2022**, *8*, 1583–1595. [CrossRef]
135. Beguin, E.; Gray, M.D.; Logan, K.A.; Nesbitt, H.; Sheng, Y.; Kamila, S.; Barnsley, L.C.; Bau, L.; McHale, A.P.; Callan, J.F.; et al. Magnetic Microbubble Mediated Chemo-Sonodynamic Therapy Using a Combined Magnetic-Acoustic Device. *J. Control. Release* **2020**, *317*, 23–33. [CrossRef] [PubMed]
136. Fan, C.-H.; Ting, C.-Y.; Lin, H.-J.; Wang, C.-H.; Liu, H.-L.; Yen, T.-C.; Yeh, C.-K. SPIO-Conjugated, Doxorubicin-Loaded Microbubbles for Concurrent MRI and Focused-Ultrasound Enhanced Brain-Tumor Drug Delivery. *Biomaterials* **2013**, *34*, 3706–3715. [CrossRef]
137. Wang, S.; Guo, X.; Xiu, W.; Liu, Y.; Ren, L.; Xiao, H.; Yang, F.; Gao, Y.; Xu, C.; Wang, L. Accelerating Thrombolysis Using a Precision and Clot-Penetrating Drug Delivery Strategy by Nanoparticle-Shelled Microbubbles. *Sci. Adv.* **2020**, *6*, eaaz8204. [CrossRef]
138. Duan, L.; Yang, F.; He, W.; Song, L.; Qiu, F.; Xu, N.; Xu, L.; Zhang, Y.; Hua, Z.; Gu, N. A Multi-Gradient Targeting Drug Delivery System Based on RGD-L-TRAIL-Labeled Magnetic Microbubbles for Cancer Theranostics. *Adv. Funct. Mater.* **2016**, *26*, 8313–8324. [CrossRef]
139. Dwivedi, P.; Kiran, S.; Han, S.; Dwivedi, M.; Khatik, R.; Fan, R.; Mangrio, F.A.; Du, K.; Zhu, Z.; Yang, C.; et al. Magnetic Targeting and Ultrasound Activation of Liposome–Microbubble Conjugate for Enhanced Delivery of Anticancer Therapies. *ACS Appl. Mater. Interfaces* **2020**, *12*, 23737–23751. [CrossRef] [PubMed]
140. Zhang, J.; Wang, S.; Deng, Z.; Li, L.; Tan, G.; Liu, X.; Zheng, H.; Yan, F. Ultrasound-Triggered Drug Delivery for Breast Tumor Therapy Through IRGD-Targeted Paclitaxel-Loaded Liposome-Microbubble Complexes. *J. Biomed. Nanotechnol.* **2018**, *14*, 1384–1395. [CrossRef] [PubMed]
141. Zhao, G.; Huang, Q.; Wang, F.; Zhang, X.; Hu, J.; Tan, Y.; Huang, N.; Wang, Z.; Wang, Z.; Cheng, Y. Targeted ShRNA-Loaded Liposome Complex Combined with Focused Ultrasound for Blood Brain Barrier Disruption and Suppressing Glioma Growth. *Cancer Lett.* **2018**, *418*, 147–158. [CrossRef]
142. Ingram, N.; McVeigh, L.E.; Abou-Saleh, R.H.; Maynard, J.; Peyman, S.A.; McLaughlan, J.R.; Fairclough, M.; Marston, G.; Valleley, E.M.A.; Jimenez-Macias, J.L.; et al. Ultrasound-Triggered Therapeutic Microbubbles Enhance the Efficacy of Cytotoxic Drugs by Increasing Circulation and Tumor Drug Accumulation and Limiting Bioavailability and Toxicity in Normal Tissues. *Theranostics* **2020**, *10*, 10973–10992. [CrossRef]
143. Charalambous, A.; Mico, V.; McVeigh, L.E.; Marston, G.; Ingram, N.; Volpato, M.; Peyman, S.A.; McLaughlan, J.R.; Wierzbicki, A.; Loadman, P.M.; et al. Targeted Microbubbles Carrying Lipid-Oil-Nanodroplets for Ultrasound-Triggered Delivery of the Hydrophobic Drug, Combretastatin A4. *Nanomed. Nanotechnol. Biol. Med.* **2021**, *36*, 102401. [CrossRef]
144. Liao, A.-H.; Chou, H.-Y.; Hsieh, Y.-L.; Hsu, S.-C.; Wei, K.-C.; Liu, H.-L. Enhanced Therapeutic Epidermal Growth Factor Receptor (EGFR) Antibody Delivery via Pulsed Ultrasound with Targeting Microbubbles for Glioma Treatment. *J. Med. Biol. Eng.* **2015**, *35*, 156–164. [CrossRef] [PubMed]
145. Kang, M.; Zhang, Y.; Jin, X.; Chen, G.; Huang, Y.; Wu, D.; Li, G.; Shan, J.; Huang, P.; Chen, J. Concurrent Treatment with Anti-DLL4 Enhances Antitumor and Proapoptotic Efficacy of a γ-Secretase Inhibitor in Gastric Cancer. *Transl. Oncol.* **2018**, *11*, 599–608. [CrossRef] [PubMed]

146. Sun, L.; Zhang, J.; Xu, M.; Zhang, L.; Tang, Q.; Chen, J.; Gong, M.; Sun, S.; Ge, H.; Wang, S.; et al. Ultrasound Microbubbles Mediated Sonosensitizer and Antibody Co-Delivery for Highly Efficient Synergistic Therapy on HER2-Positive Gastric Cancer. *ACS Appl. Mater. Interfaces* **2022**, *14*, 452–463. [CrossRef]
147. Kim, D.; Lee, S.S.; Moon, H.; Park, S.Y.; Lee, H.J. PD-L1 Targeting Immune-Microbubble Complex Enhances Therapeutic Index in Murine Colon Cancer Models. *Pharmaceuticals* **2020**, *14*, 6. [CrossRef]
148. Ma, Y.; Han, J.; Jiang, J.; Zheng, Z.; Tan, Y.; Liu, C.; Zhao, Y. Ultrasound Targeting of Microbubble-Bound Anti PD-L1 MAb to Enhance Anti-Tumor Effect of Cisplatin in Cervical Cancer Xenografts Treatment. *Life Sci.* **2020**, *262*, 118565. [CrossRef]
149. Liu, Y.; Jiang, J.; Liu, C.; Zhao, W.; Ma, Y.; Zheng, Z.; Zhou, Q.; Zhao, Y. Synergistic Anti-Tumor Effect of Anti-PD-L1 Antibody Cationic Microbubbles for Delivery of the MiR-34a Gene Combined with Ultrasound on Cervical Carcinoma. *Am. J. Transl. Res.* **2021**, *13*, 988–1005. [PubMed]
150. Li, X.; Khorsandi, S.; Wang, Y.; Santelli, J.; Huntoon, K.; Nguyen, N.; Yang, M.; Lee, D.; Lu, Y.; Gao, R.; et al. Cancer Immunotherapy Based on Image-Guided STING Activation by Nucleotide Nanocomplex-Decorated Ultrasound Microbubbles. *Nat. Nanotechnol.* **2022**, *17*, 891–899. [CrossRef] [PubMed]
151. Gao, X.; Nan, Y.; Yuan, Y.; Gong, X.; Sun, Y.; Zhou, H.; Zong, Y.; Zhang, L.; Yu, M. Gas-filled Ultrasound Microbubbles Enhance the Immunoactivity of the HSP70-MAGEA1 Fusion Protein against MAGEA1-expressing Tumours. *Mol. Med. Rep.* **2018**, *18*, 315–321. [CrossRef]
152. Jugniot, N.; Dahl, J.J.; Paulmurugan, R. Immunotheranostic Microbubbles (IMBs)—A Modular Platform for Dendritic Cell Vaccine Delivery Applied to Breast Cancer Immunotherapy. *J. Exp. Clin. Cancer Res.* **2022**, *41*, 299. [CrossRef]
153. Chomas, J.E.; Dayton, P.; May, D.; Ferrara, K. Threshold of Fragmentation for Ultrasonic Contrast Agents. *J. Biomed. Opt.* **2001**, *6*, 141. [CrossRef] [PubMed]
154. Morgan, K.E.; Allen, J.S.; Dayton, P.A.; Chomas, J.E.; Klibaov, A.L.; Ferrara, K.W. Experimental and Theoretical Evaluation of Microbubble Behavior: Effect of Transmitted Phase and Bubble Size. *IEEE Trans. Ultrason. Ferroelect. Freq. Contr.* **2000**, *47*, 1494–1509. [CrossRef] [PubMed]
155. Dayton, P.A.; Allen, J.S.; Ferrara, K.W. The Magnitude of Radiation Force on Ultrasound Contrast Agents. *J. Acoust. Soc. Am.* **2002**, *112*, 2183–2192. [CrossRef]
156. Supponen, O.; Upadhyay, A.; Lum, J.; Guidi, F.; Murray, T.; Vos, H.J.; Tortoli, P.; Borden, M. The Effect of Size Range on Ultrasound-Induced Translations in Microbubble Populations. *J. Acoust. Soc. Am.* **2020**, *147*, 3236–3247. [CrossRef] [PubMed]
157. Segers, T.; Kruizinga, P.; Kok, M.P.; Lajoinie, G.; de Jong, N.; Versluis, M. Monodisperse Versus Polydisperse Ultrasound Contrast Agents: Non-Linear Response, Sensitivity, and Deep Tissue Imaging Potential. *Ultrasound Med. Biol.* **2018**, *44*, 1482–1492. [CrossRef]
158. Wang, S.; Unnikrishnan, S.; Herbst, E.B.; Klibanov, A.L.; Mauldin, F.W.; Hossack, J.A. Ultrasound Molecular Imaging of Inflammation in Mouse Abdominal Aorta. *Invest. Radiol.* **2017**, *52*, 499–506. [CrossRef]
159. Rychak, J.J.; Klibanov, A.L.; Ley, K.F.; Hossack, J.A. Enhanced Targeting of Ultrasound Contrast Agents Using Acoustic Radiation Force. *Ultrasound Med. Biol.* **2007**, *33*, 1132–1139. [CrossRef]
160. Liu, J.; Zhang, P.; Liu, P.; Zhao, Y.; Gao, S.; Tan, K.; Liu, Z. Endothelial Adhesion of Targeted Microbubbles in Both Small and Great Vessels Using Ultrasound Radiation Force. *Mol. Imaging* **2012**, *11*, 58–66. [CrossRef]
161. Wang, S.; Wang, C.Y.; Unnikrishnan, S.; Klibanov, A.L.; Hossack, J.A.; Mauldin, F.W. Optical Verification of Microbubble Response to Acoustic Radiation Force in Large Vessels with In Vivo Results. *Investig. Radiol.* **2015**, *50*, 772–784. [CrossRef]
162. Frinking, P.J.A.; Tardy, I.; Théraulaz, M.; Arditi, M.; Powers, J.; Pochon, S.; Tranquart, F. Effects of Acoustic Radiation Force on the Binding Efficiency of BR55, a VEGFR2-Specific Ultrasound Contrast Agent. *Ultrasound Med. Biol.* **2012**, *38*, 1460–1469. [CrossRef]
163. Gessner, R.C.; Streeter, J.E.; Kothadia, R.; Feingold, S.; Dayton, P.A. An In Vivo Validation of the Application of Acoustic Radiation Force to Enhance the Diagnostic Utility of Molecular Imaging Using 3-D Ultrasound. *Ultrasound Med. Biol.* **2012**, *38*, 651–660. [CrossRef] [PubMed]
164. Navarro-Becerra, J.A.; Castillo, J.I.; Di Ruzza, F.; Borden, M.A. Monodispersity Increases Adhesion Efficiency and Specificity for Ultrasound-Targeted Microbubbles. *ACS Biomater. Sci. Eng.* **2023**, *9*, 991–1001. [CrossRef] [PubMed]
165. Feshitan, J.A.; Chen, C.C.; Kwan, J.J.; Borden, M.A. Microbubble Size Isolation by Differential Centrifugation. *J. Colloid Interface Sci.* **2009**, *329*, 316–324. [CrossRef] [PubMed]
166. Segers, T.; Lassus, A.; Bussat, P.; Gaud, E.; Frinking, P. Improved Coalescence Stability of Monodisperse Phospholipid-Coated Microbubbles Formed by Flow-Focusing at Elevated Temperatures. *Lab Chip* **2019**, *19*, 158–167. [CrossRef]
167. Segers, T.; De Rond, L.; De Jong, N.; Borden, M.; Versluis, M. Stability of Monodisperse Phospholipid-Coated Microbubbles Formed by Flow-Focusing at High Production Rates. *Langmuir* **2016**, *32*, 3937–3944. [CrossRef]
168. Dhanaliwala, A.H.; Chen, J.L.; Wang, S.; Hossack, J.A. Liquid Flooded Flow-Focusing Microfluidic Device for in Situ Generation of Monodisperse Microbubbles. *Microfluid. Nanofluid.* **2013**, *14*, 457–467. [CrossRef]
169. Khan, A.H.; Jiang, X.; Kaushik, A.; Nair, H.S.; Edirisinghe, M.; Mercado-Shekhar, K.P.; Shekhar, H.; Dalvi, S.V. Combining Ultrasound and Capillary-Embedded T-Junction Microfluidic Devices to Scale Up the Production of Narrow-Sized Microbubbles through Acoustic Fragmentation. *Langmuir* **2022**, *38*, 10288–10304. [CrossRef]
170. Gañán-Calvo, A.M.; Gordillo, J.M. Perfectly Monodisperse Microbubbling by Capillary Flow Focusing. *Phys. Rev. Lett.* **2001**, *87*, 274501. [CrossRef] [PubMed]
171. Garstecki, P.; Stone, H.A.; Whitesides, G.M. Mechanism for Flow-Rate Controlled Breakup in Confined Geometries: A Route to Monodisperse Emulsions. *Phys. Rev. Lett.* **2005**, *94*, 164501. [CrossRef]

172. Anna, S.L.; Bontoux, N.; Stone, H.A. Formation of Dispersions Using "Flow Focusing" in Microchannels. *Appl. Phys. Lett.* **2003**, *82*, 364–366. [CrossRef]
173. Wang, H.; Jiang, S.; Zhu, C.; Ma, Y.; Fu, T. Bubble Formation in T-Junctions within Parallelized Microchannels: Effect of Viscoelasticity. *Chem. Eng. J.* **2021**, *426*, 131783. [CrossRef]
174. Dixon, A.J.; Dhanaliwala, A.H.; Chen, J.L.; Hossack, J.A. Enhanced Intracellular Delivery of a Model Drug Using Microbubbles Produced by a Microfluidic Device. *Ultrasound Med. Biol.* **2013**, *39*, 1267–1276. [CrossRef] [PubMed]
175. Peyman, S.A.; Abou-Saleh, R.H.; McLaughlan, J.R.; Ingram, N.; Johnson, B.R.G.; Critchley, K.; Freear, S.; Evans, J.A.; Markham, A.F.; Coletta, P.L.; et al. Expanding 3D Geometry for Enhanced On-Chip Microbubble Production and Single Step Formation of Liposome Modified Microbubbles. *Lab Chip* **2012**, *12*, 4544. [CrossRef] [PubMed]
176. Klibanov, A.L. Ultrasound Molecular Imaging of Cancer: Design and Formulation Strategies of Targeted Contrast Agents. In *Molecular Imaging in Oncology*; Schober, O., Kiessling, F., Debus, J., Eds.; Recent Results in Cancer Research; Springer International Publishing: Cham, Switzerland, 2020; Volume 216, pp. 319–336. ISBN 978-3-030-42617-0.
177. Klibanov, A.L. Ligand-Carrying Gas-Filled Microbubbles: Ultrasound Contrast Agents for Targeted Molecular Imaging. *Bioconjugate Chem.* **2005**, *16*, 9–17. [CrossRef]
178. Sambi, M.; Bagheri, L.; Szewczuk, M.R. Current Challenges in Cancer Immunotherapy: Multimodal Approaches to Improve Efficacy and Patient Response Rates. *J. Oncol.* **2019**, *2019*, 1–12. [CrossRef] [PubMed]
179. Cruz, E.; Kayser, V. Monoclonal Antibody Therapy of Solid Tumors: Clinical Limitations and Novel Strategies to Enhance Treatment Efficacy. *BTT* **2019**, *13*, 33–51. [CrossRef]
180. Hansel, T.T.; Kropshofer, H.; Singer, T.; Mitchell, J.A.; George, A.J.T. The Safety and Side Effects of Monoclonal Antibodies. *Nat. Rev. Drug Discov.* **2010**, *9*, 325–338. [CrossRef]
181. Fares, C.M.; Van Allen, E.M.; Drake, C.G.; Allison, J.P.; Hu-Lieskovan, S. Mechanisms of Resistance to Immune Checkpoint Blockade: Why Does Checkpoint Inhibitor Immunotherapy Not Work for All Patients? *Am. Soc. Clin. Oncol. Educ. Book* **2019**, *39*, 147–164. [CrossRef]
182. Singh, S.; Hassan, D.; Aldawsari, H.M.; Molugulu, N.; Shukla, R.; Kesharwani, P. Immune Checkpoint Inhibitors: A Promising Anticancer Therapy. *Drug Discov. Today* **2020**, *25*, 223–229. [CrossRef]
183. Rohaan, M.W.; Wilgenhof, S.; Haanen, J.B.A.G. Adoptive Cellular Therapies: The Current Landscape. *Virchows Arch.* **2019**, *474*, 449–461. [CrossRef]
184. Fu, C.; Shi, G.; Liu, Y.-T. Manufacturing Anti-CD19 CAR-Tscm Cells for Immunotherapy Using Innovative Microbubble-Based Technologies for Precision Cell Processing. *Blood* **2021**, *138*, 3889. [CrossRef]
185. Lustig, A.; Manor, T.; Shi, G.; Li, J.; Wang, Y.-T.; An, Y.; Liu, Y.-T.; Weng, N. Lipid Microbubble–Conjugated Anti-CD3 and Anti-CD28 Antibodies (Microbubble-Based Human T Cell Activator) Offer Superior Long-Term Expansion of Human Naive T Cells In Vitro. *ImmunoHorizons* **2020**, *4*, 475–484. [CrossRef] [PubMed]
186. Waldmann, T.A. Cytokines in Cancer Immunotherapy. *Cold Spring Harb. Perspect. Biol.* **2018**, *10*, a028472. [CrossRef]
187. Berraondo, P.; Sanmamed, M.F.; Ochoa, M.C.; Etxeberria, I.; Aznar, M.A.; Pérez-Gracia, J.L.; Rodríguez-Ruiz, M.E.; Ponz-Sarvise, M.; Castañón, E.; Melero, I. Cytokines in Clinical Cancer Immunotherapy. *Br. J. Cancer* **2019**, *120*, 6–15. [CrossRef]
188. Qiu, Y.; Su, M.; Liu, L.; Tang, Y.; Pan, Y.; Sun, J. Clinical Application of Cytokines in Cancer Immunotherapy. *DDDT* **2021**, *15*, 2269–2287. [CrossRef]
189. Figueiredo, M.L.; Figueiredo Neto, M.; Salameh, J.W.; Decker, R.E.; Letteri, R.; Chan-Seng, D.; Emrick, T. Ligand-Mediated Targeting of Cytokine Interleukin-27 Enhances Its Bioactivity In Vivo. *Mol. Ther.-Methods Clin. Dev.* **2020**, *17*, 739–751. [CrossRef]
190. Barua, A.; Yellapa, A.; Bahr, J.M.; Adur, M.K.; Utterback, C.W.; Bitterman, P.; Basu, S.; Sharma, S.; Abramowicz, J.S. Interleukin 16- (IL-16-) Targeted Ultrasound Imaging Agent Improves Detection of Ovarian Tumors in Laying Hens, a Preclinical Model of Spontaneous Ovarian Cancer. *BioMed Res. Int.* **2015**, *2015*, 1–10. [CrossRef] [PubMed]
191. Malonis, R.J.; Lai, J.R.; Vergnolle, O. Peptide-Based Vaccines: Current Progress and Future Challenges. *Chem. Rev.* **2020**, *120*, 3210–3229. [CrossRef]
192. Amara, S.; Tiriveedhi, V. The Five Immune Forces Impacting DNA-Based Cancer Immunotherapeutic Strategy. *Int. J. Mol. Sci.* **2017**, *18*, 650. [CrossRef]
193. Saxena, M.; Balan, S.; Roudko, V.; Bhardwaj, N. Towards Superior Dendritic-Cell Vaccines for Cancer Therapy. *Nat. Biomed. Eng.* **2018**, *2*, 341–346. [CrossRef]
194. Lin, M.J.; Svensson-Arvelund, J.; Lubitz, G.S.; Marabelle, A.; Melero, I.; Brown, B.D.; Brody, J.D. Cancer Vaccines: The next Immunotherapy Frontier. *Nat. Cancer* **2022**, *3*, 911–926. [CrossRef]
195. Dewitte, H.; Van Lint, S.; Heirman, C.; Thielemans, K.; De Smedt, S.C.; Breckpot, K.; Lentacker, I. The Potential of Antigen and TriMix Sonoporation Using MRNA-Loaded Microbubbles for Ultrasound-Triggered Cancer Immunotherapy. *J. Control. Release* **2014**, *194*, 28–36. [CrossRef]
196. Song, H.-W.; Lee, H.-S.; Kim, S.-J.; Kim, H.Y.; Choi, Y.H.; Kang, B.; Kim, C.-S.; Park, J.-O.; Choi, E. Sonazoid-Conjugated Natural Killer Cells for Tumor Therapy and Real-Time Visualization by Ultrasound Imaging. *Pharmaceutics* **2021**, *13*, 1689. [CrossRef]
197. Navarro-Becerra, J.A.; Franco-Urquijo, C.A.; Ríos, A.; Escalante, B. Localized Delivery of Caveolin-1 Peptide Assisted by Ultrasound-Mediated Microbubble Destruction Potentiates the Inhibition of Nitric Oxide-Dependent Vasodilation Response. *Ultrasound Med. Biol.* **2021**, *47*, 1559–1572. [CrossRef] [PubMed]

198. Navarro-Becerra, J.A.; Caballero-Robledo, G.A.; Franco-Urquijo, C.A.; Ríos, A.; Escalante, B. Functional Activity and Endothelial-Lining Integrity of Ex Vivo Arteries Exposed to Ultrasound-Mediated Microbubble Destruction. *Ultrasound Med. Biol.* **2020**, *46*, 2335–2348. [CrossRef]
199. Chen, H.; Brayman, A.A.; Evan, A.P.; Matula, T.J. Preliminary Observations on the Spatial Correlation Between Short-Burst Microbubble Oscillations and Vascular Bioeffects. *Ultrasound Med. Biol.* **2012**, *38*, 2151–2162. [CrossRef] [PubMed]
200. Chen, H.; Brayman, A.A.; Bailey, M.R.; Matula, T.J. Blood Vessel Rupture by Cavitation. *Urol. Res.* **2010**, *38*, 321–326. [CrossRef] [PubMed]
201. Li, P.; Armstrong, W.F.; Miller, D.L. Impact of Myocardial Contrast Echocardiography on Vascular Permeability: Comparison of Three Different Contrast Agents. *Ultrasound Med. Biol.* **2004**, *30*, 83–91. [CrossRef]
202. Miller, D.L.; Lu, X.; Fabiilli, M.; Fields, K.; Dou, C. Frequency Dependence of Petechial Hemorrhage and Cardiomyocyte Injury Induced during Myocardial Contrast Echocardiography. *Ultrasound Med. Biol.* **2016**, *42*, 1929–1941. [CrossRef]
203. White, P.J.; Zhang, Y.-Z.; Power, C.; Vykhodtseva, N.; McDannold, N. Observed Effects of Whole-Brain Radiation Therapy on Focused Ultrasound Blood–Brain Barrier Disruption. *Ultrasound Med. Biol.* **2020**, *46*, 1998–2006. [CrossRef]
204. Sun, T.; Samiotaki, G.; Wang, S.; Acosta, C.; Chen, C.C.; Konofagou, E.E. Acoustic Cavitation-Based Monitoring of the Reversibility and Permeability of Ultrasound-Induced Blood-Brain Barrier Opening. *Phys. Med. Biol.* **2015**, *60*, 9079–9094. [CrossRef]
205. Anderson, N.M.; Simon, M.C. The Tumor Microenvironment. *Curr. Biol.* **2020**, *30*, R921–R925. [CrossRef] [PubMed]
206. Hernot, S.; Klibanov, A.L. Microbubbles in Ultrasound-Triggered Drug and Gene Delivery. *Adv. Drug Deliv. Rev.* **2008**, *60*, 1153–1166. [CrossRef] [PubMed]
207. Jallinoja, V.I.J.; Houghton, J.L. Current Landscape in Clinical Pretargeted Radioimmunoimaging and Therapy. *J. Nucl. Med.* **2021**, *62*, 1200–1206. [CrossRef] [PubMed]

pharmaceutics

MDPI

Review

Microbubble–Nanoparticle Complexes for Ultrasound-Enhanced Cargo Delivery

Rachel Chapla [1], Katherine T. Huynh [1,2] and Carolyn E. Schutt [1,2,*]

[1] Cancer Early Detection Advanced Research Center, Oregon Health and Science University, Portland, OR 97201, USA
[2] Department of Biomedical Engineering, Oregon Health and Science University, Portland, OR 97239, USA
* Correspondence: ibsenc@ohsu.edu

Abstract: Targeted delivery of therapeutics to specific tissues is critically important for reducing systemic toxicity and optimizing therapeutic efficacy, especially in the case of cytotoxic drugs. Many strategies currently exist for targeting systemically administered drugs, and ultrasound-controlled targeting is a rapidly advancing strategy for externally-stimulated drug delivery. In this non-invasive method, ultrasound waves penetrate through tissue and stimulate gas-filled microbubbles, resulting in bubble rupture and biophysical effects that power delivery of attached cargo to surrounding cells. Drug delivery capabilities from ultrasound-sensitive microbubbles are greatly expanded when nanocarrier particles are attached to the bubble surface, and cargo loading is determined by the physicochemical properties of the nanoparticles. This review serves to highlight and discuss current microbubble–nanoparticle complex component materials and designs for ultrasound-mediated drug delivery. Nanocarriers that have been complexed with microbubbles for drug delivery include lipid-based, polymeric, lipid–polymer hybrid, protein, and inorganic nanoparticles. Several schemes exist for linking nanoparticles to microbubbles for efficient nanoparticle delivery, including biotin–avidin bridging, electrostatic bonding, and covalent linkages. When compared to unstimulated delivery, ultrasound-mediated cargo delivery enables enhanced cell uptake and accumulation of cargo in target organs and can result in improved therapeutic outcomes. These ultrasound-responsive delivery complexes can also be designed to facilitate other methods of targeting, including bioactive targeting ligands and responsivity to light or magnetic fields, and multi-level targeting can enhance therapeutic efficacy. Microbubble–nanoparticle complexes present a versatile platform for controlled drug delivery via ultrasound, allowing for enhanced tissue penetration and minimally invasive therapy. Future perspectives for application of this platform are also discussed in this review.

Keywords: ultrasound; targeted drug delivery; microbubble; cavitation; sonoporation; nanoparticle

Citation: Chapla, R.; Huynh, K.T.; Schutt, C.E. Microbubble–Nanoparticle Complexes for Ultrasound-Enhanced Cargo Delivery. *Pharmaceutics* **2022**, *14*, 2396. https://doi.org/10.3390/pharmaceutics14112396

Academic Editors: Shashank Sirsi, James Kwan, Michael Gray and Brandon Helfield

Received: 1 September 2022
Accepted: 8 October 2022
Published: 7 November 2022

1. Introduction

An important design consideration in drug delivery systems is targeting the delivery to the intended or afflicted tissue. Systemic delivery of drugs is commonly implemented because it is cost-effective, technically straightforward, and minimally invasive. However, employing this method without a targeting element commonly results in off-target effects, sub-optimal delivery to the active site, and high uptake and clearance by the immune system. Systems for directed delivery to the target tissue protect drugs from environmental degradation and decrease off-target release, reducing the necessary administered dose for effective treatment [1,2]. In decreasing off-target delivery, targeted delivery approaches also minimize adverse effects to healthy tissue, which is particularly critical in cases of cancer chemotherapy or administration of other cytotoxic therapeutics. Many methods of targeted delivery exist, including passive, active, and stimulus-activated approaches. Additionally, targeting strategies may be combined for enhanced targeting precision.

Passive targeting consists of designing the physical properties of the particle to optimize its accumulation at the active site [3]. This type of targeting is frequently used for cancer drug delivery, where upregulated angiogenic signaling in tumor tissue promotes rapid formation of immature vasculature. This tumor vasculature is more highly branched and permeable, or "leaky" than healthy vasculature. Due to the greater permeability of these vessels, nanoparticles are more capable of extravasating into tumor tissue. Further, poor lymphatic drainage in tumor tissue permits high nanoparticle accumulation and retention [4]. This combined effect is known as the enhanced permeability and retention (EPR) effect, and much effort has been made to design therapeutic vehicle size and shape for optimal extravasation through the leaky tumor vasculature in passive targeting approaches [3,5]. Passive targeting alone does not provide tissue specificity.

Alternatively, active targeting entails designing therapeutic vehicles for enhanced delivery to a particular target [3]. Functionalization of therapeutic vehicles with targeting moieties (e.g., antibodies, peptides, nucleic acids, carbohydrates) that recognize biomolecules upregulated in cancerous cells or the extracellular tumor microenvironment is a key area of development and has shown some clinical success in targeted drug delivery [3,6–9]. Other "smart" delivery vehicles dispense their cargo in response to tumor or inflammatory tissue-induced environmental stimuli, such as a decrease in pH or increase in enzymatic activity [3,5,10,11].

Another important method of targeted delivery that affords manual control of both location and timing of drug delivery is application of an external stimulus that triggers drug release, such as light, magnetic fields, or ultrasound [11]. Both photo- and magnetic-mediated delivery have shown utility in triggering targeted cargo delivery [12–14], but each present specific limitations related to tissue penetration. Focused external magnetic fields with adequate penetration and precision for guided drug delivery can be technically difficult to assemble [11], while visible and infrared light can only penetrate ≤ 10 mm into tissue, which is limiting in terms of clinical application for many pathologies [11,15]. In contrast, ultrasound-mediated drug delivery is particularly promising because of its technically facile and noninvasive modulation and its ability to penetrate and be focused precisely to greater depths within tissues [16]. Thus, substantial research efforts are currently focused on developing and optimizing drug delivery vehicles for ultrasound-stimulated cargo release, with a particular interest in complexing nanocarriers with ultrasound-sensitive microbubbles. Many ultrasound-controlled complexes recently introduced in this growing field have demonstrated superior targeting over unstimulated vehicles, indicating great promise for future clinical translation of this targeted therapy strategy. In response to widespread interest in the growing field, this review functions to survey the state of the art for ultrasound-responsive microbubble–nanoparticle complexes for cargo delivery. We provide an in-depth exploration of the design and synthesis of each complex component, discussing design selections, complexation and linking strategies (as summarized in Figure 1), therapeutic outcomes, and areas for future growth.

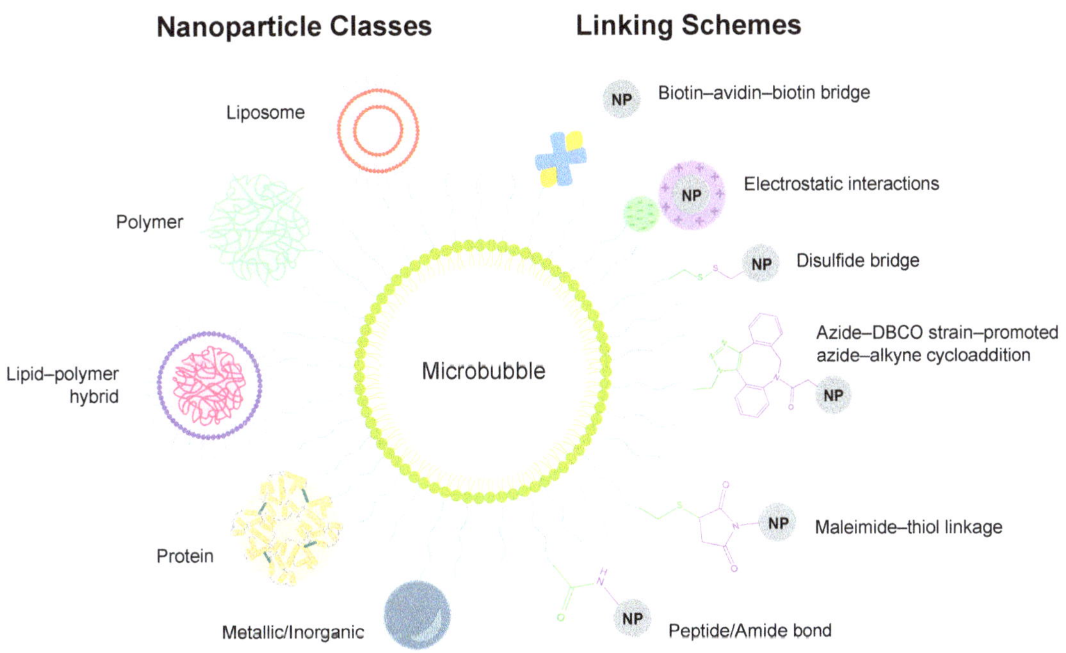

Figure 1. Microbubble–nanoparticle complex designs for ultrasound-controlled cargo delivery. Nanocarriers that have been linked to microbubbles include liposomes, polymer nanoparticles, lipid–polymer hybrid nanoparticles, protein nanoparticles, and metallic/inorganic nanoparticles. Schemes for conjugating nanoparticles to microbubbles include avidin–biotin bridge formation, electrostatic interactions, disulfide bridges, and multiple types of covalent bonds.

2. Ultrasound Stimulation

Ultrasound waves are longitudinal pressure waves of higher frequency than 20 kHz, which is above the range of human hearing [17–21]. These mechanical pressure waves cause oscillatory deformation of matter in the surrounding environment. Different tissues have varying resistance to ultrasound propagation, with more solid tissues reflecting more ultrasound than fluid tissues; thus, this technology has traditionally been implemented for diagnostic imaging to evaluate tissue density and morphology [22,23]. Diagnostic ultrasound imaging has had long-standing success, in large part due to its ability to deeply penetrate through tissue, allowing visualization of deep tissue morphology [24]. At frequencies around 1 MHz, ultrasound waves are minimally attenuated by tissue, allowing for significant tissue penetration [24,25]. This advantage, along with the ability to focus the ultrasound beam to small volumes within the body, has made ultrasound an attractive method of stimulation for other applications beyond diagnostics, including use for therapy and drug delivery [26,27]. The intensity of ultrasound waves may also be modulated for different therapeutic applications. High intensity ultrasound (100–10,000 W/cm^2) causes local increases in temperature and tissue damage by thermal ablation, which can be implemented as cancer therapy, but may cause undesired damage in many applications [28–30]. Lower intensity ultrasound (e.g., the range of 0.125–3 W/cm^2) is capable of inducing mechanical effects on local tissues and microparticles without causing temperature spiking [28]. Thus, researchers have harnessed this form of energy for local, on-demand drug delivery by designing micromaterials and complexes to release drugs in response to ultrasound stimulation.

Ultrasound Stimulation of Microbubbles

Gas-core microbubble particles are highly mechanically responsive to the oscillating ultrasound pressure waves. Due to the compressibility of the gas, these microbubbles undergo volumetric oscillations, shrinking and expanding with the compression and rarefaction phases of the ultrasound pulse [31–35]. This oscillation is known as cavitation [36]. Two types of cavitation dynamics can occur in response to ultrasound stimulation: stable cavitation and inertial cavitation (both depicted in Figure 2A). Microbubble oscillation also results in strong backscatter of the ultrasound waves, which has enabled their use as highly effective contrast agents for ultrasound imaging [37].

In the process of stable cavitation, which occurs at lower ultrasound intensities, microbubbles oscillate in diameter about their equilibrium size at the frequency of the applied ultrasound wave [38]. At higher acoustic pressures, inertial cavitation can occur, where the microbubble expands to a large enough size that, upon the next compression phase, a rapid inward rush of fluid is directed toward the center of the microbubble. This causes a forceful implosion which collapses the bubble and can free attached or encapsulated cargo. The oscillation response of the microbubble is greatest at its resonant frequency, which is determined by the physical properties of the bubble itself (e.g., size, composition, coating) [30,39]. The collapse dynamics can be influenced by proximity to a surface or to other bubbles undergoing cavitation [40]. Both stable and inertial microbubble cavitation can cause a temporary increase in permeability of nearby cell membranes known as sonoporation (Figure 2B).

As microbubbles undergo stable cavitation, their rapid expansion and contraction can have a push and pull effect on adjacent cell membranes (Figure 2B) and can also generate fluid flow, known as microstreaming (Figure 2C) [41,42]. Microstreaming exerts shear stresses upon nearby cell membranes, causing transient opening of membrane pores, e.g., sonoporation [43–45]. During inertial cavitation, the forceful collapse of the microbubble can generate radial shock waves (Figure 2D), as well as directed fluid jets (Figure 2E), caused by the asymmetric collapse and involution of the bubble, which can disrupt nearby cell membranes [44,46–49]. Several studies have observed sonoporation-induced cellular pore sizes ranging from approximately 10 nm to multiple micrometers in diameter, using scanning electron microscopy (SEM) and atomic force microscopy (AFM) (Figure 2F) [46–48]. Following the formation of transient pores, evidence of pore resealing, facilitated by lysosome-mediated exocytosis [50] and extracellular calcium ions (Ca^{2+}), has been observed [51]. The pore size generated is dependent upon ultrasound parameters, such as duration, number and localization of ultrasound pulses, bubble–cell proximity, and geometrical configuration of the microbubbles within the local 3D environment [52]. Upon cavitation-induced pore formation, local extracellular species may enter cells directly through these pores (Figure 2G) [53]. As such, sonoporation caused by ultrasound-stimulated microbubble cavitation is thought to play a key role in internalization of delivered therapeutics, as it creates the most direct route of delivery to cells. Thus, ultrasound-stimulated sonoporation may be leveraged to overcome challenges in targeted gene or therapeutic cargo transfer into cells [41].

In addition to promoting cargo delivery across cell membranes, focused ultrasound is an instrumental method for delivering cargo across the blood–brain barrier (BBB). The BBB is a neuroprotective barrier of endothelial cells joined by tight junctions impermissible to molecules larger than 400 Da [54,55]. Focused ultrasound-induced microbubble cavitation temporarily disrupts the tight junctions of the BBB without causing thermoablation of these cells, providing localized direct delivery of cargo (as large as 2000 kDa molecular weight) to brain tissue at the targeted site [55,56]. Focused ultrasound delivery is considered safer and more precise than other methods of crossing the BBB, including invasive surgeries and systemically-administered chemically-modified cargo [54,56].

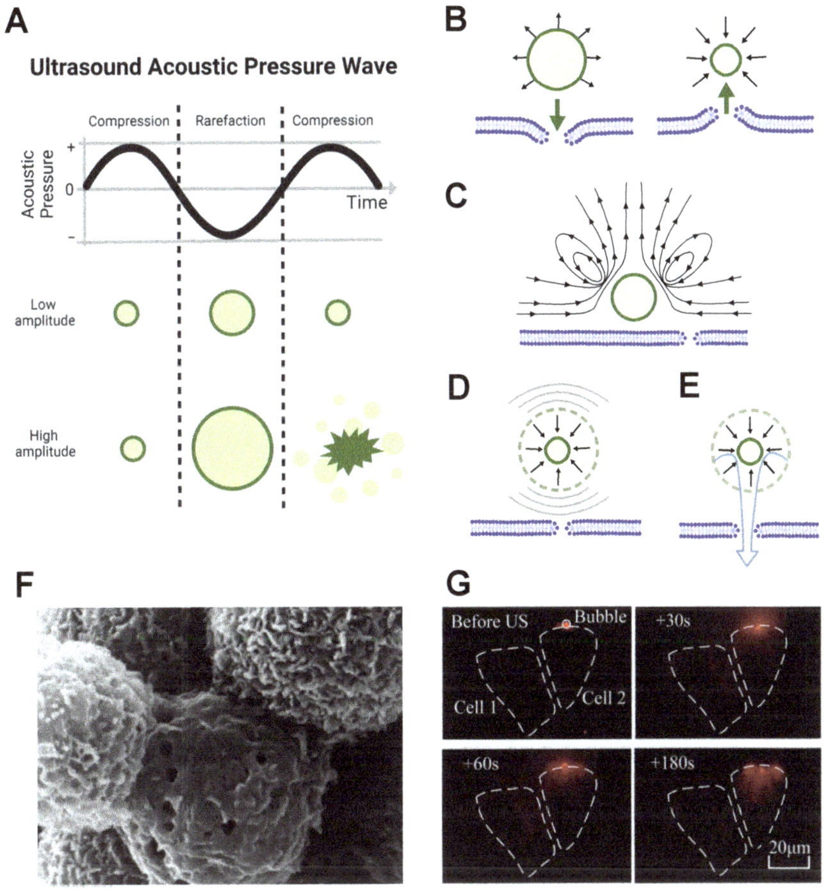

Figure 2. Microbubble cavitation induces sonoporation of local cell membranes. (**A**) During the sinusoidal alternating cycle of compression and rarefaction from the ultrasound pressure waves, gas-core microbubbles contract and expand in diameter. In stable cavitation (top row), occurring at lower acoustic pressure amplitudes, microbubbles stably oscillate between expansion and contraction, whereas at higher acoustic pressures, inertial cavitation can occur (bottom row), in which the microbubble expands and then implodes, collapsing and fragmenting the bubble. Schematic illustrations in (**A**) created with Biorender.com, and based on those from [57], Springer Nature, 2015, and [58], IvySpring International Publisher, 2012. (**B–E**) Both stable and inertial cavitation can create local dynamics which may cause sonoporation. Illustrations are based on those from [42], copyright 2017. Reproduced by permission of Taylor and Francis Group, LLC, a division of Informa plc. (**B**) Stable microbubble cavitation may cause sonoporation by pushing the cell membrane during expansion or pulling the membrane during contraction. (**C**) Stable cavitation creates local fluid microstreaming, which can also cause membrane permeation. (**D**) During inertial cavitation, resultant shockwaves from bubble implosion can create pores in cell membranes. (**E**) Bubble collapse during inertial cavitation can also cause fluid jet formation which can permeate the cell membrane. (**F**) Scanning electron microscopy images show the presence of pores in cell membranes following ultrasound exposure in the presence of microbubbles. Image reprinted from [46], copyright 2005, with permission from Elsevier. (**G**) Ultrasound stimulation in the presence of adjacent microbubbles causes cell membrane permeability, facilitating rapid cell uptake of local fluorescence marker propidium iodide (red) through pores created from sonoporation. Images reprinted from [53], Springer Nature, 2018.

Focused ultrasound-stimulated microbubble cavitation for release of therapeutic cargo is an attractive biocompatible approach for therapeutic delivery due to the deep tissue penetration and precise, noninvasive spatiotemporal control of the ultrasound stimulus. The capability to focus the acoustic energy to small mm^3-scale volumes within deep tissue also makes ultrasound an attractive stimulus to activate drug delivery vehicles. As briefly mentioned above, microbubble cavitation behavior is largely dictated by bubble physical properties. Several considerations for design of ultrasound-responsive microbubble delivery vehicles are shell composition, microbubble size, type of gas in the core, surface functionalization, and cargo loading. Below we discuss these design parameters for different delivery applications.

3. Ultrasound-Responsive Microbubble Design Parameters

Microbubbles have been designed for different therapeutic applications, including the delivery of genes or other drugs, due to their unique cavitation behavior. The general microbubble design consists of an outer protective layer, or shell, and a hydrophobic gas core. The types of protective layers include phospholipids, that create a monolayer membrane, or layers composed of protein (e.g., albumin) or polymer [59]. The microbubble monolayer can be comprised of many different lipid types and mixtures and commonly consists of a phosphatidyl-choline, such as 1,2-dipalmitoyl-sn-glycero-3-phospho-choline (DPPC) [60–63], or 1,2-distearoyl-sn-glycero-3-phosphocholine (DSPC) [64–67]. These phospholipids self-arrange into monolayers across the gas/water interface with their hydrophobic lipid tails facing inwards toward the gas-filled core and hydrophilic heads facing outward toward the aqueous environment. To increase microbubble stability, polyethylene glycol (PEG) functionalized, or (PEG)ylated, lipids, such as PEG stearate [64], or 1,2-distearoyl-sn-glycero-3-phosphoethanolamine-N-carboxy(polyethylene glycol) (DSPE-PEG2000), can be incorporated. These hydrophilic PEG brush layers inhibit bubble coalescence and also increase the particle half-life in circulation by shielding the bubble from immune cell recognition [68]. The incorporation of cholesterol can modify the packing of phospholipids in the microbubble monolayer, which affects the physical properties of the lipid monolayer and the microbubble's interaction with ultrasound [69].

Gases implemented for the microbubble core are generally inert and hydrophobic in nature to reduce dissolution into the surrounding liquid, thereby increasing microbubble stability [70,71]. Different types of gases that can be incorporated inside the phospholipid shell include: octafluoropropane [72], perfluorobutane [60], perfluorohexane [73,74] sulfur hexafluoride [75], and nitrogen [76]. Other gases may also be incorporated for different therapeutic strategies, such as oxygen to improve treatment of tumors [77,78].

In order to synthesize lipid coated microbubbles, the phospholipid mixture and gas components are combined under conditions that favor micelle self-assembly, including probe sonication, microfluidics, and high shear emulsification [79–83]. Probe sonication introduces energy into a lipid solution that breaks up lipid micelles while simultaneously pulling gas from above the liquid surface into the solution, forming microbubbles that are then coated by the free lipids before they reform micelles. This results in the formation of a monolayer of the lipid coating and stabilizes the gas core. Standard bulk production methods utilize ultrasonic emulsification to synthesize high volumes of microbubbles, but result in a broad size distribution [84]. Conversely, microfluidic production methods provide tighter size distributions, but operate at a much lower production rate [85]. Hybrid systems of these methods that achieve high production rate and more uniform size distributions may provide a novel fabrication platform for higher yield production [86]. The size distribution of the microbubbles is important because the microbubble resonance frequency is dependent on the bubble size [49], with one to ten microns in diameter effectively oscillating with ultrasound frequencies in the low MHz [87].

While lipid-based microbubbles have been well-established for drug and gene delivery applications, the capacity of the monolayer shell for housing cargo is low. Though less common, other macromolecules may be used to form the microbubble shell. Polymer-based microbubbles are advantageous for drug delivery due to their ability to form thicker crosslinked layers which enhance bubble stability. These polymers can achieve thicknesses of 50–150 nm in comparison to the 3–5 nm thickness of phospholipid-based microbubbles [88]. The thicker shell provides a larger space for loading therapeutics, which are entrapped within the polymer network [89]. However, the thicker membranes can be stiffer than lipid monolayers, which affects their echogenicity, or responsiveness to ultrasound stimulation, and is an important consideration for delivery applications. Successful delivery of molecular payloads across the BBB to brain tumors has been achieved using poly(n-butyl cyanoacrylate) (PBCA)-based polymer microbubbles incorporating macromolecular FITC-dextran in the shell [88]. Polymer–surfactant shell hybrids, comprised of multiple shell layers, also provide greater cargo loading capabilities and have been employed for different applications, such as dual drug release or protection of DNA cargo from enzymatic degradation [90–92]. Protein coated microbubbles have also been employed for gene and drug delivery purposes [93]. Proteins can be covalently crosslinked via disulfide bonds to form and stabilize a microbubble shell [94]. This design has been used to effectively deliver both adenoviruses in a rat model [95] and plasmids in vitro [93] in response to ultrasound stimulation.

Microbubble Customization and Cargo Capabilities

Microbubbles are promising carrier vehicles for gene or other drug delivery, enabling spatiotemporal control over delivery through the application of a focused ultrasound trigger. Customization of the microbubble structure is critical for loading different cargos intended for a wide range of different applications.

Linking payloads to the microbubble surface can be achieved through functionalization of microbubbles with crosslinking groups or by imparting electrostatic charges on the surface. Microbubble surface charge can be modified by incorporating cationic or anionic lipids into the monolayer. This enables complexation with oppositely-charged species via electrostatic interactions. The cationic lipids 1,2-Dioleoyl-3-trimethylammonium propane (DOTAP) and N-[1-(2,3-distearoyloxy)propyl]-N,N,N-trimethylammonium methylsulfate (DSTAP) are commonly used for conferring positive charge to microbubbles, allowing negatively charged nucleic acid cargo, such as cDNA, siRNA, or mRNA, to attach to the microbubble surface [60,96,97]. Attaching nucleic acids to microbubbles protects them from enzymatic degradation in circulation [98]. DNA cargo loaded onto the microbubble surface can be verified by labeling with SYBR-gold dye and quantification of the fluorescence signal [99].

Microbubble surfaces can also be chemically functionalized for the attachment of specific therapeutic cargo types. For example, PD-L1 therapeutic antibodies were conjugated to PEG-coated, N-Hydroxysuccinimide (NHS)-functionalized microbubbles using a covalent amine–NHS linkage, forming an antibody–microbubble conjugate for cancer immunotherapy applications. This delivery vehicle provided protection for the therapeutic antibodies from immune recognition due to the microbubble PEG brush layer and partial blockage of the antibody Fc region by the microbubble conjugation. This design enabled effective targeted release of the antibody by focused ultrasound [100].

Beyond modifying microbubble membranes for cargo loading, functional groups may also be used to add targeting capabilities. Microbubbles can be targeted for specific cell types. For instance, a phospholipid microbubble conjugated (via biotin–actin bridging) with anti-CD4 was successfully targeted to CD4 positive lymphocytes that carried the microbubbles to specific regions within the body [101]. Attaching multiple ligands on the same microbubble may allow for increased specificity to cellular targets, such as to regions of atherosclerosis through conjugation of anti-VCAM-1 and anti-ICAM-1 antibodies and synthetic polymeric sialyl Lewis X [102]. Microbubbles can also be modified to leverage multiple modalities, combining imaging with drug release capabilities. For example, incorporating iron oxide nanoparticles with microbubbles allows for dual MRI and ultrasound imaging and also allows ultrasound-mediated delivery of doxorubicin [92].

Current methods of cargo-loading directly onto microbubbles have shown success in delivery, and increasing cargo capacity further can potentially improve therapeutic outcomes. The addition of cargo-carrying nanoparticles onto the microbubble surface is a promising area of research for increasing the loading capacity of microbubbles and expanding the types of payload that can be carried. For example, hydrophilic drugs are difficult to load directly into the microbubble lipid monolayer, but can be carried inside liposome or polymeric nanoparticles, which can be easily attached to the microbubble surface. These attached nanoparticles also increase the loading capacity beyond what can be achieved by attaching the payload directly to the limited surface area of the microbubble itself. Here we detail the range of microbubble–nanoparticle complexes that have been designed to deliver cargo in response to ultrasound stimulation.

4. Nanoparticle Carriers and Microbubble–Nanoparticle Complexes

Combining microbubbles with specialized carriers for cargo retention and controlled delivery provides greater molecule-loading capabilities than microbubbles, alone and expands the library of cargo types deliverable by ultrasound stimulation. Cargo vehicles can confer, or enhance, solubility to hydrophobic or amphiphilic drugs, improving biodistribution and delivery efficiency [103]. These particles improve drug safety and efficacy by preventing off-target payload release and protecting therapeutic cargo from harsh environments that may denature the molecules and diminish their therapeutic function [103,104]. Physical and chemical characteristics of carrier particles determine the class of cargo and mode of delivery, and adding these particles to ultrasound-responsive microbubbles can provide additional layers of targeting for greater precision of delivery [105].

The size of cargo-carrying delivery particles has been proven critical to both their function and biodistribution. Nanoparticles, in particular, have been widely employed in drug delivery due to the advantages of their size. Particles between 10–200 nm in diameter show minimal renal clearance and immune clearance in vivo [106], resulting in effective passive targeting of tumor tissue via the EPR effect [103]. To further evade opsonization and phagocytic clearance, nanoparticles may be surface-modified with a stealth layer, such as a PEG brush coating. This effectively increases circulation time as required for enhanced accumulation at the tumor site via the EPR effect [107]. Additionally, due to their size, nanoscale carriers may be internalized by cells via endocytosis. Particles between 10 and 60 nm present the optimal size for cell uptake in vitro based on size alone, and surface modifications may enhance uptake as well [106,108]. Nanoparticles have also been designed to undergo endosomal escape upon cellular uptake [109–111]. Finally, the high aspect ratio (surface to volume ratio) of nanoparticles can be advantageous for solubility, cargo loading, and contact-mediated interactions with cells [112,113].

Combining nanocarriers with microbubbles confers many of the unique advantages of nanoparticles to ultrasound-responsive carriers. In early systems, microbubbles and nanoparticles were co-administered or delivered to tissues in succession, showing some success in improved delivery [114,115]. For efficient microbubble cavitation-stimulated delivery directly to cells, however, the bubbles must be in close proximity to nanoparticles [116]. Accordingly, microbubble–nanoparticle delivery systems have demonstrated greater delivery efficacy when the two components are linked together [117–119]. Thus, in recent years, many strategies have been employed for physically and chemically linking nanoparticles to microbubbles.

A variety of ultrasound-responsive microbubble–nanoparticle complex formulations exist for different cargo delivery applications, with many recent developments and innovations in design and therapeutic application. Thus, we below describe and discuss the different existing designs for these linked microbubble–nanoparticle complexes.

First, the type of the nanoparticle component has a critical influence on the cargo and delivery capabilities. As such, we have divided types of microbubble–nanoparticle complexes by nanoparticle class, which are outlined in Table 1. Second, several schemes exist by which the microbubbles and nanoparticles are connected, which vary in design and strength; thus, we then discuss the linking schema and how they affect the complex formation.

Table 1. Nanoparticles implemented in microbubble–nanoparticle complexes.

Type of Nanoparticle	Schematic	Notable Characteristics
Liposome	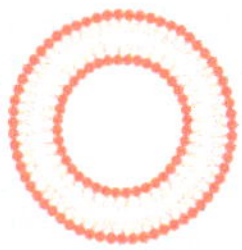	Support endocytosis of cargo [120–122], can house multiple cargo types simultaneously [123], stability influenced by environmental factors [124,125]
Polymer		Customizable properties [126], designed for enhanced stability [127], controlled cargo release [126–130], endosomal escape [109–111,128]
Lipid–polymer hybrid	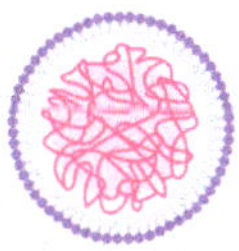	Can house multiple classes of cargo at once and deliver cargos on different timescales [121,131], high stability, due to multilayer design, promotes sustained drug release and cellular uptake [121,132,133]
Protein		Biologically-derived nanoparticles, inherent biofunctional groups allow for complexing with microbubbles, loading therapeutics, or targeting receptors [134–139]
Metallic/Inorganic		High stability, facile functionalization [140], can be light-/magnetic-responsive [105], can have some toxicity issues [105]

4.1. Nanoparticle Classes for Microbubble–Nanoparticle Complexes

4.1.1. Microbubble–Liposome Complexes

Liposomes as Nanocarriers

One nanocarrier class commonly complexed with microbubbles is liposomes. Liposomes, depicted in Table 1, are vesicles composed of phospholipids assembled into an amphiphilic bi-layered membrane encircling an aqueous interior, which can hold hydrophilic cargo [141,142]. Loading molecules into liposomes for delivery provides protection to the cargo from the biological environment, reduces off-target tissue exposure to the payload, increases drug circulation time, and facilitates passive delivery via EPR [143].

The organization of the molecules comprising liposomes is uniquely beneficial for cargo delivery. Their double-layered phospholipid membrane is similar in structure and curvature to biological membranes, rendering them highly biocompatible and capable of direct interaction with cells. Upon contact with the cell membrane, liposomes have been shown to fuse with cellular plasma membranes for direct delivery of liposomal cargo into the cell [120–122]. Liposome membrane structure also enables simultaneous incorporation of multiple classes of cargo with different properties. In addition to housing cargo within its hydrophilic interior, hydrophobic cargo may also be contained in the hydrophobic compartment between lipid layers of the membrane [123]. Delivering multiple classes of therapeutics in combination is a promising approach for treatment of therapeutic-resistant cancers. Furthermore, strategies for loading cargo into liposomes are technically straightforward.

Liposomes may be modified and functionalized [142,144]; while typically phospholipid- and cholesterol-based [141], liposomes may vary widely in design for different applications. Unilamellar liposomes for drug delivery can vary in size from ~100 nm to 800 nm in diameter, and are typically designed to be around 100 nm for optimal EPR extravasation from tumor vasculature [122,144]. Their surface functionalization can be tuned to escape immune recognition, attach charged cargo, release drug in response to a stimulus, or actively target biological molecules, which may also promote cellular uptake [122,142].

Liposomes provide excellent delivery of different types of cargo, but do not have effective echogenicity alone, due to their fluid-filled structure, and are more capable of US-stimulated delivery when coupled to highly echogenic microbubbles [122,125]. Linking liposomes with microbubbles confers many of the design advantages of liposomes to ultrasound-responsive microbubbles.

Microbubble–Liposome Complexes for Cargo Delivery

Many microbubble–liposome complex designs exist. Liposomal cargo housing capabilities expand the range of therapeutics that can be delivered by microbubbles, as evidenced by the studies outlined in Figure 3. As both lipid microbubbles and liposomes contain outer phospholipid layers, they are typically linked together by incorporating lipids functionalized with complementary moieties into their shells. This allows for a variety of conjugation strategies [145].

Early microbubble–liposome complexes were characterized by Kheirolomoom et al., who evaluated binding strategies and liposome parameters to maximize liposome loading efficiency onto each decafluorobutane-filled microbubble [122]. Kheirolomoom additionally evaluated liposome formulation and size for stability and complexation with microbubbles, with 5% biotinylation of PEG surface chains and 100 nm liposome diameter were found to provide efficient stability and microbubble binding. The liposome membranes were comprised of PEGylated lipids, biotinylated + PEGylated lipids, and 22-(*N*-(7-Nitrobenz-2-Oxa-1,3-Diazol-4-yl)Amino)-23,24-Bisnor-5-Cholen-3β-Ol (NBD) fluorescent cholesterol. These seminal microbubble–liposome complexes were shown to oscillate in response to ultrasound insonation, and bound liposomes were released upon ultrasound-mediated bubble oscillation or disruption [122]. Finally, ultrasound-controlled liposome delivery to PC-3 prostate cancer cells was evaluated in vitro. NBD was internalized by cells when delivered from the microbubble–liposome complex by ultrasound stimulation, but no effective transfer into cells was demonstrated by liposomes alone, microbubbles alone, or complexes without ultrasound treatment [122]. This study established the effectiveness of microbubble–liposome complexes for targeted delivery applications.

As targeting is critical for safe cancer therapies, microbubble–liposome complexes have been evaluated for chemotherapeutic delivery. Liposomes loaded with hydrophobic mitotic inhibitor paclitaxel (PTX) were conjugated to perfluoropropane-filled microbubbles via biotin–avidin linking. This early study evaluated cell uptake of cargo as a function of ultrasound parameters, finding that higher intensity and longer stimulation times increased uptake. Administration of this complex under ultrasound stimulus resulted in significantly reduced cell viability of 4T1 breast tumor cells in vitro in comparison to PTX-liposomes alone under identical ultrasound parameters. In a 4T1 breast carcinoma flank mouse model, micro-complex PTX delivery via ultrasound significantly inhibited tumor growth and showed high PTX accumulation in the tumor and low accumulation in the liver, indicating effective targeted treatment [146]. The same group later implemented this complex to deliver cancer therapeutic molecule doxorubicin (DOX) to DOX-resistant

MCF-7/ADR human breast cancer cells in vitro, ameliorating therapeutic resistance more effectively than verapamil, a drug currently used to reverse multidrug resistance. DOX delivery from the microbubble–liposome complex under ultrasound stimulation resulted in the highest cellular DOX uptake, lowest efflux from cells over time, and largest decrease in cell viability compared to delivery by complexes without ultrasound, liposomes with ultrasound, and liposomes plus free verapamil with ultrasound. The higher DOX delivery corresponded to high intercellular reactive oxygen species (ROS) production, greater γ-H2AX expression, and more apoptosis [147]. Another class of therapeutics requiring targeted delivery is that of platinum-based drugs, which are effective in cancer treatment, but are toxic and highly reactive and can cause damaging off-target effects. In one complex, iproplatin was loaded into the aqueous cores of azide-tagged liposomes, which were then covalently linked to DBCO-functionalized sulfur hexafluoride-filled microbubbles. Passive drug leakage from liposomes was evaluated, finding 19% iproplatin remaining within the liposomes after six days in solution [148]. When the complexes were administered to MCF7 breast cancer cells in vitro, ultrasound stimulation enhanced cell uptake of iproplatin from the microbubble–liposome constructs, while cell uptake of iproplatin from liposomes alone with applied sonication was negligible [148]. In another covalently-linked complex, DOX-loaded maleimide-functionalized liposomes were combined with a mixture of lipids, some of which presented thiol groups. Upon mixing with hydrophobic gas, the lipids self-assembled into liposome-loaded microbubbles. This covalent maleimide–thiol linkage withstood buffer and plasma conditions, with liposomes remaining bonded to microbubbles in the absence of ultrasound stimulation. When activated by ultrasound in vitro, these complexes showed highly effective cytotoxicity to melanoma cells, with roughly the same cell viability at 5.4 μg/mL as DOX-loaded liposomes alone showed at 27 μg/μL (~60% viability). Delivery of free DOX at 0.5 μg/mL resulted in ~90% cell viability, while delivery of 0.5 μg/mL delivered from the carrier complex by ultrasound resulted in ~65% cell viability. These findings suggest that ultrasound-controlled delivery can enhance therapeutic efficiency and allow for lower administered doses of harsh therapeutic molecules, increasing overall safety [149]. In all the above studies, complexing liposomes with microbubbles and stimulating with ultrasound resulted in microbubble cavitation-induced enhancement of cell uptake and therapeutic efficacy of chemotherapeutic small molecules.

Due to their ability to protect fragile cargo in circulation, microbubble–liposome complexes have also been implemented in gene delivery. Therapeutic plasmids can be effective in upregulating pro-healing pathways and repressing pro-tumor or pathological marker expression in many disease states, but they are highly sensitive to degradation from free nucleases in circulation and have demonstrated poor accumulation at the target site due to denaturation [150]. Viral delivery vectors are effective carriers, but they can cause immune reactions and mutagenesis [151,152]. Liposomal delivery serves as a safer alternative for delivery of therapeutic nucleic acids.

Nucleic acid delivery from microbubble–liposome complexes has shown particularly promising results in treatments for liver fibrosis, for which microbubble–cationic–liposome complexes have been developed for electrostatic linkage of negatively-charged genetic material to the liposome surfaces. In one study of hepatocyte growth factor (HGF) gene delivery as a therapy for liver fibrosis, pCDH-HGF plasmid was loaded onto the cationic liposome surface of microbubble–liposome complexes. Liposome charge was imparted by cationic cholesterol DC-chol incorporation in the lipid layer, and biotin–avidin–biotin bridges linked the liposomes to the microbubbles. These complexes were administered intravenously in a rat bile duct ligation model of liver fibrosis, and ultrasound stimulation was locally applied. Enhanced echo signals from ultrasound imaging confirmed microbubble–liposome complex accumulation at the target site. This therapy reversed the fibrotic effects of the ligation as measured by local hydroxyproline content [151]. A second study targeted another fibrosis-related gene, connective tissue growth factor (CTGF). Delivery of silencing miCTGF RNA by cationic microbubble–liposome complexes downregulated CTGF gene expression in HSC-T6 cells in vitro and significantly ameliorated fibrotic factors CTGF, TGFβ1, collagen I, and α-SMA expression in the rat liver in a fibrosis model in vivo. This therapy method effectively inhibited fibrotic progression and prevented development into cirrhosis [153]. Effective gene delivery by microbubble–liposome complexes in vivo shows promise for targeted genetic therapy.

Liposomes can support the delivery of multiple classes of cargo at once, thus, facilitating multidrug therapy for improved therapeutic outcomes. Demonstrating the expanded cargo loading capabilities of the microbubble–liposome complex, one study delivered a known synergistic combination of hydrophilic and hydrophobic drugs to a BxPC-3 human pancreatic tumor model in severe combined immunodeficient (SCID) mice. Combinatorial therapies implemented in pancreatic cancer treatment often cause off-target toxicity and adverse systemic effects, presenting a need for improved targeted therapies. In this design, hydrophobic drug irinotecan was incorporated into the lipid membrane layer of a perfluorobutane-filled microbubble, while hydrophilic oxaliplatin was loaded into the aqueous liposome core, and the two carriers were linked together via the biotin–avidin–biotin bridge (Schematic shown in Figure 3A). In vivo delivery of this complex under ultrasound exposure significantly impeded tumor growth in vivo without causing systemic toxicity, resulting in tumors 136% smaller than those treated with the complex without ultrasound and those treated with free multidrug therapy (Figure 3B). These results underscore the importance of both the microcarrier complex and the ultrasound stimulation in effective delivery for efficacious treatment of drug-resistant pancreatic tumors [154]. In another multi-drug delivery study for breast cancer therapy, DOX was dissolved in the aqueous liposome core and therapeutic RNA was incorporated in the lipid layers by including a plasmid–protamine complex in the lipid film during liposome synthesis. A green fluorescent protein (GFP) tester plasmid was first implemented to evaluate cell transfection and then was replaced by therapeutic siRNA siSTAT3 (signal transducer and activator of transcription 3, cell proliferation-promoting transcription factor). Liposomes were conjugated to microbubbles via a Thiol–pyridyldithiopropionate (PDP) disulfide bond and surface labeled with anti-HER2 antibodies for targeting HER2+ breast cancer cells (Figure 3C). Upon ultrasound-stimulated delivery to SkBr3 (breast cancer cells over-expressing HER2) in vitro, cells expressed GFP (while there was no expression when microcarriers were delivered without ultrasound) (Figure 3D). In the therapeutic model, ultrasound exposure enhanced cell internalization of DOX. When complexes loaded with both DOX and siSTAT3 were exposed to ultrasound, the treatment significantly decreased expression of pro-tumor and pro-mitotic factors STAT3, Cyclin D1, and c-Myc, reduced cell viability, and inhibited tumor growth (Figure 3E), indicating efficacy for cancer therapy [155].

Finally, one unique model conferred ultrasound responsiveness to liposomes not by linking the particle surfaces together, but by loading microbubbles within a large, micron-scale liposome aqueous core, along with the therapeutic cargo. Cavitation of microbubbles caused liposome rupture, releasing cargo [156]. This method eliminated the need for any linkage strategies, provided bubble stability, and ensured liposomes and microbubbles were delivered together. These large liposomes were formed at 5 μm in diameter with 1 μm loaded microbubbles and exhibited significantly greater cargo loading efficiency than conventional liposomes. Interestingly, this group encapsulated large protein molecules (IgG, 150 kD) within the liposomes, which can serve as immunotherapeutic agents. Microbubble-loaded liposomes, but not unloaded liposomes, underwent forceful rupture under ultrasound stimulation. Without stimulation, these microcarriers showed stability when incubated in a blood sample [156].

Many microbubble–liposome designs have been explored, and they present key advantages including facile fabrication and complexation, expanded cargo loading capacity, enhanced cargo protection, and direct delivery to cells. A key challenge with liposome-based vehicles, however, is that they have lower stability than some other nanoparticle types, and membrane destabilization can result in off-target release of cargo. Liposome stability and lipid layer integrity is controlled by several factors including liposome composition, lipid tail length and saturation, ether linkages, and cholesterol content [124]. Without cholesterol, liposome interactions with proteins can cause membrane instability. Lipid stability and liposome permeability are also dictated by environmental temperature, as the lipid membrane experiences enhanced permeability at the phase transition temperature of the lipids. Instability above the transition temperature leads to liquid disorder and can cause leakage [124,125]. Finally, unmodified liposomes are subject to opsonization, where the particles are tagged with opsonins for clearance by the immune system. Charged liposomes and larger liposomes are frequently opsonized [123]. Particle uptake and clearance by the reticuloendothelial system is mitigated but not entirely eliminated by the addition of PEG stealth coatings, and many liposomal formulations have high accumulation in the liver [157]. Though many design modifications can improve liposome stability, undesired effects, such as lipid oxidation or hydrolysis, membrane destabilization, aggregation, liposome fusion, and immune-based elimination, all still occur [124]. This motivates the implementation of additional classes of nanocarriers to complex with microbubbles for cavitation-based directed delivery.

4.1.2. Microbubble–Polymer Nanoparticle Complexes

Polymer Nanoparticles as Nanocarriers

Another class of nanoparticle carrier that has been complexed with microbubbles for ultrasound-controlled delivery is polymer-based nanoparticles, which are illustrated in Table 1. One of the most common and well-characterized polymer materials used for nanoparticle delivery systems is poly(lactic-co-glycolic acid) (PLGA). PLGA is stable in circulation, biocompatible, and biodegradable, resulting in products that are easily metabolized and eliminated by the body [158]. Naturally anionic PLGA nanoparticles can be loaded with hydrophobic or hydrophilic cargo during particle fabrication by using emulsion or double emulsion techniques, respectively. In the emulsion process, polymers and cargo are mixed with water and a surfactant, then sonicated to form nanoparticles [158]. As compared to PEG-coated liposomes, loaded with DOX, PEG-coated DOX-loaded PLGA nanoparticles have shown reduced off-target cardiotoxicity, while preserving anti-tumor therapeutic effects [159]. PLGA nanoparticles have been employed to deliver small hydrophobic drugs, proteins, and nucleic acids; however, loading efficiency, and in some cases, burst release of drugs, remain challenges [158].

Another synthetic polymer nanoparticle, the poly(ethylenimine) (PEI) nanocarrier, is hydrophilic and inherently positively-charged, and, as such, it is considered the gold standard for non-viral therapeutic gene delivery from polymer nanoparticles [126,160]. The PEI structure possesses large numbers of protonatable amine groups, which are separated by short alkyl spacers, resulting in a high positive charge density [160,161]. In addition to improving gene complexation efficiency, it is suggested that the positive charge of the PEIs enhances transfection, due to increased cell internalization from electrostatic interactions with cell membranes and endosomal escape, as described below [162].

Natural polymers commonly used in therapeutic delivery, such as chitosan and hyaluronic acid, provide biocompatibility and inherent bioactivity [163–165], and complexing naturally-derived polymer nanoparticles with ultrasound microbubbles for ultrasound targeting could be an area of future interest.

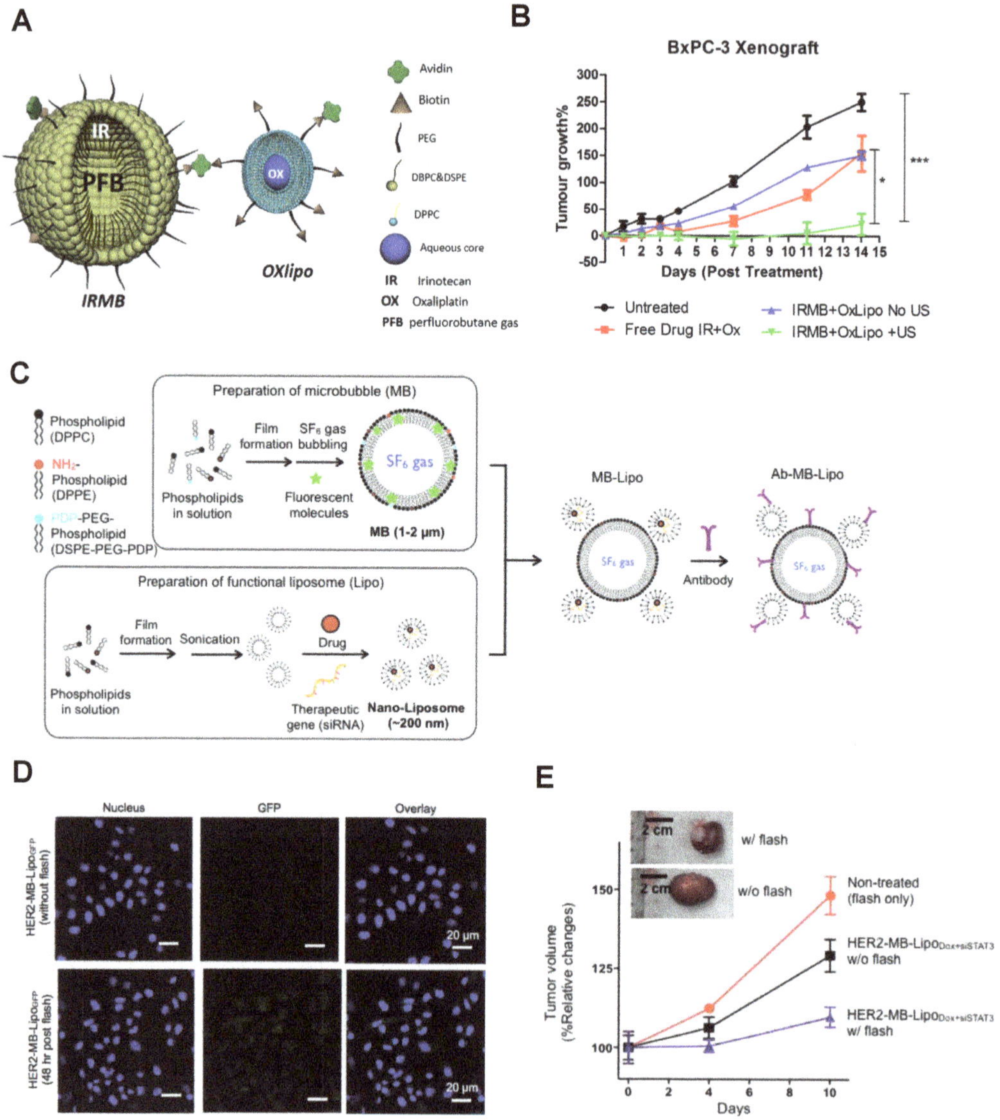

Figure 3. Microbubble–liposome complexes expand the types of cargo that can be delivered by ultrasound-induced microbubble cavitation, leading to improved therapeutic efficacy. (**A**) Microbubbles loaded with hydrophobic chemotherapeutic Irinotecan and liposomes loaded with hydrophilic chemotherapeutic Oxaliplatin were linked together by biotin–avidin bridging for a multi-drug cancer therapy approach. (**B**) Implementation of this system (IRMB + OxLipo + US, green) in vivo resulted in significantly lower tumor growth than delivery without ultrasound (blue), * $p < 0.05$, *** $p < 0.0001$. (**A**,**B**) reprinted from [154], copyright 2020, with permission from Elsevier. (**C**). Both DOX and therapeutic siRNA were loaded into liposomes, which were covalently conjugated to microbubbles and labeled with antibodies for multidrug active targeting. (**D**) Ultrasound-stimulated delivery (flash) of this complex promoted cell uptake and gene transfection (bottom row). (**E**) This therapeutic microcarrier complex design inhibited tumor growth when stimulated with ultrasound (flash) in vivo (shown in blue). (**C**–**E**) reprinted from [155], copyright 2014, Ivyspring International Publisher.

While polymer-based nanoparticle carriers provide some of the same advantages as liposomes, several characteristics of polymers are specifically advantageous to nanocarrier systems, including physical stability in biological environments, ability to confer solubility to otherwise insoluble therapeutics via complexation [127], and diversity in properties and custom design options [126]. Polymer properties vary widely, and cargo loading and delivery abilities are governed by these physicochemical properties. For example, PEI properties are influenced by polymer molecular weight and the extent of polymer branching [162]. While PEI and other cationic polymers, like chitosan, are attractive options for delivery of anionic nucleic acid cargo, PLGA nanoparticles must be modified with cationic molecules when implemented for gene delivery [126,158]. Beyond modifying particle charge for improved cargo loading, multiple polymers are commonly combined in nanoparticle fabrication to optimize properties for specific applications. Adding PEG and poly(caprolactone) to PEI counters toxicity effects, and coating PLGA with chitosan and alginate impacts nanoparticle interaction with the epithelium [126,166–170]. As with other nanocarriers, PEG chains are often added to polymer nanoparticles to reduce immunogenicity and enhance circulation time [121]. Retention of cargo within polymer nanoparticles during circulation is also influenced by the degree of polymer crosslinking, and delivery of cargo is impacted by the type of polymer crosslinking. Noncovalent crosslinking may allow more pH-sensitive, rapid cargo delivery than covalent polymer crosslinking [127,128]. For many PLGA nanocarriers, the mechanism of therapeutic release is hydrolytic polymer degradation-mediated diffusion of cargo, which is tuned by altering the molar ratio of lactic acid to glycolic acid [171].

Many polymer nanoparticle systems have been engineered for targeted drug delivery. Several designs present either polymer–polymer crosslinks or polymer–cargo linkages that are labile to environmental stimuli, such as disulfide bonds (glutathione-sensitive), ester linkages (ROS-labile and esterase-labile), or pH-labile peptide linkers [127,129]. These stimuli are associated with pathological tissue environments and can facilitate degradation of polymer linkages and release of entrapped or linkage-attached cargo. Another polymer-specific method of delivery is network swelling to release the drug from expanded pores. Swelling may occur in response to an endogenous stimulus, such as pH or temperature, which also adds a layer of targeting for controlled cargo delivery [128,130]. Stimulus-responsive swelling behavior is also a critical asset of polymer molecules that enhances direct delivery in cells. Upon endosomal internalization of nanoparticles by cells, many polymer-based nanoparticles undergo pH-responsive swelling, or pH buffering, in response to the acidic endosomal environment, resulting in endosomal swelling and rupture, and, thereby, releasing the nanoparticles directly into the cell cytoplasm [109–111,128]. This mechanism is suggested to contribute to improved gene transfection from PEI nanocarriers [111,162].

Polymer nanoparticles are also engineered for active targeting by functionalizing these carriers with biochemical compounds, such as γ-PGA and RGD, for active targeting of tumor-specific or tumor-upregulated markers [127,169]. For remotely-controlled on-demand delivery, polymer nanoparticles have been designed to release their cargo in response to light exposure or changes in temperature [13,127,172,173]. While many existing sophisticated polymer nanoparticles are designed to achieve localized drug delivery through targeting or environmentally-responsive strategies, they are not inherently responsive to ultrasound stimulation. Thus, to confer the design advantages of ultrasound-mediated therapy to different polymer nanoparticles, the nanoparticles can be complexed with microbubbles.

Microbubble–Polymer Nanoparticle Complexes for Cargo Delivery

To impart on-demand delivery of DOX from PLGA nanoparticles, DOX-loaded microbubble–PLGA–nanoparticle complexes were developed. PLGA nanoparticles were fabricated and loaded with DOX by the double emulsion–solvent evaporation technique. PLGA polymer COOH end groups were then covalently linked to free-amine-functionalized, perfluoropropane-filled, microbubbles via amide bonds. When implemented in a rabbit liver tumor model, delivery of DOX from this complex, via ultrasound stimulation, resulted in inhibition of tumor growth, improved survival, and, importantly, limited off-target effects [174]. In a different study further investigating this same ultrasound-responsive complex, delivery of DOX-loaded PLGA nanoparticles to rabbit tumor models reduced tumor growth, increasing apoptosis of tumor cells via increased Bax expression and decreased Bcl-2 expression [175]. The double emulsion–solvent evaporation technique of loading DOX into the nanoparticles by encapsulating within the polymer structure during nanoparticle formation was very efficient in both of the above cases, with >85% encapsulation efficiency [174,175].

PEI-based nanoparticles for nucleic acid delivery have also been coupled to microbubbles for ultrasound-controlled delivery. One delivery complex incorporating PEI was designed for multileveled targeting of ovarian cancer stem cells, which are an appropriate candidate cell type for on-demand targeted delivery because they are not fast replicating and are. Thus. not as responsive to systemically administered therapies designed for highly proliferative cell populations. In this design, PEI–PEG-based nanocarriers were fabricated with disulfide bonds linking the two polymer types together. Polymer nanoparticles were loaded with plasmid DNA, via electrostatic complexation, and functionalized with biotin for conjugation to perfluoropropane gas-filled microbubbles via biotin–avidin–biotin linkage [176]. Conjugating PEG to positively-charged PEI via disulfide bonds provides an environmentally-sensitive steric shield, which protects the charged nanoparticle from immune responses in circulation, but is cleaved in glutathione-rich and acidic environments, such as tumors, exposing the cationic nanoparticle for cell uptake. This temporary stealth coating imparts tumor-specific targeted delivery. Ultrasound delivery of these particles enhanced the delivery of the DNA payload to the targeted tissues while reducing off-target delivery [176,177].

Ultrasound-mediated delivery of GFP plasmid (pGFP) from PEI–PEG nanocarriers to ovarian cancer stem cells in vitro resulted in significantly greater transfection efficiency than cationic commercial transfection agent lipofectamine. When evaluated in vivo, this delivery method also resulted in enhanced transfection efficiency, which was confined to the targeted area, and caused less off-target effects, as compared to lipofectamine treatment in a mouse ovarian tumor model [176]. Upon application of this system for therapeutic delivery of antitumorigenic genes (short hairpin RNA for aldehyde dehydrogenase 1) to ovarian cancer stem cells in vitro, cells showed higher rates of apoptosis than those treated with lipofectamine. Mechanisms contributing to efficacy included enhanced endocytosis and endosomal escape of the plasmids delivered by ultrasound-stimulated PEG–PEI nanocarriers [178]. Future directions for this model include evaluation of therapeutic efficacy in vivo.

PEG has been shown to consistently confer biocompatibility and increased circulation time to nanoparticles by forming a neutral, hydrophilic shield from opsonization [179]. When implemented instead as the base of a polymer nanocarrier, PEG can be modified to optimize other properties for cargo loading and delivery. To form a carrier for delivering sodium borocaptate (BSH), for targeted boron neutron capture therapy of glioblastoma multiforme, PEG was complexed with the polymer segment poly(chloromethylstyrene) (PCMS) to form PEG-b-PCMS. Anionic drug BSH linked to the chloromethyl groups of the PCMS to form the BSH-loaded polymer nanoparticle–drug complex PEG-b-poly((closododecaboranyl)thiomethylstyrene) (PEG-b-PMBSH). PEG-b-PMBSH nanoparticles were electrostatically complexed with cationic microbubbles for ultrasound-mediated BBB penetration and on-demand tumor targeting. This system was delivered in a mouse cranial glioblastoma model, where ultrasound-induced microbubble cavitation caused BBB permeation and rapid boron accumulation in the tumor tissue [180]. This study demonstrated efficient targeted delivery and motivates future work in evaluating therapeutic efficacy of this complex for clinical translation.

One microbubble–polymer nanoparticle complex combining PLGA and PEI was designed to treat renal interstitial fibrosis. PLGA nanoparticles were simultaneously formed and loaded with therapeutic rosiglitazone (RSG, which is an agonist for the anti-inflammatory and anti-fibrotic transcription factor PPARγ) via a double emulsion preparation before complexing with PEI (also via electrostatic linking). As RSG has poor water solubility, PLGA was an apt choice for forming a water-soluble prodrug for effective delivery [181]. The blended polymer nanoparticles presented sufficient positive charge for electrostatic conjugation to commercially available sulfur hexafluoride-filled anionic SonoVue® microbubbles to impart ultrasound responsiveness. This ultrasound-sensitive complex is particularly promising for kidney-related diseases, because ultrasound microbubble destruction has been shown to safely enable renal interstitial capillary permeation for therapeutic delivery. As expected, ultrasound sonication in vitro enhanced nanoparticle uptake in mouse SV40-MES-13 cells. In a rat unilateral ureteral obstruction (UUO) model, intravenous therapeutic complex delivery via ultrasound stimulation effectively reversed PPARγ downregulation, and profibrotic TGFβ, αSMA, and Collagen I expression enhancement caused by UUO. Neither micro-complexes without ultrasound nor therapeutic nanoparticles with ultrasound but no microbubbles were as effective as the microbubble–nanoparticle ultrasound condition, which also attenuated fibrotic collagen deposition [181].

Polymer nanoparticles present many advantages in drug delivery applications and have been successfully utilized in ultrasound-controlled targeted delivery. Development toward effective delivery strategies has also motivated design of a new class of nanoparticles made by combining both lipids and polymers to impart the beneficial properties of both material classes to the delivery vehicles.

4.1.3. Microbubble-Lipid-Polymer Hybrid Nanoparticle Complexes

Lipid-Polymer–Hybrid Nanoparticles as Nanocarriers

As outlined above, liposomes and polymer-based nanoparticles both have advantageous characteristics resulting from their individual inherent chemical properties. A more recently developed nanocarrier design combines these structural components into lipid–polymer–hybrid nanoparticles (LPHNs), depicted in Table 1. Most commonly organized as a polymer core with a lipid outer layer (Figure 4A), this class of molecule combines the functionality of polymer nanoparticles and lipid-based nanoparticles for multimodal delivery [121,131]. Cargo resides within the inner polymer core, and the lipid layer acts as a barrier to protect and stabilize the core and reduce leakage. An outermost PEG coating may be added to reduce immune clearance of the hybrid particles [121,131]. Like other nanoparticles, LPHNs may be surface-functionalized with moieties for active tissue targeting or for ameliorating the complement activation pathway [131,182].

LPHNs are a particularly advantageous class of vehicle for cancer drug delivery. Simultaneous combinatorial drug therapy has shown promise in treating therapeutic-resistant tumors, and LPHNs present several approaches for loading different types of cargo with varying properties, such as hydrophobic and hydrophilic drugs, charged and uncharged molecules, etc., depending upon the nanoparticle fabrication method used. Beyond the multidrug loading capabilities of liposomes alone, these hybrid nanoparticles can be designed to deliver drugs at different timescales or in response to different stimuli [121].

The fabrication process parameters of LPHNs (including lipid to polymer mass ratio and the manner and order in which the molecules are combined) determine the nanoparticles' physical properties, such as size, loading efficiency, and release kinetics. For example, the outer lipid coating must cover the nanoparticle core completely without gaps to prevent off-target cargo leakage. This coverage, and resulting release kinetics, are determined by the nanoparticle fabrication conditions. As such, a major area of research in the current field of LPHN development for drug delivery is investigating the effects of the preparation and assembly methods on nanoparticle properties and function [121,131,132]. When compared to polymer and lipid-only nanoparticles, drug delivery via LPHNs has shown significantly enhanced sustained drug release and cellular uptake in vitro, and improved delivery to cells in vivo, which indicates their potential for targeted delivery (Figure 4B,C) [121,131,133].

These results are likely attributable to increased sustained drug release and greater carrier stability [133]. Though current studies are limited, localized on-demand delivery of LPHNs can be improved by complexing these nanoparticles with microbubbles to make them ultrasound-responsive.

Microbubble-LPHN Complexes for Cargo Delivery

Brain tumors are difficult to treat due to the protective BBB inhibiting drug delivery. Glioblastoma is resistant to the chemotherapeutic molecule temozolomide (TMZ) by expression of O6-methylguanine-DNA methyltransferase (MGMT), despite TMZ having BBB-penetrative properties. This necessitates delivery of additional therapeutics. A promising approach to increase effectiveness of TMZ treatment is to inhibit MGMT expression via gene therapy, which must be precisely controlled to prevent off-target effects. The high stability of the LPHNs makes them an attractive carrier for this application. Yang et al. elegantly designed a microbubble-LPHN complex with multiple layers of targeting to safely and precisely control delivery of CRISPR/Cas9 plasmids for gene editing to glioblastoma cells in vivo. Here, a PLGA nanoparticle core was complexed with the MGMT-targeting CRISPR/Cas9 plasmids, then combined with biotin- and cRGD-functionalized lipids and self-assembled into LPHNs. The nanoparticles were then bound to perfluoropropane-filled microbubbles by biotin–avidin linking. Encapsulation within the LPHNs protected plasmids from enzymatic degradation and also facilitated sustained plasmid release from the nanoparticles. The functionalization of cRGD served to target the $\alpha5\beta3$ surface receptor upregulated in tumor tissue and was correlated with higher T98G glioblastoma cell uptake in vitro. In a T98G orthotopic xenograft model in vivo, focused ultrasound-mediated microbubble cavitation caused BBB opening, and microbubble–LPHN complexes functionalized with cRGD showed significantly greater accumulation in the brain than control LPHN–bubble complexes without active targeting components (Figure 4D). When co-administered with TMZ, ultrasound exposure with microbubble–cRGD–LPHN complexes loaded with Cas9/MGMT plasmids inhibited tumor growth and considerably extended survival period than control conditions TMZ alone, TMZ + cRGD-LPHN–microbubble complexes without ultrasound, and TMZ + LPHN–microbubble complexes with ultrasound (Figure 4E). In this system, each targeting facet enhanced precision of delivery. The ultrasound stimulus-controlled nanoparticle transit across the BBB, the PLGA–core–LPHN facilitated sustained therapeutic release, and the cRGD moiety provided active molecular targeting to tumor cells. Through effective targeted delivery of gene inhibitors against glioblastoma therapy resistance, this technology overcame TMZ resistance to impede tumor growth [183] (Figure 4E).

Another unique design merged properties of LPHNs and microbubbles into a single carrier. This system combined the echogenicity of lipid microbubbles with the stability and cargo-loading capabilities of polymers. The complex was made via a water/oil/water double emulsion process with ammonium bicarbonate, which decomposed to form the air-filled gas core. The final bubble diameter was 800 nm, with a 30–60 nm thick lipid-PLGA shell. These microbubbles showed greater echogenicity than PLGA-only microbubbles, and were shown to be stable at acidic pH but to undergo cavitation under ultrasound exposure. DOX was loaded into the carriers by dissolving it in ammonium bicarbonate before microbubble formation (Figure 4F). Ultrasound stimulation of the microbubbles facilitated controlled DOX delivery, causing reduced 4T1 tumor cell viability in vitro and targeted accumulation, tumor growth inhibition, and longer survival in a 4T1 tumor model in vivo, indicating the efficacy of this system (Figure 4G) [184].

While few studies currently exist that have complexed LPHNs with microbubbles for ultrasound-mediated cargo delivery, the therapeutic outcomes of the above work indicate that this technology is promising and the field is open for more complex LPHN-based delivery designs.

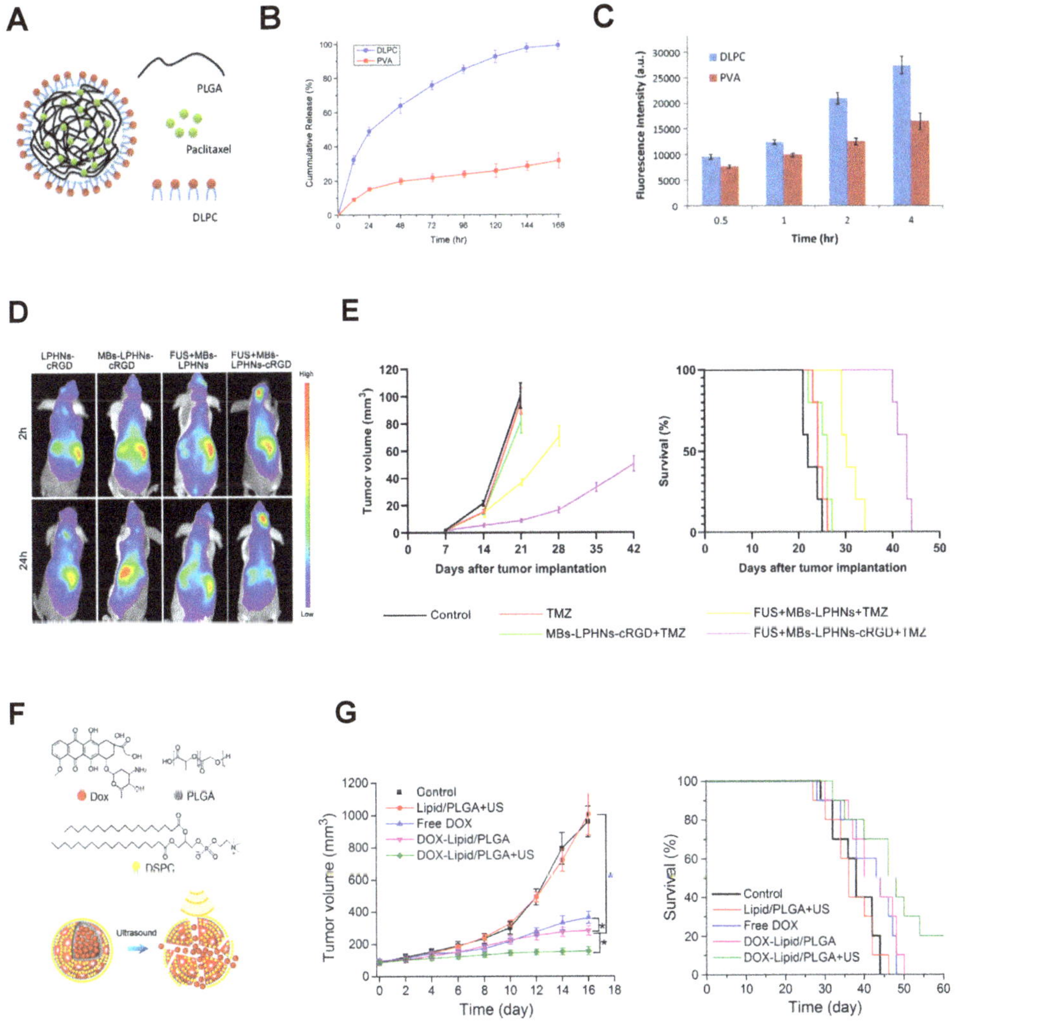

Figure 4. Lipid–polymer hybrid nanoparticles (LPHNs) demonstrate advantages in stability and cargo delivery, and are promising nanocarriers for ultrasound-mediated delivery. (**A**) LPHN composed of a polymer core which houses cargo and a lipid outer layer. (**B**) Cargo release magnitude over time was greater from the LPHNs (DLPC, blue) than from polymer nanoparticles (PVA, red). (**C**) Cargo uptake by cells was greater over time when delivered by LPHNs (DLPC, blue) compared to polymer carriers (PVA, red). (**A–C**) reprinted from [133], copyright 2010, with permission from Elsevier. (**D**) Adding ultrasound-responsiveness to LPHNs via microbubble conjugation improved biodistribution and targeting in vivo and enhanced cargo accumulation at the target location in the brain (FUS + MBs-LPHNs-cRGD). (**E**) Using this complex design for the delivery of temozolomide was correlated with inhibition of tumor growth and extended survival (purple). (**D**,**E**) reprinted from [183]: *International Journal of NanoMedicine 2021:16 185–199'* Originally published by and used with permission from Dove Medical Press Ltd. (**F**) Unique model: a microbubble composed of a lipid–polymer hybrid shell, loaded with DOX. (**G**) Ultrasound-stimulated cargo delivery from this system (DOX-Lipid/PLGA + US, green) also resulted in significantly inhibited tumor growth in vivo and greater survival, * $p < 0.05$. (**F**,**G**) Reprinted or adapted with permission from [184]. Copyright 2019 American Chemical Society.

4.1.4. Microbubble–Protein Nanoparticle Complexes

Protein Nanoparticles as Nanocarriers

Another type of nanoparticle implemented in ultrasound-responsive complexes is the protein-based nanoparticle, illustrated in Table 1. As biologically-derived molecules, albumin nanoparticles have inherent bioactivity which contributes to targeting capabilities. Other biologically-derived protein nanoparticles do exist, including gelatin, elastin, gliadin, silk sericin, and others. However, albumin is the most abundant plasma protein, and, thus, it is widely employed for drug delivery, and albumin-based prodrugs are in clinical study [134,135]. Protein nanoparticles are fabricated by emulsion techniques, or, more commonly, desolvation, where the proteins are precipitated out of an organic solution to form nanoprecipitate particles. The addition of glutaraldehyde after precipitation crosslinks albumin molecules together, providing stability to the particles. Fabrication parameters, such as the amount of desolvation solution, pH, and crosslinker concentration, all influence the final albumin nanoparticle size [185,186].

Like many other nanoparticle carriers, albumin nanoparticles are 50–200 nm in diameter, facilitating extravasation into tumor tissue via the EPR effect, and they can be functionalized for targeted delivery [186]. However, as natural biological molecules with protein structure, they differ significantly in other ways. Similar to many polymers, proteins are biocompatible and biodegradable, and elicit a limited immune reaction. Unlike most synthetic nanomaterials, however, albumin's surface innately presents multiple functional groups, including thiol, amino, and carboxyl groups [134]. This intrinsic property is advantageous for linking the nanoparticles to microbubbles or active targeting moieties. The albumin surface site cysteine-34 binds drugs and prodrugs, and free fatty acid binding sites also exist on the surface [135]. The amphiphilic structure of albumin also promotes cargo attachment via hydrophobic interactions. Many common therapeutics, including penicillin and benzodiazepines, bind to the surface of albumin. While active targeting moieties, such as peptides, antibodies, and aptamers, may be conjugated to these surface groups, albumin itself has high binding affinity for tumor endothelial cell surface receptor gp60, resulting in internalization of the nanoparticle [136]. Albumin also binds to several other surface receptors highly expressed in tumor and inflammatory environments, including the glioblastoma-expressed protein SPARC (secreted protein acidic and rich in cysteine) [137–139]. These inherent binding capabilities promote precision delivery and accumulation of albumin-based carriers.

Albumin nanoparticles are stable at a range of pH conditions (pH 4–9) and temperatures up to 60 °C, and also have a prolonged circulation time, which is critical for effective EPR-mediated therapy. Albumin has also been shown to accumulate in areas of inflammation, supporting its versatility in therapeutic potential for different pathologies [135]. Albumin nanoparticles can form prodrugs by binding directly to the cargo to impart greater solubility and to extend circulation time. This complexation is particularly effective for short therapeutic peptides with short circulation times [134,136]. While native albumin is nonimmunogenic, modified albumin may trigger immune responses and can be coated with a stealth PEG layer for improved circulation and biocompatibility [137]. One potential challenge of utilizing albumin nanoparticles is aggregation of albumin-covered microbubble complexes due to amine–NHS interactions between adjacent particles. This issue has been ameliorated by adjusting the concentration of drug loaded onto the surface of the nanoparticles to sterically block these interactions [187].

Microbubble–Protein–Nanoparticle Complexes for Cargo Delivery

In addition to their advantages for active targeting, the functional groups present on the protein nanoparticle surface are an excellent asset for linking the particles to microbubbles to create ultrasound-controlled delivery systems. A common design scheme consists of covalently linking NHS-functionalized microbubbles to free amine groups on the albumin surface, forming a strong bond.

For enhanced BBB permeation and delivery, NHS-functionalized, sulfur hexafluoride gas-filled lipid microbubbles were covalently linked to albumin nanoparticles by NHS–amine amide bond formation. To evaluate albumin complex delivery, NHS-functionalized Cy5.5 dye molecules were also bonded to the albumin particles. These microbubble–albumin complexes were used to evaluate ultrasound stimulation parameters for safe BBB permeation and delivery of the particles. At optimized settings, ultrasound stimulation resulted in 1.5× greater nanoparticle accumulation in the brain than that of complexes administered without ultrasound [188].

Another approach for cargo loading of therapeutic microbubble–albumin complexes involved linking DOX to the albumin nanoparticle surface via electrostatic interactions, resulting in sustained release of DOX from the nanoparticles over an extended time period. This loading strategy was pH-dependent with higher release profiles at acidic pH, which may add another targeting capability to nanoparticles for greater cargo release in tumor environments. Sulfur hexafluoride gas-filled NHS-microbubbles were again linked to the albumin surface via covalent amide bonds, and it was demonstrated that addition of the albumin nanoparticles did not adversely affect echogenicity of the microbubbles. In a VX2 liver tumor rabbit model, more tumor growth inhibition was observed with ultrasound-controlled delivery of these therapeutic particles, using both intra-vascular and more invasive intra-arterial injections, than with intra-arterial delivery of free DOX with ultrasound [187].

While several designs have loaded DOX via electrostatic interactions, one group instead thiolated the nanoparticles during desolvation fabrication for DOX attachment. The nanoparticles were then linked to NHS-functionalized perfluoropropane-filled microbubbles via amide linkages. The complexes were dispersed in Lipiodol (oil phase) to form an emulsion. This emulsion was then delivered to a VX2 rabbit liver carcinoma model, serving as an improved ultrasound-targeted delivery method for trans-arterial chemoembolization (TACE) compared to a conventional TACE formulation of DOX in Lipiodol. This ultrasound-based delivery had a greater inhibitory effect on tumor growth than traditional therapeutic methods [189].

4.1.5. Microbubble–Metallic–Nanoparticle Complexes and Inorganic Nanoparticle Complexes

Metallic and Inorganic Nanoparticles in Therapeutic Delivery

One class of nanoparticles that is in early stages of complexation with microbubbles for delivery is inorganic and metallic nanoparticles. This class, illustrated in Table 1, includes gold, iron oxide, and silica-based nanoparticles, among others, and, like polymer nanoparticles, they can be designed with a range of properties for specific applications. In addition to drug delivery purposes, these particles themselves can serve as therapeutic, theranostic, or targeting agents. Some metallic nanoparticles are inherently responsive to external stimuli. Gold nanoparticles (AuNPs) are responsive to light stimulation and iron oxide nanoparticle accumulation can be directed by magnetic fields [105]. Metallic nanoparticles are favorable drug delivery agents because they are stable and can be easily functionalized for biochemical interactions [140]. While effective, inorganic nanoparticles present some toxicity and solubility issues, making them excellent candidates for microbubble complexation for ultrasound-targeted delivery [105]. Current systems combining metallic nanoparticles with microbubbles for delivery vary widely in both design and application.

Microbubble–Metallic–Nanoparticle Complexes and Inorganic Nanoparticle Complexes for Therapeutic Targeting

A few systems exist in which AuNPs are delivered in combination with microbubbles, though many do not complex these components together into a single microparticle, instead administering the nanoparticles and microbubbles independently. In each of these approaches, ultrasound-mediated microbubble cavitation has promoted delivery of nanoparticles to specified regions. An example of this is copper-alloyed-gold nanocluster delivery to the brainstem. The nanoclusters were delivered intranasally and the microbubbles were administered systemically. Focused ultrasound was applied to the brainstem, resulting in enhanced cluster accumulation in the targeted region [190]. In another study, microbubble cavitation enhanced chitosan-coated AuNP delivery to the inner ear by inducing permeability of the round window membrane within the ear [140]. Similarly, microbubble-assisted BBB permeation enhanced the passage of silica-coated gold nanorods into the brain [191]. These results all indicate the potential for targeted drug delivery and future enhanced therapeutic efficiency of therapeutic AuNPs. For all of these approaches, delivery may be enhanced with complexation of the AuNPs to the microbubbles for ensured proximity during bubble cavitation.

In many current systems, inorganic nanoparticles delivered by microbubbles serve solely as imaging contrast agents, particularly in the case of iron oxide nanoparticles [192–195]. However, some metallic nanoparticle properties allow them to perform therapeutic roles as well. Inorganic nanoparticles have been implemented as delivery devices for antibiotic agents. Gold and silver nanoparticles promote antibiotic molecule interaction with bacterial cell walls, while some inorganic particles themselves bear antibacterial or antimicrobial properties [196–198]. Additionally, metallic nanoparticles, particularly AuNP, have unique optical properties and can convert light to heat for targeted thermoablation therapy. Finally, the facile functionalization of metallic nanoparticles, along with their stability, provides the capability to serve as drug delivery vehicles.

In recent studies where metallic nanoparticles were complexed to microbubbles, the complexation tended to differ from that of other nanoparticle classes. AuNPs have been incorporated into the microbubble shell itself, while mesoporous silica nanoparticles were either incorporated into the shell or encapsulated within the hydrophobic gas core [199–203]. In one recent study, gold nanorods were incorporated within the microbubble shell, and a PEI-therapeutic DNA prodrug was complexed to the shell. Ultrasound stimulation enhanced accumulation of the nanoparticles and gene therapy payloads at the target site in a hepatoma xenograft mouse model. This gene therapy was combined with colocalized laser-induced photothermal ablation therapy, implemented by the stimulated gold nanorods, resulting in a significant reduction in tumor growth [200]. Similarly, AuNPs incorporated into bovine serum albumin-coated microbubble surfaces accumulated within the ultrasound-targeted region of a U-87 MG xenograft mouse model. Laser irradiation of the delivered AuNPs caused targeted photothermal inhibition of tumor growth [201].

Mesoporous silica nanoparticles (MSNs) are promising drug delivery vehicles, due to their porosity and capacity for loading therapeutic genes via hydrophobic interactions. When MSNs were loaded with a reporter plasmid, complexed with PEI, and then incorporated into a microbubble phospholipid layer, they showed enhanced cell uptake and gene transfection, as well as targeted accumulation in a mouse ovarian tumor model, in response to focused ultrasound stimulation [203]. MSNs may also be modified for additional layers of targeting. In one design, MSNs loaded with the chemotherapeutic tanshinone IIA were functionalized with folate to actively target folate receptor-overexpressing cancer cells, then encapsulated within the hydrophobic core of a microbubble for ultrasound delivery. This complex induced cell apoptosis in vitro. When administered in a mouse H22 tumor model under ultrasound exposure, this complex effectively inhibited tumor growth [202].

Ultrasound-controlled delivery of microbubble–metallic nanoparticle complexes has shown enhanced targeting and efficacy in tumor treatment, and this technology can be further expanded for targeted antibacterial treatment applications.

4.2. Multi-Stimulus-Controlled Cargo Delivery from Microbubble–Nanoparticle Complexes

High precision delivery is particularly important for very aggressive treatment regimens, which can be necessary in cases of highly malignant, accelerated, or late-diagnosed cancers, such as pancreatic cancer. Many of the ultrasound-sensitive complexes discussed thus far have incorporated design components for additional passive, bio-active, or environmentally-stimulated targeting. However, recent approaches have combined multiple external on-demand stimuli for increased precision of targeting with minimal invasiveness. Multi-stimulus microcarriers span multiple classes of nanoparticle components, and they promote precise targeted delivery and high efficacy (Figure 5).

Towards delivering efficacious therapeutics for pancreatic cancer, a magnetic microbubble–liposome complex was developed to enable magnetic field-influenced accumulation of these therapeutic carriers at a designated location with ultrasound-mediated cavitation to induce targeted delivery. Magnetic liposomes were prepared by rehydrating a lipid thin film layer with an aqueous solution of citric acid-stabilized iron oxide nanoparticles, upon which the magnetic metal particles became housed in the liposome aqueous core. These maleimide-functionalized liposomes were loaded with DOX and covalently conjugated to the thiol-functionalized surfaces of perfluorocarbon-filled microbubbles (Figure 5A). Under ultrasound treatment in vitro, these novel microcarriers caused efficient cellular uptake of DOX by pancreatic tumor cells, resulting in significantly reduced cell viability. In a pancreatic tumor xenograft mouse model, a magnetic field was applied to the tumor region, and tumor growth inhibition was significantly improved when both magnetic and ultrasound stimuli were applied, demonstrating that magnetic-facilitated particle accumulation can enhance efficacy of ultrasound-controlled therapeutic delivery (Figure 5B). These results correlated with the biodistribution of magnetic + ultrasound-controlled delivery of fluorophore as a model drug in healthy mice, where the fluorophore signal was precisely concentrated to one site with magnetic accumulation of liposomes only, but fluorescence intensity was enhanced when microbubbles and ultrasound stimulation were added [204].

Another aggressive form of combinatorial cancer therapy which leverages external stimulation is chemo-photodynamic therapy (PDT). PDT is the chemical excitation of photosensitizer dye molecules by a light stimulus, resulting in local ROS production [205]. ROS cause cell death by oxidative stress, establishing an effective method for killing tumor cells and tumor vasculature. Combining this method with small molecule chemotherapy presents multiple mechanisms of cell death simultaneously for enhanced efficacy. Light stimulation provides spatial control of PDT; however, photosensitizer distribution can be improved by ultrasound-mediated delivery, in part because ultrasound is less scattered by tissue than light.

One elegant system imparted ultrasound targeting to combine photo- and chemical-mediated therapy for improving the safety and efficacy of this already multi-pronged therapeutic approach. The hydrophobic photosensitizer chlorin e6 (Ce6) was loaded onto the tails of the lipid monolayer of perfluoropropane gas-filled, NHS-functionalized microbubbles. To simultaneously load chemotherapeutic agents, human serum albumin (HSA) nanoparticles were formed by disulfide bond crosslinking between thiol groups on albumin molecule surfaces and loaded with DOX. The loaded nanoparticles were then covalently linked to the microbubbles by amine–NHS covalent conjugation (Figure 5C).

Ultrasound stimulation enhanced cell uptake of both DOX and Ce6 from this complex in vitro, and intracellular Ce6 distribution suggested ultrasound caused delivery through efficient sonoporation, while non-stimulated Ce6 uptake was inferred to be by endocytosis. Laser stimulation of the complex enhanced intracellular ROS production in vitro. When stimuli were combined, cell viability in vitro was significantly decreased (Figure 5D). In a mouse pancreas tumor model, these multitherapeutic complexes inhibited tumor growth significantly more effectively with both ultrasound and laser together than with either stimulus alone (Figure 5E) [206].

While the ultrasound sensitive micro-complex effectively improved chemical–PDT outcomes, potential adverse downstream effects can result from the generation of ROS in PDT. The effective cell death resulting from ROS may cause feedback effects, such as increasing tumor angiogenic signals that enhance growth, requiring an additional form of therapy to combat these effects, often gene therapy to attenuate VEGF upregulation. Further, ROS impede gene therapy efficacy, so delivery of the PDT photosensitizer and genetic material together necessitates a system for both delivery and nucleic acid protection. Current nanocarrier systems incorporating photosensitizers with VEGF inhibitors do not yet successfully undergo targeted delivery, with only 0.7% delivery of administered therapeutic molecules reaching the tumor. As such, the same research group later improved upon their ultrasound-controlled chemo–PDT by complexing the Ce6-loaded microbubble for PDT with a different nanoparticle carrier designed for optimal delivery and protection of genetic therapeutics (siRNA) to combat the unfavorable ROS effects from PDT.

The perfluoropropane microbubbles were loaded with hydrophobic photosensitizer Ce6 as before, while the therapeutic nanoparticles in this design consisted of siRNA hydrogels loaded with DOX and coated with PEI and HA polymers for protection. To synthesize these particles, anti-VEGF siRNA hydrogel nanoparticles were loaded with DOX before reacting with positively-charged PEI and HA–amine via electrostatic interactions to form a protective coating to prevent RNA denaturation. These amine-presenting nanoparticles were then covalently conjugated to Ce6-loaded, NHS-functionalized microbubbles via an amide linkage (Figure 5F). Physical separation of therapeutic components proved critical, as the siRNA was only effective in blocking the Ce6-induced increase in VEGF expression when delivered from within the separate, polymer-protected nanoparticle.

As before, ultrasound stimulation enhanced breast cancer cell uptake of DOX, Ce6, and siVEGF, and resulted in downregulation of VEGF expression and significantly decreased cell viability in vitro. In a rat cancer model, ultrasound treatment enhanced Ce6 and siRNA accumulation in tumor tissue, and siRNA reduced VEGF upregulation resulting from Ce6 + Laser (PDT) treatment, resulting in greater decreases in tumor growth in comparison to chemotherapy alone or chemo + PDT without siRNA for VEGF reduction (Figure 5G). Importantly, the combined therapy significantly attenuated upregulation of angiogenesis caused by chemo–PDT as evaluated by local blood vessel area. This combined therapy also did not cause major losses in body weight, indicating limited off-target effects. Here, ultrasound stimulation localized microbubble cavitation and rupture of the complexes to just the tumor site, successfully delivering the necessary components for DOX therapy, PDT, and angiogenesis suppression all at the designated location for highly effective therapy [207].

The above multi-stimulus systems show improved targeting and treatment efficacy in comparison to individual stimuli, and their results indicate that combining ultrasound with other on-demand stimuli is a promising future direction for ultrasound-mediated targeted delivery.

4.3. Microbubble–Nanoparticle Complex Linking Strategies

As described above, microbubbles and nanoparticles must be situated in close proximity for microbubble cavitation to cause delivery of nanoparticles to cells, and physically or chemically attaching nanoparticles to the microbubbles ensures this close interaction. As such, microbubble–nanoparticle conjugation is a critical design consideration for ultrasound-controlled nanocarrier delivery, and many linkage types have been employed, as summarized in Table 2. Linkages differ in affinity, and different conjugation strategies are influenced by the inherent properties and chemical groups present on both the microbubbles and nanoparticles themselves.

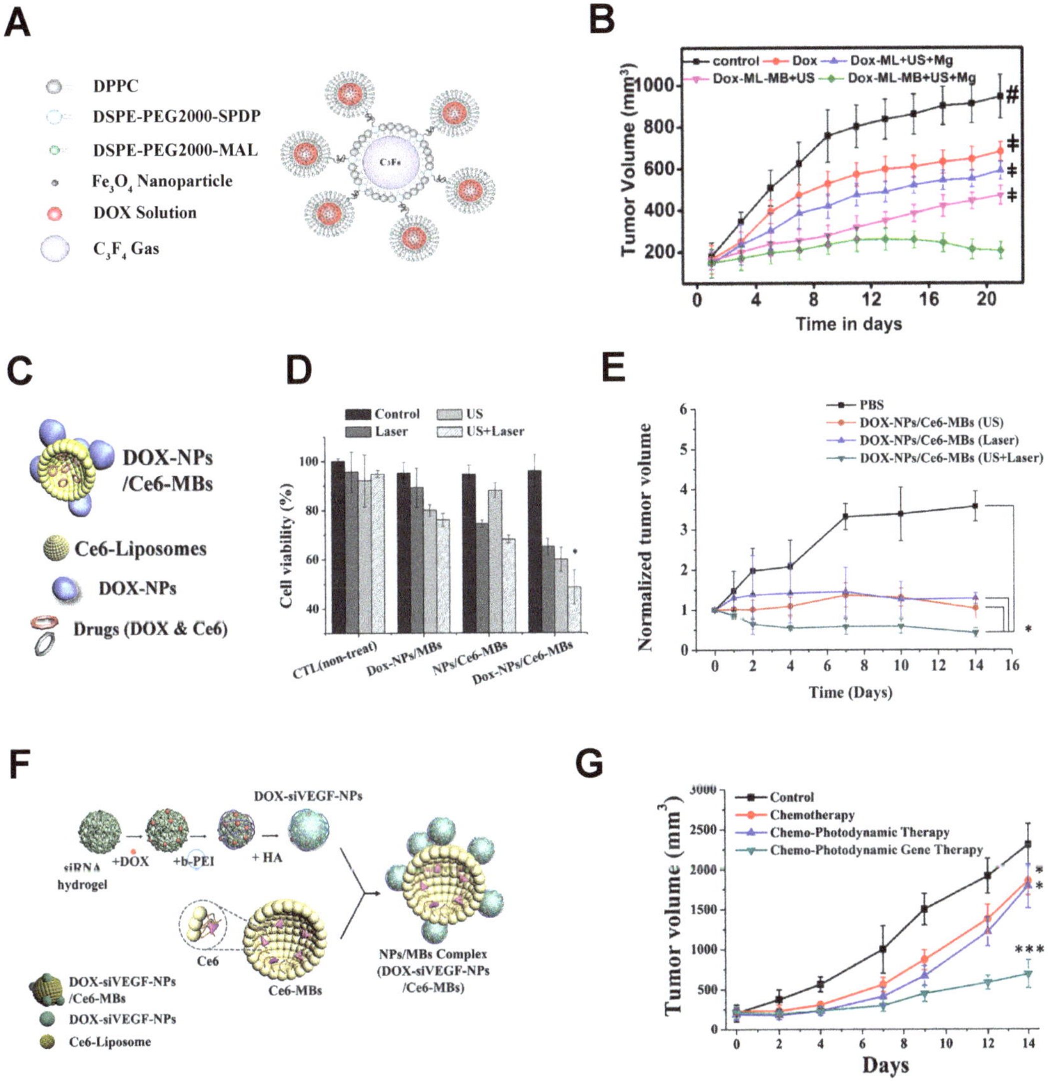

Figure 5. Combined external stimuli enhances cargo delivery from microbubble–nanoparticle complexes. (**A**). Magnetically-responsive iron oxide nanoparticle-containing liposomes were loaded with DOX and covalently linked to microbubbles. (**B**) Dual micro-complex stimulation by both magnetic field and ultrasound (DOX-ML-MB + US + Mg, green) had a greater inhibitory effect on tumor growth than either stimulus alone, ‡ $p < 0.05$, # $p < 0.001$. (**A**,**B**) Adapted with permission from [204], copyright 2020 American Chemical Society. (**C**) DOX-loaded albumin nanoparticles were linked to Ce6 photosensitizer-loaded microbubbles. (**D**) Combining ultrasound-mediated delivery of the DOX and Ce6 photosensitizer with PDT significantly improved treatment efficacy in vitro (DOX-NPs/Ce6-MBs + US + Laser) and (**E**) inhibition of tumor growth in vivo, * $p < 0.05$. (**C**–**E**) reprinted from [206], copyright 2018, with permission from Elsevier. (**F**) Polymer nanoparticles loaded with DOX and therapeutic siRNA were covalently linked to the Ce6-containing microbubbles. (**G**) The combination of chemotherapy, gene therapy, and photodynamic therapy with ultrasound targeting (green line) significantly inhibited tumor growth, * $p < 0.05$, *** $p < 0.001$. (**F**,**G**) reprinted from [207], copyright 2020, with permission from Elsevier.

A commonly-used method for particle linking harnesses the natural non-covalent binding reactions between biotin molecules and protein avidin [122]. This method is straightforward and well established. Both nanoparticles and microbubbles are functionalized with biotin, which does not affect the particle biochemical properties. The biotin groups are then linked together with soluble avidin. Avidin–biotin secondary binding, a naturally occurring reaction utilized for many lab assays, is the strongest non-covalent interaction, though its effectiveness is context-dependent [208]. It should be noted that, due to steric effects from the surface coatings of microbubbles and nanoparticles, a higher soluble avidin concentration is necessary for particle linking than for free biotin binding reactions. Additionally, avidin molecules are charged at biological pH and are, thus, highly repulsive of other avidin molecules, impeding the formation of adjacent biotin–avidin linkages. For improved binding efficiency, avidin may be replaced with neutravidin, a smaller molecule with a lower isoelectric point. Due to the size and lower repulsion between molecules, neutravidin has shown a three-fold higher binding efficiency to biotin than avidin [122,209,21 One challenge of employing this scheme for linking nanoparticles to microbubbles is that it is a time-consuming process that requires several washing steps, which may result in decreased yield. Additionally, these particles can cause immunogenic responses and some off-target binding, due to endogenous free avidin [149,208]. While employed in some recent complexes, this method was most widely used in early microbubble–nanoparticle complex designs, particularly microbubble-liposomes, perhaps due to the convenience of commercially available lipid–PEG–biotin molecules.

Electrostatic interactions have been implemented to directly link nanoparticles to microbubbles without a biotin–avidin bridge. This scheme is attractive for gene delivery complexes, as anionic nucleic acids may be complexed with cationic nanoparticles, often inherently charged polymers, for delivery. These nanoparticles bind to microbubbles with a negative surface charge (such as commercially available SonoVue® microbubbles) via electrostatic interactions [181]. Alternatively, anionic polymer nanoparticle prodrugs have been efficiently bound to cationic microbubbles, which were synthesized by incorporating cationic lipids into the microbubble lipid membrane. This secondary binding scheme works for microbubble–nanoparticle complexation, but binding stability can depend on physiological conditions. Covalent linkages are the strongest and most robust to environmental factors, making covalent bonding an appealing option for ensuring codelivery of microbubbles and nanoparticles as intact microbubble–nanoparticle complexes.

An early assessment of covalently linked microbubble–liposome complexes evaluated liposome binding efficiency and linkage stability. It was shown that liposomes attach to microbubbles via non-specific chain entanglement; however, significantly greater numbers of liposomes were loaded onto the microbubbles using covalent thiol–maleimide bonding [149]. Thiol–maleimide is an efficient Michael-type reaction that occurs very rapidly [211–213]. These covalent linkages were stable when exposed to blood plasma, suggesting their stability in circulation when injected intravenously [149]. When exposed to thiol compounds in biological environments, however, the thiol–maleimide bond may become unstable [213,214]. Similarly, disulfide bonds implemented in microbubble–liposome linkages are stable in circulation but are subject to degradation in reducing environments, which can cause microbubble–nanoparticle separation. In contrast, another "click-chemistry" reaction, used to link liposomes and microbubbles, was the strain-promoted azide–alkyne (SPAAC) DBCO–azide reaction, a relatively rapid reaction that occurs under physiological conditions and does not produce toxic byproducts. Though its reaction rate is slower than that of maleimide–thiol, it may provide greater stability in a range of biological conditions [148]. It can be seen in Table 2 that microbubble–liposome complexes can be bound via a variety of chemistries, as PEG-coated liposomes may be chemically functionalized with different reactive linking groups. Other nanoparticles, such as polymer and lipid–polymer hybrid nanoparticles, may also be modified in order to link to microbubbles. Some nanoparticles, however, inherently present chemical moieties capable of forming covalent bonds.

Two methods take advantage of existing NHS-reactive functional groups on nanoparticles. PLGA-based nanoparticle polymer chain ends contain carboxylic acid groups, which can be activated to undergo carbodiimide chemistry (COOH + NHS reaction), forming a covalent amide bond with NHS-functionalized microbubbles [208]. Similarly, albumin nanoparticles present primary amine groups, which again can form covalent bonds with NHS-functionalized microbubbles. This method of linkage was employed for all albumin nanoparticle-based complexes surveyed in this review [187–189,206,215].

Alternative complex designs implement relative spatial arrangements to combine particles with microbubbles instead of linking their surfaces together. In one design, ultrasound stimulation of microbubbles encapsulated within liposomes resulted in internal bubble fragmentation causing the liposome membrane to rupture and release the housed payload [156]. Polymer nanoparticles that self-arrange into microbubbles can be created through high-speed mixing of gas, nanoparticle solution, and a protein. These nanoparticles can be successfully delivered across the BBB and can effectively deliver drugs to specified sites [216–218]. Metallic or inorganic nanoparticles can instead be incorporated within the microbubble lipid layer, and these complexes have also demonstrated effective targeted delivery under ultrasound-induced microbubble cavitation [199–203].

Table 2. Microbubble–nanoparticle (NP) linking strategies.

Type of Linkage	Schematic	Notable Characteristics	References
Avidin–biotin bridging		Strong interactions between biotin and avidin, very commonly used, binding efficiency increased with neutravidin. Synthesis is time consuming and results in material waste. Can stimulate immunogenic reactions	Liposomes: [122,146,147,151,153,154] Polymer NP [176,178] LPHN: [183]
Electrostatic bonds	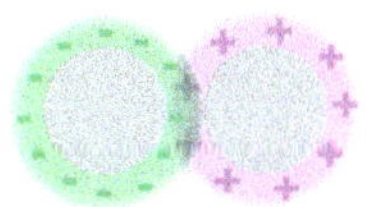	Appropriate for charged nanoparticles, used for cationic polymer nanoparticles for nonviral gene delivery. SonoVue® microbubbles designed for this linking method	Polymer NP: [180,181]
Disulfide bonds	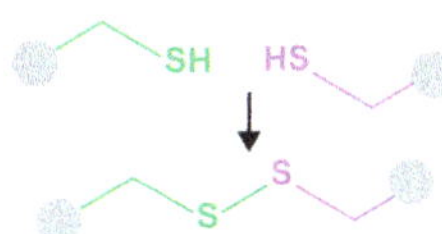	Covalent bonds that can be reversible	Liposomes: [155]
Maleimide–thiol bonds	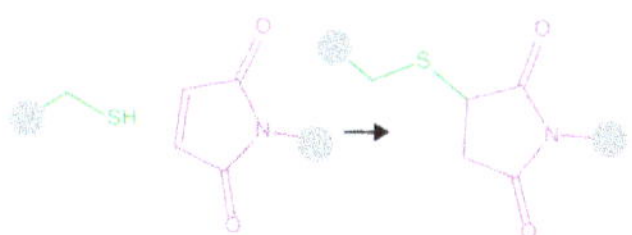	Rapid, biocompatible reaction. Cysteines in proteins have thiol groups on sidechains which may cause non-specific binding, also may cleave at physiological conditions	Liposomes: [155,204,219]
DBCO–azide SPAAC (strain-promoted alkyne–azide cycloaddition)	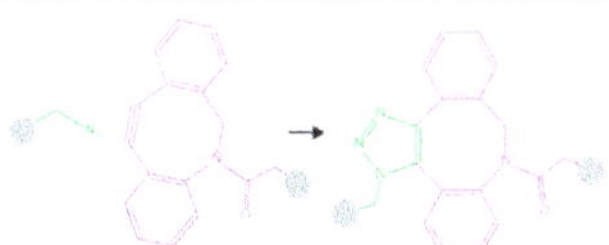	Rapid, nontoxic biorthogonal reaction forms covalent bonds that are stable under biological conditions	Liposomes: [148]

Table 2. *Cont.*

Type of Linkage	Schematic	Notable Characteristics	References
Peptide bonds	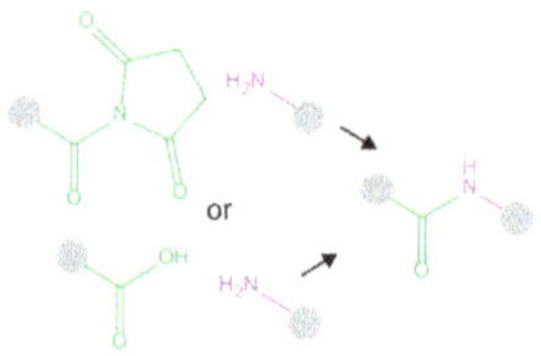	Amide bond formation between Amine and NHS or COOH requires chemical activation, forms stabile covalent linkage. Commonly implemented with albumin nanoparticles which possess surface amines	NHS: Protein NPs: [187–189,206,215] NHS: Polymer NP: [207] COOH: Polymer NP: [174,175]
Other complexation strategies	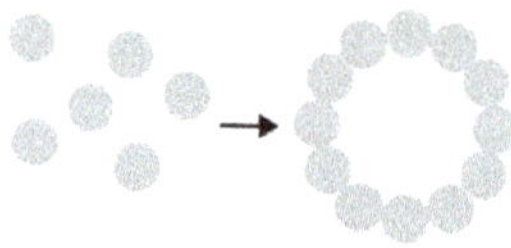	Multiple nanoparticles can be arranged around a gas compartment or embedded in the microbubble shell	Polymer NP: [216–218] Metallic/Inorganic NP: [199–203]

5. Conclusions and Future Perspectives

Due to limited efficacy of systemic drug delivery and potentially harmful off-target effects of various therapeutics, methods to target drug delivery vehicles directly to diseased tissues remain a clinically-relevant area of active development. On-demand external stimulation of cargo release provides advantageous spatiotemporal control over drug delivery. Ultrasound-stimulated targeted delivery can facilitate controlled delivery through use of applied pressure waves that cause echogenic microbubble cavitation, resulting in the release of attached nanoparticle payload. This attractive option is biocompatible and noninvasive, and the ultrasound can focus to greater penetration depths in tissues compared to other user-controlled targeting stimuli, such as light [15,16]. Ultrasound-mediated drug delivery is enhanced when ultrasound-sensitive bubbles and drug nanocarriers are combined into one complex, and many material options are available for both components.

In this article, we reviewed current designs for microbubble–nanoparticle complexes, as well as the different schemes for linking particles and microbubbles together. In designing these complexes, the class of nanoparticle used may be selected for particular applications, including fragile or highly toxic cargo, hydrophobicity or charge of cargo, or incorporation of other targeting design elements. Though variation exists within classes, properties relate to nanoparticle type. Liposomes boast facile fabrication and the ability to house both hydrophilic and hydrophobic cargo, but are less stable than other carrier types. Polymer-based nanoparticles are robust in circulation and allow customization. Lipid–polymer hybrid nanoparticles combine the advantages of both molecule types to provide protection and stability to cargo and can deliver multiple cargo types with different properties simultaneously or at different timescales. Protein nanoparticles are stable across a broad pH and temperature range and inherently present functional groups for loading cargo, linking to microbubbles, and for bioactive targeting. Metallic nanoparticles are stable and provide additional targeting layers for thermal ablation therapy or magnetic accumulation.

Like nanoparticle types and properties, microbubble–nanoparticle linkage strategies vary in technical complexity, biocompatibility, and strength. The linkage strategy depends on the application and inherent properties of each component, though both microbubbles and nanoparticles may be modified with different functional groups for specific linkages. Finally, a few systems have successfully combined the nanocarriers and ultrasound sensitive species in unique ways, such as nesting bubbles within nanocarriers [156], or forming microbubbles from assembled spheres of nanoparticles [216–218].

Across different complexation designs, nanoparticle classes, and linkages, ultrasound exposure consistently enhanced targeted delivery via increased in vitro cell uptake of cargo. These results may be largely attributed to local sonoporation, transient formation of pores in cell membranes caused by microbubble cavitation that facilitates direct cellular uptake of released cargo. Ultrasound stimulation was also correlated to improved biodistribution in animal models, showing accumulation of cargo at targeted locations and diminished off-target release and side effects. This was frequently linked to improved therapeutic functional outcomes. Other targeting components, such as nanoparticle size, active targeting moieties, pH-responsive elements, and additional external stimuli, further increased the targeting efficiency of ultrasound-responsive drug delivery complexes. These results are highly promising, and there are many avenues for further advancement of this technology.

As supported by the results described in Section 4.2, a potential future direction for optimized targeting of ultrasound-responsive nanocarrier complexes is the combination of ultrasound with other targeting elements for synergistic targeting precision. Adding ultrasound responsiveness to nanoparticles designed for passive or active targeting has shown improved spatial targeting and efficacy across many studies, and more recent experiments combining ultrasound with light or magnetic external stimuli further improved delivery efficiency and therapeutic efficacy [204,206,207]. Additional combinations of targeting elements, particularly user-controlled stimuli, can enhance precision of targeting, improving therapeutic efficacy [129]. Another future prospect is to expand the range of different nanoparticle types that can be complexed with microbubbles, as well as the amount of cargo that can be carried by these complexes. New nanoparticles, such as supramolecular structures, may improve cargo loading efficiency, leading to greater therapeutic effects, and expand the types of therapy that can be administered with targeted ultrasound [220,221]. Metallic nanoparticles, comprised of gold or iron oxide, have been incorporated within microbubbles to enhance imaging contrast [192–195]. Direct delivery of metallic nanoparticles by ultrasound from microbubble–nanoparticle complexes for targeted therapy is in early development [199]. These nanoparticles are excellent candidates for ultrasound-targeted delivery, due to presenting solubility and in vivo toxicity complications as free particles [105]. Microbubble–nanoparticle complexes may also be implemented to deliver therapeutic gases to targeted tissues, or to deliver other molecules to specific tissue environments to alter the local extracellular matrix via crosslinking or degradation, rather than targeting resident cells directly. The results outlined in this review highlight the potential of this targeting strategy and motivate future work in translational applications towards clinical trials and clinical implementation. Currently, microbubbles are clinically approved for use as contrast agents [222,223] and some nanoparticles (liposomes, PLGA nanoparticles) have been implemented in the clinic [120,224]. A few studies have tested microbubbles co-delivered with free nanoparticles or free drug for cargo delivery in large animal studies and clinical trials, finding limited adverse effects and some therapeutic efficacy [225–230], but linked microbubble–nanoparticle complexes have not yet been widely tested as drug carriers in large animal models or clinical trials. In vitro results suggest that complexed microbubble–nanoparticle platforms can enhance therapeutic delivery efficacy over component co-delivery [117–119]. A small number of large animal studies of liposome–microbubble complexes show effective circulation, but therapeutic efficacy was not evaluated [231]. Thus, long term future work consists of moving ultrasound-responsive microbubble–nanoparticle complexes into additional large animal biosafety and efficacy studies before use in clinical trials. One consideration that should be noted for clinical implementation is the limited circulation time for the microbubble–nanoparticle complexes, due to their micron-scale size, which is in the order of <10 min and affects the window during which the complexes can be activated [232,233]. Activation of these particles necessitates training of skilled ultrasound operators to successfully aim and precisely stimulate the targeted tissue. Tailored protocols with adjusted ultrasound parameters must be developed for ultrasound stimulation for this specific application [234]. A major economic feasibility advantage of these drug delivery systems is that they use ultrasound technology and equipment that is already present in many clinics. Overall, ultrasound-directed drug delivery by microbubble–nanoparticle complexes is a valuable advancing technology that has the potential to enhance therapeutic efficacy and improve patient clinical outcomes.

Author Contributions: Conceptualization, C.E.S. and R.C.; information curation, R.C. and K.T.H.; writing—original draft preparation, R.C. and K.T.H.; writing—review and editing, R.C. and C.E.S.; original schematics, R.C.; figure preparation, R.C. and K.T.H.; supervision, C.E.S. All authors have read and agreed to the published version of the manuscript.

Funding: This work was supported by funding (Exploratory 2022-1537, Full 2021-1308) from the Cancer Early Detection Advanced Research Center at Oregon Health & Science University's Knight Cancer Institute.

Conflicts of Interest: The authors declare no conflict of interest.

References

1. Torchilin, V.P. Passive and active drug targeting: Drug delivery to tumors as an example. *Handb. Exp. Pharmacol.* **2010**, *197*, 3–53. [CrossRef]
2. Poon, K.A.; Flagella, K.; Beyer, J.; Tibbitts, J.; Kaur, S.; Saad, O.; Yi, J.H.; Girish, S.; Dybdal, N.; Reynolds, T. Preclinical safety profile of trastuzumab emtansine (T-DM1): Mechanism of action of its cytotoxic component retained with improved tolerability. *Toxicol. Appl. Pharmacol.* **2013**, *273*, 298–313. [CrossRef] [PubMed]
3. Liu, D.; Auguste, D.T. Cancer targeted therapeutics: From molecules to drug delivery vehicles. *J. Control. Release* **2015**, *219*, 632–643. [CrossRef] [PubMed]
4. Danhier, F.; Feron, O.; Preat, V. To exploit the tumor microenvironment: Passive and active tumor targeting of nanocarriers for anti-cancer drug delivery. *J. Control. Release* **2010**, *148*, 135–146. [CrossRef] [PubMed]
5. Kamaly, N.; Xiao, Z.; Valencia, P.M.; Radovic-Moreno, A.F.; Farokhzad, O.C. Targeted polymeric therapeutic nanoparticles: Design, development and clinical translation. *Chem. Soc. Rev.* **2012**, *41*, 2971–3010. [CrossRef] [PubMed]
6. Banerjee, A.; Pathak, S.; Subramanium, V.D.; Dharanivasan, G.; Murugesan, R.; Verma, R.S. Strategies for targeted drug delivery in treatment of colon cancer: Current trends and future perspectives. *Drug Discov. Today* **2017**, *22*, 1224–1232. [CrossRef] [PubMed]
7. Srinivasarao, M.; Low, P.S. Ligand-Targeted Drug Delivery. *Chem. Rev.* **2017**, *117*, 12133–12164. [CrossRef] [PubMed]
8. Hudis, C.A. Trastuzumab—Mechanism of action and use in clinical practice. *N. Eng. J. Med.* **2007**, *357*, 39–51. [CrossRef]
9. Wiley, D.T.; Webster, P.; Gale, A.; Davis, M.E. Transcytosis and brain uptake of transferrin-containing nanoparticles by tuning avidity to transferrin receptor. *Proc. Natl. Acad. Sci. USA* **2013**, *110*, 8662–8667. [CrossRef]
10. Shahriari, M.; Zahiri, M.; Abnous, K.; Taghdisi, S.M.; Ramezani, M.; Alibolandi, M. Enzyme responsive drug delivery systems in cancer treatment. *J. Control. Release* **2019**, *308*, 172–189. [CrossRef] [PubMed]
11. Mura, S.; Nicolas, J.; Couvreur, P. Stimuli-responsive nanocarriers for drug delivery. *Nat. Mater.* **2013**, *12*, 991–1003. [CrossRef] [PubMed]
12. Ding, Z.; He, K.; Duan, Y.; Shen, Z.; Cheng, J.; Zhang, G.; Hu, J. Photo-degradable micelles for co-delivery of nitric oxide and doxorubicin. *J. Mater. Chem. B* **2020**, *8*, 7009–7017. [CrossRef] [PubMed]
13. Schutt, C.; Ibsen, S.; Zahavy, E.; Aryal, S.; Kuo, S.; Esener, S.; Berns, M.; Esener, S. Drug Delivery Nanoparticles with Locally Tunable Toxicity Made Entirely from a Light-Activatable Prodrug of Doxorubicin. *Pharm. Res.* **2017**, *34*, 2025–2035. [CrossRef] [PubMed]
14. Oliveira, H.; Perez-Andres, E.; Thevenot, J.; Sandre, O.; Berra, E.; Lecommandoux, S. Magnetic field triggered drug release from polymersomes for cancer therapeutics. *J. Control. Release* **2013**, *169*, 165–170. [CrossRef] [PubMed]
15. Stolik, S.; Delgado, J.A.; Perez, A.; Anasagasti, L. Measurement of the penetration depths of red and near infrared light in human "ex vivo" tissues. *J. Photochem. Photobiol. B* **2000**, *57*, 90–93. [CrossRef]
16. Qian, X.; Zheng, Y.; Chen, Y. Micro/Nanoparticle-Augmented Sonodynamic Therapy (SDT): Breaking the Depth Shallow of Photoactivation. *Adv. Mater.* **2016**, *28*, 8097–8129. [CrossRef]
17. Pitt, W.G.; Husseini, G.A.; Staples, B.J. Ultrasonic drug delivery—A general review. *Expert Opin. Drug Deliv* **2004**, *1*, 37–56. [CrossRef] [PubMed]
18. Gelmi, A.; Schutt, C.E. Stimuli-Responsive Biomaterials: Scaffolds for Stem Cell Control. *Adv. Healthc. Mater.* **2021**, *10*, e2001125. [CrossRef] [PubMed]
19. de Leon, A.; Perera, R.; Nittayacharn, P.; Cooley, M.; Jung, O.; Exner, A.A. Ultrasound Contrast Agents and Delivery Systems in Cancer Detection and Therapy. *Adv. Cancer Res.* **2018**, *139*, 57–84. [CrossRef] [PubMed]
20. Seah, B.C.; Teo, B.M. Recent advances in ultrasound-based transdermal drug delivery. *Int. J. Nanomed.* **2018**, *13*, 7749–7763. [CrossRef] [PubMed]
21. Rodríguez Valiente, A.; Trinidad, A.; García Berrocal, J.; Górriz, C.; Ramírez Camacho, R. Extended high-frequency (9–20 kHz) audiometry reference thresholds in 645 healthy subjects. *Int. J. Audiol.* **2014**, *53*, 531–545. [CrossRef] [PubMed]
22. Abu-Zidan, F.M.; Hefny, A.F.; Corr, P. Clinical ultrasound physics. *J. Emerg. Trauma Shock* **2011**, *4*, 501–503. [CrossRef] [PubMed]
23. Aldrich, J.E. Basic physics of ultrasound imaging. *Crit. Care Med.* **2007**, *35*, S131–S137. [CrossRef]
24. Chan, V.; Perlas, A. Basics of ultrasound imaging. In *Atlas of Ultrasound-Guided Procedures in Interventional Pain Management*; Springer: Berlin/Heidelberg, Germany, 2011; pp. 13–19.
25. Sirsi, S.R.; Borden, M.A. State-of-the-art materials for ultrasound-triggered drug delivery. *Adv. Drug Deliv. Rev.* **2014**, *72*, 3–14. [CrossRef] [PubMed]

26. Mitragotri, S. Healing sound: The use of ultrasound in drug delivery and other therapeutic applications. *Nat. Rev. Drug Discov.* **2005**, *4*, 255–260. [CrossRef]
27. Baek, H.; Pahk, K.J.; Kim, H. A review of low-intensity focused ultrasound for neuromodulation. *Biomed. Eng. Lett.* **2017**, *7*, 135–142. [CrossRef] [PubMed]
28. Elhelf, I.A.S.; Albahar, H.; Shah, U.; Oto, A.; Cressman, E.; Almekkawy, M. High intensity focused ultrasound: The fundamentals, clinical applications and research trends. *Diagn. Interv. Imaging* **2018**, *99*, 349–359. [CrossRef] [PubMed]
29. Phenix, C.P.; Togtema, M.; Pichardo, S.; Zehbe, I.; Curiel, L. High intensity focused ultrasound technology, its scope and applications in therapy and drug delivery. *J. Pharm. Sci.* **2014**, *17*, 136–153. [CrossRef]
30. Chowdhury, S.M.; Abou-Elkacem, L.; Lee, T.; Dahl, J.; Lutz, A.M. Ultrasound and microbubble mediated therapeutic delivery: Underlying mechanisms and future outlook. *J. Control. Release* **2020**, *326*, 75–90. [CrossRef] [PubMed]
31. Qin, S.; Caskey, C.F.; Ferrara, K.W. Ultrasound contrast microbubbles in imaging and therapy: Physical principles and engineering. *Phys. Med. Biol.* **2009**, *54*, R27–R57. [CrossRef] [PubMed]
32. Zheng, H.; Dayton, P.A.; Caskey, C.; Zhao, S.; Qin, S.; Ferrara, K.W. Ultrasound-Driven Microbubble Oscillation and Translation Within Small Phantom Vessels. *Ultrasound Med. Biol.* **2007**, *33*, 1978–1987. [CrossRef] [PubMed]
33. Schutt, C.E.; Ibsen, S.D.; Benchimol, M.J.; Hsu, M.J.; Esener, S.C. Manipulating nanoscale features on the surface of dye-loaded microbubbles to increase their ultrasound-modulated fluorescence output. *Small* **2014**, *10*, 3316–3324. [CrossRef] [PubMed]
34. Schutt, C.; Benchimol, M.; Hsu, M.; Esener, S. *Ultrasound-Modulated Fluorescent Contrast Agent for Optical Imaging through Turbid Media*; SPIE: Philadelphia, PA, USA, 2011; Volume 8165.
35. Schutt, C.E.; Ibsen, S.; Benchimol, M.; Hsu, M.; Esener, S. Optical detection of harmonic oscillations in fluorescent dye-loaded microbubbles ensonified by ultrasound. *Opt. Lett.* **2015**, *40*, 2834–2837. [CrossRef] [PubMed]
36. Humphrey, V.F. Ultrasound and matter—Physical interactions. *Prog. Biophys. Mol. Biol.* **2007**, *93*, 195–211. [CrossRef] [PubMed]
37. Chong, W.K.; Papadopoulou, V.; Dayton, P.A. Imaging with ultrasound contrast agents: Current status and future. *Abdom. Radiol.* **2018**, *43*, 762–772. [CrossRef] [PubMed]
38. Fan, Z.; Chen, D.; Deng, C.X. Characterization of the dynamic activities of a population of microbubbles driven by pulsed ultrasound exposure in sonoporation. *Ultrasound Med. Biol.* **2014**, *40*, 1260–1272. [CrossRef]
39. de Jong, N.; Emmer, M.; van Wamel, A.; Versluis, M. Ultrasonic characterization of ultrasound contrast agents. *Med. Biol. Eng. Comput.* **2009**, *47*, 861–873. [CrossRef]
40. Schutt, C.E.; Ibsen, S.D.; Thrift, W.; Esener, S.C. The influence of distance between microbubbles on the fluid flow produced during ultrasound exposure. *J. Acoust. Soc. Am.* **2014**, *136*, 3422. [CrossRef]
41. He, J.; Liu, Z.; Zhu, X.; Xia, H.; Gao, H.; Lu, J. Ultrasonic Microbubble Cavitation Enhanced Tissue Permeability and Drug Diffusion in Solid Tumor Therapy. *Pharmaceutics* **2022**, *14*, 1642. [CrossRef]
42. Postema, M. *Fundamentals of Medical Ultrasonics*; CRC Press: Boca Raton, FL, USA, 2011.
43. Wu, J. Shear stress in cells generated by ultrasound. *Prog. Biophys. Mol. Biol.* **2007**, *93*, 363–373. [CrossRef]
44. Doinikov, A.A.; Bouakaz, A. Theoretical investigation of shear stress generated by a contrast microbubble on the cell membrane as a mechanism for sonoporation. *J. Acoust. Soc. Am.* **2010**, *128*, 11–19. [CrossRef] [PubMed]
45. Wu, J.; Nyborg, W.L. Ultrasound, cavitation bubbles and their interaction with cells. *Adv. Drug Deliv. Rev.* **2008**, *60*, 1103–1116. [CrossRef] [PubMed]
46. Mehier-Humbert, S.; Bettinger, T.; Yan, F.; Guy, R.H. Plasma membrane poration induced by ultrasound exposure: Implication for drug delivery. *J. Control. Release* **2005**, *104*, 213–222. [CrossRef]
47. Deng, C.X.; Sieling, F.; Pan, H.; Cui, J. Ultrasound-induced cell membrane porosity. *Ultrasound Med. Biol.* **2004**, *30*, 519–526. [CrossRef] [PubMed]
48. Schlicher, R.K.; Radhakrishna, H.; Tolentino, T.P.; Apkarian, R.P.; Zarnitsyn, V.; Prausnitz, M.R. Mechanism of intracellular delivery by acoustic cavitation. *Ultrasound Med. Biol.* **2006**, *32*, 915–924. [CrossRef] [PubMed]
49. Helfield, B. A Review of Phospholipid Encapsulated Ultrasound Contrast Agent Microbubble Physics. *Ultrasound Med. Biol.* **2019**, *45*, 282–300. [CrossRef]
50. Yang, F.; Gu, N.; Chen, D.; Xi, X.; Zhang, D.; Li, Y.; Wu, J. Experimental study on cell self-sealing during sonoporation. *J. Control. Release* **2008**, *131*, 205–210. [CrossRef]
51. Zhou, Y.; Shi, J.; Cui, J.; Deng, C.X. Effects of extracellular calcium on cell membrane resealing in sonoporation. *J. Control. Release* **2008**, *126*, 34–43. [CrossRef]
52. Fan, Z.; Kumon, R.E.; Deng, C.X. Mechanisms of microbubble-facilitated sonoporation for drug and gene delivery. *Ther. Deliv.* **2014**, *5*, 467–486. [CrossRef]
53. Wang, M.; Zhang, Y.; Cai, C.; Tu, J.; Guo, X.; Zhang, D. Sonoporation-induced cell membrane permeabilization and cytoskeleton disassembly at varied acoustic and microbubble-cell parameters. *Sci. Rep.* **2018**, *8*, 3885. [CrossRef]
54. Arif, W.M.; Elsinga, P.H.; Gasca-Salas, C.; Versluis, M.; Martinez-Fernandez, R.; Dierckx, R.; Borra, R.J.H.; Luurtsema, G. Focused ultrasound for opening blood-brain barrier and drug delivery monitored with positron emission tomography. *J. Control. Release* **2020**, *324*, 303–316. [CrossRef] [PubMed]
55. Chen, H.; Konofagou, E.E. The size of blood-brain barrier opening induced by focused ultrasound is dictated by the acoustic pressure. *J. Cereb. Blood Flow Metab.* **2014**, *34*, 1197–1204. [CrossRef] [PubMed]

56. Meng, Y.; Hynynen, K.; Lipsman, N. Applications of focused ultrasound in the brain: From thermoablation to drug delivery. *Nat. Rev. Neurol.* **2021**, *17*, 7–22. [CrossRef]
57. Ibsen, S.; Tong, A.; Schutt, C.; Esener, S.; Chalasani, S.H. Sonogenetics is a non-invasive approach to activating neurons in Caenorhabditis elegans. *Nat. Commun.* **2015**, *6*, 8264. [CrossRef] [PubMed]
58. Wrenn, S.P.; Dicker, S.M.; Small, E.F.; Dan, N.R.; Mleczko, M.; Schmitz, G.; Lewin, P.A. Bursting bubbles and bilayers. *Theranostics* **2012**, *2*, 1140–1159. [CrossRef] [PubMed]
59. Upadhyay, A.; Dalvi, S.V. Microbubble formulations: Synthesis, stability, modeling and biomedical applications. *Ultrasound Med. Biol.* **2019**, *45*, 301–343. [CrossRef]
60. Dewitte, H.; Van Lint, S.; Heirman, C.; Thielemans, K.; De Smedt, S.C.; Breckpot, K.; Lentacker, I. The potential of antigen and TriMix sonoporation using mRNA-loaded microbubbles for ultrasound-triggered cancer immunotherapy. *J. Control. Release* **2014**, *194*, 28–36. [CrossRef]
61. Bae, Y.J.; Yoon, Y.I.; Yoon, T.J.; Lee, H.J. Ultrasound-Guided Delivery of siRNA and a Chemotherapeutic Drug by Using Microbubble Complexes: In Vitro and In Vivo Evaluations in a Prostate Cancer Model. *Korean J. Radiol.* **2016**, *17*, 497–508. [CrossRef]
62. Chen, S.; Chen, J.; Meng, X.L.; Shen, J.S.; Huang, J.; Huang, P.; Pu, Z.; McNeill, N.H.; Grayburn, P.A. ANGPTL8 reverses established adriamycin cardiomyopathy by stimulating adult cardiac progenitor cells. *Oncotarget* **2016**, *7*, 80391–80403. [CrossRef]
63. Fu, Y.Y.; Zhang, L.; Yang, Y.; Liu, C.W.; He, Y.N.; Li, P.; Yu, X. Synergistic antibacterial effect of ultrasound microbubbles combined with chitosan-modified polymyxin B-loaded liposomes on biofilm-producing Acinetobacter baumannii. *Int. J. Nanomed.* **2019**, *14*, 1805–1815. [CrossRef]
64. Kilroy, J.P.; Klibanov, A.L.; Wamhoff, B.R.; Bowles, D.K.; Hossack, J.A. Localized in vivo model drug delivery with intravascular ultrasound and microbubbles. *Ultrasound Med. Biol.* **2014**, *40*, 2458–2467. [CrossRef] [PubMed]
65. Bioley, G.; Lassus, A.; Terrettaz, J.; Tranquart, F.; Corthesy, B. Long-term persistence of immunity induced by OVA-coupled gas-filled microbubble vaccination partially protects mice against infection by OVA-expressing Listeria. *Biomaterials* **2015**, *57*, 153–160. [CrossRef] [PubMed]
66. Dong, Y.; Li, J.; Li, P.; Yu, J. Ultrasound Microbubbles Enhance the Activity of Vancomycin Against Staphylococcus epidermidis Biofilms In Vivo. *J. Ultrasound Med.* **2018**, *37*, 1379–1387. [CrossRef] [PubMed]
67. Goyal, A.; Yu, F.T.H.; Tenwalde, M.G.; Chen, X.; Althouse, A.; Villanueva, F.S.; Pacella, J.J. Inertial Cavitation Ultrasound with Microbubbles Improves Reperfusion Efficacy When Combined with Tissue Plasminogen Activator in an In Vitro Model of Microvascular Obstruction. *Ultrasound Med. Biol.* **2017**, *43*, 1391–1400. [CrossRef] [PubMed]
68. Ferrara, K.W.; Borden, M.A.; Zhang, H. Lipid-shelled vehicles: Engineering for ultrasound molecular imaging and drug delivery. *Acc. Chem. Res.* **2009**, *42*, 881–892. [CrossRef] [PubMed]
69. Langeveld, S.A.G.; Beekers, I.; Collado-Lara, G.; van der Steen, A.F.W.; de Jong, N.; Kooiman, K. The Impact of Lipid Handling and Phase Distribution on the Acoustic Behavior of Microbubbles. *Pharmaceutics* **2021**, *13*, 119. [CrossRef] [PubMed]
70. Rossi, S.; Szíjjártó, C.; Gerber, F.; Waton, G.; Krafft, M.P. Fluorous materials in microbubble engineering science and technology—Design and development of new bubble preparation and sizing technologies. *J. Fluor. Chem.* **2011**, *132*, 1102–1109. [CrossRef]
71. Schutt, E.G.; Klein, D.H.; Mattrey, R.M.; Riess, J.G. Injectable microbubbles as contrast agents for diagnostic ultrasound imaging: The key role of perfluorochemicals. *Angew. Chem. Int. Ed. Eng.* **2003**, *42*, 3218–3235. [CrossRef] [PubMed]
72. Sarkar, K.; Katiyar, A.; Jain, P. Growth and dissolution of an encapsulated contrast microbubble: Effects of encapsulation permeability. *Ultrasound Med. Biol.* **2009**, *35*, 1385–1396. [CrossRef] [PubMed]
73. Abou-Saleh, R.H.; Peyman, S.A.; Johnson, B.R.; Marston, G.; Ingram, N.; Bushby, R.; Coletta, P.L.; Markham, A.F.; Evans, S.D. The influence of intercalating perfluorohexane into lipid shells on nano and microbubble stability. *Soft Matter* **2016**, *12*, 7223–7230. [CrossRef]
74. Shi, D.; Wallyn, J.; Nguyen, D.-V.; Perton, F.; Felder-Flesch, D.; Bégin-Colin, S.; Maaloum, M.; Krafft, M.P. Microbubbles decorated with dendronized magnetic nanoparticles for biomedical imaging: Effective stabilization via fluorous interactions. *Beilstein J. Nanotechnol.* **2019**, *10*, 2103–2115. [CrossRef]
75. Horsley, H.; Owen, J.; Browning, R.; Carugo, D.; Malone-Lee, J.; Stride, E.; Rohn, J.L. Ultrasound-activated microbubbles as a novel intracellular drug delivery system for urinary tract infection. *J. Control. Release* **2019**, *301*, 166–175. [CrossRef] [PubMed]
76. Dixon, A.J.; Li, J.; Rickel, J.-M.R.; Klibanov, A.L.; Zuo, Z.; Hossack, J.A. Efficacy of Sonothrombolysis Using Microbubbles Produced by a Catheter-Based Microfluidic Device in a Rat Model of Ischemic Stroke. *Ann. Biomed. Eng.* **2019**, *47*, 1012–1022. [CrossRef] [PubMed]
77. Fix, S.M.; Papadopoulou, V.; Velds, H.; Kasoji, S.K.; Rivera, J.N.; Borden, M.A.; Chang, S.; Dayton, P.A. Oxygen microbubbles improve radiotherapy tumor control in a rat fibrosarcoma model—A preliminary study. *PLoS ONE* **2018**, *13*, e0195667. [CrossRef]
78. Song, L.; Wang, G.; Hou, X.; Kala, S.; Qiu, Z.; Wong, K.F.; Cao, F.; Sun, L. Biogenic nanobubbles for effective oxygen delivery and enhanced photodynamic therapy of cancer. *Acta BioMater.* **2020**, *108*, 313–325. [CrossRef] [PubMed]
79. Hall, R.L.; Juan-Sing, Z.D.; Hoyt, K.; Sirsi, S.R. Formulation and Characterization of Chemically Cross-linked Microbubble Clusters. *Langmuir* **2019**, *35*, 10977–10986. [CrossRef]
80. Pancholi, K.; Stride, E.; Edirisinghe, M. Dynamics of bubble formation in highly viscous liquids. *Langmuir* **2008**, *24*, 4388–4393. [CrossRef] [PubMed]

81. Mahalingam, S.; Meinders, M.B.; Edirisinghe, M. Formation, stability, and mechanical properties of bovine serum albumin stabilized air bubbles produced using coaxial electrohydrodynamic atomization. *Langmuir* **2014**, *30*, 6694–6703. [CrossRef]
82. Mahalingam, S.; Raimi-Abraham, B.T.; Craig, D.Q.; Edirisinghe, M. Formation of protein and protein-gold nanoparticle stabilized microbubbles by pressurized gyration. *Langmuir* **2015**, *31*, 659–666. [CrossRef]
83. Lee, M.; Lee, E.Y.; Lee, D.; Park, B.J. Stabilization and fabrication of microbubbles: Applications for medical purposes and functional materials. *Soft Matter* **2015**, *11*, 2067–2079. [CrossRef]
84. Stride, E.; Segers, T.; Lajoinie, G.; Cherkaoui, S.; Bettinger, T.; Versluis, M.; Borden, M. Microbubble agents: New directions. *Ultrasound Med. Biol.* **2020**, *46*, 1326–1343. [CrossRef] [PubMed]
85. Pulsipher, K.W.; Hammer, D.A.; Lee, D.; Sehgal, C.M. Engineering theranostic microbubbles using microfluidics for ultrasound imaging and therapy: A review. *Ultrasound Med. Biol.* **2018**, *44*, 2441–2460. [CrossRef]
86. Carugo, D.; Browning, R.J.; Iranmanesh, I.; Messaoudi, W.; Rademeyer, P.; Stride, E. Scaleable production of microbubbles using an ultrasound-modulated microfluidic device. *J. Acoust. Soc. Am.* **2021**, *150*, 1577. [CrossRef]
87. Khismatullin, D.B. Resonance frequency of microbubbles: Effect of viscosity. *J. Acoust. Soc. Am.* **2004**, *116*, 1463–1473. [CrossRef]
88. Koczera, P.; Appold, L.; Shi, Y.; Liu, M.; Dasgupta, A.; Pathak, V.; Ojha, T.; Fokong, S.; Wu, Z.; Van Zandvoort, M. PBCA-based polymeric microbubbles for molecular imaging and drug delivery. *J. Control. Release* **2017**, *259*, 128–135. [CrossRef] [PubMed]
89. Barmin, R.A.; Dasgupta, A.; Bastard, C.; De Laporte, L.; Rütten, S.; Weiler, M.; Kiessling, F.; Lammers, T.; Pallares, R.M. Engineering the Acoustic Response and Drug Loading Capacity of PBCA-Based Polymeric Microbubbles with Surfactants. *Mol. Pharm.* **2022**, *19*, 3256–3266. [CrossRef] [PubMed]
90. Lentacker, I.; De Geest, B.G.; Vandenbroucke, R.E.; Peeters, L.; Demeester, J.; De Smedt, S.C.; Sanders, N.N. Ultrasound-responsive polymer-coated microbubbles that bind and protect DNA. *Langmuir* **2006**, *22*, 7273–7278. [CrossRef] [PubMed]
91. Nahire, R.; Haldar, M.K.; Paul, S.; Mergoum, A.; Ambre, A.H.; Katti, K.S.; Gange, K.N.; Srivastava, D.; Sarkar, K.; Mallik, S. Polymer-coated echogenic lipid nanoparticles with dual release triggers. *Biomacromolecules* **2013**, *14*, 841–853. [CrossRef] [PubMed]
92. Niu, C.; Wang, Z.; Lu, G.; Krupka, T.M.; Sun, Y.; You, Y.; Song, W.; Ran, H.; Li, P.; Zheng, Y. Doxorubicin loaded superparamagnetic PLGA-iron oxide multifunctional microbubbles for dual-mode US/MR imaging and therapy of metastasis in lymph nodes. *Biomaterials* **2013**, *34*, 2307–2317. [CrossRef]
93. Teupe, C.; Richter, S.; Fisslthaler, B.; Randriamboavonjy, V.; Ihling, C.; Fleming, I.; Busse, R.; Zeiher, A.M.; Dimmeler, S. Vascular gene transfer of phosphomimetic endothelial nitric oxide synthase (S1177D) using ultrasound-enhanced destruction of plasmid-loaded microbubbles improves vasoreactivity. *Circulation* **2002**, *105*, 1104–1109. [CrossRef] [PubMed]
94. Grinstaff, M.W.; Suslick, K.S. Air-filled proteinaceous microbubbles: Synthesis of an echo-contrast agent. *Proc. Natl. Acad. Sci. USA* **1991**, *88*, 7708–7710. [CrossRef]
95. Shohet, R.V.; Chen, S.; Zhou, Y.T.; Wang, Z.; Meidell, R.S.; Unger, R.H.; Grayburn, P.A. Echocardiographic destruction of albumin microbubbles directs gene delivery to the myocardium. *Circulation* **2000**, *101*, 2554–2556. [CrossRef]
96. De Temmerman, M.L.; Dewitte, H.; Vandenbroucke, R.E.; Lucas, B.; Libert, C.; Demeester, J.; De Smedt, S.C.; Lentacker, I.; Rejman, J. mRNA-Lipoplex loaded microbubble contrast agents for ultrasound-assisted transfection of dendritic cells. *Biomaterials* **2011**, *32*, 9128–9135. [CrossRef] [PubMed]
97. Lentacker, I.; De Smedt, S.C.; Demeester, J.; Van Marck, V.; Bracke, M.; Sanders, N.N. Lipoplex-Loaded Microbubbles for Gene Delivery: A Trojan Horse Controlled by Ultrasound. *Adv. Funct. Mater.* **2007**, *17*, 1910–1916. [CrossRef]
98. Sirsi, S.; Borden, M. Microbubble compositions, properties and biomedical applications. *Bubble Sci. Eng. Technol.* **2009**, *1*, 3–17. [CrossRef] [PubMed]
99. Xie, A.; Wu, M.D.; Cigarroa, G.; Belcik, J.T.; Ammi, A.; Moccetti, F.; Lindner, J.R. Influence of DNA-microbubble coupling on contrast ultrasound–mediated gene transfection in muscle and liver. *J. Am. Soc. Echocardiogr.* **2016**, *29*, 812–818. [CrossRef] [PubMed]
100. Kim, D.; Lee, S.S.; Moon, H.; Park, S.Y.; Lee, H.J. PD-L1 Targeting Immune-Microbubble Complex Enhances Therapeutic Index in Murine Colon Cancer Models. *Pharmaceuticals* **2020**, *14*, 6. [CrossRef]
101. Xie, Y.; Chen, Y.; Zhang, L.; Wu, M.; Deng, Z.; Yang, Y.; Wang, J.; Lv, Q.; Zheng, H.; Xie, M.; et al. Ultrasound Molecular Imaging of Lymphocyte-endothelium Adhesion Cascade in Acute Cellular Rejection of Cardiac Allografts. *Transplantation* **2019**, *103*, 1603–1611. [CrossRef] [PubMed]
102. Yan, F.; Sun, Y.; Mao, Y.; Wu, M.; Deng, Z.; Li, S.; Liu, X.; Xue, L.; Zheng, H. Ultrasound Molecular Imaging of Atherosclerosis for Early Diagnosis and Therapeutic Evaluation through Leucocyte-like Multiple Targeted Microbubbles. *Theranostics* **2018**, *8*, 1879–1891. [CrossRef] [PubMed]
103. Kumari, P.; Ghosh, B.; Biswas, S. Nanocarriers for cancer-targeted drug delivery. *J. Drug Target* **2016**, *24*, 179–191. [CrossRef]
104. de Matos, M.B.C.; Deckers, R.; van Elburg, B.; Lajoinie, G.; de Miranda, B.S.; Versluis, M.; Schiffelers, R.; Kok, R.J. Ultrasound-Sensitive Liposomes for Triggered Macromolecular Drug Delivery: Formulation and In Vitro Characterization. *Front. Pharmacol.* **2019**, *10*, 1463. [CrossRef] [PubMed]
105. Mitchell, M.J.; Billingsley, M.M.; Haley, R.M.; Wechsler, M.E.; Peppas, N.A.; Langer, R. Engineering precision nanoparticles for drug delivery. *Nat. Rev. Drug Discov.* **2021**, *20*, 101–124. [CrossRef]
106. Hoshyar, N.; Gray, S.; Han, H.; Bao, G. The effect of nanoparticle size on in vivo pharmacokinetics and cellular interaction. *Nanomedicine* **2016**, *11*, 673–692. [CrossRef]

107. Alexis, F.; Pridgen, E.; Molnar, L.K.; Farokhzad, O.C. Factors affecting the clearance and biodistribution of polymeric nanoparticles. *Mol. Pharm.* **2008**, *5*, 505–515. [CrossRef] [PubMed]
108. Zhang, S.; Gao, H.; Bao, G. Physical Principles of Nanoparticle Cellular Endocytosis. *ACS Nano* **2015**, *9*, 8655–8671. [CrossRef] [PubMed]
109. Smith, S.A.; Selby, L.I.; Johnston, A.P.R.; Such, G.K. The Endosomal Escape of Nanoparticles: Toward More Efficient Cellular Delivery. *Bioconjug. Chem.* **2019**, *30*, 263–272. [CrossRef]
110. Cupic, K.I.; Rennick, J.J.; Johnston, A.P.; Such, G.K. Controlling endosomal escape using nanoparticle composition: Current progress and future perspectives. *Nanomedicine* **2019**, *14*, 215–223. [CrossRef] [PubMed]
111. Ma, D. Enhancing endosomal escape for nanoparticle mediated siRNA delivery. *Nanoscale* **2014**, *6*, 6415–6425. [CrossRef]
112. Banerjee, A.; Qi, J.; Gogoi, R.; Wong, J.; Mitragotri, S. Role of nanoparticle size, shape and surface chemistry in oral drug delivery. *J. Control. Release* **2016**, *238*, 176–185. [CrossRef]
113. Zhang, L.; Gu, F.X.; Chan, J.M.; Wang, A.Z.; Langer, R.S.; Farokhzad, O.C. Nanoparticles in medicine: Therapeutic applications and developments. *Clin. Pharmacol. Ther.* **2008**, *83*, 761–769. [CrossRef]
114. Yao, L.; Song, Q.; Bai, W.; Zhang, J.; Miao, D.; Jiang, M.; Wang, Y.; Shen, Z.; Hu, Q.; Gu, X.; et al. Facilitated brain delivery of poly (ethylene glycol)-poly (lactic acid) nanoparticles by microbubble-enhanced unfocused ultrasound. *Biomaterials* **2014**, *35*, 3384–3395. [CrossRef]
115. Treat, L.H.; McDannold, N.; Vykhodtseva, N.; Zhang, Y.; Tam, K.; Hynynen, K. Targeted delivery of doxorubicin to the rat brain at therapeutic levels using MRI-guided focused ultrasound. *Int. J. Cancer* **2007**, *121*, 901–907. [CrossRef]
116. Ibsen, S.; Schutt, C.E.; Esener, S. Microbubble-mediated ultrasound therapy: A review of its potential in cancer treatment. *Drug Des. Devel. Ther.* **2013**, *7*, 375–388. [CrossRef] [PubMed]
117. Snipstad, S.; Hanstad, S.; Bjorkoy, A.; Morch, Y.; de Lange Davies, C. Sonoporation Using Nanoparticle-Loaded Microbubbles Increases Cellular Uptake of Nanoparticles Compared to Co-Incubation of Nanoparticles and Microbubbles. *Pharmaceutics* **2021**, *13*, 640. [CrossRef]
118. De Cock, I.; Lajoinie, G.; Versluis, M.; De Smedt, S.C.; Lentacker, I. Sonoprinting and the importance of microbubble loading for the ultrasound mediated cellular delivery of nanoparticles. *Biomaterials* **2016**, *83*, 294–307. [CrossRef]
119. Cool, S.K.; Geers, B.; Roels, S.; Stremersch, S.; Vanderperren, K.; Saunders, J.H.; De Smedt, S.C.; Demeester, J.; Sanders, N.N. Coupling of drug containing liposomes to microbubbles improves ultrasound triggered drug delivery in mice. *J. Control. Release* **2013**, *172*, 885–893. [CrossRef] [PubMed]
120. Torchilin, V.P. Recent advances with liposomes as pharmaceutical carriers. *Nat. Rev. Drug Discov.* **2005**, *4*, 145–160. [CrossRef] [PubMed]
121. Hadinoto, K.; Sundaresan, A.; Cheow, W.S. Lipid-polymer hybrid nanoparticles as a new generation therapeutic delivery platform: A review. *Eur. J. Pharm. Biopharm.* **2013**, *85*, 427–443. [CrossRef] [PubMed]
122. Kheirolomoom, A.; Dayton, P.A.; Lum, A.F.; Little, E.; Paoli, E.E.; Zheng, H.; Ferrara, K.W. Acoustically-active microbubbles conjugated to liposomes: Characterization of a proposed drug delivery vehicle. *J. Control. Release* **2007**, *118*, 275–284. [CrossRef] [PubMed]
123. Sercombe, L.; Veerati, T.; Moheimani, F.; Wu, S.Y.; Sood, A.K.; Hua, S. Advances and Challenges of Liposome Assisted Drug Delivery. *Front. Pharmacol.* **2015**, *6*, 286. [CrossRef]
124. Guimaraes, D.; Cavaco-Paulo, A.; Nogueira, E. Design of liposomes as drug delivery system for therapeutic applications. *Int. J. Pharm.* **2021**, *601*, 120571. [CrossRef]
125. Moussa, H.G.; Martins, A.M.; Husseini, G.A. Review on triggered liposomal drug delivery with a focus on ultrasound. *Curr. Cancer Drug Targets* **2015**, *15*, 282–313. [CrossRef]
126. Jiang, X.; Abedi, K.; Shi, J. Polymeric nanoparticles for RNA delivery. In *Reference Module in Materials Science and Materials Engineering*; Elsevier: Amsterdam, The Netherlands, 2021.
127. Hershberger, K.K.; Gauger, A.J.; Bronstein, L.M. Utilizing Stimuli Responsive Linkages to Engineer and Enhance Polymer Nanoparticle-Based Drug Delivery Platforms. *ACS Appl. Biol. Mater.* **2021**, *4*, 4720–4736. [CrossRef]
128. Deirram, N.; Zhang, C.; Kermaniyan, S.S.; Johnston, A.P.R.; Such, G.K. pH-Responsive Polymer Nanoparticles for Drug Delivery. *Macromol. Rapid Commun.* **2019**, *40*, e1800917. [CrossRef]
129. Lopez Ruiz, A.; Ramirez, A.; McEnnis, K. Single and Multiple Stimuli-Responsive Polymer Particles for Controlled Drug Delivery. *Pharmaceutics* **2022**, *14*, 421. [CrossRef]
130. Karimi, M.; Sahandi Zangabad, P.; Ghasemi, A.; Amiri, M.; Bahrami, M.; Malekzad, H.; Ghahramanzadeh Asl, H.; Mahdieh, Z.; Bozorgomid, M.; Ghasemi, A.; et al. Temperature-Responsive Smart Nanocarriers for Delivery Of Therapeutic Agents: Applications and Recent Advances. *ACS Appl. Mater. Interfaces* **2016**, *8*, 21107–21133. [CrossRef]
131. Mukherjee, A.; Waters, A.K.; Kalyan, P.; Achrol, A.S.; Kesari, S.; Yenugonda, V.M. Lipid-polymer hybrid nanoparticles as a next-generation drug delivery platform: State of the art, emerging technologies, and perspectives. *Int. J. Nanomed.* **2019**, *14*, 1937–1952. [CrossRef]
132. Siewert, C.D.; Haas, H.; Cornet, V.; Nogueira, S.S.; Nawroth, T.; Uebbing, L.; Ziller, A.; Al-Gousous, J.; Radulescu, A.; Schroer, M.A.; et al. Hybrid Biopolymer and Lipid Nanoparticles with Improved Transfection Efficacy for mRNA. *Cells* **2020**, *9*, 2034. [CrossRef]

133. Liu, Y.; Pan, J.; Feng, S.S. Nanoparticles of lipid monolayer shell and biodegradable polymer core for controlled release of paclitaxel: Effects of surfactants on particles size, characteristics and in vitro performance. *Int. J. Pharm.* **2010**, *395*, 243–250. [CrossRef]
134. Lohcharoenkal, W.; Wang, L.; Chen, Y.C.; Rojanasakul, Y. Protein Nanoparticles as Drug Delivery Carriers for Cancer Therapy. *BioMed Res. Int.* **2014**, *2014*, 180549. [CrossRef]
135. Kratz, F. Albumin as a drug carrier: Design of prodrugs, drug conjugates and nanoparticles. *J. Control. Release* **2008**, *132*, 171–183. [CrossRef] [PubMed]
136. Gregory, J.V.; Kadiyala, P.; Doherty, R.; Cadena, M.; Habeel, S.; Ruoslahti, E.; Lowenstein, P.R.; Castro, M.G.; Lahann, J. Systemic brain tumor delivery of synthetic protein nanoparticles for glioblastoma therapy. *Nat. Commun.* **2020**, *11*, 5687. [CrossRef] [PubMed]
137. Spada, A.; Emami, J.; Tuszynski, J.A.; Lavasanifar, A. The Uniqueness of Albumin as a Carrier in Nanodrug Delivery. *Mol. Pharm.* **2021**, *18*, 1862–1894. [CrossRef] [PubMed]
138. Hawkins, M.J.; Soon-Shiong, P.; Desai, N. Protein nanoparticles as drug carriers in clinical medicine. *Adv. Drug Deliv. Rev.* **2008**, *60*, 876–885. [CrossRef]
139. Park, C.R.; Jo, J.H.; Song, M.G.; Park, J.Y.; Kim, Y.H.; Youn, H.; Paek, S.H.; Chung, J.K.; Jeong, J.M.; Lee, Y.S.; et al. Secreted protein acidic and rich in cysteine mediates active targeting of human serum albumin in U87MG xenograft mouse models. *Theranostics* **2019**, *9*, 7447–7457. [CrossRef]
140. Lin, Y.C.; Shih, C.P.; Chen, H.C.; Chou, Y.L.; Sytwu, H.K.; Fang, M.C.; Lin, Y.Y.; Kuo, C.Y.; Su, H.H.; Hung, C.L.; et al. Ultrasound Microbubble-Facilitated Inner Ear Delivery of Gold Nanoparticles Involves Transient Disruption of the Tight Junction Barrier in the Round Window Membrane. *Front. Pharmacol.* **2021**, *12*, 689032. [CrossRef]
141. Akbarzadeh, A.; Rezaei-Sadabady, R.; Davaran, S.; Joo, S.W.; Zarghami, N.; Hanifehpour, Y.; Samiei, M.; Kouhi, M.; Nejati-Koshki, K. Liposome: Classification, preparation, and applications. *Nanoscale Res. Lett.* **2013**, *8*, 102. [CrossRef]
142. Yang, B.; Song, B.P.; Shankar, S.; Guller, A.; Deng, W. Recent advances in liposome formulations for breast cancer therapeutics. *Cell Mol. Life Sci.* **2021**, *78*, 5225–5243. [CrossRef]
143. Golombek, S.K.; May, J.N.; Theek, B.; Appold, L.; Drude, N.; Kiessling, F.; Lammers, T. Tumor targeting via EPR: Strategies to enhance patient responses. *Adv. Drug Deliv. Rev.* **2018**, *130*, 17–38. [CrossRef]
144. Chinnagounder Periyasamy, P.; Leijten, J.C.H.; Dijkstra, P.J.; Karperien, M.; Post, J.N. Nanomaterials for the Local and Targeted Delivery of Osteoarthritis Drugs. *J. Nanomater.* **2012**, *2012*, 673968. [CrossRef]
145. Malik, R.; Pancholi, K.; Melzer, A. Microbubble-liposome conjugate: Payload evaluation of potential theranostic vehicle. *Nanobiomedicine* **2016**, *3*, 1849543516670806. [CrossRef] [PubMed]
146. Yan, F.; Li, L.; Deng, Z.; Jin, Q.; Chen, J.; Yang, W.; Yeh, C.K.; Wu, J.; Shandas, R.; Liu, X.; et al. Paclitaxel-liposome-microbubble complexes as ultrasound-triggered therapeutic drug delivery carriers. *J. Control. Release* **2013**, *166*, 246–255. [CrossRef] [PubMed]
147. Deng, Z.; Yan, F.; Jin, Q.; Li, F.; Wu, J.; Liu, X.; Zheng, H. Reversal of multidrug resistance phenotype in human breast cancer cells using doxorubicin-liposome-microbubble complexes assisted by ultrasound. *J. Control. Release* **2014**, *174*, 109–116. [CrossRef] [PubMed]
148. Browning, R.; Thomas, N.; Marsh, L.K.; Tear, L.R.; Owen, J.; Stride, E.; Farrer, N.J. Ultrasound-Triggered Delivery of Iproplatin from Microbubble-Conjugated Liposomes. *Chem. Open* **2021**, *10*, 1170–1176. [CrossRef]
149. Geers, B.; Lentacker, I.; Sanders, N.N.; Demeester, J.; Meairs, S.; De Smedt, S.C. Self-assembled liposome-loaded microbubbles: The missing link for safe and efficient ultrasound triggered drug-delivery. *J. Control. Release* **2011**, *152*, 249–256. [CrossRef]
150. Kanasty, R.; Dorkin, J.R.; Vegas, A.; Anderson, D. Delivery materials for siRNA therapeutics. *Nat. Mater.* **2013**, *12*, 967–977. [CrossRef]
151. Zhang, S.H.; Wen, K.M.; Wu, W.; Li, W.Y.; Zhao, J.N. Efficacy of HGF carried by ultrasound microbubble-cationic nano-liposomes complex for treating hepatic fibrosis in a bile duct ligation rat model, and its relationship with the diffusion-weighted MRI parameters. *Clin. Res. Hepatol. Gastroenterol.* **2013**, *37*, 602–607. [CrossRef]
152. Thomas, C.E.; Ehrhardt, A.; Kay, M.A. Progress and problems with the use of viral vectors for gene therapy. *Nat. Rev. Genet.* **2003**, *4*, 346–358. [CrossRef]
153. Yang, D.; Gao, Y.H.; Tan, K.B.; Zuo, Z.X.; Yang, W.X.; Hua, X.; Li, P.J.; Zhang, Y.; Wang, G. Inhibition of hepatic fibrosis with artificial microRNA using ultrasound and cationic liposome-bearing microbubbles. *Gene Ther.* **2013**, *20*, 1140–1148. [CrossRef]
154. Gao, J.; Nesbitt, H.; Logan, K.; Burnett, K.; White, B.; Jack, I.G.; Taylor, M.A.; Love, M.; Callan, B.; McHale, A.P.; et al. An ultrasound responsive microbubble-liposome conjugate for targeted irinotecan-oxaliplatin treatment of pancreatic cancer. *Eur. J. Pharm. Biopharm.* **2020**, *157*, 233–240. [CrossRef]
155. Yoon, Y.I.; Kwon, Y.S.; Cho, H.S.; Heo, S.H.; Park, K.S.; Park, S.G.; Lee, S.H.; Hwang, S.I.; Kim, Y.I.; Jae, H.J.; et al. Ultrasound-mediated gene and drug delivery using a microbubble-liposome particle system. *Theranostics* **2014**, *4*, 1133–1144. [CrossRef] [PubMed]
156. Ibsen, S.; Benchimol, M.; Simberg, D.; Schutt, C.; Steiner, J.; Esener, S. A novel nested liposome drug delivery vehicle capable of ultrasound triggered release of its payload. *J. Control. Release* **2011**, *155*, 358–366. [CrossRef] [PubMed]
157. Ross, C.; Taylor, M.; Fullwood, N.; Allsop, D. Liposome delivery systems for the treatment of Alzheimer's disease. *Int. J. Nanomed.* **2018**, *13*, 8507–8522. [CrossRef]

158. Danhier, F.; Ansorena, E.; Silva, J.M.; Coco, R.; Le Breton, A.; Preat, V. PLGA-based nanoparticles: An overview of biomedical applications. *J. Control. Release* **2012**, *161*, 505–522. [CrossRef] [PubMed]
159. Park, J.; Fong, P.M.; Lu, J.; Russell, K.S.; Booth, C.J.; Saltzman, W.M.; Fahmy, T.M. PEGylated PLGA nanoparticles for the improved delivery of doxorubicin. *Nanomedicine* **2009**, *5*, 410–418. [CrossRef]
160. Patnaik, S.; Gupta, K.C. Novel polyethylenimine-derived nanoparticles for in vivo gene delivery. *Expert Opin. Drug Deliv.* **2013**, *10*, 215–228. [CrossRef]
161. Wang, X.; Niu, D.; Hu, C.; Li, P. Polyethyleneimine-Based Nanocarriers for Gene Delivery. *Curr. Pharm. Des.* **2015**, *21*, 6140–6156. [CrossRef]
162. Zakeri, A.; Kouhbanani, M.A.J.; Beheshtkhoo, N.; Beigi, V.; Mousavi, S.M.; Hashemi, S.A.R.; Karimi Zade, A.; Amani, A.M.; Savardashtaki, A.; Mirzaei, E.; et al. Polyethylenimine-based nanocarriers in co-delivery of drug and gene: A developing horizon. *Nano Rev. Exp.* **2018**, *9*, 1488497. [CrossRef]
163. Wong, K.H.; Lu, A.; Chen, X.; Yang, Z. Natural Ingredient-Based Polymeric Nanoparticles for Cancer Treatment. *Molecules* **2020**, *25*, 3620. [CrossRef]
164. Choi, K.Y.; Saravanakumar, G.; Park, J.H.; Park, K. Hyaluronic acid-based nanocarriers for intracellular targeting: Interfacial interactions with proteins in cancer. *Colloids Surf. B Biointerfaces* **2012**, *99*, 82–94. [CrossRef]
165. Fan, Y.; Liu, Y.; Wu, Y.; Dai, F.; Yuan, M.; Wang, F.; Bai, Y.; Deng, H. Natural polysaccharides based self-assembled nanoparticles for biomedical applications—A review. *Int. J. Biol. Macromol.* **2021**, *192*, 1240–1255. [CrossRef] [PubMed]
166. Davoudi, Z.; Peroutka-Bigus, N.; Bellaire, B.; Jergens, A.; Wannemuehler, M.; Wang, Q. Gut Organoid as a New Platform to Study Alginate and Chitosan Mediated PLGA Nanoparticles for Drug Delivery. *Marine Drugs* **2021**, *19*, 282. [CrossRef]
167. Ghasemiyeh, P.; Mohammadi-Samani, S. Polymers Blending as Release Modulating Tool in Drug Delivery. *Front. Mater.* **2021**, *8*, 752813. [CrossRef]
168. Chen, D.; Han, S.; Zhu, Y.; Hu, F.; Wei, Y.; Wang, G. Kidney-targeted drug delivery via rhein-loaded polyethyleneglycol-co-polycaprolactone-co-polyethylenimine nanoparticles for diabetic nephropathy therapy. *Int. J. Nanomed.* **2018**, *13*, 3507–3527. [CrossRef] [PubMed]
169. Yan, J.; Su, T.; Cheng, F.; Cao, J.; Zhang, H.; He, B. Multifunctional nanoparticles self-assembled from polyethylenimine-based graft polymers as efficient anticancer drug delivery. *Colloids Surf. B Biointerfaces* **2017**, *155*, 118–127. [CrossRef]
170. Glodde, M.; Sirsi, S.R.; Lutz, G.J. Physiochemical properties of low and high molecular weight poly(ethylene glycol)-grafted poly(ethylene imine) copolymers and their complexes with oligonucleotides. *Biomacromolecules* **2006**, *7*, 347–356. [CrossRef] [PubMed]
171. Kumari, A.; Yadav, S.K.; Yadav, S.C. Biodegradable polymeric nanoparticles based drug delivery systems. *Colloids Surf. B Biointerfaces* **2010**, *75*, 1–18. [CrossRef] [PubMed]
172. He, X.; Yang, X.; Li, D.; Cao, Z. Red and NIR Light-Responsive Polymeric Nanocarriers for On-Demand Drug Delivery. *Curr. Med. Chem.* **2020**, *27*, 3877–3887. [CrossRef]
173. Zhao, W.; Zhao, Y.; Wang, Q.; Liu, T.; Sun, J.; Zhang, R. Remote Light-Responsive Nanocarriers for Controlled Drug Delivery: Advances and Perspectives. *Small* **2019**, *15*, e1903060. [CrossRef] [PubMed]
174. Gong, Y.; Wang, Z.; Dong, G.; Sun, Y.; Wang, X.; Rong, Y.; Li, M.; Wang, D.; Ran, H. Low-intensity focused ultrasound mediated localized drug delivery for liver tumors in rabbits. *Drug Deliv.* **2016**, *23*, 2280–2289. [CrossRef] [PubMed]
175. Chen, K.; Zhang, L. Effect of drug-loaded microbubbles combined with ultrasound on the apoptosis of cancer cells and the expression of Bax and Bcl-2 in a rabbit VX2 liver tumor model. *Biosci. Rep.* **2019**, *39*, BSR20181144. [CrossRef] [PubMed]
176. Liufu, C.; Li, Y.; Tu, J.; Zhang, H.; Yu, J.; Wang, Y.; Huang, P.; Chen, Z. Echogenic PEGylated PEI-Loaded Microbubble As Efficient Gene Delivery System. *Int. J. Nanomed.* **2019**, *14*, 8923–8941. [CrossRef] [PubMed]
177. Lei, Y.; Wang, J.; Xie, C.; Wagner, E.; Lu, W.; Li, Y.; Wei, X.; Dong, J.; Liu, M. Glutathione-sensitive RGD-poly(ethylene glycol)-SS-polyethylenimine for intracranial glioblastoma targeted gene delivery. *J. Gene Med.* **2013**, *15*, 291–305. [CrossRef] [PubMed]
178. Liufu, C.; Li, Y.; Lin, Y.; Yu, J.; Du, M.; Chen, Y.; Yang, Y.; Gong, X.; Chen, Z. Synergistic ultrasonic biophysical effect-responsive nanoparticles for enhanced gene delivery to ovarian cancer stem cells. *Drug Deliv.* **2020**, *27*, 1018–1033. [CrossRef] [PubMed]
179. Knop, K.; Hoogenboom, R.; Fischer, D.; Schubert, U.S. Poly(ethylene glycol) in drug delivery: Pros and cons as well as potential alternatives. *Angew. Chem. Int. Ed. Eng.* **2010**, *49*, 6288–6308. [CrossRef] [PubMed]
180. Fan, C.H.; Wang, T.W.; Hsieh, Y.K.; Wang, C.F.; Gao, Z.; Kim, A.; Nagasaki, Y.; Yeh, C.K. Enhancing Boron Uptake in Brain Glioma by a Boron-Polymer/Microbubble Complex with Focused Ultrasound. *ACS Appl. Mater. Interfaces* **2019**, *11*, 11144–11156. [CrossRef] [PubMed]
181. Wei, S.; Xu, C.; Zhang, Y.; Shi, Z.; Wu, M.; Yang, B. Ultrasound Assisted a Peroxisome Proliferator-Activated Receptor (PPAR)gamma Agonist-Loaded Nanoparticle-Microbubble Complex to Attenuate Renal Interstitial Fibrosis. *Int. J. Nanomed.* **2020**, *15*, 7315–7327. [CrossRef] [PubMed]
182. Salvador-Morales, C.; Zhang, L.; Langer, R.; Farokhzad, O.C. Immunocompatibility properties of lipid-polymer hybrid nanoparticles with heterogeneous surface functional groups. *Biomaterials* **2009**, *30*, 2231–2240. [CrossRef] [PubMed]
183. Yang, Q.; Zhou, Y.; Chen, J.; Huang, N.; Wang, Z.; Cheng, Y. Gene Therapy for Drug-Resistant Glioblastoma via Lipid-Polymer Hybrid Nanoparticles Combined with Focused Ultrasound. *Int. J. Nanomed.* **2021**, *16*, 185–199. [CrossRef]
184. Chen, Y.; Liang, Y.; Jiang, P.; Li, F.; Yu, B.; Yan, F. Lipid/PLGA Hybrid Microbubbles as a Versatile Platform for Noninvasive Image-Guided Targeted Drug Delivery. *ACS Appl. Mater. Interfaces* **2019**, *11*, 41842–41852. [CrossRef]

185. Langer, K.; Balthasar, S.; Vogel, V.; Dinauer, N.; von Briesen, H.; Schubert, D. Optimization of the preparation process for human serum albumin (HSA) nanoparticles. *Int. J. Pharm.* **2003**, *257*, 169–180. [CrossRef]
186. Jain, A.; Singh, S.K.; Arya, S.K.; Kundu, S.C.; Kapoor, S. Protein Nanoparticles: Promising Platforms for Drug Delivery Applications. *ACS Biomater. Sci. Eng.* **2018**, *4*, 3939–3961. [CrossRef]
187. Lee, J.H.; Moon, H.; Han, H.; Lee, I.J.; Kim, D.; Lee, H.J.; Ha, S.W.; Kim, H.; Chung, J.W. Antitumor Effects of Intra-Arterial Delivery of Albumin-Doxorubicin Nanoparticle Conjugated Microbubbles Combined with Ultrasound-Targeted Microbubble Activation on VX2 Rabbit Liver Tumors. *Cancers* **2019**, *11*, 581. [CrossRef] [PubMed]
188. Ha, S.W.; Hwang, K.; Jin, J.; Cho, A.S.; Kim, T.Y.; Hwang, S.I.; Lee, H.J.; Kim, C.Y. Ultrasound-sensitizing nanoparticle complex for overcoming the blood-brain barrier: An effective drug delivery system. *Int. J. Nanomed.* **2019**, *14*, 3743–3752. [CrossRef] [PubMed]
189. Kim, D.; Lee, J.H.; Moon, H.; Seo, M.; Han, H.; Yoo, H.; Seo, H.; Lee, J.; Hong, S.; Kim, P.; et al. Development and evaluation of an ultrasound-triggered microbubble combined transarterial chemoembolization (TACE) formulation on rabbit VX2 liver cancer model. *Theranostics* **2021**, *11*, 79–92. [CrossRef] [PubMed]
190. Ye, D.; Zhang, X.; Yue, Y.; Raliya, R.; Biswas, P.; Taylor, S.; Tai, Y.C.; Rubin, J.B.; Liu, Y.; Chen, H. Focused ultrasound combined with microbubble-mediated intranasal delivery of gold nanoclusters to the brain. *J. Control. Release* **2018**, *286*, 145–153. [CrossRef]
191. Hartman, R.K.; Hallam, K.A.; Donnelly, E.M.; Emelianov, S.Y. Photoacoustic imaging of gold nanorods in the brain delivered via microbubble-assisted focused ultrasound: A tool for in vivo molecular neuroimaging. *Laser Phys. Lett.* **2019**, *16*, 025603. [CrossRef]
192. Morch, Y.; Hansen, R.; Berg, S.; Aslund, A.K.; Glomm, W.R.; Eggen, S.; Schmid, R.; Johnsen, H.; Kubowicz, S.; Snipstad, S.; et al. Nanoparticle-stabilized microbubbles for multimodal imaging and drug delivery. *Contrast Media Mol. Imaging* **2015**, *10*, 356–366. [CrossRef] [PubMed]
193. Yang, F.; Li, Y.; Chen, Z.; Zhang, Y.; Wu, J.; Gu, N. Superparamagnetic iron oxide nanoparticle-embedded encapsulated microbubbles as dual contrast agents of magnetic resonance and ultrasound imaging. *Biomaterials* **2009**, *30*, 3882–3890. [CrossRef] [PubMed]
194. Yang, F.; Li, L.; Li, Y.; Chen, Z.; Wu, J.; Gu, N. Superparamagnetic nanoparticle-inclusion microbubbles for ultrasound contrast agents. *Phys. Med. Biol.* **2008**, *53*, 6129–6141. [CrossRef]
195. Liu, Z.; Lammers, T.; Ehling, J.; Fokong, S.; Bornemann, J.; Kiessling, F.; Gatjens, J. Iron oxide nanoparticle-containing microbubble composites as contrast agents for MR and ultrasound dual-modality imaging. *Biomaterials* **2011**, *32*, 6155–6163. [CrossRef]
196. Gao, W.; Chen, Y.; Zhang, Y.; Zhang, Q.; Zhang, L. Nanoparticle-based local antimicrobial drug delivery. *Adv. Drug Deliv. Rev.* **2018**, *127*, 46–57. [CrossRef]
197. Rai, M.; Yadav, A.; Gade, A. Silver nanoparticles as a new generation of antimicrobials. *Biotechnol. Adv.* **2009**, *27*, 76–83. [CrossRef]
198. Jones, N.; Ray, B.; Ranjit, K.T.; Manna, A.C. Antibacterial activity of ZnO nanoparticle suspensions on a broad spectrum of microorganisms. *FEMS Microbiol. Lett.* **2008**, *279*, 71–76. [CrossRef]
199. Barmin, R.A.; Rudakovskaya, P.G.; Gusliakova, O.I.; Sindeeva, O.A.; Prikhozhdenko, E.S.; Maksimova, E.A.; Obukhova, E.N.; Chernyshev, V.S.; Khlebtsov, B.N.; Solovev, A.A.; et al. Air-Filled Bubbles Stabilized by Gold Nanoparticle/Photodynamic Dye Hybrid Structures for Theranostics. *Nanomaterials* **2021**, *11*, 415. [CrossRef]
200. Wang, L.; Lu, H.; Gao, Q.; Yuan, C.; Ding, F.; Li, J.; Zhang, D.; Ou, X. A multifunctional theranostic contrast agent for ultrasound/near infrared fluorescence imaging-based tumor diagnosis and ultrasound-triggered combined photothermal and gene therapy. *Acta Biomater.* **2019**, *99*, 373–386. [CrossRef]
201. Yoon, Y.I.; Pang, X.; Jung, S.; Zhang, G.; Kong, M.; Liu, G.; Chen, X. Smart Gold Nanoparticle-Stabilized Ultrasound Microbubbles as Cancer Theranostics. *J. Mater. Chem. B* **2018**, *6*, 3235–3239. [CrossRef]
202. Lv, Y.; Cao, Y.; Li, P.; Liu, J.; Chen, H.; Hu, W.; Zhang, L. Ultrasound-Triggered Destruction of Folate-Functionalized Mesoporous Silica Nanoparticle-Loaded Microbubble for Targeted Tumor Therapy. *Adv. Healthc. Mater.* **2017**, *6*, 1700354. [CrossRef]
203. Du, M.; Chen, Y.; Tu, J.; Liufu, C.; Yu, J.; Yuan, Z.; Gong, X.; Chen, Z. Ultrasound Responsive Magnetic Mesoporous Silica Nanoparticle-Loaded Microbubbles for Efficient Gene Delivery. *ACS BioMater. Sci. Eng.* **2020**, *6*, 2904–2912. [CrossRef]
204. Dwivedi, P.; Kiran, S.; Han, S.; Dwivedi, M.; Khatik, R.; Fan, R.; Mangrio, F.A.; Du, K.; Zhu, Z.; Yang, C.; et al. Magnetic Targeting and Ultrasound Activation of Liposome-Microbubble Conjugate for Enhanced Delivery of Anticancer Therapies. *ACS Appl. Mater. Interfaces* **2020**, *12*, 23737–23751. [CrossRef]
205. Kwiatkowski, S.; Knap, B.; Przystupski, D.; Saczko, J.; Kedzierska, E.; Knap-Czop, K.; Kotlinska, J.; Michel, O.; Kotowski, K.; Kulbacka, J. Photodynamic therapy—Mechanisms, photosensitizers and combinations. *Biomed. Pharm.* **2018**, *106*, 1098–1107. [CrossRef]
206. Lee, H.; Han, J.; Shin, H.; Han, H.; Na, K.; Kim, H. Combination of chemotherapy and photodynamic therapy for cancer treatment with sonoporation effects. *J. Control. Release* **2018**, *283*, 190–199. [CrossRef]
207. Jang, Y.; Kim, D.; Lee, H.; Jang, H.; Park, S.; Kim, G.E.; Lee, H.J.; Kim, H.J.; Kim, H. Development of an ultrasound triggered nanomedicine-microbubble complex for chemo-photodynamic-gene therapy. *Nanomedicine* **2020**, *27*, 102194. [CrossRef]
208. Jain, A.; Cheng, K. The principles and applications of avidin-based nanoparticles in drug delivery and diagnosis. *J. Control. Release* **2017**, *245*, 27–40. [CrossRef]

209. Peyman, S.A.; Abou-Saleh, R.H.; McLaughlan, J.R.; Ingram, N.; Johnson, B.R.G.; Critchley, K.; Freear, S.; Evans, J.A.; Markham, A.F.; Coletta, P.L.; et al. Expanding 3D geometry for enhanced on-chip microbubble production and single step formation of liposome modified microbubbles. *Lab Chip* **2012**, *12*, 4544–4552. [CrossRef]
210. Nele, V.; Schutt, C.E.; Wojciechowski, J.P.; Kit-Anan, W.; Doutch, J.J.; Armstrong, J.P.K.; Stevens, M.M. Ultrasound-Triggered Enzymatic Gelation. *Adv. Mater.* **2020**, *32*, e1905914. [CrossRef]
211. Northrop, B.H.; Frayne, S.H.; Choudhary, U. Thiol–maleimide "click" chemistry: Evaluating the influence of solvent, initiator, and thiol on the reaction mechanism, kinetics, and selectivity. *Polym. Chem.* **2015**, *6*, 3415–3430. [CrossRef]
212. Saito, F.; Noda, H.; Bode, J.W. Critical evaluation and rate constants of chemoselective ligation reactions for stoichiometric conjugations in water. *ACS Chem. Biol.* **2015**, *10*, 1026–1033. [CrossRef]
213. Madl, C.M.; Heilshorn, S.C. Bioorthogonal Strategies for Engineering Extracellular Matrices. *Adv. Funct. Mater.* **2018**, *28*, 1706046. [CrossRef]
214. Baldwin, A.D.; Kiick, K.L. Tunable degradation of maleimide-thiol adducts in reducing environments. *Bioconjug. Chem.* **2011**, *22*, 1946–1953. [CrossRef]
215. Lee, S.; Kim, J.H.; Moon, H.; Lee, H.J.; Han, J.K. Combined treatment of sorafenib and doxorubicin-loaded microbubble-albumin nanoparticle complex for hepatocellular carcinoma: A feasibility study. *PLoS ONE* **2020**, *15*, e0243815. [CrossRef]
216. Baghirov, H.; Snipstad, S.; Sulheim, E.; Berg, S.; Hansen, R.; Thorsen, F.; Morch, Y.; Davies, C.L.; Aslund, A.K.O. Ultrasound-mediated delivery and distribution of polymeric nanoparticles in the normal brain parenchyma of a metastatic brain tumour model. *PLoS ONE* **2018**, *13*, e0191102. [CrossRef]
217. Eggen, S.; Fagerland, S.M.; Morch, Y.; Hansen, R.; Sovik, K.; Berg, S.; Furu, H.; Bohn, A.D.; Lilledahl, M.B.; Angelsen, A.; et al. Ultrasound-enhanced drug delivery in prostate cancer xenografts by nanoparticles stabilizing microbubbles. *J. Control. Release* **2014**, *187*, 39–49. [CrossRef]
218. Aslund, A.K.O.; Berg, S.; Hak, S.; Morch, Y.; Torp, S.H.; Sandvig, A.; Wideroe, M.; Hansen, R.; de Lange Davies, C. Nanoparticle delivery to the brain—By focused ultrasound and self-assembled nanoparticle-stabilized microbubbles. *J. Control. Release* **2015**, *220*, 287–294. [CrossRef]
219. Park, S.H.; Yoon, Y.I.; Moon, H.; Lee, G.H.; Lee, B.H.; Yoon, T.J.; Lee, H.J. Development of a novel microbubble-liposome complex conjugated with peptide ligands targeting IL4R on brain tumor cells. *Oncol. Rep.* **2016**, *36*, 131–136. [CrossRef]
220. Jiang, X.; He, C.; Lin, W. Supramolecular metal-based nanoparticles for drug delivery and cancer therapy. *Curr. Opin. Chem. Biol.* **2021**, *61*, 143–153. [CrossRef]
221. Bukhari, S.Z.; Zeth, K.; Iftikhar, M.; Rehman, M.; Usman Munir, M.; Khan, W.S.; Ihsan, A. Supramolecular lipid nanoparticles as delivery carriers for non-invasive cancer theranostics. *Curr. Res. Pharmacol. Drug Discov.* **2021**, *2*, 100067. [CrossRef]
222. Frinking, P.; Segers, T.; Luan, Y.; Tranquart, F. Three Decades of Ultrasound Contrast Agents: A Review of the Past, Present and Future Improvements. *Ultrasound Med. Biol.* **2020**, *46*, 892–908. [CrossRef]
223. Jangjou, A.; Meisami, A.H.; Jamali, K.; Niakan, M.H.; Abbasi, M.; Shafiee, M.; Salehi, M.; Hosseinzadeh, A.; Amani, A.M.; Vaez, A. The promising shadow of microbubble over medical sciences: From fighting wide scope of prevalence disease to cancer eradication. *J. Biomed. Sci.* **2021**, *28*, 49. [CrossRef]
224. Ding, D.; Zhu, Q. Recent advances of PLGA micro/nanoparticles for the delivery of biomacromolecular therapeutics. *Mater. Sci. Eng. C Mater. Biol. Appl.* **2018**, *92*, 1041–1060. [CrossRef]
225. Kooiman, K.; Roovers, S.; Langeveld, S.A.G.; Kleven, R.T.; Dewitte, H.; O'Reilly, M.A.; Escoffre, J.M.; Bouakaz, A.; Verweij, M.D.; Hynynen, K.; et al. Ultrasound-Responsive Cavitation Nuclei for Therapy and Drug Delivery. *Ultrasound Med. Biol.* **2020**, *46*, 1296–1325. [CrossRef] [PubMed]
226. Wang, Y.; Li, Y.; Yan, K.; Shen, L.; Yang, W.; Gong, J.; Ding, K. Clinical study of ultrasound and microbubbles for enhancing chemotherapeutic sensitivity of malignant tumors in digestive system. *Chin. J. Cancer Res.* **2018**, *30*, 553–563. [CrossRef] [PubMed]
227. Kotopoulis, S.; Dimcevski, G.; Helge Gilja, O.; Hoem, D.; Postema, M. Treatment of human pancreatic cancer using combined ultrasound, microbubbles, and gemcitabine: A clinical case study. *Med. Phys.* **2013**, *40*, 072902. [CrossRef] [PubMed]
228. Dimcevski, G.; Kotopoulis, S.; Bjanes, T.; Hoem, D.; Schjott, J.; Gjertsen, B.T.; Biermann, M.; Molven, A.; Sorbye, H.; McCormack, E.; et al. A human clinical trial using ultrasound and microbubbles to enhance gemcitabine treatment of inoperable pancreatic cancer. *J. Control. Release* **2016**, *243*, 172–181. [CrossRef]
229. Di Ianni, T.; Bose, R.J.C.; Sukumar, U.K.; Bachawal, S.; Wang, H.; Telichko, A.; Herickhoff, C.; Robinson, E.; Baker, S.; Vilches-Moure, J.G.; et al. Ultrasound/microbubble-mediated targeted delivery of anticancer microRNA-loaded nanoparticles to deep tissues in pigs. *J. Control. Release* **2019**, *309*, 1–10. [CrossRef] [PubMed]
230. Kumar, S.U.; Telichko, A.V.; Wang, H.; Hyun, D.; Johnson, E.G.; Kent, M.S.; Rebhun, R.B.; Dahl, J.J.; Culp, W.T.N.; Paulmurugan, R. Acoustically Driven Microbubbles Enable Targeted Delivery of microRNA-Loaded Nanoparticles to Spontaneous Hepatocellular Neoplasia in Canines. *Adv. Ther.* **2020**, *3*, 2000120. [CrossRef]
231. Dewitte, H.; Vanderperren, K.; Haers, H.; Stock, E.; Duchateau, L.; Hesta, M.; Saunders, J.H.; De Smedt, S.C.; Lentacker, I. Theranostic mRNA-loaded microbubbles in the lymphatics of dogs: Implications for drug delivery. *Theranostics* **2015**, *5*, 97–109. [CrossRef]

232. Hyvelin, J.M.; Gaud, E.; Costa, M.; Helbert, A.; Bussat, P.; Bettinger, T.; Frinking, P. Characteristics and Echogenicity of Clinical Ultrasound Contrast Agents: An In Vitro and In Vivo Comparison Study. *J. Ultrasound Med.* **2017**, *36*, 941–953. [CrossRef] [PubMed]
233. Willmann, J.K.; Cheng, Z.; Davis, C.; Lutz, A.M.; Schipper, M.L.; Nielsen, C.H.; Gambhir, S.S. Targeted microbubbles for imaging tumor angiogenesis: Assessment of whole-body biodistribution with dynamic micro-PET in mice. *Radiology* **2008**, *249*, 212–219. [CrossRef]
234. Snipstad, S.; Vikedal, K.; Maardalen, M.; Kurbatskaya, A.; Sulheim, E.; Davies, C.L. Ultrasound and microbubbles to beat barriers in tumors: Improving delivery of nanomedicine. *Adv. Drug Deliv. Rev.* **2021**, *177*, 113847. [CrossRef]

pharmaceutics

MDPI

Review

The Evolution and Recent Trends in Acoustic Targeting of Encapsulated Drugs to Solid Tumors: Strategies beyond Sonoporation

Arvin Honari and Shashank R. Sirsi *

Department of Bioengineering, Erik Johnson School of Engineering, The University of Texas at Dallas, Richardson, TX 75080, USA; arvin.honari@utdallas.edu
* Correspondence: shashank.sirsi@utdallas.edu

Abstract: Despite recent advancements in ultrasound-mediated drug delivery and the remarkable success observed in pre-clinical studies, no delivery platform utilizing ultrasound contrast agents has yet received FDA approval. The sonoporation effect was a game-changing discovery with a promising future in clinical settings. Various clinical trials are underway to assess sonoporation's efficacy in treating solid tumors; however, there are disagreements on its applicability to the broader population due to long-term safety issues. In this review, we first discuss how acoustic targeting of drugs gained importance in cancer pharmaceutics. Then, we discuss ultrasound-targeting strategies that have been less explored yet hold a promising future. We aim to shed light on recent innovations in ultrasound-based drug delivery including newer designs of ultrasound-sensitive particles specifically tailored for pharmaceutical usage.

Keywords: ultrasound; ultrasound targeting; ultrasound drug delivery; sonoporation; microbubbles; acoustic droplet vaporization

Citation: Honari, A.; Sirsi, S.R. The Evolution and Recent Trends in Acoustic Targeting of Encapsulated Drugs to Solid Tumors: Strategies beyond Sonoporation. *Pharmaceutics* **2023**, *15*, 1705. https://doi.org/10.3390/pharmaceutics15061705

Academic Editor: Barbara Stella

Received: 13 April 2023
Revised: 29 May 2023
Accepted: 2 June 2023
Published: 10 June 2023

1. Introduction

Modern drug-based cancer therapy is rooted in early revolutionary research carried out by physicians in the late 1800s and early 1900s. Drug targeting, a central concept in modern drug delivery, was introduced in the early 1900s by Paul Ehrlich, a German physician and a Nobel Prize winner [1]. He established the basic principles of precision medicine by defining successful drug delivery as a "magic bullet" that removes pathogens without causing deleterious side effects [2,3]. His research primarily focused on the chemical modification of drugs to enhance their binding affinity to specific pathogens, leading to their effective elimination. Ehrlich is known as the father of chemotherapy, as his search for the magic bullet led to the discovery of arsenic acid derivatives used for treating syphilis and cancer [4,5]. Despite his success in developing successful drug compounds, variable dosing and uncontrolled side effects remained ever-present issues. Thus, Ehrlich advised against intravenous injections of his formidable discoveries [2]. Although substantial progress has been made in reducing drug toxicity in healthy tissues since the start of the twentieth century, harmful side effects from systemic delivery remain a major concern today. To address this issue, various approaches have been proposed to create the ideal drug-delivery platform that dispenses the appropriate dosage of a drug in a controlled manner to the necessary tissue for an extended period. Encapsulation of active agents into nanomedicines is rapidly becoming a popular approach to constructing such an ideal system. The focus of this review is on the recent development of ultrasonic methods to more effectively deliver nanomedicine cargo to the target tissue.

2. Drug Encapsulation and Nanomedicine

The encapsulation of drugs has been around for well over a century, dating back to when gelatin capsules were developed. Encapsulation is the process of encasing medicines within a protective shell to withstand certain hostile surroundings and prevent premature release of drug content before arriving at the target site [6]. Drug encapsulation led to the development of drug carriers that have evolved tremendously in the past few decades [7]. Drug carriers gradually became a tool to improve the pharmacokinetics of drugs by controlling their water solubility, stability, or delivery rates [8]. While earlier works focused mainly on drug encapsulation for oral intake (typically on the macro scale), newer strategies are developed to improve drug delivery using nanomedicines introduced systemically.

The remarkable discovery of enhanced permeability and retention (EPR) by Maeda et al. revolutionized the creation of nanomedicines, specifically for targeting tumors. EPR targeting is a method that employs abnormal tumor structure to deliver drugs solely to tumor tissue [9,10]. Although preclinical data were promising, utilizing the EPR effect in clinics was only partially successful. The reasons behind EPR clinical challenges have been studied extensively in the literature. It is mainly related to the exaggeration of the EPR effect in preclinical tumor models and the heterogeneous nature of this effect [11,12]. Relying solely on EPR targeting has proved insufficient in clinical practice [13,14].

While nanomedicine alone did not emerge as a comprehensive remedy for cancer, it has achieved significant milestones. Firstly, it made pharmaceutical scientists look more carefully at cancer biology and better identify and circumvent the biological barriers against drug delivery [15–18]. Second, nanomedicine paved the way for designing many drug carrier formulations that successfully encapsulate bioactive agents [19,20]. These particles provided a platform for novel targeting and delivery techniques. Many scientists are further manipulating successful drug carriers developed in the nanomedicine era to improve drug delivery to a target tissue. One good example is designing the lipid nanoparticle formulation used in FDA-approved COVID-19 vaccines [21–23]. It is important to note that classic nanomedicine is still an active area of research and novel nanoparticle designs are still being investigated or modified to overcome clinical limitations. A few articles have reviewed recent advances in nanomedicine and particle designs [24,25]. However, nanomedicine has branched into major categories as various strategies have been explored to account for EPR targeting limitations. One of these promising strategies is pharmacological and physical co-treatment, which is the focus of this review. In the following, we delve deeper into solid tumor drug delivery after the EPR era. This review thoroughly explores the promising field of pharmacological and physical co-treatment using ultrasound (US) techniques.

2.1. Nanomedicine beyond the EPR Effect

Combination therapies, localized enhancement of vascular permeability, and ligand targeting are some approaches that have been examined in depth to minimize the requirements for the EPR effect. These strategies have been well documented in many studies [26–31]. In this review, our focus is on tumor-selective drug release using acoustic methods. This field of research is still in its infancy and has yet to be tested in clinical trials. Drug carriers can be made sensitive to endogenous stimuli specific to tumor tissues or exogenous energies to release their payload locally in tumors. Intrinsic stimuli targeting uses tumor anomalies, such as lower PH, hypoxia, or abnormal enzyme activities, to trigger drug release from carriers [32–34]. In contrast, external targeting applies physical stimuli such as heat, magnetism, electricity, or ultrasound locally to tumor tissue and triggers drug release from a stimuli-sensitive carrier [35–37]. Heat-sensitive particles, such as temperature-sensitive liposomes (TSLs) Thermodox®, are being explored for this purpose [38–40] and have reached clinical trials [41]. Thermosensitive drug vehicles have delivered positive results, but they have faced their own set of challenges, which include low heat resolution, the potential of energy dispersion, and difficulties with administering hyperthermia within clinical settings [42]. Although this drug-delivery approach is quite promising, better options for targeting deep tissue have been devised by altering the external stimulus, such as radiofrequency ablation or ultrasound [43]. The advent of active targeting using external stimuli led to the development of a more comprehensive drug delivery strategy called image-guided drug delivery (IGDD). IGDD is an umbrella term for using an imaging modality to guide drug vehicles to the target tissue, trigger the drug release, and visualize the location and the amount of drug release [44]. IGDD could give physicians a platform to rigorously control drug release and adjust the amount of drug unloading by manipulating the external energy source. Heat-triggered drug release cannot constitute an IGDD platform due to the lack of a feedback system. Various imaging modalities were studied to build a real-time IGDD platform to trigger and measure drug release. Among different medical imaging modalities, ultrasound is famous for its cost-effectiveness, high penetration depth, and spatiotemporal precision.

2.2. Improving Delivery by Drug Uncaging

Through investigation, it became clear that one of the key concerns early on in nanomedicine was a lack of drug release from carriers [45]. This posed a problem for liposomal doxorubicin and cisplatin liposomes, two treatments whose efficacy showed only minor improvement compared to their free counterparts [46–48]. One of the reasons for this failure was the lack of a release mechanism from the liposomes to make the drugs bioavailable [49,50]. To combat this issue and increase success rates, researchers sought innovative ways to trigger drug unloading upon a stimulus, thus opening up an entirely new field dubbed "drug uncaging" [45,51,52]. Drug uncaging, on-demand drug release, or triggering drug release all refer to the same concept: drug unloading upon excitement with a stimulus. For the rest of this review, we use the term drug uncaging for consistency. Utilizing external energies such as light, heat, or ultrasound to destabilize drug carriers is widely used for drug uncaging [53]. Despite its precision in applications, the role of ultrasound has yet to be fully explored. In this article, we delve into how therapeutic drugs are transported within tumors and evaluate the potential benefits that could arise from using ultrasound as an effective means of uncaging medicine.

3. Drug Transport in Tumor Tissue

3.1. Convection and Diffusion

The two main transport mechanisms for drug delivery to tumors are convection and diffusion [54]. Convection is the process by which fluids move from one area to another due to differences in pressure or temperature. It is often used in drug-delivery systems as it can direct a drug toward its intended target with greater accuracy than diffusion alone. Diffusion, on the other hand, refers to the process by which molecules move from an area of higher chemical potential (partial molar Gibbs free energy) to an area of lower chemical potential [55]. In drug delivery, the concentration gradient of a molecule influences the chemical potential, and typically, active molecules transport from an area of low concentration to an area of high concentration across tissues. This process occurs naturally throughout our bodies but can be enhanced with certain techniques such as ultrasound. Diffusion and convection always coexist in vivo. However, drug and tissue properties dictate which mechanism is the dominant form of transformation. There is a dimensionless number called the Péclet number (Pe), named after a physicist, that predicts which transport mechanism is the primary form. The Péclet number is the ratio of the convective flux to the diffusive flux, in which for $Pe > 1$, the primary transport mechanism is convection, and for $Pe < 1$, it is diffusion [55,56]. A drug's molecular weight (MW) significantly impacts its rate and type of transport [57]. Convective flux is directly related to the MW through the retardation coefficient. Diffusion coefficients are inversely proportional to the hydrodynamic size, demonstrating that smaller molecules have an easier time diffusing. Therefore, larger drugs primarily move through convection, while small drugs typically move by diffusion (Figure 1). As the MW increases, the Pe number also increases, which can affect how drugs interact with cell membranes and ultimately how they are transported. Thus, it is critical to consider the MW when attempting to understand the transport process of a drug [57].

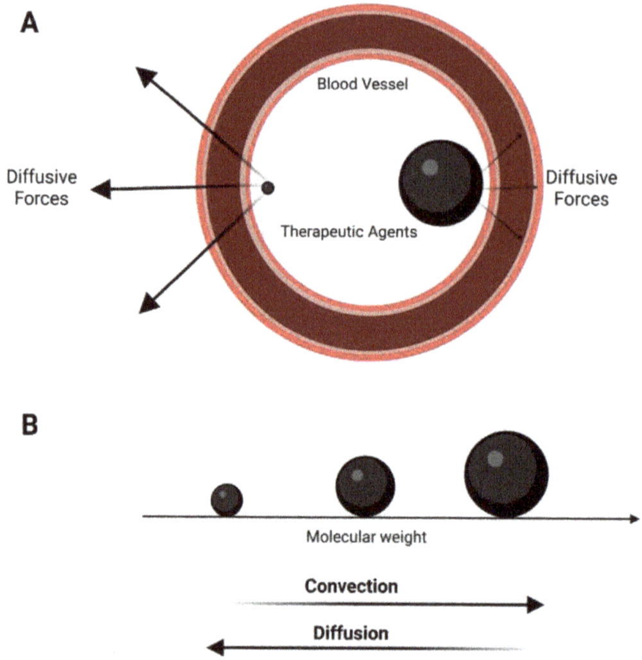

Figure 1. Two major drug transport mechanisms in blood vessels. (**A**) Small therapeutic agents are subjected to stronger diffusive forces than larger ones. (**B**) Molecular weight is directly related to convection and inversely with diffusion.

3.2. Small Molecule Chemotherapy Delivery

Chemotherapeutic molecules have the difficult task of penetrating tumors to reach and destroy the source of cancer. The vascular endothelium is the main barrier that chemotherapeutics must cross to mount an effective attack on tumor tissue. Fickian diffusion is one of the primary means by which drugs diffuse past the vascular endothelium into tumor tissue. Because diffusion is not particularly efficient due to its concentration-gradient-based nature, it requires high drug concentrations in the plasma. Furthermore, the lack of tissue specificity of small molecule chemotherapeutics poses a significant challenge to drug development. Off-target toxicity is a significant concern due to the potential for unwanted side effects that translates into a limitation in dosage in clinical settings.

3.3. Drug Encapsulation: Advantages and Challenges with Nanoparticle Drug Delivery

Encapsulated drugs in nano or micron-size carriers have a Pe number higher than 1 (Pe >> 1), indicating that their transportation is reliant on convective forces and very low in diffusion [58]. However, free drugs usually have Péclet numbers lower than or close to 1 (Pe << 1) and are higher in diffusion. This is one of the primary reasons for the necessity of drug encapsulation. Free forms are drugs rapidly diffuse into the tissue at the injection site before reaching the target. Moreover, most drugs have high lipid solubilities and are inclined to escape from the bloodstream into the tissue, leading to a low circulation half-life. Drug encapsulation drastically improves lipid-soluble drugs' circulation time by increasing solubility in aqueous solutions [59]. Despite the numerous advantages of drug encapsulation, it can reduce drug bioavailability in target cells and tissues, especially solid tumors. This is because convective forces inside a solid tumor are minimal due to its dense microenvironment, high interstitial pressures, and poor lymphatic drainage, rendering bulk flow negligible [60–62]. Consequently, most encapsulated drugs stay on the periphery of the tumor and cannot penetrate deep into the tumor bulk [15]. Solid tumor physiological abnormalities and their effects on drug delivery have been thoroughly characterized in the literature [16,63].

Slow convection flow is an established barrier to drug delivery, especially at the center of the tumor. Manual convective flow increase is one solution to this problem and has been explored extensively [64,65]. One of the key ideas to increase the convective flow in the vessels is to use ultrasound waves to open the endothelial cells to enhance the transport of the encapsulated drug (sonoporation effect) [66]. Sound waves can also affect therapeutic molecules. It was found that therapeutic molecules can be dislocated in the direction of ultrasound wave propagation. This effect, coined acoustic streaming, can potentially be used to externally control and guide drug molecules for precise spatial delivery to a target tissue or manually enforce drug penetration in solid tumors [67,68].

Another key area of focus for circumventing slow convection is drug uncaging, a promising solution for overcoming large molecule transport obstacles. This approach was initially examined with thermosensitive liposomes [69,70]. Fast drug release creates an intense concentration gradient of small molecule bioavailable drugs, thus changing the dominant mode of transport to diffusive forces. This concept was demonstrated by Manzoor et al., who showed intravascular drug discharge stimulated from temperature-sensitive liposomes in vivo [71], resulting in enhanced uptake levels. The results of the study indicated that intravascular drug uncaging led to a 2-fold increase in penetration depth and increased the duration of contact between drugs and tumor cells by 11–17 fold. This was explained as being due to slower wash-out rates resulting from deeper drug penetration. Other research has likewise proved this idea further, demonstrating a diminished washout rate when employing intravascular drug uncaging [72].

Despite the successful attempt to improve drug internalization, controling thermal drug uncaging can be difficult, especially at greater tissue depths where heat dissipation is a significant problem [42,73]. Additionally, the quick release of drugs can be difficult to manage due to the conflict between the speedy release of particles in vivo and shelf-life stability. An optimal system should be able to instantly unlock its contents yet retain constancy when stored at room temperature. The following section focuses on fresh explorations of using ultrasound to uncage drugs from nanoparticles—a method that resolves some troubles that arise from thermal unpacking. Here, we define ultrasound-mediated drug delivery as any pharmaceutic system that uses ultrasound to enhance the delivery of active molecules to diseased tissue. In this review, we focus on ultrasound-mediated drug delivery that uses ultrasound-sensitive particles, namely microbubbles (MBs) and phase-shift contrast agents (PCAs). These two types of ultrasound-sensitive particles have evolved as ultrasound contrast agents and have been extensively characterized in the literature and used for drug release purposes [74–76]. In the following section, we briefly explain ultrasound-sensitive particles, their behavior in the US field, and their application in drug-delivery systems.

4. Ultrasound and Its Contrast Agents

Ultrasound, which consists of inaudibly high-frequency sound waves, has been used for medical imaging for many years. It is primarily used for echocardiography, obstetrics, and abdominal imaging [77,78]. Ultrasound is a safe and cost-effective imaging technique suitable for visualizing the anatomy of deep tissues [79] and blood flow using Doppler imaging [80]. Although ultrasound biomedical research has historically revolved around imaging, the discovery of ultrasound contrast agents has opened new research avenues [81–83]. The first ultrasound contrast agent (UCA) was serendipitously discovered in the 1960s when two physicians accidentally imaged bubbles in the bloodstream via a US machine. Gramiak and Shah observed a bright ultrasound signal at the injection site while administering saline to a patient. This group discovered that the intense signal was due to the formation of gas bubbles while shaking the saline solution, suggesting that small gas bubbles serve as ultrasound contrast agents [84]. This discovery subsequently led to the development of surfactant-stabilized microbubbles (MBs) for contrast-enhanced imaging. Ultrasound contrast agents have seen a surge in popularity since the 1990s when the Food and Drug

Administration (FDA) first approved their use in contrast-enhanced ultrasound imaging. This approval opened up a wide range of possibilities for drug delivery, enabling targeted and localized treatment of diseases by a technique known as "sonopermeation" [85,86], which is discussed in a subsequent section. In more recent studies, UCAs are being used as efficient drug carriers, enabling researchers to deliver the payload directly to the source of disease in an enhanced and controlled manner [87,88]. Currently, two main types of UCAs, MBs and phase-shift contrast agents (PCAs), are extensively studied for pharmaceutical applications. MBs and UCAs can be loaded with drugs and release their cargo upon ultrasound excitation due to the mechanical cavitation effect. Recent review studies have explored the use of MBs or droplets alone as drug carriers [89–91].

Microbubbles and Phase-Shift Contrast Agents

Microbubbles (MBs) are gas spheres stabilized within a phospholipid, protein, or polymeric shell. MBs go through volumetric oscillation in an ultrasound field (Figure 2). This oscillation displaces the liquid medium surrounding the microbubble and produces a secondary ultrasound field that can disrupt surrounding particles or membranes [92,93]. Thus, MBs are US point scatterers used extensively as contrast agents for US imaging [94]. The response of US contrast agents in an ultrasound field has been thoroughly studied [95–97]. Microbubbles are available commercially in several formulations. The first microbubble formulation approved by the FDA was Albunex, consisting of an air bubble entrapped in an albumin shell [98]. Subsequent generations of microbubbles were designed using lipid shells to improve elasticity and fluorocarbon gases for enhanced stability [99]. From several available microbubble formulations widely used in the clinic, one can mention Definity® (Waterloo, ON, USA), Sonovue® (Milan, Italy), and Optison® (St. Louis, MO, USA).

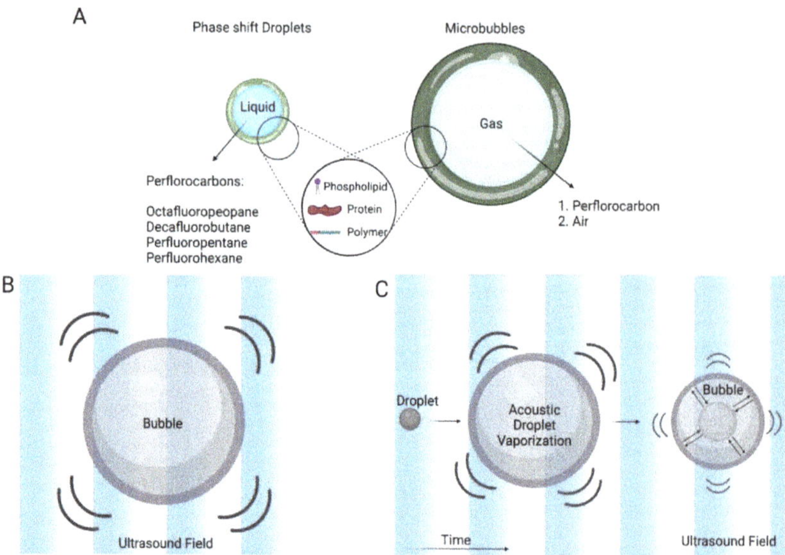

Figure 2. Microbubbles and phase-shift contrast agents' structure. (**A**) Ultrasound-sensitive agents are encapsulated in a shell for improved stability (**B**) Microbubbles expand and contract in an ultrasound field. (**C**) Phase-shift droplets go through vaporization when excited with ultrasound. They overexpand past their final resting bubble size and then relax at a smaller resting size, where they oscillate as a microbubble.

Phase-shift contrast agents (PCA) are a relatively newer class of particles made of superheated perfluorocarbon liquids. The perfluorocarbons remain in the liquid phase due to the surface tension of the polymer or phospholipid shell and can be vaporized upon US excitation or ambient temperature changes [100]. This phenomenon is called acoustic droplet vaporization (ADV). Upon vaporization, the droplet overexpands (nearly five times) past its final bubble size and then relaxes at a smaller resting size, where it oscillates as an MB [101–103] (Figure 2). This vaporization event dislocates the liquid medium and induces high levels of shear force on nearby boundaries or surrounding particles [104–106]. Phase-shift droplets have been used for US imaging and US-mediated drug delivery [107]. Following this short introductory section, we delve deeper into ultrasound contrast agents and their most significant accomplishments in drug delivery. We further focus on novel strategies for future endeavors in the field.

5. Ultrasound-Mediated Drug Delivery

To fully understand the benefits of ultrasound agents in drug delivery, it is critical to determine the biological effects of microbubble oscillation and droplet vaporization in the body. Kooiman et al. recently conducted a review to describe and classify the physical and chemical impacts of ultrasound reactive particles on biological tissue [108]. Ultrasound activation of particles can cause various thermal, chemical, and physical changes. However, ultrasound pharmaceutical research is mainly focused on sonoporation (also called 'sonopermeation'), which has proven highly advantageous.

Sonoporation is the most well-known phenomenon in ultrasound drug delivery and consists of "sono" and "poration", which means inducing gaps using sound energy. Sonoporation in biology means inducing small pores in endothelial cells by ultrasound waves [96] (Figure 3). The sonoporation effect was classically proposed to improve the EPR effect locally in the tumor by generating capillary gaps. Numerous studies have established that MB cavitation or droplet vaporization can improve intracellular drug uptake [106,109,110]. Following the success of sonoporation in preclinical research, studies were conducted to explore the biological mechanisms involved [111,112]. Further studies discovered that sonoporation could induce gaps between confluent cells or the cellular membrane [113], and it became increasingly favored in brain delivery applications due to its potential to open theblood–brain barrier (BBB) [114,115]. Brain sonoporation reached clinical trials and is currently being evaluated as a therapy method for various diseases, such as amyotrophic lateral sclerosis (ALS), Alzheimer's disease, and glioblastoma (ClinicalTrials.gov Identifiers: NCT04118764, NCT02343991, NCT05615623) [116–118]. Sonoporation has also been utilized to enhance the effectiveness of standard chemotherapeutics for treating inoperable pancreatic ductal adenocarcinoma (PDAC) patients in clinical trials (ClinicalTrials.gov Identifier: NCT04821284). Although sonoporation has been successful in preliminary studies and showed positive results in clinical trials, the long-term consequences of this technique are still largely unknown [119]. A few studies have demonstrated the disruption in cell homeostasis as a consequence of sonoporation [120], suggesting that further investigation is essential to completely comprehend the potential adverse effects of sonoporation. A particular study showed adverse effects from opening a specific brain site in rats [121], and a few review papers have warned about ubiquitous clinical BBB opening before studying long-term consequences [122,123]. As Przystupski et al. explored in a recent review, sonoproration provokes various biological responses that are not limited to the cell membrane and include intracellular and intercellular activities. Sonoporation alters the structural and physical integrity of affected cells. It is imperative that it also alters the cell's function, either temporarily or, at high doses, permanently, an issue that is more critical in the brain. The BBB is one of the most sophisticated and crucial parts of the brain's evolution history that prevents the brain's exposure to toxins. Even temporary disruption in the integrity of the BBB might prove detrimental by exposing the most vulnerable part of the human body to pre-existing toxins, such as viruses or bacteria [124,125].

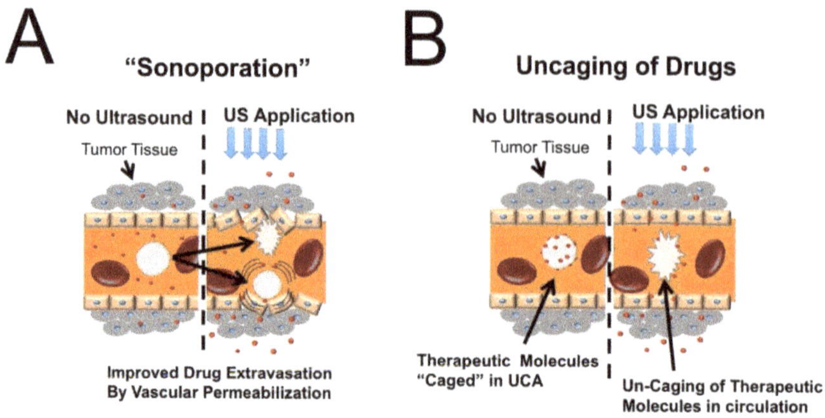

Figure 3. Comparison between drug uncaging and sonoporation strategies (**A**) Sonoporation induces small gaps between endothelial cells to enhance drug convection through diseased tissue. (**B**) Drug uncaging is designed to release drugs rapidly in the bloodstream and create a diffusion gradient for drug transport across the endothelial membrane.

Furthermore, there are arguments in the scientific community that inertial cavitation from microbubbles might enhance metastasis possibilities [126,127]. Concerns about the safety of the sonoporation effect gave rise to an attempt to pursue alternative ultrasound delivery approaches in which the integrity of blood vessels stays intact. Following the tumor vascular normalization idea for cancer therapy, first introduced by Rakesh Jain [128,129], drug delivery to endothelial cells gained popularity. Consequently, ultrasound-mediated drug delivery branched into various techniques to deliver therapeutics without the sonoporation effect. This idea was especially paramount in brain drug delivery as sonoporation is probably most practiced for opening the BBB [115,130,131].

Materials for Ultrasonic Drug Uncaging

Ultrasonic drug uncaging is only achievable when drugs are embedded in microbubbles/droplets, either on their surface, within the outer shell, or inside the core. The drug loaded onto ultrasound contrast agents (UCAs) rapidly releases after the shell material destabilizes upon ultrasound excitation. The mechanisms by which drug release is triggered depend on ultrasound parameters, and multiple mechanisms have been proposed in the literature, which are not necessarily mutually exclusive. Three major mechanisms are cavitation (which applies high shear forces to the surrounding particles and membranes), thermal ultrasonic effects, and acoustic streaming, resulting in increased frequency and intensity of particle collisions that eventually lead to drug release. The primary mechanism of release is still debated [74]. All the drug vehicles discussed here are designed to utilize these ultrasonic effects to destabilize their structures and release their cargo, in contrast to conventional ultrasonic drug vehicles (such as MBs or PCAs) that directly use ultrasound cavitation as a release mechanism (Figure 3).

One of the primary challenges associated with drug loading solely on MBs or PCAs is the restricted surface area for loading. To enhance payload and target tissue specificity, numerous studies have aimed to develop high-capacity drug-loaded ultrasound contrast agents [132–134]. One approach to address the drug capacity issue is to use polymeric microbubbles with thick shells [135]. El-Sherif et al. were the first to develop a polymeric MB formulation [136]. Later, the feasibility of loading high payloads of chemotherapeutics on polymer-based MBs and triggering drug release upon US exposure was demonstrated [137–140]. More recently, more accessible methods for generating polymeric MBs with potential drug-release applications have been proposed [141,142]. Another approach to improving drug loading on MBs is linking drugs/nanoparticles to the bubbles' surface (see Chapla et al. [143] for a comprehensive account of drug-loaded microbubble–nanoparticle complexes). The first attempt in this front was conjugating MBs to liposomes, conducted by Kheirolomoom et al. [144] and Lentacker et al. [145]. Other groups showed the feasibility of drug release upon ultrasound excitation using MB–liposome constructs [146–148], and they further used it for drug delivery in preclinical tumor models [149–153]. Another novel approach to improving loading capacity is using phase-change droplets with a liquid perfluorocarbon core for loading drugs. Rapoport et al. were the first group to uncage drugs using PCAs after co-loading chemotherapeutics and perfluorocarbons in micelles [154]. This group used drug-loaded (doxorubicin and paclitaxel) polymer nanodroplets to target tumors passively via the EPR effect and triggered drug release upon ultrasound exposure [155,156]. Droplets became a successful alternative for MBs in drug-delivery applications, and many groups showed successful drug delivery to target tissues in preclinical models [157–161]. One of the limitations of PCAs as drug carriers is the limited solubility of drugs in the perfluorocarbon core. To address this issue, our group conjugated droplets to liposomes for the first time to improve droplets' loading capabilities and demonstrated drug uncaging upon ultrasound application in vitro and in vivo [162].

6. Emerging Areas with Ultrasonic Drug Uncaging

Research on the development of drug-loaded ultrasound contrast agents typically serves to augment sonoporation rather than supplant it. However, this area of ultrasonic drug delivery is starting to branch out and develop in new directions. In recent years, a new field has emerged to deliver drugs directly into target sites without harming the vascular tissue, particularly applicable when treating brain diseases where even short-term BBB opening may be considered unsafe or undesired during therapeutic intervention. Airan et al. first introduced and investigated this idea when they uncaged propofol in a rat's brain [163]. Delivery was carried out by propofol uncaging from polymeric PCAs and was confirmed by monitoring the brain activity via EEG measurements. Intact BBB integrity was confirmed by MRI and histology evaluations. This work was the first attempt to maintain blood vessel integrity after drug uncaging [164]. More recently, Lea-Banks et al. efficiently delivered a barbiturate anesthetic to the mouse brain by vaporizing lipid PCAs with no collateral damage to the BBB, confirmed by MRI and histology [165], and they also showed sub-millimeter precision in the delivery [166].

Microbubbles have also been used for non-invasive drug uncaging. In a recent study, Ozdas et al. demonstrated that low-intensity ultrasound applied to a liposome–microbubbles construct improves localized delivery of an encapsulated $GABA_A$ receptor agonist (muscimol) without compromising blood-brain barrier integrity [167]. In another study, Gorick et al. successfully transfected plasmids to endothelial cells in a mouse brain using low-intensity ultrasound and plasmid-loaded MBs [168]. This study showed no BBB opening after US exposure using MRI measurements. Ultrasound-mediated brain drug uncaging without affecting BBB integrity is becoming increasingly popular, especially for gene delivery.

6.1. Incorporating Cavitation Nuclei in Drug-Delivery Vehicles

As discussed earlier, one of the primary limitations of using conventional US contrast agents as a drug carrier is the low drug capacity. One of the key ideas to address this issue is incorporating ultrasound-sensitive agents inside conventional drug carriers to sensitize them to US waves (Figure 4).

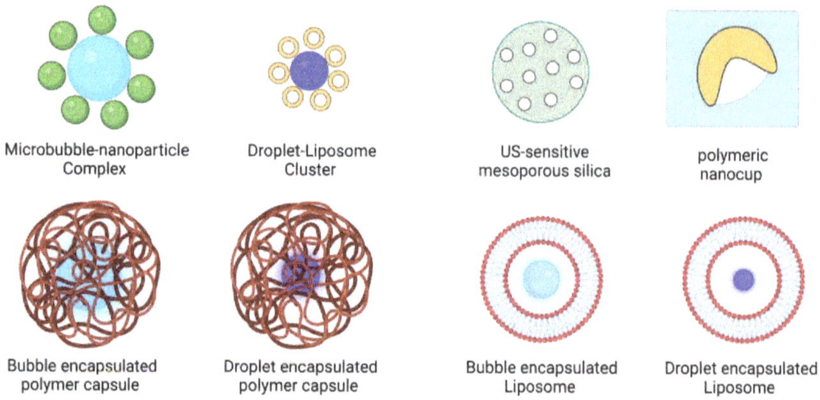

Figure 4. Ultrasound-sensitive particles designed for drug-release applications. Combining ultrasound contrast agents with various drug carriers results in novel ultrasound-sensitive drug vehicles that trigger release upon ultrasound radiation.

6.2. Microbubble-Nested Drug Carriers

One of the first attempts on this front was encapsulating lipid-shelled microbubbles inside polymeric microcapsules by Wrenn et al. [169]. Although the motivation for this study was primarily to give lipid shell MBs a longer circulation time and shelf stability [170], this group showed some advantages in drug delivery [169]. This group also encapsulated PCAs in microcapsules for potential drug-delivery applications but mainly investigated their imaging applications [171]. Encapsulating bubbles inside liposomes by the same group was the first attempt to nest cavitation nuclei in conventional drug carriers solely for drug-delivery applications. This group reported surrogate drug release in vitro from bubble-nested liposomes upon ultrasound excitation and bubble cavitation [172]. Ibsen et al. also entrapped lipid-shelled microbubbles in liposomes for drug-delivery applications [173]. More recently, Batchelor et al. nested nanobubbles in liposomes to reduce the construct's size in the submicron range. This group demonstrated drug uncaging upon activation by high-intensity focused ultrasound (HIFU). It is worth noting that this study reports nested nanobubbles to be a mixture of perfluorocarbon gas and droplet [174].

6.3. Droplet-Nested Drug Carriers

An alternative approach for sensitizing liposomes to US was nesting PCAs inside them. A few groups encapsulated perfluorocarbon droplets in liposomes and demonstrated successful drug uncaging in vitro [175,176]. The limitation of the abovementioned studies is the lack of in vivo evidence for the feasibility of drug uncaging upon US excitation from ultrasound-sensitive particles nested in drug carriers. Further studies need to address this issue.

6.4. Nesting Cavitation Nuclei in Solid Drug Carriers

Embedding cavitation nuclei in dug vehicles was also pursued in solid drug vehicles. Solid particles can stabilize or nucleate gas within pores or crevices on their surface or mantle to produce acoustically active agents [177]. This phenomenon has been used for drug delivery applications, mainly by entrapping gas bubbles in silica particles (mesoporous silica) and polymeric particles [178]. Mesoporous silica particles have been widely used for drug uncaging [179,180]. Kim et al. did one of the first studies on this front when they delivered ibuprofen from mesoporous silica upon ultrasound activation in vitro [181]. This strategy was further proven helpful in preclinical tumor models [182].

The first approach to embed cavitation nuclei in polymeric particles was developing cup-shaped polymers that can entrap small gas bubbles in their structure. Kwan et al. developed nano-sized polymeric cups to entrap gas bubbles and further demonstrated their ultrasound activity and drug-release capabilities [183]. In another attempt, the same group developed porous PLGA microparticles to trap gas bubbles and potential ultrasound-mediated drug-delivery applications [184]. More recently, our group has developed a facile method to nucleate bubbles inside polymeric microcapsules to produce ultrasound-sensitive drug carriers. We successfully nucleated bubbles in polylactic acid microcapsules by resuspending them in a propylene glycol/glycerol solution. We characterized the formulated particles and demonstrated surrogate drug loading on them. This method can drastically reduce the time and energy of conventional methods [141]. A recent paper by Sabuncu et al. reviewed gas-stabilizing solid particles for pharmaceutic applications [185].

7. Materials for Ultrasound-Sensitive Drug Carriers

In the previous sections, we discussed various structures scientists developed to trigger drug release upon ultrasound excitation. In this section, we provide a brief summary of different materials used to create these structures and compare their advantages and disadvantages (Table 1).

Table 1. Materials used in ultrasound-sensitive drug carrier designs.

<table>
<tr><th>Structure</th><th colspan="2">Materials</th><th>Pro</th><th>Con</th></tr>
<tr><td rowspan="4">UCA nested particles</td><td rowspan="2">Nested UCA Shell Material</td><td>Phospholipid (DSPC, DPPC)</td><td rowspan="4">1. High drug-loading capacity
2. Long circulation time
3. Suitable for encapsulating both hydrophobic and hydrophilic agents</td><td rowspan="4">1. Typically micro-size range
2. Limited ultrasound sensitivity</td></tr>
<tr><td>No shell</td></tr>
<tr><td rowspan="2">Encapsulation material</td><td>Polymer (PLGA, PLA, PBCA)</td></tr>
<tr><td>Phospholipid (DSPC, DPPC, DMPC)</td></tr>
<tr><td rowspan="4">UCA-Particle complex</td><td>UCA</td><td>Phospholipid (DSPC, DPPC)</td><td rowspan="4">1. Improved ultrasound sensitivity
2. Improved drug release</td><td rowspan="4">Limited circulation time compared to polymers</td></tr>
<tr><td rowspan="3">Drug carrier</td><td>Phospholipid (various PCs)</td></tr>
<tr><td>Polymer (PLA, PLGA, PEG)</td></tr>
<tr><td>Miscellaneous material (metallic particles, proteins, etc.)</td></tr>
<tr><td>Coned shape particles</td><td colspan="2">Air-entrapped polymer (PLA, PLGA)</td><td>1. Biocompatible
2. Long circulation time
3. Biodegradability</td><td>1. Large size
2. Limited ultrasound sensitivity</td></tr>
<tr><td>Mesoporous particles</td><td colspan="2">Air-entrapped silica</td><td>1. Long circulation time
2. Biocompatible</td><td>Limited ultrasound sensitivity</td></tr>
</table>

7.1. Lipids

Phospholipids are commonly used for designing drug carriers and have shown great success in clinical applications, making them one of the primary choices for drug carrier design [186]. In the field of nanomedicine, phospholipids have evolved to be biocompatible and effective in encapsulating active molecules. Phosphatidylcholines (PCs), a class of phospholipids, have been widely used for this purpose [187]. In many of the newer carrier structures discussed in this paper, phospholipids are the predominant material. They are extensively employed in producing complexes of UCAs with nanoparticles [143]. Phospholipid bilayers are also used to encapsulate UCAs, providing stability and increased drug-loading capacity by generating nested UCAs [172,173]. Phospholipids mimic cellular bilayers and possess desirable biocompatibility. Additionally, their elasticity make

UCAs from phospholipids respond well to ultrasound signals. However, the drug-loading capacity of phospholipid particles can be limited [188].

7.2. Polymers

Polymers are another primary class of materials used in ultrasound-mediated drug delivery. They offer chemical flexibility to drug carriers, allowing modification of their mechanical and physical properties to suit specific biological environments. This chemical flexibility is particularly advantageous in microbubble design, as altering the mechanical properties of the microbubble shell enables control of their response to ultrasound [189]. Polymers also provide the carrier with the unique ability to be biodegradable [190].

Among various polymer formulations, polyesters are attractive polymers extensively used in drug delivery due to their biocompatibility and biodegradability. Poly(lactic-co-glycolic acid) (PLGA), a copolymer of two polyesters, poly(lactic acid) (PLA) and poly(glycolic acid) (PGA), is one of the primary polymers used for generating microbubbles [191]. Another desirable polymer for microbubble production is poly(butyl cyanoacrylate) (PBCA), which is well tolerated by the body and is degradable [139,140,192]. Several groups, including ours, have pursued encapsulating ultrasound contrast agents in a polymeric shell for drug delivery or imaging applications [162,189]. In addition to chemical flexibility and degradability, polymers can offer in vivo stability to a drug carrier. Polymers are rigid structures capable of circulating in the bloodstream for days to months. Due to their entangled structure, they can provide better drug-loading capacity than phospholipids by offering a thicker shell material. However, one drawback of polymers is their reduced reactivity to ultrasound signals due to the lack of elasticity compared to lipids [193].

It is worth noting that proteins are also a class of materials used for designing ultrasound-sensitive particles (particularly MBs) in the past but have been a less popular choice for designing UCAs recently [194].

8. The Role of Ultrasound Parameters in Drug Vehicle Design

One major consideration for developing ultrasound drug delivery carriers is safety. Ultrasound exposure safety involves two aspects: the bioeffects of ultrasound energy that activates the drug carriers and the bioeffects of cavitation after carrier activation.

The initial designs for acoustic drug delivery with drug carriers utilized the inertial cavitation of bubbles to destabilize the carriers. The first attempt involved co-injecting MBs and drug carriers [195]. Later, MBs were bound to drug carriers to enhance drug release by increasing their proximity [143]. However, this method requires high levels of ultrasound energy, and inertial cavitation itself can be harmful to the tissue due to the strong ultrasound field or shock waves it produces. Concerns about the safety of inertial cavitation led to the use of stable cavitation as a delivery mode. However, the sonoporation effect soon overshadowed this approach, and fewer attempts were made to solely destabilize drug carriers through stable cavitation [196]. Concerns about the sonoporation effect have been extensively discussed in previous sections. The exact ultrasound parameters used in studies utilizing inertial or stable cavitation of MBs to enhance drug release and delivery vary due to different experimental conditions. For more information on ultrasound parameters, please refer to the review paper by Kooiman et al. [197].

Another approach to reduce the ultrasound energy is to use perfluorocarbon-based PCAs instead of MBs. The activation threshold of PCAs can be controlled by adjusting the boiling point of the PFC core. Theoretically, one can design a PCA that can be activated at different ultrasound energy levels by altering the chemistry of the fluorocarbon core [100,198]. However, concerns regarding the bioeffects of vaporization effects still exist.

An alternative approach to reducing the side effects of cavitation is to encapsulate the cavitation effect within a vesicle. This concept led to the nesting of cavitation nuclei within drug carriers. By placing the cavitation nuclei in a vehicle and enclosing the UCAs, the effects of cavitation can be localized and directed solely to the carrier structure, thereby reducing the bioeffects. However, concerns remain regarding the activation threshold of

UCAs when embedded in a membrane-bound vesicle. Wrenn et al. observed an increase in cavitation thresholds when MBs were nested within liposomes [199]. This concern can potentially be addressed by utilizing low-boiling-point PCAs with low activation thresholds.

More recently, attention has been given to alternative approaches for ultrasound targeting and delivery that are less invasive and can be used with low-intensity ultrasound. Other targeting modes, such as acoustic streaming or primary acoustic radiation forces, have gained attention as valuable methods for ultrasound-mediated drug delivery. The following section briefly covers recent advances in pharmaceutical uses of acoustic radiation forces.

9. Acoustic Radiation Forces in Drug Delivery

Before concluding this paper, we want to briefly define the application of acoustic radiation forces in pharmaceutics and discuss recent advances in this field. Ultrasound can offer a different targeting strategy by spatially controlling and dislocating acoustic-sensitive agents. It is known that MBs can be displaced in a liquid medium using low pressure acoustic waves due to the absorption of the sound wave momentum. This phenomenon is known as primary acoustic radiation force and is particularly advantageous in drug-delivery applications using low-intensity US waves [200,201]. Radiation forces are one of the less explored concepts in drug delivery despite its promising potential [202]. The concept was first explored in vivo by Dayton et al., where they displaced commercially available microbubbles to the wall of a blood vessel in the mouse cremaster muscle [200]. Later, Lum et al. demonstrated the feasibility of particle deposition on a tube using acoustic radiation [203].

One of the novel methods for ultrasound targeting is utilizing radiation forces to remotely implant sensitive acoustic particles in diseased tissue [142]. In a recent paper, Su et al. successfully implanted multi-cavity PLGA microparticles in a foam-cell spheroid model by ultrasound exposure [204]. The implanted particle can slowly release its cargo to the diseased tissue. Our group is trying a strategy to use acoustic radiation forces to deposit liposomes on endothelial cells, similar to the sonoprinting approach first described by Cock et al. [205]. Liposomes (or any desired drug vehicle) slowly release their cargo to the tumor tissue or the endothelial cell over time. This method uses safe, low-intensity ultrasound waves and the integrity of the endothelial cells is intact as per our hypothesis. We have shown successful lipid deposition onto neuroblastoma tumor cells in vivo.

10. Conclusions

Nanomedicine has provided hope in delivering drugs systemically, but it has only been partially successful clinically. Recent studies have shown that combination therapies, localized enhancement of vascular permeability, ligand targeting, and image-guided drug delivery hold significant promise for improved therapy. Additionally, ultrasound as an option for intravascular drug release has emerged as a cost-effective and precise technique due to its high penetration depth and spatiotemporal precision. The future of drug targeting is exciting, with more technological advancements on the horizon. We can expect to see even more precise therapies and treatments with fewer side effects in the coming years, resulting in better patient health outcomes.

Author Contributions: Conceptualization, A.H. and S.R.S.; writing—original draft preparation, A.H.; writing—review and editing, S.R.S.; Funding Acquisition, S.R.S. All authors have read and agreed to the published version of the manuscript.

Funding: This research was funded by the National Institute of Health through the NCI grant number R01CA235756.

Acknowledgments: The authors want to acknowledge the Biorender platform. The figures in this paper were created by Biorender.com.

Conflicts of Interest: The authors declare no conflict of interest.

References

1. Parascandola, J. The Theoretical Basis of Paul Ehrlich's Chemotherapy. *J. Hist. Med. Allied Sci.* **1981**, *36*, 19–43. [CrossRef] [PubMed]
2. Thorburn, A.L. Paul Ehrlich: Pioneer of Chemotherapy and Cure by Arsenic (1854–1915). *Br. J. Vener. Dis.* **1983**, *59*, 404. [CrossRef] [PubMed]
3. Strebhardt, K.; Ullrich, A. Paul Ehrlich's Magic Bullet Concept: 100 Years of Progress. *Nat. Rev. Cancer* **2008**, *8*, 473–480. [CrossRef] [PubMed]
4. Valent, P.; Groner, B.; Schumacher, U.; Superti-Furga, G.; Busslinger, M.; Kralovics, R.; Zielinski, C.; Penninger, J.M.; Kerjaschki, D.; Stingl, G.; et al. Paul Ehrlich (1854–1915) and His Contributions to the Foundation and Birth of Translational Medicine. *J. Innate Immun.* **2016**, *8*, 111–120. [CrossRef] [PubMed]
5. Ellis, H. Paul Ehrlich: Nobel Laureate and Father of Modern Chemotherapy. *Br. J. Hosp. Med.* **2015**, *76*, 483. [CrossRef]
6. Podczeck, F.; Jones, B.E. *Pharmaceutical Capsules*; Pharmaceutical Press: London, UK; Chicago, IL, USA, 2004; ISBN 0-85369-568-7.
7. Park, H.; Otte, A.; Park, K. Evolution of Drug Delivery Systems: From 1950 to 2020 and Beyond. *J. Control. Release* **2022**, *342*, 53–65. [CrossRef]
8. Abu-Thabit, N.Y.; Makhlouf, A.S.H. Historical Development of Drug Delivery Systems: From Conventional Macroscale to Controlled, Targeted, and Responsive Nanoscale Systems. In *Stimuli Responsive Polymeric Nanocarriers for Drug Delivery Applications*; Woodhead Publishing: Sawston, UK, 2018; Volume 1.
9. Matsumura, Y.; Maeda, H. A New Concept for Macromolecular Therapeutics in Cancer Chemotherapy: Mechanism of Tumoritropic Accumulation of Proteins and the Antitumor Agent Smancs. *Cancer Res.* **1986**, *46*, 6387–6392.
10. Iyer, A.K.; Khaled, G.; Fang, J.; Maeda, H. Exploiting the Enhanced Permeability and Retention Effect for Tumor Targeting. *Drug Discov. Today* **2006**, *11*, 812–818. [CrossRef]
11. Prabhakar, U.; Maeda, H.; Jain, R.K.; Sevick-Muraca, E.M.; Zamboni, W.; Farokhzad, O.C.; Barry, S.T.; Gabizon, A.; Grodzinski, P.; Blakey, D.C. Challenges and Key Considerations of the Enhanced Permeability and Retention Effect for Nanomedicine Drug Delivery in Oncology. *Cancer Res.* **2013**, *73*, 2412–2417. [CrossRef]
12. Greish, K. Enhanced Permeability and Retention Effect for Selective Targeting of Anticancer Nanomedicine: Are We There Yet? *Drug Discov. Today Technol.* **2012**, *9*, e161–e166. [CrossRef]
13. Huynh, E.; Zheng, G. Cancer Nanomedicine: Addressing the Dark Side of the Enhanced Permeability and Retention Effect. *Nanomedicine* **2015**, *10*, 1993–1995. [CrossRef] [PubMed]
14. Nakamura, Y.; Mochida, A.; Choyke, P.L.; Kobayashi, H. Nanodrug Delivery: Is the Enhanced Permeability and Retention Effect Sufficient for Curing Cancer? *Bioconjugate Chem.* **2016**, *27*, 2225–2238. [CrossRef] [PubMed]
15. Sriraman, S.K.; Aryasomayajula, B.; Torchilin, V.P. Barriers to Drug Delivery in Solid Tumors. *Tissue Barriers* **2014**, *2*, e29528. [CrossRef] [PubMed]
16. Blanco, E.; Shen, H.; Ferrari, M. Principles of Nanoparticle Design for Overcoming Biological Barriers to Drug Delivery. *Nat. Biotechnol.* **2015**, *33*, 941–951. [CrossRef] [PubMed]
17. Dewhirst, M.W.; Secomb, T.W. Transport of Drugs from Blood Vessels to Tumour Tissue. *Nat. Rev. Cancer* **2017**, *17*, 738–750. [CrossRef]
18. Kim, S.M.; Faix, P.H.; Schnitzer, J.E. Overcoming Key Biological Barriers to Cancer Drug Delivery and Efficacy. *J. Control. Release* **2017**, *267*, 15–30. [CrossRef]
19. Couvreur, P. Nanoparticles in Drug Delivery: Past, Present and Future. *Adv. Drug Deliv. Rev.* **2013**, *65*, 21–23. [CrossRef]
20. De Jong, W.H.; Borm, P.J.A. Drug Delivery and Nanoparticles:Applications and Hazards. *Int. J. Nanomed.* **2008**, *3*, 133–149. [CrossRef] [PubMed]
21. Müller, R.H.; Shegokar, R.; Keck, C.M. 20 Years of Lipid Nanoparticles (SLN and NLC): Present State of Development and Industrial Applications. *Curr. Drug Discov. Technol.* **2011**, *8*, 207–227. [CrossRef]
22. Hou, X.; Zaks, T.; Langer, R.; Dong, Y. Lipid Nanoparticles for MRNA Delivery. *Nat. Rev. Mater.* **2021**, *6*, 1078–1094. [CrossRef]
23. Wilson, B.; Geetha, K.M. Lipid Nanoparticles in the Development of MRNA Vaccines for COVID-19. *J. Drug Deliv. Sci. Technol.* **2022**, *74*, 103553. [CrossRef] [PubMed]
24. Tang, L.; Li, J.; Zhao, Q.; Pan, T.; Zhong, H.; Wang, W. Advanced and Innovative Nano-Systems for Anticancer Targeted Drug Delivery. *Pharmaceutics* **2021**, *13*, 1151. [CrossRef] [PubMed]
25. Trucillo, P. Drug Carriers: Classification, Administration, Release Profiles, and Industrial Approach. *Processes* **2021**, *9*, 470. [CrossRef]
26. Shi, J.; Kantoff, P.W.; Wooster, R.; Farokhzad, O.C. Cancer Nanomedicine: Progress, Challenges and Opportunities. *Nat. Rev. Cancer* **2017**, *17*, 20–37. [CrossRef] [PubMed]
27. Nehoff, H.; Parayath, N.N.; Domanovitch, L.; Taurin, S.; Greish, K. Nanomedicine for Drug Targeting: Strategies beyond the Enhanced Permeability and Retention Effect. *Int. J. Nanomed.* **2014**, *9*, 2539–2555. [CrossRef]
28. Moradi Kashkooli, F.; Soltani, M.; Souri, M. Controlled Anti-Cancer Drug Release through Advanced Nano-Drug Delivery Systems: Static and Dynamic Targeting Strategies. *J. Control. Release* **2020**, *327*, 316–349. [CrossRef]
29. Srinivasarao, M.; Low, P.S. Ligand-Targeted Drug Delivery. *Chem. Rev.* **2017**, *117*, 12133–12164. [CrossRef]

30. Ojha, T.; Pathak, V.; Shi, Y.; Hennink, W.E.; Moonen, C.T.W.; Storm, G.; Kiessling, F.; Lammers, T. Pharmacological and Physical Vessel Modulation Strategies to Improve EPR-Mediated Drug Targeting to Tumors. *Adv. Drug Deliv. Rev.* **2017**, *119*, 44–60. [CrossRef]
31. Bayat Mokhtari, R.; Homayouni, T.S.; Baluch, N.; Morgatskaya, E.; Kumar, S.; Das, B.; Yeger, H. Combination Therapy in Combating Cancer. *Oncotarget* **2017**, *8*, 38022–38043. [CrossRef]
32. Zhu, L.; Torchilin, V.P. Stimulus-Responsive Nanopreparations for Tumor Targeting. *Integr. Biol.* **2013**, *5*, 96–107. [CrossRef]
33. Danhier, F.; Feron, O.; Préat, V. To Exploit the Tumor Microenvironment: Passive and Active Tumor Targeting of Nanocarriers for Anti-Cancer Drug Delivery. *J. Control. Release* **2010**, *148*, 135–146. [CrossRef]
34. Li, R.; Xie, Y. Nanodrug Delivery Systems for Targeting the Endogenous Tumor Microenvironment and Simultaneously Overcoming Multidrug Resistance Properties. *J. Control. Release* **2017**, *251*, 49–67. [CrossRef] [PubMed]
35. Cheng, W.; Gu, L.; Ren, W.; Liu, Y. Stimuli-Responsive Polymers for Anti-Cancer Drug Delivery. *Mater. Sci. Eng. C* **2014**, *45*, 600–608. [CrossRef] [PubMed]
36. Mi, P. Stimuli-Responsive Nanocarriers for Drug Delivery, Tumor Imaging, Therapy and Theranostics. *Theranostics* **2020**, *10*, 4557–4588. [CrossRef] [PubMed]
37. Kauscher, U.; Holme, M.N.; Björnmalm, M.; Stevens, M.M. Physical Stimuli-Responsive Vesicles in Drug Delivery: Beyond Liposomes and Polymersomes. *Adv. Drug Deliv. Rev.* **2019**, *138*, 259–275. [CrossRef]
38. Regenold, M.; Bannigan, P.; Evans, J.C.; Waspe, A.; Temple, M.J.; Allen, C. Turning down the Heat: The Case for Mild Hyperthermia and Thermosensitive Liposomes. *Nanomed. Nanotechnol. Biol. Med.* **2022**, *40*, 102484. [CrossRef]
39. Dai, Y.; Su, J.; Wu, K.; Ma, W.; Wang, B.; Li, M.; Sun, P.; Shen, Q.; Wang, Q.; Fan, Q. Multifunctional Thermosensitive Liposomes Based on Natural Phase-Change Material: Near-Infrared Light-Triggered Drug Release and Multimodal Imaging-Guided Cancer Combination Therapy. *ACS Appl. Mater. Interfaces* **2019**, *11*, 10540–10553. [CrossRef]
40. Amin, M.; Lammers, T.; ten Hagen, T.L.M. Temperature-Sensitive Polymers to Promote Heat-Triggered Drug Release from Liposomes: Towards Bypassing EPR. *Adv. Drug Deliv. Rev.* **2022**, *189*, 114503. [CrossRef]
41. Nardecchia, S.; Sánchez-Moreno, P.; de Vicente, J.; Marchal, J.A.; Boulaiz, H. Clinical Trials of Thermosensitive Nanomaterials: An Overview. *Nanomaterials* **2019**, *9*, 191. [CrossRef]
42. Dou, Y.; Hynynen, K.; Allen, C. To Heat or Not to Heat: Challenges with Clinical Translation of Thermosensitive Liposomes. *J. Control. Release* **2017**, *249*, 63–73. [CrossRef]
43. Lyon, P.C.; Gray, M.D.; Mannaris, C.; Folkes, L.K.; Stratford, M.; Campo, L.; Chung, D.Y.F.; Scott, S.; Anderson, M.; Goldin, R.; et al. Safety and Feasibility of Ultrasound-Triggered Targeted Drug Delivery of Doxorubicin from Thermosensitive Liposomes in Liver Tumours (TARDOX): A Single-Centre, Open-Label, Phase 1 Trial. *Lancet Oncol.* **2018**, *19*, 1027–1039. [CrossRef] [PubMed]
44. Lammers, T.; Kiessling, F.; Hennink, W.E.; Storm, G. Nanotheranostics and Image-Guided Drug Delivery: Current Concepts and Future Directions. *Mol. Pharm.* **2010**, *7*, 1899–1912. [CrossRef] [PubMed]
45. Andresen, T.L.; Jensen, S.S.; Jørgensen, K. Advanced Strategies in Liposomal Cancer Therapy: Problems and Prospects of Active and Tumor Specific Drug Release. *Prog. Lipid Res.* **2005**, *44*, 68–97. [CrossRef] [PubMed]
46. Halford, S.; Yip, D.; Karapetis, C.S.; Strickland, A.H.; Steger, A.; Khawaja, H.T.; Harper, P.G. A Phase II Study Evaluating the Tolerability and Efficacy of CAELYX (Liposomal Doxorubicin, Doxil) in the Treatment of Unresectable Pancreatic Carcinoma. *Ann. Oncol.* **2001**, *12*, 1399–1402. [CrossRef] [PubMed]
47. Matsumura, Y.; Gotoh, M.; Muro, K.; Yamada, Y.; Shirao, K.; Shimada, Y.; Okuwa, M.; Matsumoto, S.; Miyata, Y.; Ohkura, H.; et al. Phase I and Pharmacokinetic Study of MCC-465, a Doxorubicin (DXR) Encapsulated in PEG Immunoliposome, in Patients with Metastatic Stomach Cancer. *Ann. Oncol.* **2004**, *15*, 517–525. [CrossRef]
48. Harrington, K.J.; Lewanski, C.R.; Northcote, A.D.; Whittaker, J.; Wellbank, H.; Vile, R.G.; Peters, A.M.; Stewart, J.S. Phase I–II Study of Pegylated Liposomal Cisplatin (SPI-077) in Patients with Inoperable Head and Neck Cancer. *Ann. Oncol.* **2001**, *12*, 493–496. [CrossRef]
49. Gabizon, A.A. Pegylated Liposomal Doxorubicin: Metamorphosis of an Old Drug into a New Form of Chemotherapy. *Cancer Investig.* **2001**, *19*, 424–436. [CrossRef]
50. Charrois, G.J.R.; Allen, T.M. Multiple Injections of Pegylated Liposomal Doxorubicin: Pharmacokinetics and Therapeutic Activity. *J. Pharmacol. Exp. Ther.* **2003**, *306*, 1058–1067. [CrossRef]
51. Mura, S.; Nicolas, J.; Couvreur, P. Stimuli-Responsive Nanocarriers for Drug Delivery. *Nat. Mater.* **2013**, *12*, 991–1003. [CrossRef]
52. Mirvakili, S.M.; Langer, R. Wireless On-Demand Drug Delivery. *Nat. Electron.* **2021**, *4*, 464–477. [CrossRef]
53. Torchilin, V.P. Multifunctional, Stimuli-Sensitive Nanoparticulate Systems for Drug Delivery. *Nat. Rev. Drug Discov.* **2014**, *13*, 813–827. [CrossRef] [PubMed]
54. Jang, S.H.; Wientjes, M.G.; Lu, D.; Au, J.L.S. Drug Delivery and Transport to Solid Tumors. *Pharm. Res.* **2003**, *20*, 1337–1350. [CrossRef] [PubMed]
55. Rapp, B.E. Chapter 9—Fluids. In *Microfluidics: Modelling, Mechanics and Mathematics*; Rapp, B.E., Ed.; Elsevier: Oxford, UK, 2017; pp. 243–263, ISBN 978-1-4557-3141-1.
56. Prichard, R.; Gibson, M.; Joseph, C.; Strasser, W. Chapter 13—A Review of Fluid Flow in and around the Brain, Modeling, and Abnormalities. In *Multiscale Biomechanical Modeling of the Brain*; Prabhu, R., Horstemeyer, M., Eds.; Academic Press: Cambridge, MA, USA, 2022; pp. 209–238, ISBN 978-0-12-818144-7.

57. Swabb, E.A.; Wei, J.; Gullino, P.M. Diffusion and Convection in Normal and Neoplastic Tissues. *Cancer Res.* **1974**, *34*, 2814–2822.
58. Wientjes, M.G.; Yeung, B.Z.; Lu, Z.; Wientjes, M.G.; Au, J.L.S. Predicting Diffusive Transport of Cationic Liposomes in 3-Dimensional Tumor Spheroids. *J. Control. Release* **2014**, *192*, 10–18. [CrossRef] [PubMed]
59. Attama, A.A.; Momoh, M.A.; Builders, P.F. Chapter 5—Lipid Nanoparticulate Drug Delivery Systems: A Revolution in Dosage Form Design and Development. In *Recent Advances in Novel Drug Carrier Systems*; Ali Demir Sezer, Ed.; IntechOpen: Rijeka, Croatia, 2012.
60. Baxter, L.T.; Jain, R.K. Transport of Fluid and Macromolecules in Tumors. I. Role of Interstitial Pressure and Convection. *Microvasc. Res.* **1989**, *37*, 77–104. [CrossRef] [PubMed]
61. Khawar, I.A.; Kim, J.H.; Kuh, H.-J. Improving Drug Delivery to Solid Tumors: Priming the Tumor Microenvironment. *J. Control. Release* **2015**, *201*, 78–89. [CrossRef]
62. Munson, J.; Shieh, A. Interstitial Fluid Flow in Cancer: Implications for Disease Progression and Treatment. *Cancer Manag. Res.* **2014**, *6*, 317–328. [CrossRef]
63. Zhou, Y.; Chen, X.; Cao, J.; Gao, H. Overcoming the Biological Barriers in the Tumor Microenvironment for Improving Drug Delivery and Efficacy. *J. Mater. Chem. B* **2020**, *8*, 6765–6781. [CrossRef]
64. Stine, C.A.; Munson, J.M. Convection-Enhanced Delivery: Connection to and Impact of Interstitial Fluid Flow. *Front. Oncol.* **2019**, *9*, 966. [CrossRef]
65. Carmeliet, P.; Jain, R.K. Principles and Mechanisms of Vessel Normalization for Cancer and Other Angiogenic Diseases. *Nat. Rev. Drug Discov.* **2011**, *10*, 417–427. [CrossRef]
66. Chowdhury, S.M.; Abou-Elkacem, L.; Lee, T.; Dahl, J.; Lutz, A.M. Ultrasound and Microbubble Mediated Therapeutic Delivery: Underlying Mechanisms and Future Outlook. *J. Control. Release* **2020**, *326*, 75–90. [CrossRef] [PubMed]
67. Raghavan, R. Theory for Acoustic Streaming in Soft Porous Matter and Its Applications to Ultrasound-Enhanced Convective Delivery. *J. Ther. Ultrasound* **2018**, *6*, 6. [CrossRef] [PubMed]
68. Wu, J. Acoustic Streaming and Its Applications. *Fluids* **2018**, *3*, 108. [CrossRef]
69. Weinstein, J.; Magin, R.; Yatvin, M.; Zaharko, D. Liposomes and Local Hyperthermia: Selective Delivery of Methotrexate to Heated Tumors. *Science* **1979**, *204*, 188–191. [CrossRef] [PubMed]
70. Yatvin, M.; Weinstein, J.; Dennis, W.; Blumenthal, R. Design of Liposomes for Enhanced Local Release of Drugs by Hyperthermia. *Science* **1978**, *202*, 1290–1293. [CrossRef]
71. Manzoor, A.A.; Lindner, L.H.; Landon, C.D.; Park, J.-Y.; Simnick, A.J.; Dreher, M.R.; Das, S.; Hanna, G.; Park, W.; Chilkoti, A.; et al. Overcoming Limitations in Nanoparticle Drug Delivery: Triggered, Intravascular Release to Improve Drug Penetration into Tumors. *Cancer Res.* **2012**, *72*, 5566–5575. [CrossRef] [PubMed]
72. ten Hagen, T.L.M.; Dreher, M.R.; Zalba, S.; Seynhaeve, A.L.B.; Amin, M.; Li, L.; Haemmerich, D. Drug Transport Kinetics of Intravascular Triggered Drug Delivery Systems. *Commun. Biol.* **2021**, *4*, 920. [CrossRef]
73. Borys, N.; Dewhirst, M.W. Drug Development of Lyso-Thermosensitive Liposomal Doxorubicin: Combining Hyperthermia and Thermosensitive Drug Delivery. *Adv. Drug Deliv. Rev.* **2021**, *178*, 113985. [CrossRef]
74. Sirsi, S.R.; Borden, M.A. State-of-the-Art Materials for Ultrasound-Triggered Drug Delivery. *Adv. Drug Deliv. Rev.* **2014**, *72*, 3–14. [CrossRef]
75. Frinking, P.; Segers, T.; Luan, Y.; Tranquart, F. Three Decades of Ultrasound Contrast Agents: A Review of the Past, Present and Future Improvements. *Ultrasound Med. Biol.* **2020**, *46*, 892–908. [CrossRef]
76. Lea-Banks, H.; O'Reilly, M.A.; Hynynen, K. Ultrasound-Responsive Droplets for Therapy: A Review. *J. Control. Release* **2019**, *293*, 144–154. [CrossRef] [PubMed]
77. Wells, P.N.T. Ultrasound Imaging. *Phys. Med. Biol.* **2006**, *51*, R83. [CrossRef] [PubMed]
78. Jensen, J.A. Medical Ultrasound Imaging. *Prog. Biophys. Mol. Biol.* **2007**, *93*, 153–165. [CrossRef] [PubMed]
79. Wells, P.N.T.; Liang, H.-D. Medical Ultrasound: Imaging of Soft Tissue Strain and Elasticity. *J. R. Soc. Interface* **2011**, *8*, 1521–1549. [CrossRef]
80. Shung, K.K. *Diagnostic Ultrasound: Imaging and Blood Flow Measurements*; CRC Press: Boca Raton, FL, USA, 2015; ISBN 1-4665-8265-0.
81. Borden, M.A.; Dayton, P.A.; Slagle, C.; Walmer, R.W. Chapter 35—Ultrasound Contrast Agents. In *Molecular Imaging*, 2nd ed.; Ross, B.D., Gambhir, S.S., Eds.; Academic Press: Cambridge, MA, USA, 2021; pp. 639–653, ISBN 978-0-12-816386-3.
82. Wang, Y.; Cong, H.; Wang, S.; Yu, B.; Shen, Y. Development and Application of Ultrasound Contrast Agents in Biomedicine. *J. Mater. Chem. B* **2021**, *9*, 7633–7661. [CrossRef]
83. Paefgen, V.; Doleschel, D.; Kiessling, F. Evolution of Contrast Agents for Ultrasound Imaging and Ultrasound-Mediated Drug Delivery. *Front. Pharmacol.* **2015**, *6*, 197. [CrossRef]
84. Gramiak, R.; Shah, P.M. Echocardiography of the Aortic Root. *Investig. Radiol.* **1968**, *3*, 356–366. [CrossRef]
85. Bouakaz, A.; Zeghimi, A.; Doinikov, A.A. Sonoporation: Concept and Mechanisms. *Adv. Exp. Med. Biol.* **2016**, *880*, 175–189. [CrossRef]
86. Liang, H.-D.; Tang, J.; Halliwell, M. Sonoporation, Drug Delivery, and Gene Therapy. *Proc. Inst. Mech. Eng. H* **2010**, *224*, 343–361. [CrossRef]
87. Tsutsui, J.M.; Xie, F.; Porter, R.T. The Use of Microbubbles to Target Drug Delivery. *Cardiovasc. Ultrasound* **2004**, *2*, 23. [CrossRef]

88. Ferrara, K.; Pollard, R.; Borden, M. Ultrasound Microbubble Contrast Agents: Fundamentals and Application to Gene and Drug Delivery. *Annu. Rev. Biomed. Eng.* **2007**, *9*, 415–447. [CrossRef] [PubMed]
89. Fan, C.-H.; Ho, Y.-J.; Lin, C.-W.; Wu, N.; Chiang, P.-H.; Yeh, C.-K. State-of-the-Art of Ultrasound-Triggered Drug Delivery from Ultrasound-Responsive Drug Carriers. *Expert Opin. Drug Deliv.* **2022**, *19*, 997–1009. [CrossRef] [PubMed]
90. Ho, Y.-J.; Huang, C.-C.; Fan, C.-H.; Liu, H.-L.; Yeh, C.-K. Ultrasonic Technologies in Imaging and Drug Delivery. *Cell. Mol. Life Sci.* **2021**, *78*, 6119–6141. [CrossRef] [PubMed]
91. Entzian, K.; Aigner, A. Drug Delivery by Ultrasound-Responsive Nanocarriers for Cancer Treatment. *Pharmaceutics* **2021**, *13*, 1135. [CrossRef]
92. Skyba, D.M.; Price, R.J.; Linka, A.Z.; Skalak, T.C.; Kaul, S. Direct in Vivo Visualization of Intravascular Destruction of Microbubbles by Ultrasound and Its Local Effects on Tissue. *Circulation* **1998**, *98*, 290–293. [CrossRef] [PubMed]
93. Sirsi, S.R.; Borden, M.A. Microbubble Compositions, Properties and Biomedical Applications. *Bubble Sci. Eng. Technol.* **2009**, *1*, 3–17. [CrossRef]
94. Blomley, M.J.; Cooke, J.C.; Unger, E.C.; Monaghan, M.J.; Cosgrove, D.O. Microbubble Contrast Agents: A New Era in Ultrasound. *BMJ* **2001**, *322*, 1222–1225. [CrossRef]
95. Leighton, T.G.; Apfel, R.E. The Acoustic Bubble. *J. Acoust. Soc. Am.* **1994**, *96*, 2616. [CrossRef]
96. Sboros, V. Response of Contrast Agents to Ultrasound. *Adv. Drug Deliv. Rev.* **2008**, *60*, 1117–1136. [CrossRef]
97. Cosgrove, D. Ultrasound Contrast Agents: An Overview. *Eur. J. Radiol.* **2006**, *60*, 324–330. [CrossRef]
98. Keller, M.W.; Glasheen, W.; Kaul, S. Albunex: A Safe and Effective Commercially Produced Agent for Myocardial Contrast Echocardiography. *J. Am. Soc. Echocardiogr.* **1989**, *2*, 48–52. [CrossRef] [PubMed]
99. Wu, S.-Y.; Chen, C.C.; Tung, Y.-S.; Olumolade, O.O.; Konofagou, E.E. Effects of the Microbubble Shell Physicochemical Properties on Ultrasound-Mediated Drug Delivery to the Brain. *J. Control. Release* **2015**, *212*, 30–40. [CrossRef] [PubMed]
100. Sheeran, P.S.; Dayton, P.A. Phase-Change Contrast Agents for Imaging and Therapy. *Curr. Pharm. Des.* **2012**, *18*, 2152–2165. [CrossRef] [PubMed]
101. Mountford, P.A.; Smith, W.S.; Borden, M.A. Fluorocarbon Nanodrops as Acoustic Temperature Probes. *Langmuir* **2015**, *31*, 10656–10663. [CrossRef]
102. Mountford, P.A.; Thomas, A.N.; Borden, M.A. Thermal Activation of Superheated Lipid-Coated Perfluorocarbon Drops. *Langmuir* **2015**, *31*, 4627–4634. [CrossRef] [PubMed]
103. Sheeran, P.S.; Luois, S.; Dayton, P.A.; Matsunaga, T.O. Formulation and Acoustic Studies of a New Phase-Shift Agent for Diagnostic and Therapeutic Ultrasound. *Langmuir* **2011**, *27*, 10412–10420. [CrossRef] [PubMed]
104. Seda, R.; Li, D.S.; Fowlkes, J.B.; Bull, J.L. Characterization of Bioeffects on Endothelial Cells under Acoustic Droplet Vaporization. *Ultrasound Med. Biol.* **2015**, *41*, 3241–3252. [CrossRef]
105. Kang, S.-T.; Lin, Y. C.; Yeh, C.-K. Mechanical Bioeffects of Acoustic Droplet Vaporization in Vessel-Mimicking Phantoms. *Ultrason. Sonochem.* **2014**, *21*, 1866–1874. [CrossRef]
106. Fix, S.M.; Novell, A.; Yun, Y.; Dayton, P.A.; Arena, C.B. An Evaluation of the Sonoporation Potential of Low-Boiling Point Phase-Change Ultrasound Contrast Agents in Vitro. *J. Ther. Ultrasound* **2017**, *5*, 7. [CrossRef]
107. Loskutova, K.; Grishenkov, D.; Ghorbani, M. Review on Acoustic Droplet Vaporization in Ultrasound Diagnostics and Therapeutics. *BioMed Res. Int.* **2019**, *2019*, 9480193. [CrossRef]
108. Kooiman, K.; Roovers, S.; Langeveld, S.A.G.; Kleven, R.T.; Dewitte, H.; O'Reilly, M.A.; Escoffre, J.-M.; Bouakaz, A.; Verweij, M.D.; Hynynen, K.; et al. Ultrasound-Responsive Cavitation Nuclei for Therapy and Drug Delivery. *Ultrasound Med. Biol.* **2020**, *46*, 1296–1325. [CrossRef]
109. Lammertink, B.H.A.; Bos, C.; Deckers, R.; Storm, G.; Moonen, C.T.W.; Escoffre, J.-M. Sonochemotherapy: From Bench to Bedside. *Front. Pharmacol.* **2015**, *6*, 138. [CrossRef]
110. Rich, J.; Tian, Z.; Huang, T.J. Sonoporation: Past, Present, and Future. *Adv. Mater. Technol.* **2022**, *7*, 2100885. [CrossRef] [PubMed]
111. Lentacker, I.; De Cock, I.; Deckers, R.; De Smedt, S.C.; Moonen, C.T.W. Understanding Ultrasound Induced Sonoporation: Definitions and Underlying Mechanisms. *Adv. Drug Deliv. Rev.* **2014**, *72*, 49–64. [CrossRef] [PubMed]
112. Qin, P.; Han, T.; Yu, A.C.H.; Xu, L. Mechanistic Understanding the Bioeffects of Ultrasound-Driven Microbubbles to Enhance Macromolecule Delivery. *J. Control. Release* **2018**, *272*, 169–181. [CrossRef] [PubMed]
113. Helfield, B.; Chen, X.; Watkins, S.C.; Villanueva, F.S. Biophysical Insight into Mechanisms of Sonoporation. *Proc. Natl. Acad. Sci. USA* **2016**, *113*, 9983–9988. [CrossRef]
114. O'Reilly, M.A.; Hynynen, K. Ultrasound Enhanced Drug Delivery to the Brain and Central Nervous System. *Int. J. Hyperth.* **2012**, *28*, 386–396. [CrossRef]
115. Dasgupta, A.; Liu, M.; Ojha, T.; Storm, G.; Kiessling, F.; Lammers, T. Ultrasound-Mediated Drug Delivery to the Brain: Principles, Progress and Prospects. *Drug Discov. Today Technol.* **2016**, *20*, 41–48. [CrossRef]
116. Mainprize, T.; Lipsman, N.; Huang, Y.; Meng, Y.; Bethune, A.; Ironside, S.; Heyn, C.; Alkins, R.; Trudeau, M.; Sahgal, A.; et al. Blood-Brain Barrier Opening in Primary Brain Tumors with Non-Invasive MR-Guided Focused Ultrasound: A Clinical Safety and Feasibility Study. *Sci. Rep.* **2019**, *9*, 321. [CrossRef]
117. Abrahao, A.; Meng, Y.; Llinas, M.; Huang, Y.; Hamani, C.; Mainprize, T.; Aubert, I.; Heyn, C.; Black, S.E.; Hynynen, K.; et al. First-in-Human Trial of Blood–Brain Barrier Opening in Amyotrophic Lateral Sclerosis Using MR-Guided Focused Ultrasound. *Nat. Commun.* **2019**, *10*, 4373. [CrossRef]

118. Meng, Y.; Pople, C.B.; Lea-Banks, H.; Abrahao, A.; Davidson, B.; Suppiah, S.; Vecchio, L.M.; Samuel, N.; Mahmud, F.; Hynynen, K.; et al. Safety and Efficacy of Focused Ultrasound Induced Blood-Brain Barrier Opening, an Integrative Review of Animal and Human Studies. *J. Control. Release* **2019**, *309*, 25–36. [CrossRef] [PubMed]
119. Krut, Z.; Gazit, D.; Gazit, Z.; Pelled, G. Applications of Ultrasound-Mediated Gene Delivery in Regenerative Medicine. *Bioengineering* **2022**, *9*, 190. [CrossRef]
120. Duan, X.; Zhou, Q.; Wan, J.M.F.; Yu, A.C.H. Sonoporation Generates Downstream Cellular Impact after Membrane Resealing. *Sci. Rep.* **2021**, *11*, 5161. [CrossRef] [PubMed]
121. Kovacs, Z.I.; Tu, T.-W.; Sundby, M.; Qureshi, F.; Lewis, B.K.; Jikaria, N.; Burks, S.R.; Frank, J.A. MRI and Histological Evaluation of Pulsed Focused Ultrasound and Microbubbles Treatment Effects in the Brain. *Theranostics* **2018**, *8*, 4837–4855. [CrossRef] [PubMed]
122. Jung, O.; Thomas, A.; Burks, S.R.; Dustin, M.L.; Frank, J.A.; Ferrer, M.; Stride, E. Neuroinflammation Associated with Ultrasound-Mediated Permeabilization of the Blood-Brain Barrier. *Trends Neurosci.* **2022**, *45*, 459–470. [CrossRef] [PubMed]
123. Gandhi, K.; Barzegar-Fallah, A.; Banstola, A.; Rizwan, S.B.; Reynolds, J.N.J. Ultrasound-Mediated Blood-Brain Barrier Disruption for Drug Delivery: A Systematic Review of Protocols, Efficacy, and Safety Outcomes from Preclinical and Clinical Studies. *Pharmaceutics* **2022**, *14*, 833. [CrossRef] [PubMed]
124. Przystupski, D.; Ussowicz, M. Landscape of Cellular Bioeffects Triggered by Ultrasound-Induced Sonoporation. *Int. J. Mol. Sci.* **2022**, *23*, 11222. [CrossRef]
125. Furusawa, Y.; Hassan, M.A.; Zhao, Q.-L.; Ogawa, R.; Tabuchi, Y.; Kondo, T. Effects of Therapeutic Ultrasound on the Nucleus and Genomic DNA. *Ultrason. Sonochem.* **2014**, *21*, 2061–2068. [CrossRef]
126. Yang, Y.; Li, Q.; Guo, X.; Tu, J.; Zhang, D. Mechanisms Underlying Sonoporation: Interaction between Microbubbles and Cells. *Ultrason. Sonochem.* **2020**, *67*, 105096. [CrossRef]
127. Izadifar, Z.; Babyn, P.; Chapman, D. Mechanical and Biological Effects of Ultrasound: A Review of Present Knowledge. *Ultrasound Med. Biol.* **2017**, *43*, 1085–1104. [CrossRef]
128. Jain, R.K. Normalization of Tumor Vasculature: An Emerging Concept in Antiangiogenic Therapy. *Science* **2005**, *307*, 58–62. [CrossRef] [PubMed]
129. Mo, S.; Coussios, C.-C.; Seymour, L.; Carlisle, R. Ultrasound-Enhanced Drug Delivery for Cancer. *Expert Opin. Drug Deliv.* **2012**, *9*, 1525–1538. [CrossRef] [PubMed]
130. Burgess, A.; Hynynen, K. Drug Delivery across the Blood-Brain Barrier Using Focused Ultrasound. *Expert Opin. Drug Deliv.* **2014**, *11*, 711–721. [CrossRef]
131. Chen, C.C.; Sheeran, P.S.; Wu, S.-Y.; Olumolade, O.O.; Dayton, P.A.; Konofagou, E.E. Targeted Drug Delivery with Focused Ultrasound-Induced Blood-Brain Barrier Opening Using Acoustically-Activated Nanodroplets. *J. Control. Release* **2013**, *172*, 795–804. [CrossRef] [PubMed]
132. Omata, D.; Unga, J.; Suzuki, R.; Maruyama, K. Lipid-Based Microbubbles and Ultrasound for Therapeutic Application. *Adv. Drug Deliv. Rev.* **2020**, *154–155*, 236–244. [CrossRef]
133. Al-Jawadi, S.; Thakur, S.S. Ultrasound-Responsive Lipid Microbubbles for Drug Delivery: A Review of Preparation Techniques to Optimise Formulation Size, Stability and Drug Loading. *Int. J. Pharm.* **2020**, *585*, 119559. [CrossRef]
134. Tinkov, S.; Winter, G.; Coester, C.; Bekeredjian, R. New Doxorubicin-Loaded Phospholipid Microbubbles for Targeted Tumor Therapy: Part I—Formulation Development and in-Vitro Characterization. *J. Control. Release* **2010**, *143*, 143–150. [CrossRef]
135. Fokong, S.; Theek, B.; Wu, Z.; Koczera, P.; Appold, L.; Jorge, S.; Resch-Genger, U.; van Zandvoort, M.; Storm, G.; Kiessling, F.; et al. Image-Guided, Targeted and Triggered Drug Delivery to Tumors Using Polymer-Based Microbubbles. *J. Control. Release* **2012**, *163*, 75–81. [CrossRef]
136. El-Sherif, D.M.; Wheatley, M.A. Development of a Novel Method for Synthesis of a Polymeric Ultrasound Contrast Agent. *J. Biomed. Mater. Res. Part A* **2003**, *66A*, 347–355. [CrossRef]
137. Cochran, M.C.; Eisenbrey, J.; Ouma, R.O.; Soulen, M.; Wheatley, M.A. Doxorubicin and Paclitaxel Loaded Microbubbles for Ultrasound Triggered Drug Delivery. *Int. J. Pharm.* **2011**, *414*, 161–170. [CrossRef]
138. Koczera, P.; Appold, L.; Shi, Y.; Liu, M.; Dasgupta, A.; Pathak, V.; Ojha, T.; Fokong, S.; Wu, Z.; van Zandvoort, M.; et al. PBCA-Based Polymeric Microbubbles for Molecular Imaging and Drug Delivery. *J. Control. Release* **2017**, *259*, 128–135. [CrossRef] [PubMed]
139. Liu, M.; Dasgupta, A.; Koczera, P.; Schipper, S.; Rommel, D.; Shi, Y.; Kiessling, F.; Lammers, T. Drug Loading in Poly(Butyl Cyanoacrylate)-Based Polymeric Microbubbles. *Mol. Pharm.* **2020**, *17*, 2840–2848. [CrossRef] [PubMed]
140. Barmin, R.A.; Dasgupta, A.; Bastard, C.; De Laporte, L.; Rütten, S.; Weiler, M.; Kiessling, F.; Lammers, T.; Pallares, R.M. Engineering the Acoustic Response and Drug Loading Capacity of PBCA-Based Polymeric Microbubbles with Surfactants. *Mol. Pharm.* **2022**, *19*, 3256–3266. [CrossRef] [PubMed]
141. Honari, A.; Kapilavaih, P.S.; Akter, N.; Sirsi, S.R. Remote Loading of Gas Bubbles into Polylactic Acid Microcapsules Creates Acoustically Active Janus Particles. *ACS Appl. Polym. Mater.* **2022**, *4*, 773–780. [CrossRef] [PubMed]
142. Peng, Y.; Seekell, R.P.; Cole, A.R.; Lamothe, J.R.; Lock, A.T.; van den Bosch, S.; Tang, X.; Kheir, J.N.; Polizzotti, B.D. Interfacial Nanoprecipitation toward Stable and Responsive Microbubbles and Their Use as a Resuscitative Fluid. *Angew. Chem. Int. Ed.* **2018**, *57*, 1271–1276. [CrossRef] [PubMed]

143. Chapla, R.; Huynh, K.T.; Schutt, C.E. Microbubble-Nanoparticle Complexes for Ultrasound-Enhanced Cargo Delivery. *Pharmaceutics* **2022**, *14*, 2396. [CrossRef]
144. Kheirolomoom, A.; Dayton, P.A.; Lum, A.F.H.; Little, E.; Paoli, E.E.; Zheng, H.; Ferrara, K.W. Acoustically-Active Microbubbles Conjugated to Liposomes: Characterization of a Proposed Drug Delivery Vehicle. *J. Control. Release* **2007**, *118*, 275–284. [CrossRef]
145. Lentacker, I.; De Smedt, S.C.; Demeester, J.; Van Marck, V.; Bracke, M.; Sanders, N.N. Lipoplex-Loaded Microbubbles for Gene Delivery: A Trojan Horse Controlled by Ultrasound. *Adv. Funct. Mater.* **2007**, *17*, 1910–1916. [CrossRef]
146. Klibanov, A.L.; Shevchenko, T.I.; Raju, B.I.; Seip, R.; Chin, C.T. Ultrasound-Triggered Release of Materials Entrapped in Microbubble–Liposome Constructs: A Tool for Targeted Drug Delivery. *J. Control. Release* **2010**, *148*, 13–17. [CrossRef]
147. Geers, B.; Lentacker, I.; Sanders, N.N.; Demeester, J.; Meairs, S.; De Smedt, S.C. Self-Assembled Liposome-Loaded Microbubbles: The Missing Link for Safe and Efficient Ultrasound Triggered Drug-Delivery. *J. Control. Release* **2011**, *152*, 249–256. [CrossRef]
148. Vandenbroucke, R.E.; Lentacker, I.; Demeester, J.; De Smedt, S.C.; Sanders, N.N. Ultrasound Assisted SiRNA Delivery Using PEG-SiPlex Loaded Microbubbles. *J. Control. Release* **2008**, *126*, 265–273. [CrossRef] [PubMed]
149. Yoon, Y.I.; Kwon, Y.-S.; Cho, H.-S.; Heo, S.-H.; Park, K.S.; Park, S.G.; Lee, S.-H.; Hwang, S.I.; Kim, Y.I.; Jae, H.J.; et al. Ultrasound-Mediated Gene and Drug Delivery Using a Microbubble-Liposome Particle System. *Theranostics* **2014**, *4*, 1133–1144. [CrossRef] [PubMed]
150. Yan, F.; Li, L.; Deng, Z.; Jin, Q.; Chen, J.; Yang, W.; Yeh, C.-K.; Wu, J.; Shandas, R.; Liu, X.; et al. Paclitaxel-Liposome-Microbubble Complexes as Ultrasound-Triggered Therapeutic Drug Delivery Carriers. *J. Control. Release* **2013**, *166*, 246–255. [CrossRef] [PubMed]
151. Gao, J.; Nesbitt, H.; Logan, K.; Burnett, K.; White, B.; Jack, I.G.; Taylor, M.A.; Love, M.; Callan, B.; McHale, A.P.; et al. An Ultrasound Responsive Microbubble-Liposome Conjugate for Targeted Irinotecan-Oxaliplatin Treatment of Pancreatic Cancer. *Eur. J. Pharm. Biopharm.* **2020**, *157*, 233–240. [CrossRef]
152. Huang, F.-Y.; Lei, J.; Sun, Y.; Yan, F.; Chen, B.; Zhang, L.; Lu, Z.; Cao, R.; Lin, Y.-Y.; Wang, C.-C.; et al. Induction of Enhanced Immunogenic Cell Death through Ultrasound-Controlled Release of Doxorubicin by Liposome-Microbubble Complexes. *Oncoimmunology* **2018**, *7*, e1446720. [CrossRef]
153. Burke, C.W.; Alexander, E., 4th; Timbie, K.; Kilbanov, A.L.; Price, R.J. Ultrasound-Activated Agents Comprised of 5FU-Bearing Nanoparticles Bonded to Microbubbles Inhibit Solid Tumor Growth and Improve Survival. *Mol. Ther.* **2014**, *22*, 321–328. [CrossRef]
154. Rapoport, N.; Gao, Z.; Kennedy, A. Multifunctional Nanoparticles for Combining Ultrasonic Tumor Imaging and Targeted Chemotherapy. *J. Natl. Cancer Inst.* **2007**, *99*, 1095–1106. [CrossRef]
155. Rapoport, N. Drug-Loaded Perfluorocarbon Nanodroplets for Ultrasound-Mediated Drug Delivery. *Adv. Exp. Med. Biol.* **2016**, *880*, 221–241. [CrossRef]
156. Rapoport, N. Phase-Shift, Stimuli-Responsive Perfluorocarbon Nanodroplets for Drug Delivery to Cancer: Phase-Shift Perfluorocarbon Nanoemulsions. *WIREs Nanomed. Nanobiotechnol.* **2012**, *4*, 492–510. [CrossRef]
157. Ji, G.; Yang, J.; Chen, J. Preparation of Novel Curcumin-Loaded Multifunctional Nanodroplets for Combining Ultrasonic Development and Targeted Chemotherapy. *Int. J. Pharm.* **2014**, *466*, 314–320. [CrossRef]
158. Zhang, W.; Shi, Y.; Abd Shukor, S.; Vijayakumaran, A.; Vlatakis, S.; Wright, M.; Thanou, M. Phase-Shift Nanodroplets as an Emerging Sonoresponsive Nanomaterial for Imaging and Drug Delivery Applications. *Nanoscale* **2022**, *14*, 2943–2965. [CrossRef] [PubMed]
159. Guo, R.; Xu, N.; Liu, Y.; Ling, G.; Yu, J.; Zhang, P. Functional Ultrasound-Triggered Phase-Shift Perfluorocarbon Nanodroplets for Cancer Therapy. *Ultrasound Med. Biol.* **2021**, *47*, 2064–2079. [CrossRef]
160. Dong, W.; Wu, P.; Zhou, D.; Huang, J.; Qin, M.; Yang, X.; Wan, M.; Zong, Y. Ultrasound-Mediated Gene Therapy of Hepatocellular Carcinoma Using Pre-MicroRNA Plasmid-Loaded Nanodroplets. *Ultrasound Med. Biol.* **2020**, *46*, 90–107. [CrossRef] [PubMed]
161. Cao, Y.; Chen, Y.; Yu, T.; Guo, Y.; Liu, F.; Yao, Y.; Li, P.; Wang, D.; Wang, Z.; Chen, Y.; et al. Drug Release from Phase-Changeable Nanodroplets Triggered by Low-Intensity Focused Ultrasound. *Theranostics* **2018**, *8*, 1327–1339. [CrossRef]
162. Honari, A.; Merillat, D.A.; Bellary, A.; Ghaderi, M.; Sirsi, S.R. Improving Release of Liposome-Encapsulated Drugs with Focused Ultrasound and Vaporizable Droplet-Liposome Nanoclusters. *Pharmaceutics* **2021**, *13*, 609. [CrossRef] [PubMed]
163. Airan, R. Neuromodulation with Nanoparticles. *Science* **2017**, *357*, 465. [CrossRef] [PubMed]
164. Airan, R.D.; Meyer, R.A.; Ellens, N.P.K.; Rhodes, K.R.; Farahani, K.; Pomper, M.G.; Kadam, S.D.; Green, J.J. Noninvasive Targeted Transcranial Neuromodulation via Focused Ultrasound Gated Drug Release from Nanoemulsions. *Nano Lett.* **2017**, *17*, 652–659. [CrossRef]
165. Lea-Banks, H.; O'Reilly, M.A.; Hamani, C.; Hynynen, K. Localized Anesthesia of a Specific Brain Region Using Ultrasound-Responsive Barbiturate Nanodroplets. *Theranostics* **2020**, *10*, 2849–2858. [CrossRef]
166. Lea-Banks, H.; Hynynen, K. Sub-Millimetre Precision of Drug Delivery in the Brain from Ultrasound-Triggered Nanodroplets. *J. Control. Release* **2021**, *338*, 731–741. [CrossRef]
167. Ozdas, M.S.; Shah, A.S.; Johnson, P.M.; Patel, N.; Marks, M.; Yasar, T.B.; Stalder, U.; Bigler, L.; von der Behrens, W.; Sirsi, S.R.; et al. Non-Invasive Molecularly-Specific Millimeter-Resolution Manipulation of Brain Circuits by Ultrasound-Mediated Aggregation and Uncaging of Drug Carriers. *Nat. Commun.* **2020**, *11*, 4929. [CrossRef]

168. Gorick, C.M.; Mathew, A.S.; Garrison, W.J.; Thim, E.A.; Fisher, D.G.; Copeland, C.A.; Song, J.; Klibanov, A.L.; Miller, G.W.; Price, R.J. Sonoselective Transfection of Cerebral Vasculature without Blood–Brain Barrier Disruption. *Proc. Natl. Acad. Sci. USA* **2020**, *117*, 5644–5654. [CrossRef] [PubMed]
169. Wrenn, S.; Dicker, S.; Small, E.; Mleczko, M. Controlling Cavitation for Controlled Release. In Proceedings of the 2009 IEEE International Ultrasonics Symposium, Rome, Italy, 20–23 September 2009; pp. 104–107.
170. Wallace, N.; Dicker, S.; Lewin, P.; Wrenn, S.P. Influence of Nesting Shell Size on Brightness Longevity and Resistance to Ultrasound-Induced Dissolution during Enhanced B-Mode Contrast Imaging. *Ultrasonics* **2014**, *54*, 2099–2108. [CrossRef] [PubMed]
171. Cimorelli, M.; Angel, B.; Fafarman, A.; Kohut, A.; Andrien, B.; Barrett, K.; Wrenn, S. Introducing a Nested Phase Change Agent with an Acoustic Response That Depends on Electric Field: A Candidate for Myocardial Perfusion Imaging and Drug Delivery. *Appl. Acoust.* **2018**, *138*, 9–17. [CrossRef]
172. Wallace, N.; Wrenn, S.P. Ultrasound Triggered Drug Delivery with Liposomal Nested Microbubbles. *Ultrasonics* **2015**, *63*, 31–38. [CrossRef]
173. Ibsen, S.; Benchimol, M.; Simberg, D.; Schutt, C.; Steiner, J.; Esener, S. A Novel Nested Liposome Drug Delivery Vehicle Capable of Ultrasound Triggered Release of Its Payload. *J. Control. Release* **2011**, *155*, 358–366. [CrossRef]
174. Batchelor, D.V.B.; Abou-Saleh, R.H.; Coletta, P.L.; McLaughlan, J.R.; Peyman, S.A.; Evans, S.D. Nested Nanobubbles for Ultrasound-Triggered Drug Release. *ACS Appl. Mater. Interfaces* **2020**, *12*, 29085–29093. [CrossRef]
175. Javadi, M.; Pitt, W.G.; Belnap, D.M.; Tsosie, N.H.; Hartley, J.M. Encapsulating Nanoemulsions Inside ELiposomes for Ultrasonic Drug Delivery. *Langmuir* **2012**, *28*, 14720–14729. [CrossRef]
176. de Matos, M.B.C.; Deckers, R.; van Elburg, B.; Lajoinie, G.; de Miranda, B.S.; Versluis, M.; Schiffelers, R.; Kok, R.J. Ultrasound-Sensitive Liposomes for Triggered Macromolecular Drug Delivery: Formulation and In Vitro Characterization. *Front. Pharmacol.* **2019**, *10*, 1463. [CrossRef]
177. Hayward, A.T.J. The Role of Stabilized Gas Nuclei in Hydrodynamic Cavitation Inception. *J. Phys. D Appl. Phys.* **1970**, *3*, 574. [CrossRef]
178. Thomas, R.G.; Jonnalagadda, U.S.; Kwan, J.J. Biomedical Applications for Gas-Stabilizing Solid Cavitation Agents. *Langmuir* **2019**, *35*, 10106–10115. [CrossRef]
179. Manzano, M.; Vallet-Regí, M. Ultrasound Responsive Mesoporous Silica Nanoparticles for Biomedical Applications. *Chem. Commun.* **2019**, *55*, 2731–2740. [CrossRef] [PubMed]
180. Lin, F.-C.; Xie, Y.; Deng, T.; Zink, J.I. Magnetism, Ultrasound, and Light-Stimulated Mesoporous Silica Nanocarriers for Theranostics and Beyond. *J. Am. Chem. Soc.* **2021**, *143*, 6025–6036. [CrossRef] [PubMed]
181. Kim, H.-J.; Matsuda, H.; Zhou, H.; Honma, I. Ultrasound-Triggered Smart Drug Release from a Poly(Dimethylsiloxane)—Mesoporous Silica Composite. *Adv. Mater.* **2006**, *18*, 3083–3088. [CrossRef]
182. Paris, J.L.; de la Torre, P.; Victoria Cabañas, M.; Manzano, M.; Grau, M.; Flores, A.I.; Vallet-Regí, M. Vectorization of Ultrasound-Responsive Nanoparticles in Placental Mesenchymal Stem Cells for Cancer Therapy. *Nanoscale* **2017**, *9*, 5528–5537. [CrossRef]
183. Kwan, J.J.; Myers, R.; Coviello, C.M.; Graham, S.M.; Shah, A.R.; Stride, E.; Carlisle, R.C.; Coussios, C.C. Ultrasound-Propelled Nanocups for Drug Delivery. *Small* **2015**, *11*, 5305–5314. [CrossRef]
184. Su, X.; Gupta, I.; Jonnalagadda, U.S.; Kwan, J.J. Complementary Effects of Porosigen and Stabilizer on the Structure of Hollow Porous Poly(Lactic-Co-Glycolic Acid) Microparticles. *ACS Appl. Polym. Mater.* **2020**, *2*, 3696–3703. [CrossRef]
185. Sabuncu, S.; Yildirim, A. Gas-Stabilizing Nanoparticles for Ultrasound Imaging and Therapy of Cancer. *Nano Converg.* **2021**, *8*, 39. [CrossRef]
186. Li, J.; Wang, X.; Zhang, T.; Wang, C.; Huang, Z.; Luo, X.; Deng, Y. A Review on Phospholipids and Their Main Applications in Drug Delivery Systems. *Asian J. Pharm. Sci.* **2015**, *10*, 81–98. [CrossRef]
187. Singh, R.P.; Gangadharappa, H.V.; Mruthunjaya, K. Phospholipids: Unique Carriers for Drug Delivery Systems. *J. Drug Deliv. Sci. Technol.* **2017**, *39*, 166–179. [CrossRef]
188. Ferrara, K.W.; Borden, M.A.; Zhang, H. Lipid-Shelled Vehicles: Engineering for Ultrasound Molecular Imaging and Drug Delivery. *Acc. Chem. Res.* **2009**, *42*, 881–892. [CrossRef]
189. Peng, Y.; Peng, C.; Nguyen, T.; Sun, T.; Porter, T.; McDannold, N.; Kheir, J.N.; Polizzotti, B.D. Engineering Caged Microbubbles for Controlled Acoustic Cavitation and Pressure Sensing. *ACS Mater. Lett.* **2021**, *3*, 978–987. [CrossRef]
190. Prajapati, S.K.; Jain, A.; Jain, A.; Jain, S. Biodegradable Polymers and Constructs: A Novel Approach in Drug Delivery. *Eur. Polym. J.* **2019**, *120*, 109191. [CrossRef]
191. Urbánek, T.; Jäger, E.; Jäger, A.; Hrubý, M. Selectively Biodegradable Polyesters: Nature-Inspired Construction Materials for Future Biomedical Applications. *Polymers* **2019**, *11*, 1061. [CrossRef] [PubMed]
192. Wei, P.; Cornel, E.J.; Du, J. Ultrasound-Responsive Polymer-Based Drug Delivery Systems. *Drug Deliv. Transl. Res.* **2021**, *11*, 1323–1339. [CrossRef] [PubMed]
193. Zhang, A.; Jung, K.; Li, A.; Liu, J.; Boyer, C. Recent Advances in Stimuli-Responsive Polymer Systems for Remotely Controlled Drug Release. *Prog. Polym. Sci.* **2019**, *99*, 101164. [CrossRef]
194. Rudakovskaya, P.G.; Barmin, R.A.; Kuzmin, P.S.; Fedotkina, E.P.; Sencha, A.N.; Gorin, D.A. Microbubbles Stabilized by Protein Shell: From Pioneering Ultrasound Contrast Agents to Advanced Theranostic Systems. *Pharmaceutics* **2022**, *14*, 1236. [CrossRef]

195. Hernot, S.; Klibanov, A.L. Microbubbles in Ultrasound-Triggered Drug and Gene Delivery. *Adv. Drug Deliv. Rev.* **2008**, *60*, 1153–1166. [CrossRef]
196. Tharkar, P.; Varanasi, R.; Wong, W.S.F.; Jin, C.T.; Chrzanowski, W. Nano-Enhanced Drug Delivery and Therapeutic Ultrasound for Cancer Treatment and Beyond. *Front. Bioeng. Biotechnol.* **2019**, *7*, 324. [CrossRef]
197. Kooiman, K.; Vos, H.J.; Versluis, M.; de Jong, N. Acoustic Behavior of Microbubbles and Implications for Drug Delivery. *Adv. Drug Deliv. Rev.* **2014**, *72*, 28–48. [CrossRef]
198. Shakya, G.; Fajrial, A.K.; Ding, X.; Borden, M.A. Effect of Thermal History and Hydrocarbon Core Size on Perfluorocarbon Endoskeletal Droplet Vaporization. *Langmuir* **2022**, *38*, 2634–2641. [CrossRef]
199. Wallace, N.; Dicker, S.; Lewin, P.; Wrenn, S.P. Inertial Cavitation Threshold of Nested Microbubbles. *Ultrasonics* **2015**, *58*, 67–74. [CrossRef] [PubMed]
200. Dayton, P.; Klibanov, A.; Brandenburger, G.; Ferrara, K. Acoustic Radiation Force in Vivo: A Mechanism to Assist Targeting of Microbubbles. *Ultrasound Med. Biol.* **1999**, *25*, 1195–1201. [CrossRef] [PubMed]
201. Zhao, S.; Borden, M.; Bloch, S.H.; Kruse, D.; Ferrara, K.W.; Dayton, P.A. Radiation-Force Assisted Targeting Facilitates Ultrasonic Molecular Imaging. *Mol. Imaging* **2004**, *3*, 135–148. [CrossRef]
202. Ciancia, S.; Cafarelli, A.; Zahoranova, A.; Menciassi, A.; Ricotti, L. Pulsatile Drug Delivery System Triggered by Acoustic Radiation Force. *Front. Bioeng. Biotechnol.* **2020**, *8*, 317. [CrossRef] [PubMed]
203. Lum, A.F.H.; Borden, M.A.; Dayton, P.A.; Kruse, D.E.; Simon, S.I.; Ferrara, K.W. Ultrasound Radiation Force Enables Targeted Deposition of Model Drug Carriers Loaded on Microbubbles. *J. Control. Release* **2006**, *111*, 128–134. [CrossRef]
204. Su, X.; Rakshit, M.; Das, P.; Gupta, I.; Das, D.; Pramanik, M.; Ng, K.W.; Kwan, J. Ultrasonic Implantation and Imaging of Sound-Sensitive Theranostic Agents for the Treatment of Arterial Inflammation. *ACS Appl. Mater. Interfaces* **2021**, *13*, 24422–24430. [CrossRef]
205. De Cock, I.; Lajoinie, G.; Versluis, M.; De Smedt, S.C.; Lentacker, I. Sonoprinting and the Importance of Microbubble Loading for the Ultrasound Mediated Cellular Delivery of Nanoparticles. *Biomaterials* **2016**, *83*, 294–307. [CrossRef]

pharmaceutics

Review

Cavitation-Mediated Immunomodulation and Its Use with Checkpoint Inhibitors

Matilde Maardalen *, Robert Carlisle and Constantin Coussios

Institute of Biomedical Engineering, Department of Engineering Science, University of Oxford, Oxford OX1 3PJ, UK
* Correspondence: matilde.maardalen@eng.ox.ac.uk

Abstract: The promotion of anti-tumour immune responses can be an effective route to the complete remission of primary and metastatic tumours in a small proportion of patients. Hence, researchers are currently investigating various methods to further characterise and enhance such responses to achieve a beneficial impact across a wider range of patients. Due to its non-invasive, non-ionising, and targetable nature, the application of ultrasound-mediated cavitation has proven to be a popular method to enhance the delivery and activity of immune checkpoint inhibitors. However, to optimise this approach, it is important to understand the biological and physical mechanisms by which cavitation may promote anti-tumour immune responses. Here, the published literature relating to the role that cavitation may play in modulating anti-tumour immunity is therefore assessed.

Keywords: ultrasound; cavitation; cancer; immunotherapy; checkpoint inhibitor

Citation: Maardalen, M.; Carlisle, R.; Coussios, C. Cavitation-Mediated Immunomodulation and Its Use with Checkpoint Inhibitors. *Pharmaceutics* **2023**, *15*, 2110. https://doi.org/10.3390/pharmaceutics15082110

Academic Editor: Xiangyang Shi

Received: 15 June 2023
Revised: 31 July 2023
Accepted: 5 August 2023
Published: 9 August 2023

1. Cancer Immunology

The interplay between tumours and the immune system was suggested in the mid-19th century when Rudolf Virchow hypothesised that cancers arise at sites of chronic inflammation [1]. A few decades later, Paul Ehrlich proposed that the immune system plays a critical role in protecting the host from cancer development [2]. Indeed, it is now clear that the immune system plays a dual role in cancer progression [3]. Thus, to best harness the immune system's anti-tumour role it is essential to understand how a tumour-specific immune response may arise, how the tumour evades it, and how therapeutic intervention can most effectively overcome such evasion.

1.1. Cancer-Immunity Cycle

Chen and Mellman described the cancer-immunity cycle—events necessary for an anti-tumour immune response to lead to effective killing of cancer cells [4]. Firstly, Tumour-associated antigens (TAAs) or Tumour-specific antigens (TSAs) need to be released and captured by tissue-resident Antigen-presenting cells (APCs) such as Dendritic cells (DCs). The phagocytosed antigens are processed into small peptides and presented on Major histocompatibility complex (MHC) I or II [5]. Additionally, immunogenic signals such as proinflammatory cytokines released by dying tumour cells are essential to induce DC maturation and turn them into potent APCs. Activated DCs increase their synthesis of MHC molecules and begin to express the co-stimulatory molecules CD80 and CD86 on their surface [4].

Next, DCs migrate to lymph nodes, along with their cargo, to activate an adaptive immune response. DCs present the captured antigen on MHC I or II to naive CD8+ or CD4+ T cells, respectively [6]. Exogenous antigens are presented on MHC II molecules but also on MHC I molecules via a process called cross-presentation. The cross-presentation of exogenous antigens is critical for anti-tumour immunity [7]. Binding to the peptide/MHC complex and co-stimulatory molecules on the DC surface stimulates the priming and activation of effector and memory T cell responses against the antigen. The activation of CD8+ T cells with specificity for TAAs or TSAs leads to cytotoxic reactions that cause tumour cell death, while activated CD4+ T cells produce cytokines. Activated effector T cells traffic to and infiltrate the tumour. Here, they bind to cancer cells via the T cell receptor presented on T cells and the antigen bound to the MHC molecule on cancer cells, and eventually kill the cancer cells. This releases additional TAAs or TSAs, and the cycle is repeated and amplified, leading to effective clearance of the primary site, potential abscopal effects against other deposits, and a durable defence against recurrence [4].

Each step in the cancer-immunity cycle is an important contributor to effective killing of cancer cells. Hence, when one or more of these steps are blocked it may lead to an insufficient anti-tumour immune response.

1.2. Evading Immune Recognition

In cancer patients, the cancer-immunity cycle does not perform optimally. Indeed, the progression of tumour growth is dependent on an escape from anti-tumour immune responses. Such an escape is achieved by hiding from immune cells, excluding them, or down-regulating their activity.

Tumour cells can hide from immune cells by altering their surface molecule expression. This may be achieved by down-regulating the expression of MHC molecules on tumour cells, particularly MHC I. This limits the ability of Cytotoxic T lymphocytes (CTLs) to recognise tumour cells and initiate apoptosis [8]. However, the absence of MHC molecules often leads to Natural killer (NK) cell activation, but these innate immune cells have no immunological memory and will therefore not provide a sustained anti-tumour immune response [9]. Furthermore, the tumours that have normal expression of MHC molecules often lack co-stimulatory molecules such as CD80 or CD86, which abrogates the activation of T lymphocytes. Lastly, cancer cells are often unrecognised by CTLs because tumour variants with low TAA or TSA production have been selected [10].

The irregular tumour vasculature, elevated interstitial fluid pressure, and dense extracellular matrix combine to restrict the movement of immune cells into and throughout the tumour [11]. Indeed, the abnormal blood and lymphatic vasculature in tumours function as both physical and functional barriers: compressed blood vessels limit the tumour infiltration of cytotoxic immune cells, whereas dysfunctional lymphatic vessels hinder APCs from migrating to lymph nodes and activating T cells [12].

Tumour cells also deploy multiple mechanisms to remodel the Tumour microenvironment (TME) into an immunosuppressive phenotype. Firstly, the TME contains different immune suppressive cells such as Myeloid-derived suppressor cells (MDSCs) and Regulatory T (Treg) cells. MDSCs migrate from the tumour to the lymph nodes, spleen, and peripheral vessels, where they inhibit activation of tumour-specific T lymphocytes [13]. Treg lymphocytes suppress CD4+ and CD8+ T cells by producing specific cytokines such as Interleukin (IL)-10, IL-4, and IL-13. Moreover, Treg cells can inhibit subpopulations of B lymphocytes and NK cells [10]. Another mechanism involves cancer cells modulating immune checkpoint pathways to evade attack by immune cells [14,15]. Immune checkpoint pathways are important for maintaining self-tolerance by modulating T cell function and protecting surrounding tissue by suppressing immune responses, thereby preventing auto-immunity in healthy individuals. Many immune checkpoints are initiated by ligand–receptor interactions; thus, tumour cells can alter this interaction to their benefit. There is an array of different types of immune checkpoint pathways and mediators, in which the most well characterised are Programmed cell death protein 1 (PD-1), Programmed death ligand 1 (PD-L1), and Cytotoxic T lymphocyte-associated protein 4 (CTLA-4). PD-1 is a cell surface receptor presented on T cells that binds to PD-L1 and PD-L2. The interaction of PD-1 with PD-L1 blocks T cell responses to prevent excessive T cell activation [14]. Hence, tumours expressing PD-L1 or PD-L2 ligands can suppress the T cell-mediated anti-tumour immune response. CTLA-4 is important in the early stages of the immune response. This molecule is expressed on tumour cells and competes with CD28 in binding to CD80 and CD86 ligands on APCs. Consequently, further T cell activation is inhibited [15].

In summary, tumours have multiple methods to evade immune recognition. Consequently, there are several ways in which an anti-tumour immune response may be reactivated. Indeed, immunotherapy is a modality that includes a range of treatment methods, including (1) cancer vaccines, which aim to activate CTLs [16]; (2) CAR T cells, which are patient-derived T cells with a chimeric antigen receptor that are engineered to recognise and bind specific antigens on the surface of cancer cells [17]; and (3) Immune checkpoint inhibitors (ICIs), which interact with immune checkpoints, thereby preventing immune tolerance [18]. The effectiveness of these treatments has been proven in a range of cancers [19–21]. In fact, the use of ICIs demonstrated that 12 out of 12 patients with rectal cancer had a clinical complete response [22]. However, it is notable that the greatest ICI successes have come in patient populations that have been carefully selected for an immune "hot" rather than "cold" phenotype [22]. For more widespread impact, methods to optimise such approaches are needed, and driving tumours towards a "hot" phenotype is thought to be an effective way of achieving this. Recent research suggests that Ultrasound (US)-mediated cavitation can result in the immunomodulation of the TME, thus partly reversing tumour immune evasion. Hence, this may be a promising strategy to promote tumour-specific immune responses and enhance the efficacy of ICIs.

2. Cavitation-Mediated Immunomodulation

Several studies have served to demonstrate that radiation or mechanically induced tumour cell damage can be used in combination with cancer immunotherapies to provide improved outcomes [23–25]. Indeed, the approach of combining non-ionising sources, such as US, to thermally ablate tumour tissue has garnered recent attention [26–28]. However, tumour antigens released from thermally ablated tumours are often denatured and may not be effectively presented to the immune system [29]. Additionally, thermal diffusion can damage surrounding healthy tissue, causing adverse effects [30].

US can also be used to induce non-ablative mechanical effects in tumours. Acoustic cavitation is used to describe the expansion and collapse of gas bubbles in response to the alternating compressional and rarefactional US wave. Notably, providing agents to nucleate such acoustic cavitation, i.e., by adding pre-existing gas bubbles—referred to henceforth as Cavitation agents (CAs), can lower the pressure amplitude required to achieve cavitation and consequently reduce the energy deposition within tissue. The volumetric changes can lead to various mechanical phenomena that can impact the surroundings. First, the gentle oscillation of CAs will cause momentum transfer from the bubble to the surrounding liquid, called microstreaming [31]. This can result in shear stresses impacting on cell membranes [32]. More violent oscillation and the collapse of CAs near a surface can produce shock waves and microjets, leading to cellular injury and tissue destruction [31,32]. The scale of the membrane disruption and damage, and the underlying mutations to apoptotic pathways within the cells, will determine the consequent route to death (apoptosis or necrosis) and the resulting immune response. Hence, due to its potential, studies have investigated the immunological effects of acoustic cavitation, which are summarised in Figure 1. Figure 1 highlights that cavitation events may modulate tumour immunity in three ways by the destructive release of TSAs, TAAs, and Damage-associated molecular patterns (DAMPs), by vascular modification, and lastly by altering the cellular activity of cells important for immune responses.

This review therefore focuses on how acoustic cavitation, used without the intention of causing heating, may promote anti-tumour immune responses. The evidence regarding the role that cavitation may play in the immune outputs, identified in Figure 1, is assessed. Details of the treatment protocols and US parameters of the papers mentioned in this section can be found in Table 1.

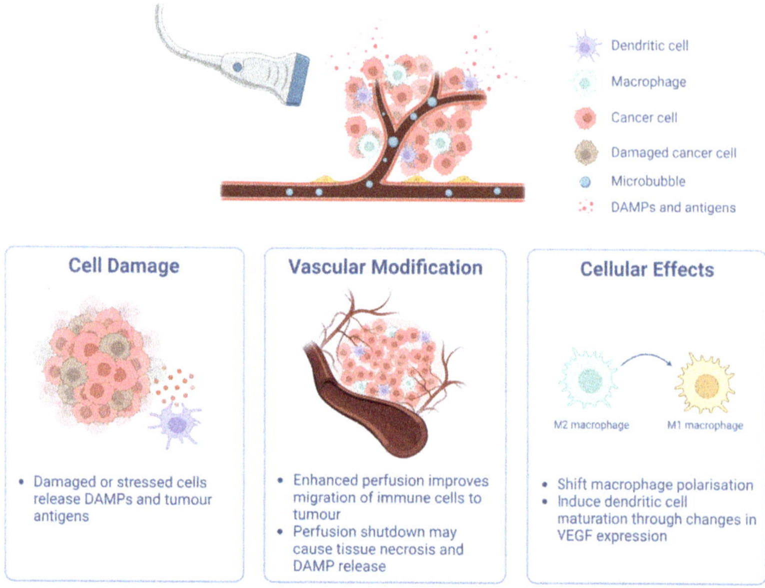

Figure 1. Schematic overview of how ultrasound-mediated cavitation may affect different parts of the tumour. It has been proposed that cavitation can cause cell damage and stress, which consequently releases Damage-associated molecular patterns (DAMPs) and tumour antigens. Moreover, it has been suggested that depending on the type, duration, and intensity of cavitation, tumour perfusion can either be increased [33] or blood flow can be shut down [34,35]. Either situation may lead to increased anti-tumour immunity as a result of improved migration and infiltration of immune effector molecules and cells or due to increased tissue necrosis and Damage-associated molecular pattern (DAMP) release. Lastly, ultrasound-mediated cavitation has been shown to transform M2 macrophages to M1 macrophages [36] and additionally induce dendritic cell maturation by altering expression of Vascular endothelial growth factor (VEGF) [37]. Figure created with BioRender.

2.1. Release of Tumour-Associated Antigens and Damage-Associated Molecular Patterns

The release of TAAs, TSAs, and DAMPs is an important part of activating an anti-tumour immune response. DAMP molecules, including High-mobility-group box 1 (HMGB1), Heat shock proteins (HSPs), calreticulin, and Adenosine triphosphate (ATP), are danger signals that are often released when cells are exposed to stress. These danger signals can promote an immune response by initiating various mechanisms such as stimulating neutrophils or monocytes to produce and secrete proinflammatory cytokines and chemokines, and up-regulate co-stimulatory molecules [38]. In regards to anti-tumour immune responses, cancer cells undergoing immunogenic cell death may expose DAMPs on their surface, which helps DCs engulf the cancer cells. This then leads to the presentation of tumour antigens followed by CTL responses [39]. Thus, the release of DAMPs ultimately leads to the activation of anti-tumour immunity. Researchers have therefore sought to characterise how cavitation may cause the release of DAMPs and tumour antigens, and subsequent immune responses in tumours.

Liu et al. investigated immunological changes induced by US+CA in a colon tumour (CT26) murine model [40]. Immune cell populations were studied at days 1, 3, and 18 after treatment and compared to untreated tumours. The analyses revealed a significant ($p < 0.05$) increase in percentage of CD8+ T cells among CD45+ leukocyte cells across all three days; however, the percentage of CD4+Foxp3+ cells (Tregs) relative to CD45+ leukocytes remained stable. The authors confirmed, by immunohistochemical staining, that US+CA triggered the production of HSP60, although the level was not quantified. A link was therefore suggested between the release of this DAMP and the increased

CTL infiltration observed. Additionally, these authors measured temperature in a tissue-mimicking phantom using similar conditions as in vivo, and no increase was observed upon US+CA treatment, indicating that the release of HSP60 was most likely due to mechanical stress and not heating [40].

In a similar study carried out by Joiner et al., pancreatic tumours (KPC) were either untreated or treated with US+CA and excised 2 or 15 days after treatment [41]. A non-significant ($p = 0.053$) 1.6-fold increase of the HMGB1 level, as measured by immunohistochemistry, was detected 2 days after US+CA treatment compared to untreated tumours. However, no significant difference in cell death marker cleaved caspase-3 or HSP70 was observed between the two groups. Furthermore, tumour draining lymph nodes exhibited a significant ($p < 0.05$, $p < 0.005$, and $p < 0.005$, respectively) increase in the number of F4/80+CD11b+ cells (macrophages), CD8+ cells (CTLs), and CD4+ cells (T helper cells) 2 days after treatment. It was concluded that this was a transient immune response due to the majority of these effects subsiding 15 days post-treatment. Notably, US-only or CA-only controls were not included in this study [41].

Furthermore, Hu et al. assessed the release of both TAAs and DAMPs after US+CA, US-only, or CA-only treatment, as well as no treatment [42]. RM1-OVA cells were used to permit measurement of the release of Ovalbumin (OVA) and DAMPs in vitro. At both 0 and 4 h after US+CA treatment, HMGB1, calreticulin, and HSPA2 levels were increased compared to the other treatment groups, though the scale of this increase was not quantified. ATP release was significantly ($p = 0.004$) higher for the US+CA-treated cells. This DAMP release was attributed to the increase in tumour cell necrosis observed both in vitro and in vivo in the US+CA group. The RNAseq data of the RM1 tumour models confirmed that the TME exhibited a high level of pathways related to 'acute inflammation response' and 'cell death', suggesting that US+CA can cause tumour cell injury. Hu et al. also investigated and compared the immune microenvironment in vivo after US+CA versus no treatment. They showed a significant ($p < 0.05$, $p < 0.005$, $p < 0.005$, $p < 0.005$, and $p < 0.005$, respectively) increase in CD45+ cells (leukocytes), CD11c+ cells (DCs), CD11b+ (monocytes), 49b+ cells (NK cells), and CD68+ cells (macrophages) both 1 day and 4 days after treatment with US+CA. Further, it was concluded that CD8+ T cells were the main contributors to the US+CA treatment effects. This was due to the fact that the tumour volume after treatment with US+CA with the addition of CD4+ and CD8+ T cell-depleting antibodies was similar to tumours treated with US+CA plus only CD8+ T cell-depleting antibodies, but both groups had a 3-fold higher tumour volume than US+CA-treated tumours that did not receive T cell depleting antibodies [42].

Wu et al. divided breast tumours (4T1) into the following treatment groups: untreated, US-only, CA-only, and US+CA [43]. They detected a significantly ($p < 0.05$) higher (1.2–1.5-fold) expression of HMGB1 and a significantly ($p < 0.05$) lower ATP level in vitro after US+CA treatment compared to the three other groups and reported an increase in calreticulin expression in the US+CA group in vivo, indicative of DAMP release. A significant ($p < 0.01$ and $p < 0.05$, respectively) increase in IL-12 and Tumour necrosis factor alpha (TNF-α) levels was also reported [43].

In contrast to the other studies, Huang et al. treated lung (LL/2) or colon (CT26) tumour cells with CA only or US+CA and detected no difference in levels of calreticulin, HMGB1, or ATP in vitro 4 h after treatment [44]. There was no significant difference in the percentage of CD80+CD86+ cells (mature DCs) or Interferon gamma (IFN-γ)-secretion in mature DCs after mature DCs were incubated with supernatants from tumour cells treated with US+CA or CA [44].

Overall, several papers confirmed that US+CA can cause DAMP release as well as TAA release. The type of DAMP and amount of DAMP release detected seem to vary between experiments, but this could be due to both different US parameters and treatment protocols. As a consequence of TAA, TSA, and DAMP release, leukocytes may be recruited to tumours and exert their anti-tumour effect [45]. Neither Liu et al. nor Joiner et al.

investigated the mechanism underlying DAMP release; however, Hu et al. suggested it was due to tumour cell necrosis, which they observed after treatment.

2.2. Modulation of Tumour Vasculature and Perfusion

Tumour vasculature, and the level of perfusion it provides, is important for the presentation of TAAs and TSAs by APCs, and the migration of immune cells such as CTLs into and throughout the tumour [46]. Hence, the impact of US+CA on tumour vasculature and the corresponding immune cell infiltration has been probed in recent papers [33–35].

Bulner et al. showed that anti-vascular US-stimulated CAs decreased colon tumour (CT26) growth compared to tumours receiving CAs alone and, perhaps as a consequence of restricted perfusion, observed no significant increase in T cell infiltration [34]. A significant ($p < 0.05$) shutdown of blood flow in tumours followed by necrosis was observed by contrast-enhanced US imaging. Specifically, the peak enhancement intensity was reduced by 88 $\pm$ 3.6% relative to CA with no US treatment. Flow cytometry analysis examined the change in CD45+ cells (leukocytes), CD45+CD8+ cells (CTLs), CD45+CD4+ cells (T helper cells), and CD45+CD4+Foxp3+ cells (Tregs) in tumours both 3 and 7 days after the initial treatment day. There was no significant difference between the US+CA group and the CA group. Nonetheless, there was a significant ($p < 0.05$) increase (1.5-fold) in the number of cells in tumour-draining lymph nodes in mice with US+CA-treated tumours 3 days after the initial treatment day, indicating some immune response was established [34].

Contrarily to Bulner et al., Hunt et al. observed an increase in intratumoural immune cells 24 h after anti-vascular US+CA treatment of melanoma tumours (K1735) compared to tumours treated with US-only [35]. Hunt et al. observed a 70% reduction of blood perfusion accompanied by subsequent necrosis. Histopathological analysis suggested that there was a significant ($p < 0.01$) increase in CD45+ cells (leukocytes). More specifically, the number of CD45+ cells per High power field (HPF) was 50 $\pm$ 7 in US+CA-treated tumours and 15 $\pm$ 1 in US-treated tumours. Likewise, there was a significant ($p < 0.01$) increase in CD3+ cells (T cells) with 60 $\pm$ 7 cells/HPF and 11 $\pm$ 1 cells/HPF in US+CA-treated and US-treated tumours, respectively. However, there was no difference in B220+ cells (B cells) between the two groups [35].

Similarly, Wu et al. used contrast enhanced US imaging to report significantly ($p < 0.05$) decreased blood perfusion that was not recovered 24 h later [43]. A significant ($p < 0.01$) upregulation of the CD11c+CD80+CD86+ cell (matured DC) proportion in tumour draining lymph nodes was detected after US+CA treatment compared to untreated tumours (26.40 $\pm$ 3.70% vs. 13.03 $\pm$ 2.03%). In addition, there was a significant ($p < 0.01$) increase in infiltration of CD3+CD8+ cells (CTLs) in tumours after US+CA treatment as measured by flow cytometry [43].

Li et al. looked at the effect of US+CA on tumour blood perfusion and the corresponding immune responses in a preclinical model of colon cancer (MC38) [33]. Mice were injected with CAs and tumours were exposed to a Peak negativepressure (PNP) of either 0.8 MPa or 2.4 MPa. Vasodilation of tumour microvessels and consequently enhanced blood perfusion, as measured by contrast enhanced US, was observed for the 0.8 MPa-treated tumours, while poor blood perfusion, microvascular damage, and hemorrhage in tumour tissues were present in those treated with 2.4 MPa. Further, the percentage of infiltrating CD8+ T cells in tumours 24 h after treatment was 32.35% or 19.43% of the total number of lymphocytes for 0.8 MPa or 2.4 MPa, respectively.

All four studies by Bulner, Hunt, Wu, and Li reported worse blood perfusion after US+CA, although for Li et al. this was only the case for the highest pressure treatment (2.4 MPa). Both Bulner et al. and Hunt et al. assessed and confirmed necrosis—a type of immunogenic cell death that can activate an inflammatory response [47]. Despite this, Bulner et al. did not observe increased tumour infiltration of immune cells. On the other hand, both Hunt et al. and Wu et al. confirmed immune cell infiltration. The use of different US parameters and CAs between studies makes it difficult to compare the reported results, particularly considering that Hunt et al. suggested that thermal changes could play a role. Yet, none of the studies looked into temperature elevation, which could provide valuable information.

Moreover, it has previously been shown that different US parameters promote different anti-vascular effects [48], which could explain the difference in detected immune response. Indeed, tumour perfusion recovery after a 30 min treatment with a PNP of 2.6 MPa and CA has been reported, but with US+CA at a PNP of 4.8 MPa the perfusion was reduced even after 24 h [48]. Hence, it could be that the US parameters used by Hunt et al. and Wu et al. allowed for anti-vascular effects to be reversed within their set time frame, thus explaining the enhanced immune response observed in these tumours despite the initial damage imparted. In contrast, in the studies by Bulner et al. and Li et al., tumour perfusion may not have been restored within their time frame, which may have hindered immune cell trafficking [12]. To gain a better understanding of the presence of tumour-infiltrating immune cells or the lack thereof, it would be valuable for future studies to determine blood perfusion before excising the tumours.

2.3. Cellular Effects of Cavitation

US-mediated cavitation may also have a direct or indirect effect on cells important for anti-tumour immunity. These can either be immune cells [36,37] or cells that can suppress immune responses [49].

Lin et al. showed that US+CA could redirect Tumour-associated macrophage (TAM) polarisation and investigated its role in vessel normalisation in a pancreatic tumour model (SW1990) [36]. There are two types of TAMs: M1, which is associated with promoting inflammation and anti-tumour activity, and M2, which typically enhances tumour progression [50]. Lin et al. measured changes in F4/80+ cells (macrophages) by imaging stained tissue (immunohistochemistry) and calculating the percentage of positive F4/80+ area over the total image area. There was a significant ($p < 0.05$) increase in the F4/80+ area after US+CA treatment compared to untreated tumours (29.87 ± 1.16% vs. 20.87 ± 1.17%). Moreover, there was a significant ($p < 0.001$) 2-fold increase in F4/80+CD86+ cells (M1 macrophages) among F4/80+ cells in US+CA-treated tumours compared to untreated tumours, suggesting a shift in TAM polarisation after treatment.

It has previously been demonstrated that shifting from the M2 to the M1 phenotype can normalise tumour vessels and enhance anti-tumour immunity [51]. Indeed, Lin et al. observed, by contrast enhanced US imaging, a 1.5 fold higher perfusion intensity in the US+CA group than that of the untreated group. To examine the relationship between vessel normalisation and TAM polarisation, they treated tumours with clodronate liposomes to remove TAMs. Firstly, TAM depletion reduced tumour growth in the untreated group, indicating that TAMs were predominantly M2 macrophages. Moreover, the vessel perfusion was improved in TAM-depleted tumours, but there was no significant difference between the untreated and treated group. Thus, it was concluded that the induction of vessel normalisation primarily relied on converting M2-type macrophages to M1-type macrophages [36].

Moreover, Zhang et al. looked at how US+CA may affect VEGF expression and the following immune activation [37]. VEGF can interfere with immune cell migration indirectly by promoting a vascular immune barrier, and more directly by interacting with Vascular endothelial growth factor receptor (VEGFR) on effector T cells, Tregs, DCs, and MDSCs; it can also affect the immune cell phenotype and function [52,53]. Zhang et al. observed that the murine prostate cancer cells (RM-1) exposed to US+CA had significantly ($p < 0.05$) lower VEGF expression, as assessed by western blotting, than those exposed to US-only, CA-only, or untreated cells (32.7 ± 4.9% vs. 47.9 ± 5.9%, 71.7 ± 6.6%, and 74.2 ± 4.7%, respectively). These cells were then co-cultured with CD11c+ DCs and CD8a+ T lymphocytes, and it was demonstrated that there were slight but significant ($p < 0.05$) increases in the amounts of CD11c+ cells (22.2 ± 0.9% vs. 18.1 ± 0.4%) and CD8a+ T cells (34.0 ± 1.5% vs. 31.5 ± 0.8%) when co-cultured with US+CA-treated RM-1 cells than untreated RM-1 cells. Thus, it was concluded that the inhibition of VEGF expression in RM-1 cells caused by US+CA can promote DC and CTL proliferation [37].

Table 1. Summary of treatment protocol, ultrasound (US) parameters, and main findings of the papers mentioned in Section 2.

Author	Treatment Protocol	US Parameters	Main Findings
Liu et al. [40]	CT26 tumours were divided into three treatment groups: (1) untreated, (2) US-only, and (3) US+CA. SonoVue MBs were used as CAs. Bolus of CAs was injected immediately before US exposure. Mice were sacrificed 1, 3, and 18 days after treatment.	F [1] = 0.5 MHz, PRF [2] = 1 Hz, PL [3] = 100 ms, and PNP [4] = 0.6 MPa and 1.4 MPa. Exposure time: 20 s per spot, and 9–12 spots were sonicated to cover the entire tumour. Transducer details: element diameter of 64 mm and radius curvature of 55 mm.	Tumour growth was inhibited for both pressures, albeit it was greater for the higher pressure. The percentage of CD45+CD8+ T cells increased in tumours treated with PNP of 1.4 MPa compared to untreated tumours across all three time points.
Joiner et al. [41]	Mice with KPC tumours were divided into two groups: (1) untreated and (2) US+CA. Lipid MBs with C_4F_{10} in the core were used as CAs. CAs were infused for the entire duration of US treatment. Tumours were excised either 2 or 15 days after treatment.	F = 1 MHz, PRF = 100 Hz, PL = 1 ms, and PNP = 0.5 MPa. Exposure time: 7–10 min depending on tumour size. Transducer details: eight-element annular array, 80 mm focus, 1 mm × 1 cm focal spot.	Significant reduction in tumour growth after treatment. The number of CD4+ T cells, CD8+ T cells, and Ly6C-F4/80+CD11b+ macrophages in lymph nodes was significantly higher in treated tumours 2 days after treatment. However, the observed immune response was likely transient due to no significant difference in these immune cell populations being observed 15 days after treatment. A substantial increase in HMGB1 was measured in treated tumours. It was suggested that stable cavitation was the dominant bubble behaviour in this study.
Hu et al. [42]	RM1, MC38, and B19 tumours were divided into four groups: (1) untreated, (2) US+CA, (3) aPD-1, and (4) US+CA+aPD-1. The CA was a lipid nanobubble with C_3F_8 as core gas. US was applied to tumours 5 min after intravenously administering the CAs. aPD-1 was administered intraperitoneally into mice once every 3 days during the treatment period for a total of four doses.	F = 1 MHz, PRF = 100 Hz, PL = 6 ms, and I [5] = 1 W/cm^2. Exposure time: 30 s. Transducer details: collimated beam and an effective probe radiation area of 2 cm^2.	US+CA and US+CA+aPD-1 substantially inhibited tumour growth compared to untreated controls. These two groups had significantly more CD44+CD8+ cells compared to untreated tumours, but the difference was greater for the US+CA+aPD-1 group. There were significantly more granzyme B and IFN-γ-secreting CD8+ T cells in the combination therapy group and the US+CA group compared to the untreated group. They also reported TAA release and DAMP release in vitro when RM-1 cells were treated with US+CA compared to the untreated, US-only, and CA-only groups.

Table 1. *Cont.*

Author	Treatment Protocol	US Parameters	Main Findings
Wu et al. [43]	For the first study, 4T1 tumours were divided into one of the following groups: (1) no treatment, (2) US, (3) CA, or (4) US+CA. Lipid MBs with C_3F_8 as core gas were used as CAs. Tumours were exposed to US immediately after the injection of lipid MBs on days 0, 1, 2, 3, and 4. In another in vivo, they divided the tumour-bearing mice into these four groups: (1) control, (2) aPD-L1, (3) US+CA, and (4) US+CA+aPD-L1. The treatment strategy was similar to the first in vivo, but aPD-L1 was injected intravenously on days 1, 4, and 7. Mice were sacrificed on day 11 after treatment.	F = 1 MHz, PRF = not specified, PL = not specified, I = 3.0 W/cm^2, and duty cycle = 50%. Exposure time: 5 min. Transducer details: diameter of 1 cm, focal length of 1.5 cm, and focus area of 0.4 cm^2.	In the first study, a substantial difference was reported in tumour volume after treating with US+CA compared to the untreated, US-only, and CA-only groups. The tumour blood perfusion was blocked even 24 h after treatment. There was a significant increase in CD11c+CD80+CD86+ cells (matured DCs) and CD3+CD8+ T cells, as well as the level of IL-12 and TNF-α cytokines. In the second study, it was observed that the tumour growth was even more inhibited by the US+CA+aPD-L1 and there was a remarkable increase in activated CD8+ T cell infiltration compared to untreated tumours and aPD-L1-treated tumours.
Huang et al. [44]	LL/2 and CT26 tumour cells and tumours were treated with either (1) CA or (2) US+CA. The CA was a lipid MB with C_3F_8 in the core. CAs with or without US exposure were administered every 3 days, for a total of 6 treatments (18 days), and mice were observed for 28 days. US exposure occurred 1 min after CA administration.	F = 2.25 MHz, PRF = 1 Hz, PL = 10 ms, and PNP = 1.9 MPa. Exposure time: 10 min. Transducer details: diameter of 20 mm and focal length of 50 mm.	Tumour cells treated with only CA or US+CA did not show induced translocation of calreticulin or Erp57, or release of HMGB1 or ATP in vitro. There was no substantial difference in CD80+CD86+ cells (mature DCs) or IFN-γ-secreting cells in vivo between the two groups.
Bulner et al. [34]	Mice with CT26 tumours were randomised into these groups: (1) CA, (2) aPD-1, (3) US+CA, and (4) US+CA+aPD-1. The CAs were MBs consisting of lipids encapsulating C_3F_8 gas. US exposure commenced immediately after intravenous injection of CAs. For the acute experiments, animals sacrificed at day 3 received US+CA or CA treatment at day 3 with or without aPD-1 at day 0. Animals sacrificed at day 7 received US+CA or CA treatment with or without aPD-1 at day 0, 3, and 6. For the longitudinal experiment, the treatment schedule was similar but aPD-1 was administered on day 9 and 12 as well, and mice were sacrificed at day 30.	F = 1 MHz, PRF = 100 Hz, PL = 0.1 ms, and PNP = 1.65 MPa. Exposure time: 50 pulses were repeated at 20 s interval for a duration of 2 min. Transducer details: spherically focused, 3.75 cm diameter, 15 cm focal length, and 1.05 cm −6 dB beam width.	The improved tumour growth inhibition was attributed to the shutdown of blood flow due to no evidence supporting a T cell-dependent mechanism (CD45+CD8+ cells and CD45+CD4+ cells). However, the re-challenge experiment suggested an engagement of adaptive memory response. Passive cavitation detection detected broadband noise.

Table 1. *Cont.*

Author	Treatment Protocol	US Parameters	Main Findings
Hunt et al. [35]	Animals with K1735 tumours were initially divided into three treatment groups: (1) 3 min US exposure, (2) US+CA with 1 min exposure, and (3) US+CA with 3 min exposure. Definity MBs were used as CAs. Tumours were insonated immediately after intravenous injection of CAs. Mice were sacrificed 24 h after treatment.	F = 3 MHz, PRF = continuous, PL = continuous, PNP = 0.22 MPa, and exposure time: 1 min or 3 min. Transducer details: unfocused, power level 3, and spatial average intensity 2.3 W/cm^2.	A significant shutdown of blood flow after both US+CA treatments compared to US-treated tumours was reported. There was a significant increase in the mean number of CD45+ cells and CD3+ cells after US+CA (3 min) compared to untreated tumours.
Li et al. [33]	MC38 tumour-bearing mice were divided into the following groups for the perfusion study: (1) US+CA at 0.8 MPa, (2) US+CA at 2.4 MPa, (3) US at 2.4 MPa, and (4) untreated control. For the second study, the groups were: (1) untreated, (2) US+CA at 0.8 MPa, (3) aPD-L1, and (4) US+CA+aPD-L1 at 0.8 MPa. Sonazoid MBs were used as CAs. The CAs were slowly injected through tail vein during US exposure. For the perfusion study, mice were sacrificed 24 h after treatment. For the combination therapy, the mice were injected with aPD-L1 on days 4, 7, 10, and 13, US+CA treatment was performed 24 h after each aPD-L1 administration, and mice were sacrificed 24 h after the final treatment.	F = 4 MHz, PRF = 1 kHz, PL = 4.5 µs, and PNP = 0.8 MPa and 2.4 MPa. Exposure time: 1 s on and 1 s off for 10 min. Transducer details: phased focus.	The 0.8 MPa exposure enhanced perfusion substantially, whereas the 2.4 MPa reduced blood perfusion. There was a significant increase in CD8+ T cells for the 0.8 MPa exposure compared to the untreated tumours and US+CA treatment at 2.4 MPa. Moreover, US+CA+aPD-L1 had significantly better therapeutic effect and more CD8+ T cells than US+CA and aPD-L1 only. Additionally, the US+CA+aPD-L1 boosted IFN-γ and granzyme B secretion.
Lin et al. [36]	Mice with SW1990 tumours were randomised into (1) untreated and (2) US+CA groups. MBs from Bracco were used as CAs. Tumours were treated 5 days per week. The CAs were slowly injected via the tail vein.	F = 1 MHz, PRF = 1 kHz, PL = 0.2 ms, and I = 1.2 W/cm^2. Exposure time: 2 min five times. Transducer details: not specified.	It was demonstrated that the induction of vessel normalisation by US+CA mainly relied on shifting TAM polarisation from M2-type to M1-type.
Zhang et al. [37]	For the VEGF expression experiment, RM-1 cells were divided into the following groups: (1) US, (2) CA, (3) US+CA, and (4) untreated. The CAs were SonoVue MBs. To detect DC and T lymphocyte phenotype, RM-1 cells either treated with US+CA or untreated were co-cultured with DCs and T lymphocytes.	F = 800 kHz, PRF = 1 Hz, PL = 0.5 s, and $ISATA$ [6] = 360 mW/cm^2. Exposure time: 30 s. Transducer details: cylindrical probe with a diameter of 13 mm.	VEGF expression was significantly decreased after US+CA treatment compared to the other groups. There was a significant increase in CD11c+ DCs and CD8a+ T cells in the US+CA group compared to the untreated group.

Table 1. *Cont.*

Author	Treatment Protocol	US Parameters	Main Findings
Tan et al. [49]	Spleens of LLC tumour-bearing mice were in the first study divided into two groups: (1) US+CA and (2) untreated. The CAs were Sonazoid MBs. Diluted CA of 0.1 mL was injected at the first 100 s, 0.02 mL per 100 s for three times, and 0.02 mL per 50 s at the rest of the treatment. Tumours and spleens were excised 24 h after treatment. In the second study, mice were divided into the following groups: (1) US+CA, (2) aPD-L1, (3) US+CA+aPD-L1, and (4) untreated. Spleens were treated once every 3 days for a total of 3 times with US+CA, and aPD-L1 was injected intraperitoneally on the following day. Spleens and tumours were excised on day 12.	F = 5 MHz, PRF = 500 Hz, PL = 1.3 µs, PNP = 2.2 MPa, and 2.3 MPa. Exposure time: transmitting and intermittent time of 0.1 s, and total duration 600 s. Transducer details: linear array probe.	In the first study, the results showed a significant reduction in splenic CECs and an increase in CD8+ T cells when treated with US+CA compared to untreated spleens. There was no substantial difference in CD11b+Gr1+ cells (MDSCs), CD11b+CD11c+ cells (DCs), CD11b+F4/80+ cells (macrophages), or B220+ cells (B cells). In the second study, tumour growth was only inhibited when US+CA was combined with aPD-L1. The US+CA+aPD-L1 treatment demonstrated a significant increase in number of IFN-γ-producing CD8+ T cells and CD4+ T cells.

[1] Frequency, [2] pulse repetition frequency, [3] pulse length, [4] peak negative pressure, [5] intensity, and [6] spatial average time average intensity.

In another study, Tan et al. exposed spleens rather than tumours to US+CA in a lewis lung cancer model [49]. The reason for this was that immature red blood cells called CD71+ erythroid progenitor cells (CECs), which expand in the spleen, suppress immune responses [54,55]. Immunosuppression can induce tumour immune evasion to promote tumour growth. Additionally, CECs also modulate T cells via the PD-L1/PD-1 pathway and express genes encoding immune checkpoint molecules [56]. Thus, by targeting and consequently reducing the CECs residing in the spleen, the anti-tumour immune response might be enhanced. Indeed, Tan et al. detected a significant ($p = 0.006$) decrease in CD71+TER119+ cells in the spleen, and the percentage of CD8+ T cells was significantly ($p = 0.003$) increased after US+CA treatment compared to untreated spleens. However, there was no difference in tumour growth between these two groups. Likewise, there was no difference in the number of CD11b+Gr1+ cells (MDSCs), CD11b+CD11c+ cells (DCs), CD11b+F4/80+ cells (macrophages), or B220+ cells (B cells) in the spleen [49]. It is unclear how US+CA would impact specifically on the intended cell targets within the spleen without also disrupting the general immune cell profile.

Altogether, US-mediated cavitation can contribute to the immunomodulation of tumours through a variety of mechanisms: releasing DAMPs and TAAs, altering tumour perfusion, and modifying cellular activity. These changes to the tumour immune microenvironment might produce a TME more favourable for ICI therapies.

3. Cavitation-Enhanced Checkpoint Inhibitor Therapy

In addition to the mounting evidence that US-mediated cavitation can promote an anti-tumour immune response without drug addition, there is also gathering evidence that combining US+CA with immunotherapeutic drugs may provide an additional benefit to the treatment. Cancer immunotherapy, particularly the use of ICIs, is considered to be one of the most promising strategies to treat cancer [57], and thus combining US-mediated cavitation with ICIs could be a powerful treatment strategy. Cavitation has the potential to both improve the drug delivery of ICIs and, as previously discussed, engage the immune system.

3.1. Improved Delivery

A drawback of ICIs is their relatively large size (~150 kDa), which limits their intratumoural delivery [58] and consequently their therapeutic efficacy [59]. US-mediated cavitation has been shown to improve the intratumoural delivery of drugs [60]. In fact, Grundy et al. observed a 2.1–3.6-fold increase in cetuximab concentration in murine tumours targeted with US+CA compared to CA-treated tumours [61]. Moreover, Li et al. reported a significant ($p < 0.0001$) 1.37-fold increase in concentration of the aPD-L1 antibody in the tumour when co-administered with US+CA compared to aPD-L1 monotherapy [33]. In addition, Kim et al. measured a significant ($p < 0.01$) increased intratumoural staining of aPD-L1 after administering aPD-L1-coated CA and exposing to US compared to aPD-L1-coated CA without US, free aPD-L1 monotherapy, or US+aPD-L1 without CA [62]. Thus, there are data showing that US+CA can increase the tumour uptake of therapeutic antibodies. Even though the poor response to ICI in some patients might be due to poor delivery, the heterogeneous tumour immune environment is also an important consideration for ICI efficacy.

3.2. Favourable Tumour Immune Microenvironment

It has been reported that having a sufficient CD8+ T cell presence within the TME favours tumour susceptibility to PD-1-based therapy [63]. In other words, "hot" tumours, defined as highly T cell-infiltrated tumours, are associated with a better treatment response. In contrast, "cold" tumours that typically have poor T cell infiltration limit the efficacy of ICIs [64]. Converting a "cold" tumour into a "hot" tumour and successfully activating T cells could be essential for sufficient therapeutic effect. As discussed in Section 2, US-mediated cavitation can contribute to making the tumour immune microenvironment more favourable for ICIs.

Indeed, Li et al. observed a significant ($p < 0.05$ and $p < 0.001$, respectively) increase in tumour growth inhibition when treated with US+CA+aPD-L1 compared to US+CA or aPD-L1 alone. There was a significant ($p < 0.05$) increase in the percentage of CD8+ T cells, as well as a significant ($p < 0.01$) increase in IFN-γ and granzyme B secretion in tumours receiving US+CA+aPD-L1 [33]. Similarly, Hu et al. reported substantial tumour growth inhibition in three different tumour models when tumours were treated with US+CA+aPD-1 compared to US+CA or aPD-1 alone [42]. All tumour models were characteristically "cold" tumours, but after US+CA+aPD-1 treatment they observed significantly ($p < 0.05$) more CD44+CD8+ cells and the presence of T cell effector functions compared to monotherapies. The US+CA+aPD-1 group also exhibited a high expression of genes related to 'immune response', 'T cell activation', and 'antigen processing and presentation' [42]. Wu et al. reported greater tumour growth inhibiton in US+CA+aPD-L1-treated tumours than in the US+CA group and aPD-L1 group, and the combination therapy resulted in a substantial increase in activated CD8+ T cell infiltration [43].

Tan et al. treated spleens with US+CA+aPD-L1 and showed that tumour volume was significantly ($p < 0.001$) reduced compared to tumours in the US+CA, aPD-L1 alone, or untreated groups [49]. The results demonstrated that the number of IFN-γ-producing CD8+ T cells and CD4+ T cells significantly ($p = 0.0134$ and $p = 0.0082$, respectively) increased in US+CA+aPD-L1-treated spleens compared to untreated spleens. Tan et al. hypothesised that US+CA prevented CECs from inhibiting CD8+ T cell proliferation in the spleen, so more cytotoxic T cells in the peripheral circulation system could benefit from aPD-L1 [49]. Notably, although the spleen was located by a small animal US system, cavitation was not mapped or quantified to ensure that cavitation was confined to the spleen alone.

Bulner et al. also observed enhanced tumour growth inhibition for mice with tumours treated with US+CA+aPD-1 compared to both the US+CA group and aPD-1-only group, but a clear T cell-dependent mechanism was not detected [34]. However, there was a significant ($p < 0.05$) increase in the number of cells in the tumour draining lymph node 7 days after US+CA+aPD L1 treatment compared to US+CA treatment but not aPD-1 treatment. The cells within the lymph node were not characterised. Additionally, one mouse that exhibited complete regression was subjected to a re-challenge experiment with the same tumour cell line, and no tumour was present on US imaging after 90 days, suggesting an adaptive immune response prevented tumour growth upon re-challenge. Hence, a potential contribution of anti-tumour immune responses should not be excluded.

3.3. *Reduced Adverse Effects*

Lastly, US and CAs may limit the impact of drug in non-target tissue as well as reducing Immune-related adverse effects (irAEs). An increasing number of studies on irAEs are being reported with approximately 10–20% of patients treated with aPD-L1 showing irAEs [62,65]. Kim et al. showed that US and CAs coated with aPD-L1 significantly ($p < 0.05$) reduced tumour volume compared to US+aPD-L1 without CA, aPD-L1-coated CA without US, or free aPD-L1. Moreover, it was reported that 18/20 mice that received the aPD-L1-coated CA without US survived 15 days post-treatment, whereas 9/20 mice survived after administration of free aPD-L1 [62]. Hence, incorporating ICIs into CAs may minimise ICI toxicities.

There are multiple ways in which US-mediated cavitation can improve ICI therapy. Although US+CA has the potential to improve the delivery of ICIs, the efficacy of ICIs is not necessarily improved. Similarly, though reducing the adverse effects of ICIs is important, enhancing the therapeutic efficacy is still desirable. Thus, if the aim is to enhance the anti-tumour effect of ICIs, converting "cold" tumours into "hot" tumours by US-mediated cavitation is an attractive strategy.

4. Discussion

It is clear from the reported data that US-mediated cavitation can affect anti-tumour immunity in several ways, even without the administration of additional therapeutics, although the responses are mostly transient. Furthermore, evidence suggests that cavitation can contribute to enhancing the efficacy of ICIs. However, gaining a better understanding of how cavitation increases the therapeutic effect of ICIs is essential for both improving upon and developing new treatment approaches.

With the sparse data available, there are likely still multiple immunological effects of cavitation that remain undefined. Thus, to achieve the most optimal effect of US-mediated cavitation in combination with ICIs, it would be valuable to further explore what immunological effects cavitation may induce. Previous studies have begun to profile the negative and positive effects of cytokines and chemokines on prognosis following ICI treatment [66,67], and there are some studies indicating that cavitation can affect cytokine presence [43]. Defining the impact of the type and duration of cavitation on the release of chemokines and cytokines could help link cavitation outputs to potential ICI efficacy. Similarly, assessing the effect of cavitation on cancer-associated fibroblasts or NK cells could be interesting since they both have important roles in anti-tumour immunity [68,69].

More importantly, whether there is a direct relationship between the oscillation regime and the biological and immunological effects remains to be elucidated. It is currently unclear whether the gentle oscillation or violent collapse of bubbles is more favourable to achieve certain immunological effects like DAMP release, immune cell migration, and change in cellular activity such as TAM polarisation. It is suggested that the microjets that arise from the violent collapse of bubbles can cause irreversible cell damage and necrotic cell death [70], which would consequently result in the release of DAMPs and TAAs. However, stressed cells are also known to release DAMPs [71], and thus shear stress generated by microstreaming from oscillating bubbles can also explain the observed DAMP release. Furthermore, Matsuura reported the detection of broadband noise while observing a reduction in perfusion [72], suggesting that violent collapse might be responsible for the shutdown of blood flow. However, the type of oscillation regime has not been confirmed for the enhanced tumour perfusion observed by Li et al. [33].

In regards to tumour vasculature, there are advantages and disadvantages to both vascular effects caused by cavitation. First, anti-vascular effects can cause tissue necrosis to activate immune responses and produce numerous antigens to induce APC maturation [73]. However, vascular disruption will also restrict the access of immune cells and therapeutic drugs. Thus, it is important for future studies to determine the long-term vascular effects of cavitation and assess blood flow immediately before tumour excision. Additionally, it is important to consider the sequencing and timing of US exposure and drug administration. On the other hand, cavitation that causes enhanced perfusion has the disadvantage of potentially not eliciting a strong immune response due to lack of DAMP and TAA release. In these situations, it is useful to know which bubble behaviour causes which vascular effect to ensure that the desired effect is achieved.

Notably, bubble behaviour is not only dependent on US parameters such as frequency, PNP, pulse repetition frequency, and sonication duration but also the characteristics of the CAs, such as their size, composition, and concentration [74]. Additionally, the tumour type, location, and environment will affect the bubble oscillation and the subsequent physical mechanisms. Thus, passive cavitation detection, i.e., monitoring cavitation behaviour, is an essential tool in terms of uncovering which bubble behaviour is dominant and desirable for certain biological and immunological effects [32,75–77]. Yet, only one paper in Section 2 assessed cavitation levels, allowing Bulner et al. to report which acoustic emissions were dominant [34], although Joiner et al. confirmed the presence of CAs in tumours after treatment by contrast enhanced imaging. Thus, it would be beneficial if future studies included passive cavitation detection to provide more details about the oscillation regime.

Moreover, a better understanding of the level, type, and intratumoural location of cavitation events will not only be of vital importance in optimising the efficacy of this approach but also its safety as it progresses to the clinic. Indeed, technology that allows a clinician to verify and map the level and type of cavitation taking place during treatment of a patient with US+CA+ICI will be of huge benefit. However, it is noteworthy that the recorded frequency content should be interpreted with care: although the violent oscillation of bubbles (inertial cavitation) emits broadband signals, the range of bubble sizes is continually changing, and this can also produce acoustic emissions with a wide range of frequencies [78]. Moreover, harmonic frequencies may be produced both by non-linear propagation and oscillating bubbles [79].

5. Conclusions

This review attempts to give an overview of some of the biological effects that cavitation may induce to promote an anti-tumour immune response and how cavitation may be combined with checkpoint inhibitors. The aforementioned papers suggest different ways US-mediated cavitation can enhance immune cell presence and activity in tumours, including DAMP and tumour antigen release, vascular modification, and the alteration of cellular activity. However, due to the limited number of studies and each of them either using different tumour models and tumour stages, treatment protocol, immune cell markers, or US parameters, it is difficult to draw any specific conclusions on how cavitation may reverse immune escape.

Compiling additional reproducible studies would be beneficial and improve the understanding of cavitation-mediated immunomodulation. Moreover, most studies do not report which oscillation regime is dominant or fully define the level and location of cavitation. By gaining a better understanding of which physical mechanisms trigger the desired anti-tumour immune responses, it may be easier to choose US parameters that ICIs and other immunotherapeutic drugs will benefit from the most and help achieve consistent results. A particular emphasis on how cavitation may turn "cold" tumours "hot" and thereby widen the clinical response to ICIs will be key.

Author Contributions: Conceptualisation, M.M.; information curation, M.M.; writing—original draft preparation, M.M.; writing—review and editing, R.C.; original schematics, M.M.; figure preparation, M.M.; supervision, R.C. and C.C.; funding acquisition, M.M., R.C. and C.C. All authors have read and agreed to the published version of the manuscript.

Funding: We thank Donald Porteus for funding this research and Aker Scholarship awarded to Matilde Maardalen.

Conflicts of Interest: The authors declare no conflict of interest.

Abbreviations

ATP	Adenosine triphosphate
APC	Antigen-presenting cell
aPD-L1	Anti-programmed death ligand 1
aPD-1	Anti-programmed cell death protein 1
CA	Cavitation agent
CEC	CD71+ erythroid progenitor cell
CTL	Cytotoxic T lymphocyte
CTLA-4	Cytotoxic T lymphocyte-associated protein 4
DAMP	Damage-associated molecular pattern
DC	Dendritic cell
HSP	Heat shock protein
HMGB1	High-mobility-group box 1
HPF	High power field
ICI	Immune checkpoint inhibitor

irAE	Immune-related adverse effect
IFN-γ	Interferon gamma
IL	Interleukin
MHC	Major histocompatibility complex
MB	Microbubble
MDSC	Myeloid-derived suppressor cell
NK	Natural killer
OVA	Ovalbumin
PNP	Peak negativepressure
PD-L1	Programmed death ligand 1
PD-1	Programmed cell death protein 1
Treg	Regulatory T
TNF-β	Transforming growth factor beta
TAA	Tumour-associated antigen
TAM	Tumour-associated macrophage
TME	Tumour microenvironment
TNF-α	Tumour necrosis factor alpha
TSA	Tumour-specific antigen
US	Ultrasound
VEGF	Vascular endothelial growth factor
VEGFR	Vascular endothelial growth factor receptor

References

1. Virchow, R. An address on the value of pathological experiments. *Br. Med. J.* **1881**, *2*, 198. [CrossRef] [PubMed]
2. Ehrlich, P. Ueber den Jetzigen Stand der Karzinomforschung. 1909; Volume 5, pp. 117–164. Available online: https://www.pei.de/SharedDocs/Downloads/DE/institut/veroeffentlichungen-von-paul-ehrlich/1906-1914/1909-karzinomforschung.pdf?__blob=publicationFile&v=2 (accessed on 10 June 2023)
3. Zamarron, B.F.; Chen, W. Dual roles of immune cells and their factors in cancer development and progression. *Int. J. Biol. Sci.* **2011**, *7*, 651. [CrossRef] [PubMed]
4. Chen, D.S.; Mellman, I. Oncology meets immunology: The cancer-immunity cycle. *Immunity* **2013**, *39*, 1–10. [CrossRef] [PubMed]
5. Savina, A.; Amigorena, S. Phagocytosis and antigen presentation in dendritic cells. *Immunol. Rev.* **2007**, *219*, 143–156. [CrossRef]
6. Marshall, J.S.; Warrington, R.; Watson, W.; Kim, H.L. An introduction to immunology and immunopathology. *Allergy Asthma Clin. Immunol.* **2018**, *14*, 1–10. [CrossRef]
7. Wculek, S.K.; Cueto, F.J.; Mujal, A.M.; Melero, I.; Krummel, M.F.; Sancho, D. Dendritic cells in cancer immunology and immunotherapy. *Nat. Rev. Immunol.* **2020**, *20*, 7–24. [CrossRef]
8. Garrido, F.; Cabrera, T.; Concha, A.; Glew, S.; Ruiz-Cabello, F.; Stern, P.L. Natural history of HLA expression during tumour development. *Immunol. Today* **1993**, *14*, 491–499. [CrossRef]
9. Garcia-Lora, A.; Algarra, I.; Garrido, F. MHC class I antigens, immune surveillance, and tumor immune escape. *J. Cell. Physiol.* **2003**, *195*, 346–355. [CrossRef]
10. Lisiecka, U.; Kostro, K. Mechanisms of tumour escape from immune surveillance. *J. Vet. Res.* **2016**, *60*, 453–460. [CrossRef]
11. Gordon-Weeks, A.; Yuzhalin, A.E. Cancer extracellular matrix proteins regulate tumour immunity. *Cancers* **2020**, *12*, 3331. [CrossRef]
12. Schaaf, M.B.; Garg, A.D.; Agostinis, P. Defining the role of the tumor vasculature in antitumor immunity and immunotherapy. *Cell Death Dis.* **2018**, *9*, 115. [CrossRef]
13. Zou, W. Immunosuppressive networks in the tumour environment and their therapeutic relevance. *Nat. Rev. Cancer* **2005**, *5*, 263–274. [CrossRef]
14. Iwai, Y.; Ishida, M.; Tanaka, Y.; Okazaki, T.; Honjo, T.; Minato, N. Involvement of PD-L1 on tumor cells in the escape from host immune system and tumor immunotherapy by PD-L1 blockade. *Proc. Natl. Acad. Sci. USA* **2002**, *99*, 12293–12297. [CrossRef] [PubMed]
15. Leach, D.R.; Krummel, M.F.; Allison, J.P. Enhancement of antitumor immunity by CTLA-4 blockade. *Science* **1996**, *271*, 1734–1736. [CrossRef] [PubMed]
16. Melero, I.; Gaudernack, G.; Gerritsen, W.; Huber, C.; Parmiani, G.; Scholl, S.; Thatcher, N.; Wagstaff, J.; Zielinski, C.; Faulkner, I.; et al. Therapeutic vaccines for cancer: An overview of clinical trials. *Nat. Rev. Clin. Oncol.* **2014**, *11*, 509–524. [CrossRef]
17. Sterner, R.C.; Sterner, R.M. CAR-T cell therapy: Current limitations and potential strategies. *Blood Cancer J.* **2021**, *11*, 69. [CrossRef]

18. Jenkins, R.W.; Barbie, D.A.; Flaherty, K.T. Mechanisms of resistance to immune checkpoint inhibitors. *Br. J. Cancer* **2018**, *118*, 9–16. [CrossRef] [PubMed]
19. Kochenderfer, J.N.; Somerville, R.P.; Lu, T.; Shi, V.; Bot, A.; Rossi, J.; Xue, A.; Goff, S.L.; Yang, J.C.; Sherry, R.M.; et al. Lymphoma remissions caused by anti-CD19 chimeric antigen receptor T cells are associated with high serum interleukin-15 levels. *J. Clin. Oncol.* **2017**, *35*, 1803. [CrossRef]
20. Nemunaitis, J. GVAX (GMCSF gene modified tumor vaccine) in advanced stage non small cell lung cancer. *J. Control. Release* **2003**, *91*, 225–231. [CrossRef]
21. Nemunaitis, J. Vaccines in cancer: GVAX®, a GM-CSF gene vaccine. *Expert Rev. Vaccines* **2005**, *4*, 259–274. [CrossRef]
22. Cercek, A.; Lumish, M.; Sinopoli, J.; Weiss, J.; Shia, J.; Lamendola-Essel, M.; El Dika, I.H.; Segal, N.; Shcherba, M.; Sugarman, R.; et al. PD-1 blockade in mismatch repair–deficient, locally advanced rectal cancer. *N. Engl. J. Med.* **2022**, *386*, 2363–2376. [CrossRef] [PubMed]
23. Lhuillier, C.; Rudqvist, N.P.; Elemento, O.; Formenti, S.C.; Demaria, S. Radiation therapy and anti-tumor immunity: Exposing immunogenic mutations to the immune system. *Genome Med.* **2019**, *11*, 1–10. [CrossRef] [PubMed]
24. Hu, Z.I.; Ho, A.Y.; McArthur, H.L. Combined radiation therapy and immune checkpoint blockade therapy for breast cancer. *Int. J. Radiat. Oncol. Biol. Phys.* **2017**, *99*, 153–164. [CrossRef] [PubMed]
25. Abe, S.; Nagata, H.; Crosby, E.J.; Inoue, Y.; Kaneko, K.; Liu, C.X.; Yang, X.; Wang, T.; Acharya, C.R.; Agarwal, P.; et al. Combination of ultrasound-based mechanical disruption of tumor with immune checkpoint blockade modifies tumor microenvironment and augments systemic antitumor immunity. *J. Immunother. Cancer* **2022**, *10*, e003717. [CrossRef] [PubMed]
26. Wu, F.; Wang, Z.B.; Lu, P.; Xu, Z.L.; Chen, W.Z.; Zhu, H.; Jin, C.B. Activated anti-tumor immunity in cancer patients after high intensity focused ultrasound ablation. *Ultrasound Med. Biol.* **2004**, *30*, 1217–1222. [CrossRef]
27. Xia, J.Z.; Xie, F.L.; Ran, L.F.; Xie, X.P.; Fan, Y.M.; Wu, F. High-intensity focused ultrasound tumor ablation activates autologous tumor-specific cytotoxic T lymphocytes. *Ultrasound Med. Biol.* **2012**, *38*, 1363–1371. [CrossRef] [PubMed]
28. Yuan, S.M.; Li, H.; Yang, M.; Zha, H.; Sun, H.; Li, X.R.; Li, A.F.; Gu, Y.; Duan, L.; Luo, J.Y.; et al. High intensity focused ultrasound enhances anti-tumor immunity by inhibiting the negative regulatory effect of miR-134 on CD86 in a murine melanoma model. *Oncotarget* **2015**, *6*, 37626. [CrossRef]
29. Hu, Z.; Yang, X.Y.; Liu, Y.; Sankin, G.N.; Pua, E.C.; Morse, M.A.; Lyerly, H.K.; Clay, T.M.; Zhong, P. Investigation of HIFU-induced anti-tumor immunity in a murine tumor model. *J. Transl. Med.* **2007**, *5*, 1–11. [CrossRef]
30. Hoogenboom, M.; Eikelenboom, D.; den Brok, M.H.; Heerschap, A.; Fütterer, J.J.; Adema, G.J. Mechanical high-intensity focused ultrasound destruction of soft tissue: Working mechanisms and physiologic effects. *Ultrasound Med. Biol.* **2015**, *41*, 1500–1517. [CrossRef]
31. Stride, E.; Coussios, C. Nucleation, mapping and control of cavitation for drug delivery. *Nat. Rev. Phys.* **2019**, *1*, 495–509. [CrossRef]
32. Kooiman, K.; Roovers, S.; Langeveld, S.A.; Kleven, R.T.; Dewitte, H.; O'Reilly, M.A.; Escoffre, J.M.; Bouakaz, A.; Verweij, M.D.; Hynynen, K.; et al. Ultrasound-responsive cavitation nuclei for therapy and drug delivery. *Ultrasound Med. Biol.* **2020**, *46*, 1296–1325. [CrossRef]
33. Li, N.; Tang, J.; Yang, J.; Zhu, B.; Wang, X.; Luo, Y.; Yang, H.; Jang, F.; Zou, J.; Liu, Z.; et al. Tumor perfusion enhancement by ultrasound stimulated microbubbles potentiates PD-L1 blockade of MC38 colon cancer in mice. *Cancer Lett.* **2021**, *498*, 121–129. [CrossRef]
34. Bulner, S.; Prodeus, A.; Gariepy, J.; Hynynen, K.; Goertz, D.E. Enhancing checkpoint inhibitor therapy with ultrasound stimulated microbubbles. *Ultrasound Med. Biol.* **2019**, *45*, 500–512. [CrossRef] [PubMed]
35. Hunt, S.J.; Gade, T.; Soulen, M.C.; Pickup, S.; Sehgal, C.M. Antivascular ultrasound therapy: Magnetic resonance imaging validation and activation of the immune response in murine melanoma. *J. Ultrasound Med.* **2015**, *34*, 275–287. [CrossRef] [PubMed]
36. Lin, L.; Du, Y.; Hao, J.; Wu, R.; Du, L. UTMD inhibits pancreatic cancer growth and metastasis by inducing macrophage polarization and vessel normalization. *Biomed. Pharmacother.* **2023**, *160*, 114322. [CrossRef] [PubMed]
37. Zhang, W.; Shou, W.D.; Xu, Y.J.; Bai, W.K.; Hu, B. Low-frequency ultrasound-induced VEGF suppression and synergy with dendritic cell-mediated anti-tumor immunity in murine prostate cancer cells in vitro. *Sci. Rep.* **2017**, *7*, 1–11. [CrossRef]
38. Gallucci, S.; Matzinger, P. Danger signals: SOS to the immune system. *Curr. Opin. Immunol.* **2001**, *13*, 114–119. [CrossRef] [PubMed]
39. Obeid, M.; Tesniere, A.; Ghiringhelli, F.; Fimia, G.M.; Apetoh, L.; Perfettini, J.L.; Castedo, M.; Mignot, G.; Panaretakis, T.; Casares, N.; et al. Calreticulin exposure dictates the immunogenicity of cancer cell death. *Nat. Med.* **2007**, *13*, 54–61. [CrossRef] [PubMed]
40. Liu, H.L.; Hsieh, H.Y.; Lu, L.A.; Kang, C.W.; Wu, M.F.; Lin, C.Y. Low-pressure pulsed focused ultrasound with microbubbles promotes an anticancer immunological response. *J. Transl. Med.* **2012**, *10*, 1–14. [CrossRef]
41. Joiner, J.B.; Kren, N.P.; Durham, P.G.; McRee, A.J.; Dayton, P.A.; Pylayeva-Gupta, Y. Low-Intensity Focused Ultrasound Produces Immune Response in Pancreatic Cancer. *Ultrasound Med. Biol.* **2022**, *48*, 2344–2353. [CrossRef]
42. Hu, J.; He, J.; Wang, Y.; Zhao, Y.; Fang, K.; Dong, Y.; Chen, Y.; Zhang, Y.; Zhang, C.; Wang, H.; et al. Ultrasound combined with nanobubbles promotes systemic anticancer immunity and augments anti-PD1 efficacy. *J. Immunother. Cancer* **2022**, *10*. [CrossRef] [PubMed]

43. Wu, N.; Cao, Y.; Liu, Y.; Zhou, Y.; He, H.; Tang, R.; Wan, L.; Wang, C.; Xiong, X.; Zhong, L.; et al. Low-intensity focused ultrasound targeted microbubble destruction reduces tumor blood supply and sensitizes anti-PD-L1 immunotherapy. *Front. Bioeng. Biotechnol.* **2023**, *11*, 1173381. [CrossRef] [PubMed]
44. Huang, F.Y.; Lei, J.; Sun, Y.; Yan, F.; Chen, B.; Zhang, L.; Lu, Z.; Cao, R.; Lin, Y.Y.; Wang, C.C.; et al. Induction of enhanced immunogenic cell death through ultrasound-controlled release of doxorubicin by liposome-microbubble complexes. *Oncoimmunology* **2018**, *7*, e1446720. [CrossRef]
45. Bianchi, M.E.; Crippa, M.P.; Manfredi, A.A.; Mezzapelle, R.; Rovere Querini, P.; Venereau, E. High-mobility group box 1 protein orchestrates responses to tissue damage via inflammation, innate and adaptive immunity, and tissue repair. *Immunol. Rev.* **2017**, *280*, 74–82. [CrossRef]
46. van der Woude, L.L.; Gorris, M.A.; Halilovic, A.; Figdor, C.G.; de Vries, I.J.M. Migrating into the tumor: A roadmap for T cells. *Trends Cancer* **2017**, *3*, 797–808. [CrossRef] [PubMed]
47. Kepp, O.; Tesniere, A.; Schlemmer, F.; Michaud, M.; Senovilla, L.; Zitvogel, L.; Kroemer, G. Immunogenic cell death modalities and their impact on cancer treatment. *Apoptosis* **2009**, *14*, 364–375. [CrossRef]
48. Liu, Z.; Gao, S.; Zhao, Y.; Li, P.; Liu, J.; Li, P.; Tan, K.; Xie, F. Disruption of tumor neovasculature by microbubble enhanced ultrasound: A potential new physical therapy of anti-angiogenesis. *Ultrasound Med. Biol.* **2012**, *38*, 253–261. [CrossRef]
49. Tan, X.; Yi, C.; Zhang, Y.; Tang, N.; Xu, Y.; Liu, Z. Ultrasound-targeted microbubble destruction alleviates immunosuppression induced by CD71+ erythroid progenitor cells and promotes PDL-1 blockade immunotherapy in the lewis lung cancer model. *Front. Oncol.* **2021**, *11*, 768222. [CrossRef]
50. Mantovani, A.; Marchesi, F.; Malesci, A.; Laghi, L.; Allavena, P. Tumour-associated macrophages as treatment targets in oncology. *Nat. Rev. Clin. Oncol.* **2017**, *14*, 399–416. [CrossRef]
51. Rolny, C.; Mazzone, M.; Tugues, S.; Laoui, D.; Johansson, I.; Coulon, C.; Squadrito, M.L.; Segura, I.; Li, X.; Knevels, E.; et al. HRG inhibits tumor growth and metastasis by inducing macrophage polarization and vessel normalization through downregulation of PlGF. *Cancer Cell* **2011**, *19*, 31–44. [CrossRef]
52. Zhang, Y.; Brekken, R.A. Direct and indirect regulation of the tumor immune microenvironment by VEGF. *J. Leukoc. Biol.* **2022**, *111*, 1269–1286. [CrossRef] [PubMed]
53. Ferrara, N.; Gerber, H.P.; LeCouter, J. The biology of VEGF and its receptors. *Nat. Med.* **2003**, *9*, 669–676. [CrossRef] [PubMed]
54. Chen, J.; Qiao, Y.D.; Li, X.; Xu, J.L.; Ye, Q.J.; Jiang, N.; Zhang, H.; Wu, X.Y. Intratumoral CD45+ CD71+ erythroid cells induce immune tolerance and predict tumor recurrence in hepatocellular carcinoma. *Cancer Lett.* **2021**, *499*, 85–98. [CrossRef] [PubMed]
55. Grzywa, T.M.; Justyniarska, M.; Nowis, D.; Golab, J. Tumor immune evasion induced by dysregulation of erythroid progenitor cells development. *Cancers* **2021**, *13*, 870. [CrossRef]
56. Sano, Y.; Yoshida, T.; Choo, M.K.; Jiménez-Andrade, Y.; Hill, K.R.; Georgopoulos, K.; Park, J.M. Multiorgan signaling mobilizes tumor-associated erythroid cells expressing immune checkpoint molecules. *Mol. Cancer Res.* **2021**, *19*, 507–515. [CrossRef]
57. Pardoll, D.M. The blockade of immune checkpoints in cancer immunotherapy. *Nat. Rev. Cancer* **2012**, *12*, 252–264. [CrossRef]
58. Dreher, M.R.; Liu, W.; Michelich, C.R.; Dewhirst, M.W.; Yuan, F.; Chilkoti, A. Tumor vascular permeability, accumulation, and penetration of macromolecular drug carriers. *J. Natl. Cancer Inst.* **2006**, *98*, 335–344. [CrossRef]
59. Dewhirst, M.W.; Secomb, T.W. Transport of drugs from blood vessels to tumour tissue. *Nat. Rev. Cancer* **2017**, *17*, 738–750. [CrossRef]
60. Chowdhury, S.M.; Abou-Elkacem, L.; Lee, T.; Dahl, J.; Lutz, A.M. Ultrasound and microbubble mediated therapeutic delivery: Underlying mechanisms and future outlook. *J. Control. Release* **2020**, *326*, 75–90. [CrossRef]
61. Grundy, M.; Bau, L.; Hill, C.; Paverd, C.; Mannaris, C.; Kwan, J.; Crake, C.; Coviello, C.; Coussios, C.; Carlisle, R. Improved therapeutic antibody delivery to xenograft tumors using cavitation nucleated by gas-entrapping nanoparticles. *Nanomedicine* **2020**, *16*, 37–50. [CrossRef]
62. Kim, D.; Lee, S.S.; Moon, H.; Park, S.Y.; Lee, H.J. PD-L1 targeting immune-microbubble complex enhances therapeutic index in murine colon cancer models. *Pharmaceuticals* **2020**, *14*, 6. [CrossRef] [PubMed]
63. Chen, P.L.; Roh, W.; Reuben, A.; Cooper, Z.A.; Spencer, C.N.; Prieto, P.A.; Miller, J.P.; Bassett, R.L.; Gopalakrishnan, V.; Wani, K.; et al. Analysis of Immune Signatures in Longitudinal Tumor Samples Yields Insight into Biomarkers of Response and Mechanisms of Resistance to Immune Checkpoint BlockadeImmune Signatures of Response to Checkpoint Blockade. *Cancer Discov.* **2016**, *6*, 827–837. [CrossRef] [PubMed]
64. Maleki Vareki, S. High and low mutational burden tumors versus immunologically hot and cold tumors and response to immune checkpoint inhibitors. *J. Immunother. Cancer* **2018**, *6*, 1–5. [CrossRef] [PubMed]
65. Macy, E. Immune-related adverse drug reactions and immunologically mediated drug hypersensitivity. *Immunol. Allergy Clin.* **2020**, *40*, 635–647. [CrossRef]
66. Benoot, T.; Piccioni, E.; De Ridder, K.; Goyvaerts, C. TNFα and immune checkpoint inhibition: Friend or foe for lung cancer? *Int. J. Mol. Sci.* **2021**, *22*, 8691. [CrossRef] [PubMed]
67. Mortezaee, K.; Majidpoor, J. Checkpoint inhibitor/interleukin-based combination therapy of cancer. *Cancer Med.* **2022**, *11*, 2934–2943. [CrossRef]
68. Liu, T.; Han, C.; Wang, S.; Fang, P.; Ma, Z.; Xu, L.; Yin, R. Cancer-associated fibroblasts: An emerging target of anti-cancer immunotherapy. *J. Hematol. Oncol.* **2019**, *12*, 1–15. [CrossRef]
69. Shimasaki, N.; Jain, A.; Campana, D. NK cells for cancer immunotherapy. *Nat. Rev. Drug Discov.* **2020**, *19*, 200–218. [CrossRef]

70. Hwang, J.H.; Tu, J.; Brayman, A.A.; Matula, T.J.; Crum, L.A. Correlation between inertial cavitation dose and endothelial cell damage in vivo. *Ultrasound Med. Biol.* **2006**, *32*, 1611–1619. [CrossRef]
71. Murao, A.; Aziz, M.; Wang, H.; Brenner, M.; Wang, P. Release mechanisms of major DAMPs. *Apoptosis* **2021**, *26*, 152–162. [CrossRef]
72. Matsuura, N.; Koonar, E.; Zhu, S.; Leung, B.; Seo, M.; Sivapalan, N.; Goertz, D. Inducing antivascular effects in tumors with ultrasound stimulated micron-sized bubbles. In Proceedings of the 2015 IEEE International Ultrasonics Symposium (IUS), Taipei, Taiwan, 21–24 October 2015; pp. 1–4.
73. Ho, Y.J.; Li, J.P.; Fan, C.H.; Liu, H.L.; Yeh, C.K. Ultrasound in tumor immunotherapy: Current status and future developments. *J. Control. Release* **2020**, *323*, 12–23. [CrossRef]
74. Martinez, P.; Bottenus, N.; Borden, M. Cavitation characterization of size-isolated microbubbles in a vessel phantom using focused ultrasound. *Pharmaceutics* **2022**, *14*, 1925. [CrossRef]
75. Gray, M.D.; Elbes, D.; Paverd, C.; Lyka, E.; Coviello, C.M.; Cleveland, R.O.; Coussios, C.C. Dual-array passive acoustic mapping for cavitation imaging with enhanced 2-D resolution. *IEEE Trans. Ultrason. Ferroelectr. Freq. Control* **2020**, *68*, 647–663. [CrossRef]
76. Choi, J.J.; Carlisle, R.C.; Coviello, C.; Seymour, L.; Coussios, C.C. Non-invasive and real-time passive acoustic mapping of ultrasound-mediated drug delivery. *Phys. Med. Biol.* **2014**, *59*, 4861. [CrossRef]
77. Graham, S.M.; Carlisle, R.; Choi, J.J.; Stevenson, M.; Shah, A.R.; Myers, R.S.; Fisher, K.; Peregrino, M.B.; Seymour, L.; Coussios, C.C. Inertial cavitation to non-invasively trigger and monitor intratumoral release of drug from intravenously delivered liposomes. *J. Control. Release* **2014**, *178*, 101–107. [CrossRef]
78. Yasui, K.; Tuziuti, T.; Lee, J.; Kozuka, T.; Towata, A.; Iida, Y. Numerical simulations of acoustic cavitation noise with the temporal fluctuation in the number of bubbles. *Ultrason. Sonochem.* **2010**, *17*, 460–472. [CrossRef]
79. Hou, G.Y.; Marquet, F.; Wang, S.; Konofagou, E.E. Multi-parametric monitoring and assessment of high-intensity focused ultrasound (HIFU) boiling by harmonic motion imaging for focused ultrasound (HMIFU): An ex vivo feasibility study. *Phys. Med. Biol.* **2014**, *59*, 1121. [CrossRef]

MDPI
St. Alban-Anlage 66
4052 Basel
Switzerland
www.mdpi.com

Pharmaceutics Editorial Office
E-mail: pharmaceutics@mdpi.com
www.mdpi.com/journal/pharmaceutics

www.ingramcontent.com/pod-product-compliance
Lightning Source LLC
Chambersburg PA
CBHW061847210926
51850CB00006B/997